Animal Viruses: Molecular Biology

Edited by:

Thomas C. Mettenleiter

and

Francisco Sobrino

Caister Academic Press

Caister Academic Press
Norfolk, UK

www.caister.com

British Library Cataloguing-in-Publication Data

A catalogue record for this book is available from the British Library
ISBN: 978-1-904455-22-6

Printed and bound in Great Britain

Contents

Other Books of Interest

Contributors

Charles C. Abrams
Institute for Animal Health
Pirbright Laboratory
Ash Road
Pirbright
Woking
Surrey GU24 0NF
UK

Fernando Almazán
Centro Nacional de Biotecnología
CSIC Department of Molecular and Cell Biology
Campus Universidad Autónoma
Cantoblanco
Darwin 3
Madrid 28049
Spain

Udeni B. R. Balasuriya*
Department of Veterinary Science
108 Maxwell. H. Gluck Equine Research Center
University of Kentucky
Lexington
KY 40546-0099
USA

Dave G. Chapman
Institute for Animal Health
Pirbright Laboratory
Ash Road
Pirbright
Woking
Surrey GU24 0NF
UK

Markus Czub*
Faculty of Veterinary Medicine
University of Calgary
Calgary
Alberta
Canada

Linda K. Dixon*
Institute for Animal Health
Pirbright Laboratory
Ash Road
Pirbright
Woking
Surrey GU24 0NF
UK

Esteban Domingo*
Centro de Biologia Molecular "Severo Ochoa"
(CSIC-UAM)
Universidad Autonoma de Madrid
Cantoblanco
28049 Madrid
Spain

Luis Enjuanes*
Centro Nacional de Biotecnología
CSIC Department of Molecular and Cell Biology
Campus Universidad Autónoma
Cantoblanco
Darwin 3
Madrid 28049
Spain

Walter Fuchs
Institute of Molecular Biology
Friedrich-Loeffler-Institut
Boddenblick 5A
17493 Greifswald-Insel Riems
Germany

Marian C. Horzinek
Spes Nostra
Haydnlaan
3723KE Bilthoven
The Netherlands

Günther M. Keil
Institute of Molecular Biology
Friedrich-Loeffler-Institut
Boddenblick 5A
17493 Greifswald-Insel Riems
Germany

Hans-Dieter Klenk*
Institute of Virology
Philipps-University
Marburg
Germany

Annette Mankertz*
National Reference Centre for Measles
Mumps and Rubella
Robert Koch-Institut
Nordufer 20
13353 Berlin
Germany

Encarnacion Martinez-Salas*
Centro de Biología Molecular Severo Ochoa
CSIC-UAM
Cantoblanco
28049 Madrid
Spain

Mikhail Matrosovich
National Institute for Medical Research
Mill Hill
London
UK

Thomas C. Mettenleiter*
Institute of Molecular Biology
Friedrich-Loeffler-Institut
Boddenblick 5A
17493 Greifswald-Insel Riems
Germany

Charlene Ranadheera
Department of Medical Microbiology
University of Manitoba
Winnipeg
Canada

Polly Roy*
Department of Infectious and Tropical Diseases
London School of Hygiene and Tropical Medicine
Keppel Street
London WC1E 7HT
UK

Till Rümenapf
Institute of Virology
Faculty of Veterinary Medicine
Justus Liebig University Giessen
Frankfurter Strasse 107 D- 35392 Giessen
Germany

Margarita Saiz
Centro de Biología Molecular Severo Ochoa
CSIC-UAM
Cantoblanco
28049 Madrid
Spain

Bevan Sawatsky
Department of Medical Microbiology
University of Manitoba
Winnipeg
Canada

Eric J. Snijder
Molecular Virology Laboratory
Department of Medical Microbiology
Center of Infectious Diseases
Leiden University Medical Center
LUMC P4-26
PO Box 9600
2300 RC Leiden
The Netherlands

Francisco Sobrino
Centro de Biología Molecular Severo Ochoa
CSIC-UAM
Cantoblanco
28049 Madrid
Spain

Isabel Sola
Centro Nacional de Biotecnología
CSIC Department of Molecular and Cell Biology
Campus Universidad Autónoma
Cantoblanco
Darwin 3
Madrid 28049
Spain

Jürgen Stech
Friedrich-Loeffler-Institut
Greifswald-Insel Riems
Germany

Heinz–Jürgen Thiel*
Institute of Virology
Faculty of Veterinary Medicine
Justus Liebig University Giessen
Frankfurter Strasse 107 D- 35392 Giessen
Germany

Hana M. Weingartl
Department of Medical Microbiology
University of Manitoba
Winnipeg
Canada

Fuquan Zhang
Institute for Animal Health
Pirbright Laboratory
Ash Road
Pirbright
Woking
Surrey GU24 0NF
UK

Sonia Zúñiga
Centro Nacional de Biotecnología
CSIC Department of Molecular and Cell Biology
Campus Universidad Autónoma
Cantoblanco
Darwin 3
Madrid 28049
Spain

*** Corresponding author**

Preface

It was in 1898 that Friedrich Loeffler and Paul Frosch reported their results on the filterability and replicative capacity of the infectious agent causing foot-and-mouth disease in cattle. In the same year, Wilhelmus Martinus Beijerinck published his findings on mosaic disease of tobacco and identified an agent with the same basic properties. With these pioneering studies they founded a new scientific discipline, virology. This was several years before the identification of the causative agent of yellow fever as the first human virus by Reed and Carroll. Thus, an infectious agent of veterinary importance stood at the basis of what is now being recognized as the wide field of 'animal virology', which also includes human viruses.

In this book, the editors purposely selected 'veterinary' viruses to demonstrate their important contribution to our understanding of the molecular basis of viral infection in general. Of course, we recognize that within the scope of a book like this not all important findings, which originated from studies of viruses in animals could be covered adequately. However, we tried to carefully select those viruses, virus genera or virus families, whose study had and still has an important impact on our understanding of viruses, their replication, their evolution and their interaction with the host. Working with 'animal' viruses has one important advantage over the study of human viruses: natural virus-host systems are experimentally accessible and those who embark on unravelling the secrets of these viruses, as they infect and cause disease in their host are not dependent on more or less artificial 'surrogate' hosts with sometimes only limited relevance for the natural infection. Thus, work on animal viruses, in the strict sense, gives a high level of reality. As in any other area of experimental research, current animal virology benefits from methodological and conceptual advances that have been emerging during the last decades in different disciplines, such as pathology, immunology, and molecular, structural and cellular biology. Only from the adequate combination of the scopes and tools these disciplines offer, can the challenging goal of understanding the molecular mechanisms responsible for host-virus interactions be achieved. Thus, special attention has been paid by the authors of the different chapters to emphasize strategic and methodological aspects of current research, as well as in providing the readers with key related references.

At the basis of animal virology was foot-and-mouth disease (FMD) virus. Although since Loeffler's and Frosch's days a lot has been learned on the infectious agent, this virus is still the most feared in animal farming. The re-emergence of old viruses and emergence of new sero- and genotypes of the virus is a constant threat to agriculture, resulting in dramatic losses either due to direct harm to the infected host, to trade restrictions or to the necessity of slaughtering animals on a large scale to contain disease outbreaks. Although numerous efforts have been undertaken since Loeffler's times, no safe live vaccine against FMD has so far been developed and prevention still relies on inactivated whole virus vaccines, which have to be produced under high biosecurity standards to avoid leakage of infectious agent from the production sites. Molecular analyses of FMD virus (the type member of the genus *Aphthovirus* within the *Picornaviridae*) yielded important insight into the role different viral proteins play for replication in cell culture and in the infected animal. Viral factors relevant for FMD pathogenesis have been identified, and new diagnostic assays have been developed that allow the differentiation between vaccinated and infected animals. Moreover, work on FMD virus has also contributed to the revolutionary concept of the quasispecies of RNA viruses, which stands at the basis of their impressive evolutionary potential. The concept of quasispecies, which results from the error-prone genome replication by RNA-dependent RNA polymerase, leads straight to the new finding that, besides increasing the fitness, i.e. the extent of virus adaptation to the environment, it can also lead to extinction. These are radical new views on an 'old' and supposedly well-known infectious agent of animals.

Within the *Flaviviridae*, the genus *Pestivirus* encompasses three animal viruses with utmost importance to agriculture: classical swine fever virus, bovine viral diarrhea virus, and border disease virus. Molecular studies elucidated several basic features of replication of these viruses and provided the basis for understanding the molecular pathogenesis of a most devastating and interesting disease in cattle, mucosal disease. This can serve as a prime example of how molecular virological studies increase our understanding of disease mechanisms. Interest in pestiviruses soared when it became clear that they are closely related to the newly discovered human hepatitis C virus. Although this agent has in the meantime been relegated to a separate family, many similarities remain at the molecular level. Thus, studies on animal pathogens can also greatly increase our knowledge on human pathogens.

The family *Arteriviridae* shares many basic features with the coronaviruses and both have been grouped together in the order *Nidovirales*. The hallmark of this group of viruses is the transcription of the viral genomes into a nested set of mRNAs. No human pathogens have so far emerged from the arteriviruses. However, they contain important animal pathogens. Besides the name-giving equine arteritis virus, a new virus has emerged in the last 15 years, basically 'out of nowhere', which has become one of the most devastating viral agents in modern pig husbandry. Porcine respiratory and reproductive syndrome virus (PRRSV) remains a challenge for pig producers, although its molecular biological characteristics are becoming increasingly known. Besides uncertainty about its origin, the nearly simultaneous appearance of two only distantly related PRRSV strains in North America and Europe is puzzling.

For a long time, coronaviruses had been considered to be of only limited importance for humans. Thus, coronavirology has largely been a niche occupied by a few dedicated researchers who primarily worked with viruses causing important diseases in animals, such as feline infectious peritonitis, porcine transmissible gastroenteritis or avian infectious bronchitis. The advent of severe acute respiratory syndrome, or SARS, radically changed this situation. Coronavirology exploded in the effort to identify and characterize the SARS agent and to find ways to prevent and/or cure this life-threatening infection. The world was on the brink of a major catastrophe; but immediate action, and a lot of luck, prevented a greater disaster. Within an incredibly short time, the molecular details of the SARS agent were unravelled. Most of this speedy success was due to the longstanding research of scientists in the veterinary field who provided the nucleus for the new surge in coronavirology. The emergence of the SARS virus in the human population, coming from an animal source, also highlighted the importance of animals in harbouring infectious agents with the potential of transfer into the human population. Molecular biology of coronaviruses is well advanced and the detailed knowledge of the replication strategy of this family of viruses, their specific enzymatic functions, as well as the intricate mechanism of genome transcription and replication can open new ways of treatment. Coronaviruses are the RNA viruses with the largest genomes and it is thought that, due to the error-rate of the RNA dependent RNA polymerase, longer RNA genomes can not be faithfully replicated, and, thus, did not appear in evolution.

One of the most challenging and fastest growing family of viruses are the *Paramyxoviridae*. This diverse family encompasses important human pathogens as well as highly relevant animal viruses. Moreover, paramyxoviruses are considered to be the family with one of the most extensive natural reservoirs, largely unknown, resulting in the continuous identification of new paramyxoviruses. Among the most interesting are the Hendra and Nipah viruses, which had been recognized as animal pathogens only in the last decade. Importantly, both have a zoonotic potential with the capacity to kill their human host. Both viruses are considered as highly pathogenic level 4 agents and they are under increased scrutiny due to their potential use as bioterrorist or biological warfare agents. Despite the difficulties of working with these viruses, significant advances have been made in understanding their molecular biology, in particular concerning receptor usage and replicative strategy. They originate from a natural reservoir in fruit bats and, just like the SARS virus, embody the constant threat to human health by these 'animal' viruses.

The impact of influenza viruses on public health is well known. Not only during pandemics but also in interpandemic periods, influenza virus infections of humans exert a significant death toll on the human population. However, it has only been recognized recently that influenza viruses, which persist in their natural avian host, can directly infect humans without further reassortment with human-adapted viruses. The current pandemic in wild and domestic birds of the highly pathogenic avian influenza strain H5N1, originating from Asia, is causing concern since it may presage, after adaptation to humans, the next, possibly devastating pandemic. Although some of the molecular mechanisms which transform a low pathogenic avian influenza into a highly pathogenic fowl plague virus are known, only little information is available on the molecular alterations required for efficient infection of and transmission between humans. Thus, a better understanding of the molecular biology of influenza viruses, the requirements for infection of different influenza virus hosts, as well as cross-species transmission, is urgently needed to better prepare mankind for the next human pandemic. It will certainly come but nobody knows when, from where and which influenza virus subtype will be involved.

There are only two families of animal viruses with double-stranded RNA genomes: the *Birnaviridae* and the *Reoviridae*. The family *Reoviridae* comprises the genus *Orbivirus* with three important animal pathogens causing bluetongue disease in sheep and cattle, enzootic hemorrhagic disease in cattle and deer, and African horse sickness. Studies on bluetongue virus (BTV) have contributed significantly to the understanding of the complex structure of reoviruses as well as their transcription and genome replication strategies. Moreover, functions of individual virally encoded proteins have been elucidated, and biotechnology has been employed to produce modern vaccines. The recent introduction of BTV into Central Europe, and its subsequent transmission by midges indigenous to this area highlights the potential of BTV to expand into hitherto unaffected regions. Setting an example, the same could also be true for the other orbiviruses.

Turning into the world of DNA virus pathogens, porcine circoviruses have first been described as contaminants of porcine cell cultures without pathogenic potential for pigs. Only recently, a second type of porcine circovirus was found to be associated with a novel, devastating pig disease designated as postweaning multisystemic wasting syndrome. Although the molecular basis for the different pathogenicity of the two porcine circovirus types has not yet been unravelled, molecular analyses of viral gene expression and genome replication yielded novel insights into the molecular properties of these agents.

Herpesviruses are arguably the most successful viral pathogens. After initial infection, they remain in their host for its life, being reactivated from the latent state in irregular intervals, thereby gaining the opportunity to infect naïve individuals throughout the lifetime of their host. Whereas human herpesviruses are being studied in numerous laboratories throughout the world, the work on the equally important animal herpesviruses is mostly focussed on a few important pathogens. Among them, the porcine pseudorabies virus is probably the best analyzed non-human herpesvirus. The study of the molecular biology of this virus has been most relevant for understanding herpesvirus replication at the molecular level, and its interaction with specific host cells, e.g. neurons. Moreover, it led to a new concept in animal disease control, which for the first time employed genetically modified live vaccines in an eradication program that is based on the serological differentiation between infected and vaccinated animals. Thus, a long-standing desire in animal disease control: the use of vaccination for reducing circulation of the infectious agent without covering continuing virus replication under a vaccination blanket, has become reality. Using this approach PrV has been eliminated from domestic pig populations in North America and large parts of Western Europe. Beyond this success, the study of animal herpesviruses in their natural hosts has provided new insights into the molecular mechanisms of latency, in particular through work on bovine herpesvirus 1. Moreover, comparative analyses of porcine, bovine and gallid herpesviruses, including studies in their natural hosts, has shed light on the similarities and differences in the molecular biological features of these different viruses.

It is rare that virus families are represented by only one single member. This is the case for the *Asfarviridae* that only contain the type member African swine fever virus. This virus, which has significant homology to poxviruses in its genetic content and replicative cycle, is not only highly interesting for basic virology but is also the reason why in large parts of Africa, pig production can not be pursued due to unacceptable losses incurred by infection with this virus. The complex genome encodes numerous gene products for modulating the host cell and evading host immune responses. So far, it has been impossible, even using all the tools of modern molecular biology, to create an efficient vaccine against this disease. However, the more detailed our understanding of the complex replication of this virus at the molecular level, the higher are the chances of success in the design of novel approaches towards stimulating protective immunity.

The editors freely concede that the contents of this book on molecular biology of animal viruses do not cover all the exciting discoveries which have been made by researchers studying viruses of veterinary importance. However, we hope that the selection of topics gives an impression on how far advanced the study of veterinary viruses is, and how valuable the results are for virology as a whole. We are very grateful to the authors of the different chapters for their excellent and timely contributions, and we appreciated the friendly interaction with them. We are particularly grateful to Marian Horzinek and Esteban Domingo for contributing a thought-provoking conclusion. Thank you all for your time and effort! We would also like to thank Anette Beidler, secretary of TCM, who very diligently copy-edited the chapters so that they conform to the required standard. Lastly, we appreciate the interest of the publisher in producing this book. Thank you, Hugh, for giving us the chance to edit it!

Thomas C. Mettenleiter
Greifswald-Insel Riems

Francisco Sobrino
Madrid

Chapter 1

Foot-and-Mouth Disease Virus

Encarnacion Martinez-Salas, Margarita Saiz and Francisco Sobrino

ABSTRACT

Foot-and-mouth disease virus (FMDV) is the prototypic member of the *Aphthovirus* genus in the *Picornaviridae* family. This picornavirus is the etiological agent of an acute systemic vesicular disease that affects cattle worldwide. Here we have addressed several aspects dealing with the molecular biology of this highly variable and transmissible virus that are relevant to understand the viral infectious cycle. Particular emphasis has been given to the peculiarities of its genome organization, as well as its control of gene expression. Soon after infection, the single stranded positive RNA that constitutes the viral genome is efficiently translated using a cap-independent mechanism driven by the internal ribosome entry site element (IRES). This process occurs concomitantly with the inhibition of cellular protein synthesis, caused by the expression of viral proteases. Processing of the viral polyprotein is achieved cotranslationally by viral encoded proteases, giving rise to the different mature viral proteins. Viral RNA as well as viral proteins interact with different components of the host cell, acting as key determinants of viral pathogenesis. In depth knowledge of the molecular basis of the viral cycle is needed to control viral pathogenesis and disease spreading.

Foot-and-mouth disease (FMD) virus (FMDV) is the etiological agent of one of the most important and highly transmissible diseases of livestock, an acute systemic vesicular disease affecting extensive areas worldwide and ranking first in the A list of infectious diseases of animals, according to the Office International des Epizooties (Sobrino et al., 2001). FMDV infects artiodactylae, mostly cattle, swine, sheep and goats (Thomson et al., 2003). In endemic areas FMD control is implemented by regular vaccination while FMD-free countries impose severe movement and trade restrictions on animals and livestock products from affected areas (Sutmoller et al., 2003).

Taxonomically, FMDV is the prototypic member of the *Aphthovirus* genus in the *Picornaviridae* family. Some picornaviruses account for important human diseases like Poliovirus (PV), the most extensively characterized member of the family at the molecular level. FMDV exhibits a high potential for genetic and antigenic variation, leading to the seven viral serotypes (A, O, C, Asia1, SAT 1, SAT 2 and SAT 3) identified on the basis of their ability to induce cross-protection in animals. The pioneer studies made by Domingo and co-workers using FMDV as a model system for the study of genetic variability in RNA viruses (Domingo et al., 1992) have provided instrumental information for the characterization of the quasispecies structure of RNA virus populations and the understanding of its biological properties. This concept, initially proposed from theoretical basis to describe sets of related but different nucleic acids during the early evolution of replicating macromolecules in earth, was experimentally first demonstrated for bacteriophage Qβ populations and relays on the observation that RNA viruses consist of multiple variants (quasispecies) in a dynamic and complex equilibrium between high mutation rate and the competitive fitness in response to selective pressures. Discussion of this important aspect of FMDV biology is beyond the scope of this chapter; the interested readers can find detailed reviews at (Domingo et al., 2006; 2004, 2001).

FMDV productively infects several primary cultures as well as established cell lines such as BHK-21 and IBRS-2 cells. Intraperitoneal inoculation of FMDV produces death in suckling mice, and this has been applied to titrate virus infectivity. Direct inoculation of *in vitro* transcribed RNA on mice has allowed the recovery of infectious virus from mutants unable to grow on cell culture (Baranowski et al., 2003). By serial passages FMDV can be adapted to guinea-pig, an animal model that has been used for immunological and tropism analyses (Francis et al., 1987; Nunez et al., 2001). FMDV can also establish persistent infections in cultured cells in which a coevolution of both, viruses (which showed increased virulence for the parental BHK-21 cells) and cells (which became progressively resistant to the initial virus), occurs along the cellular passages (de la Torre et al., 1988).

During the last years relevant information has been generated aimed to understand the molecular basis of the viral cycle and their implications on virulence, tropism and disease control. FMDV is a current threat and needs further input on research to address the molecular mechanisms responsible for viral pathogenesis and disease prophylaxis.

GENOME ORGANIZATION

As all Picornaviruses FMDV has a positive-sense RNA genome of about 8500 nucleotides (nt) (Figure 1). A single copy of the genome is enclosed within a protein capsid comprised of 60 copies of the 4 different structural proteins (VP4, VP3, VP2 and VP1). The viral RNA contains a single open reading frame (ORF) flanked by two highly structured non-coding regions (NCRs) that contain *cis*-acting structural elements involved in viral replication and gene expression (Figure 1). The RNA is polyadenilated at its 3' end. In contrast to most cellular mRNAs there is no cap structure at the 5' end of the viral genome but a small viral-encoded protein, VPg (3B), covalently linked to the 5'-terminus.

The viral RNA is infectious (de la Torre et al., 1987). This feature has allowed the construction of infectious cDNA clones (Zibert et al., 1990), which are a useful tool to study the function of different viral gene products and RNA structural motifs. The viral genome contains all the information needed to take over the host machinery and shut down cellular protein synthesis to the limit that only viral RNA and proteins are produced in the infected cell (Belsham and Martínez-Salas, 2004). The entire replication cycle occurs in the cytoplasm of cells and starts with the translation of the genomic RNA to produce the viral polyprotein that will be co-translationally processed by viral proteases, L^{pro} and $3C^{pro}$, into the mature viral products (Figure 1). One of these products is the RNA-dependent RNA polymerase $3D^{pol}$ that will generate copies of the negative-sense RNA using the positive-sense strand as a template. These negative strands will be then used as templates for the synthesis of many positive strands. Some of them will be packaged into viral capsids to generate new infectious virions.

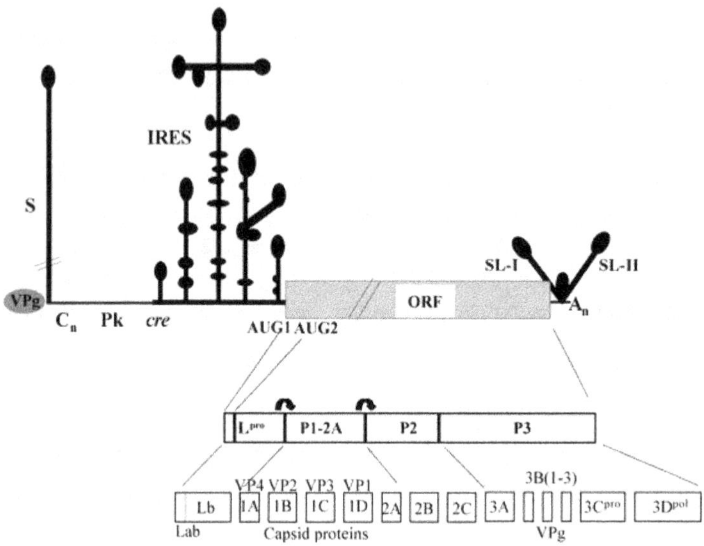

Fig. 1. Schematic representation of the FMDV genome. Relevant motifs present in the FMDV RNA have been indicated as well as the mature viral products derived from the polyprotein. See the text for details.

VIRAL 5' END. STRUCTURE AND FUNCTION

The FMDV 5'NCR is a long region of about 1300 nt including sequences required for the initiation of viral replication and translation. As we walk along this region we find several structural and functional elements. At the very 5'end of the RNA there is a region of about 360 nt named the S-fragment, predicted to form a large hairpin structure (Figure 1). The S region corresponds to the smaller of two fragments obtained after treatment of the viral RNA with oligo(dG) and RNase H. For FMDV S region it has been recently shown that is able to establish RNA/RNA interactions with the viral 3'end (Serrano et al., 2006). Moreover, two cellular factors, p116 and p47, presumably poly(rC) binding protein (PCBP), have been proved to interact with the S region in UV-crosslinking experiments using labelled S-RNA probes and cellular protein extracts. The homologous element in PV 5'end, though displaying a different secondary structure, has been shown to interact with several viral (3CD precursor) and cellular (PCBP2) proteins (Gamarnik and Andino, 1997) (Parsley et al., 1997). Poly(A) binding protein (PABP) is able to interact with PCBP and the 3'end poly(A). Some authors have suggested that those interactions play a role in the switch from translation to replication and likely in the putative circularization of the viral RNA by protein bridges between 5' and 3'ends (Gamarnik and Andino, 1998) (Herold and Andino, 2001).

Following the S region on the 5'end there is a poly(C) tract (termed C_n in Figure 1) almost homopolymeric. The presence of the polyC tract is a common feature only with most cardioviruses among the picornaviruses. The poly(C) length varies from about 80 to 400 nt, tipically around 200 in FMDV field isolates (Brown, 1972) (Harris and Brown, 1977). Studies aimed to address the role on infectivity of the length of the poly(C) tract have yielded controversial results and its significance is not clear to date.

Downstream of the poly(C) there is a region about 250 nt predicted to contain multiple pseudo-knots, 2 to 4 depending on the viral isolate (Escarmis et al., 1995). Interestingly, in cardioviruses the pseudoknot stretch is predicted to be present on the 5'side of the poly(C) tract (Martin and Palmenberg, 1996) suggesting a common function for both regions.

As mentioned above, for picornaviruses it is believed that structures present at both 5' and 3'ends of the viral RNA are required to direct the recognition of their genomes by the viral RNA-dependent RNA polymerase $3D^{pol}$. In the last few years evidence for internal RNA sequences required for replication has been shown for enterovirus, rhinovirus, and cardiovirus (McKnight and Lemon, 1998) (Goodfellow et al., 2000) (Gerber et al., 2001). This sequence element, termed as '*cis*-acting replication element" (*cre*) includes a conserved motif, AAACA, located within a loop at the end of a stable stem structure. The *cre* elements from human rhinovirus (HRV-14, HRV-2), cardioviruses and PV are within the coding regions for 1D, 2A, 1B and 2C respectively.

Cre elements can be moved without losing their function. It has been shown recently for PV and HRV-2 *cre* sequences their role as templates for the uridylylation of VPg (3B) by the viral RNA polymerase *in vitro* (Paul et al., 2000) (Gerber et al., 2001). This reaction starting by the covalent attachment of a UMP molecule to the hydroxyl group of a tyrosine in the terminal protein VPg generates the products VPgpU and/or VPgpUp that act as primers for viral RNA synthesis. This process is considered an essential step in PV RNA replication.

For FMDV, functional evidence for the presence of a *cre* in the 5′NCR, just upstream the IRES, was reported (Mason et al., 2002) (Figure 1). The element contains the AAACA motif, and replication could be restored on an RNA transcript carrying a defective *cre* by insertion of the wild type *cre* into the 3'NCR. This finding was consistent with previous work showing that FMDV mutants carrying deletions of 94% of the P1 coding region were still able to replicate efficiently (McInerney et al., 2000). The presence of a *cre* in the non-coding regions of the genome is a distinguishing feature of FMDV suggesting some differences in the replication strategies among picornaviruses. Additionally, the presence of a *ts* FMDV RNA mutant carrying a destabilising mutation in the conserved stem-loop was described. This mutation could be complemented in *trans* by other *ts* FMDV RNAs with defects in other regions of the genome (Tiley et al., 2003). It was then suggested that the *cre* should be renamed a 3B-uridylylation site.

Downstream of the *cre* element and partly overlapping with it, we find the IRES (internal ribosome entry site), present in all picornaviruses, a complex highly structured element of about 450 nt responsible for the internal cap-independent initiation of protein synthesis on the viral RNA.

FMDV IRES: PROTEIN SYNTHESIS CONTROL

Following cell entry, release and translation of the viral RNA is the first step during the viral replication cycle. In common to all picornaviruses, the genetic organization of the 5'NCR in FMDV genome is incompatible with the scanning model for several reasons. First, the 5' terminus of the genome has a virus-encoded protein, VPg, covalently linked to the terminal nucleotide. Second, the 5'NCR that varies between 700 and 1300 nt, comprises distinct regions, heavily structured. Third, the uncapped 5'NCR harbours more than 10 AUG triplets not recognized as initiator codons.

Protein synthesis in FMDV viral RNA is driven by the IRES element (Martinez-Salas et al., 2001). This region of the FMDV genome efficiently drives protein synthesis of a second cistron outside of its genetic context in both transfected cells and reticulocyte lysates (Martinez-Salas, 1999). Initiation of translation mediated by IRES elements represents an alternative to the cap-dependent mechanism used for most cellular mRNAs. The 5'end of eukaryotic mRNAs has a cap structure m^7GpppN... that is recognized by the initiation factor eIF4F. This factor comprises three polypeptides, the cap-binding protein eIF4E, the RNA helicase eIF4A and the scaffold protein eIF4G. The latter has distinct binding sites for other proteins including eIF3, the poly(A) binding protein (PABP) and Mnk-1 (an eIF4E kinase). To initiate translation, the small ribosomal subunit interacts with the eIF4F complex bound at the 5' cap and migrates along the 5'NCR until an AUG codon in appropriate context is found (Kozak, 1989). Recognition of the correct initiation codon during this scanning process is inhibited by complex RNA structure and by the presence of upstream AUG codons. In contrast, initiation of translation mediated by IRES elements involves the direct recruitment of the translational machinery to an internal position in the mRNA. IRES-dependent protein synthesis is resistant to inhibitory conditions for cap-dependent translation initiation. Such conditions occur following eIF4G cleavage, which is induced during picornavirus infection by proteases L (FMDV) or 2A (poliovirus, rhinovirus), or due to the dephosphorylation of the translational repressor 4E-BP1 in encephalomyocarditis virus (EMCV) infected cells (reviewed in (Martinez-Salas et al., 2001).

Definition of the picornavirus IRES genetic borders by deletion mapping in bicistronic constructs indicated that around 450 nt were enough to drive 5´end-independent translation initiation (Pelletier and Sonenberg, 1988). Then, comparison of different picornavirus IRES sequence revealed only small stretches of conserved sequences. According to secondary RNA structure prediction models, two groups of picornavirus IRES structures have been described, type I includes the entero-/rhinoviruses and type II, the cardio-/aphthoviruses. A third type of IRES is represented by hepatitis A, and recently a fourth group of picornavirus IRES have been demonstrated to share some similarities with hepatitis C virus (HCV) (Pisarev et al., 2004). The absence of structural conservation between different picornavirus IRES may reflect the diversity of strategies used by these elements to interact with the translation machinery.

IRES STRUCTURE-FUNCTION RELATIONSHIP

The involvement of RNA structural motifs in FMDV IRES activity has been studied in detail by a combination of RNA structure probing, mutational analysis and phylogenetic conservation. RNA viruses are variable but essential elements are conserved (Martinez-Salas and Fernandez-Miragall, 2004). Nucleotide covariation is indicative of stable secondary or tertiary structure, possibly involved in IRES activity. Independent changes are often present in loops or unstructured regions. Invariant positions accounts for region of potential interest during internal initiation. As shown in Figure 2 the FMDV IRES is organized in five structural domains. Residues of domain 1 are now recognized as the right arm of the *cre* stem-loop (Mason et al., 2002). Domain 2 is predicted to form a conserved stem-loop structure that has an invariant pyrimidine-rich sequence at its apical loop. This stem-loop is conserved with EMCV and contains the main binding site for the polypyrimidine tract binding protein (PTB) (Luz and Beck, 1991).

The central domain in the FMDV IRES, domain 3, shows two distinct regions in terms of secondary structure (Figure 2). According to RNA structure probing performed by chemical and RNase accessibility, the apical part contains stable stem-loops that form a four-way junction (Fernandez-Miragall and Martinez-Salas, 2003). Two phylogenetically conserved motifs bearing sequences GUAA and AAAA are present in the apical loops. Mutational analysis of these motifs revealed that they constitute essential regions for the activity of the IRES (Lopez de Quinto and Martinez-Salas, 1997). Nucleotide substitutions in the GUAA sequence in the FMDV IRES demonstrated that a GNRA motif is strictly required to retain 100% activity. A study of the equivalent GNRA motif in the EMCV IRES was also consistent with the functional relevance of this motif for translation initiation (Robertson et al., 1999). The GNRA motif adopts a tetraloop structure that is responsible for the organization of the adjacent stem-loops (Fernandez-Miragall and Martinez-Salas, 2003). A single A to G substitution in the fourth position of this motif led to a strong RNA reorganization, affecting several nucleotides away in the secondary structure of domain 3. The study of mutants bearing UNCG or GUAG sequences revealed lack of protection to chemical attack in native RNA at specific nucleotides relative to the parental GUAA, suggesting that the GNRA motif dictates the organization and stability of domain 3. This effect may be mediated by the interaction with distant residues bearing the sequence CACG, phylogenetically conserved (Fernandez-Miragall et al., 2006). The presence of a structural element that serves as substrate for RNase P, overlapping with the apical region of the central domain has been recently reported (Serrano et al., 2007). Thereby, a potential tRNA-like signal might reside in this IRES region.

The GNRA tetraloop is a building block that is commonly found in structured RNA molecules (Doherty et al., 2001). Thus, a similar role for these motifs within IRES elements is likely. As inferred from gel-shift assays, the central region of the FMDV IRES is unique in its ability to interact with each of the other domains, including itself and the entire IRES (Ramos and Martinez-Salas, 1999). This result opens the possibility that domain 3 acts as a scaffold structure that holds together the IRES domains. In addition, its strong dimerization capacity may determine intermolecular interactions with other IRES molecules, a result consistent with complementation data reported between defective FMDV IRES elements (Drew and Belsham, 1994). The conserved C-rich loop, also present in the central domain (Figure 2), is tolerant to nucleotide substitutions (Martinez-Salas et al., 2002) in common with the equivalent motif in PV IRES (Gamarnik et al., 2000). The C-rich loop is a candidate to interact with the poly(rC) binding proteins (PCBP) (Stassinopoulos and Belsham, 2001). However, PCBP2 binding to the EMCV or the FMDV 5′NCR is not required for IRES activity (Walter et al., 1999) consistent with the functional analysis of FMDV C-rich loop. The proximal region of the central domain contains a long stem interrupted with internal bulges (Fernandez-Miragall et al., 2006). Accordingly, second site mutation in sequences located in the base of domain 3 showed a requirement for secondary structure involving formation of a helical structure (Martinez-Salas et al., 1996). This structural element may contribute to the stability of the entire domain, as deduced from RNA probing analysis of the basal region.

As mentioned before, both primary sequence and secondary structure conservation is observed between the aphtho- and cardiovirus IRESs. At the base of domain 4, two A-rich internal bulges are conserved within the secondary structure of the different field isolates of FMDV (Figure 2). A single point substitution of A312 to U impaired IRES activity more strongly than mutations introduced in other motifs (Lopez de Quinto and Martinez-Salas, 2000). Mutations that abrogated the predicted secondary structure were detrimental for IRES activity in transfected cells. However, restoration of the secondary structure resulted in a complete recovery of IRES activity demonstrating that base-pair formation is required at these positions within the IRES for internal initiation activity. Sequence comparison of independent isolates of FMDV showed that domain 5 contains a conserved hairpin (Lopez de Quinto et al., 2001), which is also shared with the EMCV IRES. However, substitution of invariant residues did not impair FMDV IRES activity to the same extent as the mutations introduced into the A-bulge of domain 4, or the GNRA loop of domain 3.

IRES INTERACTION WITH PROTEINS

Cellular proteins play an essential role in FMDV IRES-dependent initiation (Figure 2, right panel). The canonical initiation factors eIF4A, eIF4G, eIF2 and eIF3, are required for 48S complex formation in a reconstituted 40S ribosome assay with the EMCV or FMDV IRES elements (Kolupaeva et al., 1998) (Pilipenko et al., 2000). The carboxy-terminal end of the proteolyticaly processed form of eIF4G that interacts with the EMCV and FMDV IRES contains the binding sites for eIF3 and eIF4A. In agreement with this, internal initiation promoted by these elements is highly efficient under conditions of eIF4G cleavage induced by the FMDV L protease (Martinez-Salas et al., 1993) (Belsham and Brangwyn, 1990) (Ohlmann et al., 1996).

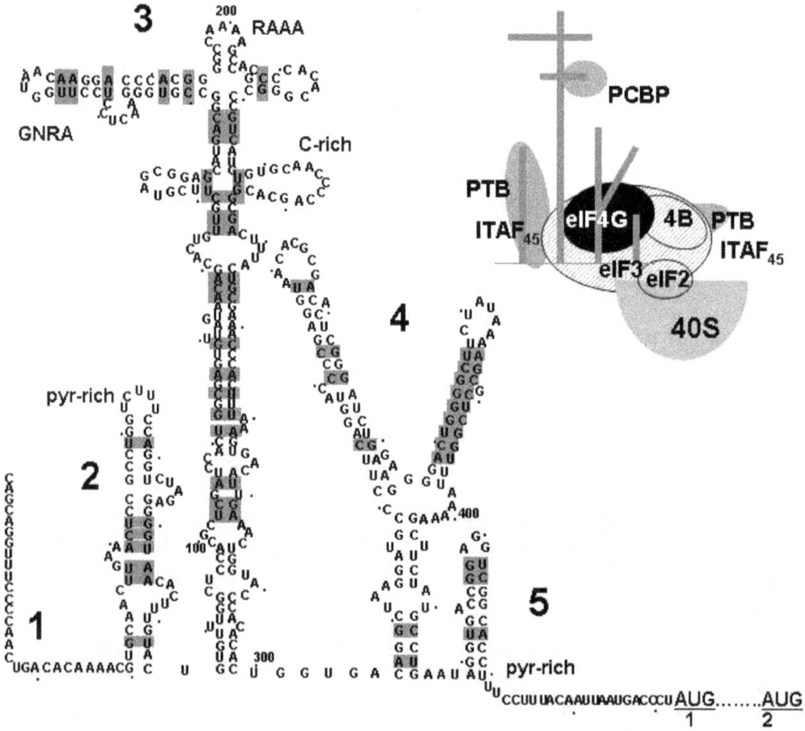

Fig. 2. Secondary structure of the FMDV IRES. The phylogenetic conservation is depicted together with RNA probing analysis according to enzymatic and chemical accessibility. Covariant nucleotides noticed in FMDV field isolates deposited in databases are depicted in grey boxes. IRES domains are numbered 1 to 5, and the functional AUG start codon are underlined. On the right panel, the binding sites of initiation factors and auxiliary proteins interacting with the FMDV IRES are schematically represented.

Using a transcript corresponding to domain 4 in a UV-crosslink assay, binding of eIF4G was as efficient as the full length IRES, strongly suggesting that no additional sites for recognition of this protein in the FMDV IRES are required (Lopez de Quinto and Martinez-Salas, 2000). Furthermore, the processed form of both eIF4GI and eIF4GII, present in cells transfected with the FMDV Lb protease, binds as efficiently to the FMDV IRES as the unprocessed protein (Lopez de Quinto et al., 2001). Disruption of the conserved structural motif at the base of domain 4 impairs eIF4G-IRES interactions. The strong correlation found between the eIF4G-IRES interaction and IRES activity in transfected cells demonstrates that eIF4G binding is an essential step in the recruitment of the translational machinery *in vivo*. In agreement with this result, the use of transcripts corresponding to the 3′ end of the FMDV IRES to deplete reticulocyte lysates lead to a strong reduction of factors required for cap-dependent translation (Stassinopoulos and Belsham, 2001). Domain 5 in the 3'region of the FMDV IRES interacts strongly with eIF4B in UV-crosslinking assays (Lopez de Quinto et al., 2001). However, mutants impaired in eIF4B binding reduced IRES activity about two to four-fold. Consistently, eIF4B only stimulates 48S complex formation on the EMCV IRES about two-fold (Pestova et al., 1996).

Related IRES elements that share secondary structure behave differently in terms of functional RNA-protein association. The FMDV IRES required binding to PTB and to the proliferation-associated factor, $ITAF_{45}$, for 48S complex formation *in vitro* in contrast to cardiovirus IRES (Pilipenko et al., 2000). Many IRES-interacting proteins contain several RNA binding motifs and display multiple sites of interaction with the IRES molecule (Hunt et al., 1999) (Walter et al., 1999) (Song et al., 2005), hence it is likely that these proteins act as RNA chaperones, directing or stabilizing the IRES tertiary folding in an active conformation. In recent years, the combination of RNA chromatography methods with mass spectrometry has allowed the identification of novel IRES-interacting factors. Understanding the role played by these factors in internal initiation would allow to generate models of IRES function, in addition to explain IRES specific interference (Reigadas et al., 2005).

IRES activity is specifically stimulated by the FMDV 3'NCR in a poly(A)-independent manner (Lopez de Quinto et al., 2002). Direct RNA-RNA interaction between the 5' and 3'NCRs as well as RNA-protein bridging is in full agreement with a functional circularization of the viral ends (Serrano et al., 2006).

IRES-PROTEIN INTERACTION: ROLE IN TISSUE TROPISM AND VIRAL PATHOGENESIS

Proteins not recognized as canonical initiation factors were found to interact with picornavirus IRES, suggesting that they could contribute to IRES-mediated cell type specificity. Additionally, viral IRES may have key roles in the regulation of viral pathogenesis in the infected organism as found in the case of poliovirus neurovirulence. A novel form of PTB appears to play an important role in the neurovirulence of Theiler's virus (Pilipenko et al., 2001). Accordingly, the GDVII virulent strain replicates in the mouse central nervous system where motor neurons have nPTB. However, involvement of cell-specific factors that facilitate conformational changes in the IRES allowing the transfer of 48S complex to a particular start codon is still an open question. Although FMDV IRES is known to specifically require $ITAF_{45}$ for initiation complex formation, the possible requirement of other factors presumably involved in tissue tropism is still under debate.

SELECTION OF THE INITIATION CODON FOR PROTEIN SYNTHESIS

Two in-frame AUG codons separated by 84 nt are present in the viral genome (Figure 2) (Sangar et al., 1987). To date, the reason why two initiation sites for protein synthesis are conserved across all FMDV isolates remains unknown. Both AUG codons are used as initiation sites for protein synthesis leading to 2 forms of the Leader protein, Lab and Lb (Figure 1), that differ in their N-termini. The Lab initiation codon was not required for virus viability, but modification of the Lb initiation codon was lethal (Cao et al., 1995). Synthesis of two forms of the L protein was observed in virus-infected cells as well as in bicistronic vectors (Lopez de Quinto and Martinez-Salas, 1999). To determine the recognition mechanism of the 2 start sites on FMDV RNA, RNA transcripts including the two FMDV initiation codons fused to a reporter gene were expressed in transfected cells (Belsham, 1992). The Lb start site was recognised preferentially in all cases. Insertion of two additional AUG codons in-frame between the authentic initiation codons led to recognition of the four AUG codons. This result suggests that FMDV IRES directed ribosomes to bind upstream of the Lab start site. The recognition of the first FMDV start site was inefficient, perhaps reflecting the suboptimal sequence context of this AUG codon in the viral RNA. Modification of the Lab start context to improve its recognition as initiation site increased the efficiency of translation at this site but surprisingly, this had no apparent effect on the efficiency of initiation at the Lb start site (Lopez de Quinto and Martinez-Salas, 1999). Hence, it was suggested that these two initiation codons could act independently. This view was supported by the observation that an antisense oligonucleotide annealed to the region of the Lab start site effectively blocked initiation at this site but had little effect on initiation at the Lb site. Thus, it may be that some ribosomes are directed by the FMDV IRES to bind to the RNA downstream of the Lab start site, or that once bound upstream of the first site they are transferred to the second one by an unknown mechanism.

VIRAL POLYPROTEIN

The viral ORF encodes a single polyprotein of about 2330 amino acids, which is undetectable as it is rapidly cleaved by viral proteases termed L^{pro} and $3C^{pro}$. Complete processing of the polyprotein yields 15 mature products (Figure 1).

The L region contains two overlapping proteins Lab and Lb, resulting from the translation initiation starting at each of the two functional AUGs of the polyprotein. The L protease is a member of the papain-like cystein proteases (Guarne et al., 1998). Both the FMDV Lab and Lb proteins have been shown to cleave the L/P1 junction within the polyprotein (Medina et al., 1993) as well as the translation initiation factor eIF4G (Devaney et al., 1988). The cleavage of eIF4G requires a very low level of the L protease and can be observed as soon as one hour postinfection (Belsham and Sonenberg, 2000). Recombinant Lb protease cleaves eIF4GI *in vitro* (Kirchweger et al., 1994) (Gradi et al., 2004) using high protease concentration compared to the low levels present in infected cells. However, although the L protease is not essential for virus viability (Piccone et al., 1995), its expression inhibits the induction of beta-interferon mRNA locking the host innate immune response (de Los Santos et al., 2006). In FMDV infected cells, cleavage of host factors PTB, PABP and eIF3a,b has been recently reported (Rodríguez-Pulido et al., 2007). While the processing of PABP is at least partially due to Lb cleavage, the involvement of 3C protease in any of these cleavages is still unknown.

Removal of L^{pro} is followed by a myristoylation reaction at the N-terminus of the P1-2A precursor encoding the structural proteins that compose the viral capsid VP1 (1D), VP2 (1B), VP3 (1C) and VP4 (1A) (Figure 1). The processing of the FMDV P1-2A precursor to 1AB, 1C and 1D is achieved by the 3C protease. The cleavage of 1AB to 1A (VP4) and 1B (VP2) occurs at the time of the RNA genome encapsidation. VP4 myristoylation is important for the assembly and/or stability of the FMDV virion. The 16-amino-acid peptide 2A has been proposed to prevent the formation of the peptide bond at the 2A/2B junction during polyprotein synthesis (Donnelly et al., 2001).

The regions P2 and P3 encode six different mature non-structural (NS) proteins, processed by $3C^{pro}$ (Figure 1). Each of them, as well as some of the processing intermediates, are involved in functions relevant to the virus cycle in infected cells. The functions of 2B and 2C proteins are poorly understood. Within PV-infected cells, the P2 proteins are found in membrane-associated viral replication complexes. The protein 2B is known to contain certain hydrophobic domains (van Kuppeveld et al., 1996) and has been related to cell permeability and protein secretion (Van kuppeveld et al., 1997) (Doedens and Kirkegaard, 1995). Picornavirus 2C protein contains helicase motifs but this activity has never been reported. Mutations that confer resistance to guanidine, an inhibitor of viral RNA replication, are located in 2C (Saunders and King, 1982). In poliovirus, the 3A protein has been proposed to be the membrane anchor for the picornavirus replication complex, it is associated to viral-induced membrane vesicles and contributes to the cytopathic effect development and inhibition of protein secretion. Changes in FMDV 3A are associated with attenuation for cattle (O'Donnell et al., 2001) as well as the ability to infect guinea pig (Nunez et al., 2001).

VPg (3B) participates in the initiation of picornavirus RNA synthesis. FMDV is the only picornavirus containing three tandem, non-identical copies of 3B protein. Viable viruses can be recovered from infectious RNAs carrying a single copy of VPg but the level of viral infectivity correlates with the number of copies present in the RNA (Falk et al., 1992). The components required for the uridylylation of each of the distinct FMDV 3B peptides by the FMDV replicase $3D^{pol}$ *in vitro* show some differences compared to PV replication. This process is strictly dependent on the presence of FMDV 3CD, the *cre* and the $3D^{pol}$ while the poly(A) tail did not support this reaction (Nayak et al., 2006).

The FMDV 3C protease belongs to the trypsin-like family of serine proteases and is responsible for most of the cleavages within the polyprotein coding sequence. The key catalytic residues of the FMDV 3C have been identified and its crystal structure has been resolved (Birtley et al., 2005; Grubman et al., 1995). Unlike PV 3C, it does not require 3D sequences for its processing activity, and its expression is toxic for cell survival (Martinez-Salas and Domingo, 1995). In addition to cleaving the viral polyprotein, the FMDV 3C also modifies several cellular proteins (Falk et al., 1990) (Li et al., 2001). The 3C protease also cleaves eIF4GI on the C-terminal side of the site generated by the expression of the L protease (Belsham et al., 2000) (Strong and Belsham, 2004). This observation is consistent with the loss of eIF4GI in cells infected by the leader-less FMDV. $3C^{pro}$ also induces the proteolytic processing of histone H3 and several host factors (Falk et al., 1990), which may be related to the inhibition of host transcription observed in infected cells.

The 3D protein is the viral RNA-dependent RNA polymerase (Martinez-Salas et al., 1985). $3D^{pol}$ is able to recognize and interact with the 3´end of both the positive- and negative-sense strands of the RNA regardless of their different sequence and structure. The mechanisms behind these processes remain to be well defined. The X-ray structure of the complex formed by FMDV $3D^{pol}$, VPg1, the substrate UTP and divalent cations, in the absence and presence of an oligoadenylate of 10 residues has been recently determined (Ferrer-Orta et al., 2006a). In both complexes, VPg fits the RNA binding cleft of the polymerase. The complex obtained in the presence of the oligoadenylate showed the product of the VPg uridylylation (VPg-UMP). Two metal ions and the catalytic aspartic acids of the polymerase active site, together with the basic residues of motif F, have been identified as participating in the priming reaction. In common with other viral RNA replicases the FMDV $3D^{pol}$ exhibits a closed-hand conformation attained by interconnection of the fingers and the thumb (Ferrer-Orta et al., 2006b).

VIRAL 3'END (STRUCTURE AND FUNCTION)

The FMDV 3'noncoding region consists of a sequence of about 100 nt predicted to be structured in two stem loops, SLI and SLII (Figure 1) (Serrano et al., 2006), and followed by a poly(A) tail genetically encoded by the virus. Experimental evidence showing the relevance of this region in the viral cycle has been accumulating recently. Thus, deletion of the 3'NCR on infectious FMDV clones abrogated replication. The infectious phenotype could not be restored by the replacement of the FMDV 3'NCR with the corresponding sequence from another swine picornavirus, Swine vesicular disease virus (SVDV) (Saiz et al., 2001). Previous reports had shown the viability of a PV RNA carrying a complete deletion of the 3'NCR but with significant defects on replication (Todd et al., 1997) (Mellits et al., 1998) (Brown et al., 2005). For cardiovirus RNA one of the three stem loops present in the 3'NCR was found to be essential for virus viability (Duque and Palmenberg, 2001).

The interaction of cellular proteins, identified as polyA-binding protein (PABP) and PCBP, with the FMDV 3'NCR has been described (Serrano et al., 2006). The dual role of this region in regulating not only replication but as well translation was proved in transfection experiments using bicistronic constructs (Lopez de Quinto et al., 2002). The presence of FMDV 3'NCR stimulated IRES-dependent translation. This effect was independent of the stimulatory effect exerted by the poly(A) tail and reached the highest levels under Lb viral protease expression.

As mentioned above, long-range RNA/RNA interactions with the 5'NCR have been recently described (Serrano et al., 2006). These interactions involve two different elements of the 5'NCR, the S fragment and the IRES. The requirements for both interactions are different. While the IRES – 3'NCR contact needs a higher order structure adopted by the complete elements, the S – 3'NCR interaction still takes place after deletion of each of the SLs separately. In contrast, the presence of the poly(A) is dispensable for IRES – 3'NCR interaction but necessary for the S – 3'NCR interaction, suggesting the relevance of a poly(A)-dependent conformation in the 3´end of the viral RNA. The role of these interactions in the viral cycle will need further work but are consistent with the 5'/3' RNA bridging model proposed for PV and other positive-strand RNA viruses involving proteins (Herold and Andino, 2001) (Isken et al., 2004) (Vende et al., 2000), and even more with recent reports showing direct cyclization of flavivirus genome (Alvarez et al., 2005) (Filomatori et al., 2006). Circularization of the PV RNA was proposed through the interaction of PABP bound to the 3' poly(A) tail with PCBPs that bind to the 5' end of the genome (Herold and Andino, 2001). PABP is a multifunctional protein that interacts among other proteins with the translation initiation factor eIF4G and the poly(rC) binding protein 2. PABP recognizes FMDV 3'UTR in UV-crosslink assays, whereas PCBP interacts with both, the S region and 3'UTR in a competitive manner (Serrano et al., 2006). Therefore, such interactions may serve to circularise the RNA.

THE VIRION: STRUCTURE AND PROPERTIES

Upon cleavage, precursor P1 renders proteins VP0 (VP4 and VP2), VP1 and VP3, which assemble into asymmetric protomers. Viral particles are produced from the assembling of twelve pentamers, each resulting from the association of five protomers, and the incorporation of a newly synthesised RNA molecule. VP2 and VP4 are generated, following RNA encapsidation, from the autocatalytic cleavage of VP0. The feasibility to derive high quality crystals from purified virus preparations has allowed determination of the three-dimensional structure of FMDV particles. The structure of FMDV virions corresponding to serotypes O, A and C, elucidated by X-ray crystallography (Acharya et al., 1989; Lea et al., 1994), has shown the main structural features common for the picornavirus family (reviewed in (Ferrer-Orta, 2004). They consist of non-enveloped icosahedral capsids, 28-30 nm in diameter, in which VP1, VP2 and VP3 are surface-oriented, whereas VP4 is internal and has a N-terminal myristic acid (Chow et al., 1987).

Unlike most picornaviruses, the FMDV capsid surface is relatively smooth, exhibiting a main protruding element, the G-H loop, in VP1. The G-H loop spans about 20 residues around positions 140-160, and can adopt different conformations without apparent alterations of the rest of the capsid. This large loop contains a highly conserved Arg-Gly-Asp (RGD) triplet, a cell-recognition site present in various extracellular proteins, which mediates interaction with cell surface integrin receptors (Fox et al., 1989).

B-cell antigenic sites are conformed by amino acid residues exposed on the surface of the capsid (Acharya et al., 1989). A major continuous neutralizing antigenic site (site A) is located in the G-H loop of VP1, as deduced from the immunogenicity in animal models of peptides spanning VP1 residues around positions 140 to 160 (Bittle et al., 1982; Pfaff et al., 1982). For serotype C, the antigenic structure of the G-H loop is complex, since different overlapping epitopes, defined by their differential ability of peptides from this region to react with panels of neutralizing monoclonal antibodies (MAbs), have been mapped within this loop (Mateu, 1995). Thus, the G-H loop is involved in both the antigenic properties of FMDV and its interaction with host cells. Additional antigenic sites involved in the induction of neutralizing antibodies sites have been identified by analysing the reactivity of viruses resistant to neutralization with MAb (MAR mutants) (reviewed in (Mateu, 2004). Site C is located at the C-terminus of VP1 and is apparently continuous and independent from the G-H loop in serotypes A and C (Lea et al., 1994; Thomas et al., 1988). In type O, its vicinity to the G-H loop in the structure of the capsid, as well as competition studies with neutralising MAbs, suggest that sites A and C conform a single site (Parry et al., 1990). Cross-neutralization assays using MAR mutants have allowed the identification of discontinuous antigenic sites, which are not mimicked by lineal peptides, contributed by residues from different capsid proteins in serotypes A and O (Baxt et al., 1989; Kitson et al., 1990; Pfaff et al., 1988; Thomas et al., 1988). These antigenic sites are located at exposed regions adjacent to each other and close to the 3-fold axis of symmetry in the capsid. In serotype C, the discontinuous site D, comprises residues involving the C-terminus of VP1, the VP3 B-B knob, and VP2 B-C loop (Lea et al., 1994).

The instability of FMDV capsids at pH below neutrality is unique among Picornaviruses (Brown, 1972). It has been proposed that this higher acidic sensitivity is due to the contribution of a cluster of His present at the pentamer interface (Curry et al., 1995).

HOST CELL INTERACTIONS

A first requirement for FMDV infection is the interaction of the virus with receptors present in host cells. Early experiments showed that a mild treatment with trypsin rendered FMDV virions that were non-infectious and did not bind tissue culture cells (Cavanagh et al., 1977) and that peptides containing the RGD motif at the G-H loop in VP1 inhibited virus cell attachment (Fox et al., 1989). Further structural and functional studies established that the RGD motif is critically involved in virus interaction with different α_V integrins (Baxt, 2004). Despite being part of an important antigenic site, the RGD triplet is highly conserved among field FMDV isolates, probably reflecting its requirement for the *in vivo* interaction with cell receptors (Duque and Baxt, 2003; Neff et al., 1998). This is supported by the failure of disease induction in susceptible animals by engineered viruses, recovered from RNA obtained from infectious clones harboring mutated versions of the RGD (Mason et al., 1994).

Expression of single human or bovine α_v integrins in cells that does not allow FMDV entry and infection has been the approach followed to identify molecules that can be used to initiate FMDV infection. Viruses that are infectious *in vivo* can enter cultured cells using $\alpha_v\beta_3$, $\alpha_v\beta_6$ and $\alpha_v\beta_8$ integrins (Jackson et al., 2004; Jackson et al., 2000; Neff et al., 1998). Integrin $\alpha_v\beta_6$ is expressed constitutively on the epithelial cells targeted by FMDV in cattle, and it has been suggested to be the major *in vivo* receptor for this virus (Monaghan et al., 2005).

Upon cell culture growth, FMDV can acquire the capacity to initiate infection via heparin sulphate (Jackson et al., 1996) by selection of amino acid replacements at a shallow depression at the junction of the three major capsid proteins (Fry et al., 1999). The RGD motif can become dispensable upon multiple passages of FMDV in cell culture, associated with amino acid substitutions at the capsid surface and the use of alternative receptors for cell entry (Baranowski et al., 2001; Baranowski et al., 2000). An alternative cell-binding site in pigs has also been reported for engineered viruses harboring lys-gly-glu (KGE) instead of RGD (Zhao et al., 2003).

The analysis of the early FMDV internalization events following receptor binding has been recently approached by using cell lines constitutively expressing or transfected with human integrin $\alpha_v\beta_6$. In both cases, tracking of viral particles by fluorescence microscopy with capsid-specific MAbs, combined with the analysis of the effect on virus entry and infection of inhibitors of endocytic pathways indicated that FMDV types A and O mainly utilize a clathrin-dependent mechanism for internalization (Berryman et al., 2005; O'Donnell et al., 2005). This was also supported by the inhibition of infected cells observed upon expression of a dominant-negative version of AP180, which is required for clathrin cage assembly (Berryman et al., 2005). The virus was rapidly taken up into the cells and colocalized with markers for early and recycling endosomes but not with a marker for lysosomes. In agreement with these results, FMDV infection is not inhibited by nocodazole, a reagent that inhibits vesicular trafficking between early and late endosomes, and hence trafficking to lysosomes. The integrin $\alpha v\beta 6$ was also seen to accumulate in early and recycling endosomes on virus entry, suggesting that the integrin serves not only as an attachment receptor but also to deliver the virus to the acidic endosome (Berryman et al., 2005).

In other Picornaviruses, such as PV, each of the NS proteins and most of their precursors are known to participate in multiple functions needed for RNA replication, particle formation and viral pathogenesis (Andino et al., 1999; Choe et al., 2005; Whitton et al., 2005). Among them, major rearrangements of the host cell membranes to generate vesicular structures where genomic RNA replication takes place have been reported (Bienz et al., 1987), and proteins 2B, 2BC, 2C and 3A are involved in these interactions with cell membranes (Aldabe and Carrasco, 1995; Cho et al., 1994; Towner et al., 1996). Relative to other Picornaviruses, differences in the membrane rearrangements induced in FMDV infected cells have been reported, such as its insensitivity to brefeldin A (O'Donnell et al., 2001), a drug that inhibits Golgi-derived coat protomer I (COPI)-coated vesicle formation and disrupts the Golgi complex. Also, collapse of cell organelles at one side of the cells has been revealed by electron microscopy (Monaghan et al., 2004). The origin of the membranes generated at the FMDV replication site is unclear (Knox et al., 2005). Partial co-localizations of the Golgi-stacks marker p58 with 3A and 2C and of calreticulin (an endoplasmic reticulum (ER) marker) and 3A have been found in confocal microscope studies of infected cell lines (Garcia-Briones et al., 2006; O'Donnell et al., 2001). In common with other Picornaviruses (Aminev et al., 2003) $3D^{pol}$ contains a nuclear localization motif and 3CD and $3D^{pol}$ are found in the nucleus of FMDV infected cells (Garcia-Briones et al., 2006), suggesting that precursor 3CD could mediate nuclear location of protease 3C and presumably contribute to the nuclear reorganization observed in infected cells.

The analysis of the effect of transient expression of viral proteins on cells has been widely used to attempt dissection of the functional role of different NS proteins. In PV and coxsackie virus, 3A is responsible for the inhibition of the anterograde traffic between the ER and the Golgi (Doedens et al., 1997) that can mediate the reduction of the secretion of proteins such as major histocompatibility complex (MHC) class I molecules (Choe et al., 2005), which is observed during Picornavirus infection (Deitz et al., 2000; Sanz-Parra et al., 1998). Such a mechanism could contribute to the evasion of the host immune system.

FMDV proteins 2B, 2C and 3A have been found associated to crude membrane extracts in infected cultured cells (Moffat et al., 2005). Unlike what is observed in other Picornaviruses, transient expression of FMDV 2BC, but not of 3A, inhibits host protein secretion in Vero cells. Transient expression of 3A did not induce major rearrangements of intracellular membranes as judged by immunofluorescence studies (Garcia-Briones et al., 2006; O'Donnell et al., 2001), and the protein appeared located at small vesicles (Moffat et al., 2005; O'Donnell et al., 2001). The sequential addition copies of the 3B coding region resulted in a relocation of the 3AB products that became progressively compact and fibrous (Garcia-Briones et al., 2006); the possible relevance of this observation in the context of FMDV infection remains to be determined.

IN VIVO PATHOGENESIS

Limited information is available on the determinants of FMDV host-range and virulence *in vivo,* reviewed in (Baxt, 2004; Mason et al., 2003). FMDV induces similar clinical signs in the wild and domestic species of cloven-hoof animals in which this virus causes a acute, systemic vesicular disease, characterized mainly by fever and vesicular lesions at mouth and upper respiratory tract epithelia (reviewed in (Donaldson, 2004). Current evidence in FMDV and other RNA viruses exhibiting a high potential for variation and adaptation (Domingo et al., 2003; Domingo, 2004) indicates that the interaction of viruses with different hosts results in a complex dynamics driven by different selective pressures, such as the requirement for cell receptors to initiate infection in different target tissues, the need to replicate to a minimal extent to allow virus progression, and the evasion of the host immune response. Indeed, naive and vaccinated ruminants can develop, upon FMDV exposure, an asymptomatic, persistent infection during which virus can be isolated from the upper respiratory track from a few weeks up to several years after infection. The mechanism underlying viral persistence is unclear, but it is likely the result of a dynamic equilibrium between the host immune response and the selection of viral antigenic variants at the mucosae of the upper respiratory tract (reviewed in (Salt, 1993)).

As discussed previously, interaction of the RGD triplet on VP1 with cell integrins is considered a requirement for viral entry. Nevertheless, integrins are expressed in a variety of tissues and species, including those considered as non-permissive for FMDV infection. Thus, additional host factors are likely to be determinant for FMDV virulence and host range. Interestingly, the accessibility of the integrin-binding site to neutralizing antibodies suggests the possibility of a co-evolution of FMDV antigenicity and receptor usage (Baranowski et al., 2001; Baranowski et al., 2000). Thus, amino acid replacements affecting either the RGD motif ($R_{141}\rightarrow G$), or adjacent positions ($L_{144}\rightarrow P$ and $L_{147}\rightarrow P$) (where L stands for leucine, and P, proline) reported to be critical for FMDV binding to some RGD-dependent integrins (Mateu et al., 1996; Pfaff, 1997), were present in FMDV type C mutants selected in cattle immunized with synthetic peptides (Taboga et al., 1997). These mutants rapidly reverted to a parental RGD context when grown on cultured cells, and this reversion was delayed by the addition of sera from immunized cattle, indicating a co-evolution of antigenicity and receptor usage in partially protected cattle (Tami et al., 2003).

The analysis of viruses derived from infectious cDNAs harbouring deletions in the L and 3B genes indicate that the integrity of these proteins is required to retain full virulence in natural hosts. Other evidences of the role of NS proteins on viral virulence and host tropism come from the analysis of viruses adapted to non-natural hosts and from naturally emerging FMDV variants with altered virulence. Thus, deletions in protein 3A were associated with attenuation for cattle of FMDV serotypes O and C, upon passages in chicken embryos (Giraudo et al., 1990). A 10-amino acid deletion together with different point mutations in 3A have been shown to contribute to the low virulence for cattle of a variant of FMDV serotype O, isolated in Taiwan in 1997 (Beard and Mason, 2000). Interestingly, this virus showed a highly virulent character for swine, which illustrates the species-specificity often associated to FMDV attenuation, and the complex interactions that take place between the host and the virus that determine the progress of the infection towards the emergence of lesions and clinical symptoms. The analysis of the molecular basis of the adaptation of a type C FMDV to the guinea-pig has also provided evidence on the role of protein 3A in the host-range of this virus. Viruses that acquired the capacity to produce fever and vesicular lesions in the guinea-pig exhibited single amino acid replacements in proteins 2C, 3A and VP1. The use of recombinant viruses harbouring these mutations revealed that the single amino acid replacement in 3A was sufficient to produce disease symptoms in guinea-pigs (Nunez et al., 2001).

Expression of different Picornavirus proteins can modulate apoptotic homeostasis leading to cell death (Buenz and Howe, 2006). The balance between proapototic and antiapopotic effects can be relevant for virus-cell interactions, particularly in the establishment of persistent infections. For FMDV, a proapoptotic effect has been recently described in cells expressing VP1 capsid protein (Peng et al., 2004).

FMDV infection and vaccination induces specific B and T lymphocyte responses in natural hosts that lead to production of neutralizing antibodies, whose levels correlate with the protection conferred (reviewed in (Collen, 1994; Doel, 2005; McCullough, 2004). In pigs, these responses seem to occur following a transient leuckopenia that mostly affect CD8+ cells (Diaz-San Segundo et al., 2006). FMDV can infect macrophage-derived cell lines following FC-receptor mediated adsorption (Baxt and Mason, 1995), which has been proposed as a mechanism for systemic viral spread (Rigden et al., 2002). FMDV is also highly sensitive to interferons, which have been proposed to mediate early protection against virus infection (Chinsangaram et al., 2001).

Vaccination to FMDV and its consequences for disease control exceeds the scope of this chapter. Current FMDV vaccines are based on chemically-inactivated viruses (Barteling, 2004; Doel, 2003). Even when these vaccines have proven useful, its use became a controversial issue because of the difficulty for an efficient serological distinction between infected and conventionally vaccinated animals when outbreaks occur in FMD-free countries (Sobrino and Domingo, 2001). Infection-specific tests based on the detection of antibodies to 3AB NS proteins are available and can contribute to reconsider emergency vaccination (Mackay, 1998). The wide genetic and antigenic diversity exhibited by FMDV in the field (seven serotypes, each showing wide antigenic and immunogenic spectra) complicates the vaccine choice and prompted the development of FMDV molecular epidemiology from nucleotide sequencing of different RNA regions (Knowles and Samuel, 2003). This scenario has led to search new safe immunogens and different strategies for vaccine delivery (reviewed in (Grubman, 2005; Rowlands, 2004; Sobrino et al., 2001). Antiviral strategies based on RNA interference have also been explored. Expression of short hairpin RNAs directed against a highly conserved sequence in the 2B nonstructural protein coding region of FMDV RNA in porcine cells induced a significant reduction in the synthesis of viral RNA and proteins, as well as a decrease in virus yield when compared to control cells (de los Santos et al., 2005). The antiviral effect was sequence specific, not attributable to induction of the interferon pathway, and effective against multiple serotypes. Thus, RNA interference may be a potential alternative control strategy to limit the spread of this highly contagious virus in livestock.

ACKNOWLEGEMENTS

This work was supported by grants BFU-2005-00948, BIO2005-07592-C02-01, CSD2006-0007 and by an Institutional grant from Fundación Ramón Areces.

REFERENCES

Acharya, R., Fry, E., Stuart, D., Fox, G., Rowlands, D., and Brown, F. (1989). The three-dimensional structure of foot-and-mouth disease virus at 2.9 A resolution. Nature *337*, 709-716.

Aldabe, R., and Carrasco, L. (1995). Induction of membrane proliferation by poliovirus proteins 2C and 2BC. Biochem. Biophys. Res. Commun. *206*, 64-76.
Alvarez, D.E., Lodeiro, M.F., Luduena, S.J., Pietrasanta, L.I., and Gamarnik, A.V. (2005). Long-range RNA-RNA interactions circularize the dengue virus genome. J. Virol. *79*, 6631-6643.
Aminev, A.G., Amineva, S.P., and Palmenberg, A.C. (2003). Encephalomyocarditis virus (EMCV) proteins 2A and 3BCD localize to nuclei and inhibit cellular mRNA transcription but not rRNA transcription. Virus. Res. *95*, 59-73.
Andino, R., Boddeker, N., Silvera, D., and Gamarnik, A.V. (1999). Intracellular determinants of picornavirus replication. Trends Microbiol. *7*, 76-82.
Baranowski, E., Molina, N., Nunez, J.I., Sobrino, F., and Saiz, M. (2003). Recovery of infectious foot-and-mouth disease virus from suckling mice after direct inoculation with in vitro-transcribed RNA. J. Virol. *77*, 11290-11295.
Baranowski, E., Ruiz-Jarabo, C.M., and Domingo, E. (2001). Evolution of cell recognition by viruses. Science *292*, 1102-1105.
Baranowski, E., Ruiz-Jarabo, C.M., Sevilla, N., Andreu, D., Beck, E., and Domingo, E. (2000). Cell recognition by foot-and-mouth disease virus that lacks the RGD integrin-binding motif: flexibility in aphthovirus receptor usage. J. Virol. *74*, 1641-1647.
Barteling, S.J. (2004). Modern inactivated Foot-and-mouth disease (FMD) vaccines: historical background and key elements in production and use. In Foot-and-mouth disease, F. Sobrino and E. Domingo, eds. (Norfolk, UK: Horizon Bioscience), pp. 305-334.
Baxt, B., and Mason, P.W. (1995). Foot-and-mouth disease virus undergoes restricted replication in macrophage cell cultures following Fc receptor-mediated adsorption. Virology *207*, 503-509.
Baxt, B., Vakharia, V., Moore, D.M., Franke, A.J., and Morgan, D.O. (1989). Analysis of neutralizing antigenic sites on the surface of type A12 foot-and-mouth disease virus. J. Virol. *63*, 2143-2151.
Baxt, B., and Rieder, E. (2004). Molecular aspects of foot-and-mouth disease virus virulence and host range: role of host cell receptors and viral factors, In Foot-and-mouth disease. Current perspectives, F. Sobrino and E. Domingo, eds. (Norfolk, UK: Horizon Bioscience), pp. 145-172.
Beard, C.W., and Mason, P.W. (2000). Genetic determinants of altered virulence of Taiwanese foot-and-mouth disease virus. J. Virol. *74*, 987-991.

Belsham, G.J. (1992). Dual initiation sites of protein synthesis on foot-and-mouth disease virus RNA are selected following internal entry and scanning of ribosomes in vivo. EMBO J. *11*, 1105-1110.

Belsham, G.J. and Martínez-Salas, E. (2004). Genome organization, translation and replication of foot-and-mouth disease virus. In Foot-and-mouth disease Current perspectives, F. Sobrino and E. Domingo, eds. (Norfolk, UK, Horizon Bioscience). pp. 19-52

Belsham, G.J., and Brangwyn, J.K. (1990). A region of the 5' noncoding region of foot-and-mouth disease virus RNA directs efficient internal initiation of protein synthesis within cells: involvement with the role of L protease in translational control. J. Virol. *64*, 5389-5395.

Belsham, G.J., McInerney, G.M., and Ross-Smith, N. (2000). Foot-and-mouth disease virus 3C protease induces cleavage of translation initiation factors eIF4A and eIF4G within infected cells. J. Virol. *74*, 272-280.

Belsham, G.J., and Sonenberg, N. (2000). Picornavirus RNA translation: roles for cellular proteins. Trends Microbiol. *8*, 330-335.

Berryman, S., Clark, S., Monaghan, P., and Jackson, T. (2005). Early events in integrin alphavbeta6-mediated cell entry of foot-and-mouth disease virus. J. Virol. *79*, 8519-8534.

Bienz, K., Egger, D., and Pasamontes, L. (1987). Association of polioviral proteins of the P2 genomic region with the viral replication complex and virus-induced membrane synthesis as visualized by electron microscopic immunocytochemistry and autoradiography. Virology *160*, 220-226.

Birtley, J.R., Knox, S.R., Jaulent, A.M., Brick, P., Leatherbarrow, R.J., and Curry, S. (2005). Crystal structure of foot-and-mouth disease virus 3C protease. New insights into catalytic mechanism and cleavage specificity. J. Biol. Chem. *280*, 11520-11527.

Bittle, J.L., Houghten, R.A., Alexander, H., Shinnick, T.M., Sutcliffe, J.G., Lerner, R.A., Rowlands, D.J., and Brown, F. (1982). Protection against foot-and-mouth disease by immunization with a chemically synthesized peptide predicted from the viral nucleotide sequence. Nature *298*, 30-33.

Brown, D.M., Cornell, C.T., Tran, G.P., Nguyen, J.H., and Semler, B.L. (2005). An authentic 3' noncoding region is necessary for efficient poliovirus replication. J. Virol. *79*, 11962-11973.

Brown, F. (1972). Structure-function relationships in foot-and-mouth disease virus. Adv. Exp. Med. Biol. *31*, 19-28.

Buenz, E.J., and Howe, C.L. (2006). Picornaviruses and cell death. Trends Microbiol. *14*, 28-36.

Cao, X., Bergmann, I.E., Fullkrug, R., and Beck, E. (1995). Functional analysis of the two alternative translation initiation sites of foot-and-mouth disease virus. J. Virol. *69*, 560-563.
Cavanagh, D., Sangar, D.V., Rowlands, D.J., and Brown, F. (1977). Immunogenic and cell attachment sites of FMDV: further evidence for their location in a single capsid polypeptide. J. Gen. Virol. *35*, 149-158.
Chinsangaram, J., Koster, M., and Grubman, M.J. (2001). Inhibition of L-deleted foot-and-mouth disease virus replication by alpha/beta interferon involves double-stranded RNA-dependent protein kinase. J. Virol. *75*, 5498-5503.
Cho, M.W., Teterina, N., Egger, D., Bienz, K., and Ehrenfeld, E. (1994). Membrane rearrangement and vesicle induction by recombinant poliovirus 2C and 2BC in human cells. Virology *202*, 129-145.
Choe, S.S., Dodd, D.A., and Kirkegaard, K. (2005). Inhibition of cellular protein secretion by picornaviral 3A proteins. Virology *337*, 18-29.
Chow, M., Newman, J.F., Filman, D., Hogle, J.M., Rowlands, D.J., and Brown, F. (1987). Myristylation of picornavirus capsid protein VP4 and its structural significance. Nature *327*, 482-486.
Collen, T. (1994). Foot-and-mouth disease (aphthovirus): viral T cell epitopes., In Cell mediated immunity in ruminants, B.M.L. Goddeeris, and W.I. Morrison, eds. (Boca Raton: CRC Press Inc.).
Curry, S., Abrams, C.C., Fry, E., Crowther, J.C., Belsham, G.J., Stuart, D.I., and King, A.M. (1995). Viral RNA modulates the acid sensitivity of foot-and-mouth disease virus capsids. J. Virol. *69*, 430-438.
de la Torre, J.C., Alarcon, B., Martinez-Salas, E., Carrasco, L., and Domingo, E. (1987). Ribavirin cures cells of a persistent infection with foot-and-mouth disease virus in vitro. J. Virol. *61*, 233-235.
de la Torre, J.C., Martinez-Salas, E., Diez, J., Villaverde, A., Gebauer, F., Rocha, E., Davila, M., and Domingo, E. (1988). Coevolution of cells and viruses in a persistent infection of foot-and-mouth disease virus in cell culture. J. Virol. *62*, 2050-2058.
de Los Santos, T., de Avila Botton, S., Weiblen, R., and Grubman, M.J. (2006). The leader proteinase of foot-and-mouth disease virus inhibits the induction of beta interferon mRNA and blocks the host innate immune response. J. Virol. *80*, 1906-1914.
de los Santos, T., Wu, Q., de Avila Botton, S., and Grubman, M.J. (2005). Short hairpin RNA targeted to the highly conserved 2B nonstructural protein coding region inhibits replication of multiple serotypes of foot-and-mouth disease virus. Virology *335*, 222-231.

Deitz, S.B., Dodd, D.A., Cooper, S., Parham, P., and Kirkegaard, K. (2000). MHC I-dependent antigen presentation is inhibited by poliovirus protein 3A. Proc. Natl. Acad. Sci. USA *97*, 13790-13795.

Devaney, M.A., Vakharia, V.N., Lloyd, R.E., Ehrenfeld, E., and Grubman, M.J. (1988). Leader protein of foot-and-mouth disease virus is required for cleavage of the p220 component of the cap-binding protein complex. J. Virol. *62*, 4407-4409.

Diaz-San Segundo, F., Salguero, F.J., de Avila, A., de Marco, M. M., Sanchez-Martin, M.A., and Sevilla, N. (2006). Selective lymphocyte depletion during the early stage of the immune response to foot-and-mouth disease virus infection in swine. J. Virol. *80*, 2369-2379.

Doedens, J.R., Giddings, T.H., Jr., and Kirkegaard, K. (1997). Inhibition of endoplasmic reticulum-to-Golgi traffic by poliovirus protein 3A: genetic and ultrastructural analysis. J. Virol. *71*, 9054-9064.

Doedens, J.R., and Kirkegaard, K. (1995). Inhibition of cellular protein secretion by poliovirus proteins 2B and 3A. EMBO J. *14*, 894-907.

Doel, T.R. (2003). FMD vaccines. Virus Res. *91*, 81-99.

Doel, T.R. (2005). Natural and vaccine induced immunity to FMD. Curr. Top. Microbiol. Immunol. *288*, 103-131.

Doherty, E.A., Batey, R.T., Masquida, B., and Doudna, J.A. (2001). A universal mode of helix packing in RNA. Nat. Struct. Biol. *8*, 339-343.

Domingo, E., Escarmis, C., Baranowski, E., Ruiz-Jarabo, C.M., Carrillo, E., Nunez, J.I., and Sobrino, F. (2003). Evolution of foot-and-mouth disease virus. Virus Res. *91*, 47-63.

Domingo, E., Escarmis, C., Martinez, M.A., Martinez-Salas, E., and Mateu, M.G. (1992). Foot-and-mouth disease virus populations are quasispecies. Curr. Top. Microbiol. Immunol. *176*, 33-47.

Domingo, E., Martin, V., Perales, C., Grande-Perez, A., Garcia-Arriaza, J., and Arias, A. (2006). Viruses as quasispecies: biological implications. Curr. Top. Microbiol. Immunol. *299*, 51-82.

Domingo, E., Mas, A., Yuste, E., Pariente, N., Sierra, S., Gutierrez-Riva, M., and Menendez-Arias, L. (2001). Virus population dynamics, fitness variations and the control of viral disease: an update. Prog. Drug. Res. *57*, 77-115.

Domingo, E., Ruiz-Jarabo, C., Arias, A., García-Arriaza, J., Escarmis, C. (2004). Quasispecies dynamics and evolution of foot-and-mouth disease virus, In Foot-and-mouth disease. Current perspectives, F. Sobrino and E. Domingo, eds. (Norfolk, UK: Horizon Bioscience), pp. 261-304

Donaldson, A. (2004). Clinical signs of foot-and-mouth disease virus, In Foot-and-mouth disease. Current perspectives, F. Sobrino and E. Domingo, eds. (Norfolk, UK: Horizon Bioscience), pp. 93-102.

Donnelly, M.L., Luke, G., Mehrotra, A., Li, X., Hughes, L.E., Gani, D., and Ryan, M.D. (2001). Analysis of the aphthovirus 2A/2B polyprotein 'cleavage' mechanism indicates not a proteolytic reaction, but a novel translational effect: a putative ribosomal 'skip'. J. Gen. Virol. *82*, 1013-1025.

Drew, J., and Belsham, G.J. (1994). trans complementation by RNA of defective foot-and-mouth disease virus internal ribosome entry site elements. J. Virol. *68*, 697-703.

Duque, H., and Baxt, B. (2003). Foot-and-mouth disease virus receptors: comparison of bovine alpha(V) integrin utilization by type A and O viruses. J. Virol. *77*, 2500-2511.

Duque, H., and Palmenberg, A.C. (2001). Phenotypic characterization of three phylogenetically conserved stem-loop motifs in the mengovirus 3' untranslated region. J. Virol. *75*, 3111-3120.

Escarmis, C., Dopazo, J., Davila, M., Palma, E.L., and Domingo, E. (1995). Large deletions in the 5'-untranslated region of foot-and-mouth disease virus of serotype C. Virus Res. *35*, 155-167.

Falk, M.M., Grigera, P.R., Bergmann, I.E., Zibert, A., Multhaup, G., and Beck, E. (1990). Foot-and-mouth disease virus protease 3C induces specific proteolytic cleavage of host cell histone H3. J. Virol. *64*, 748-756.

Falk, M.M., Sobrino, F., and Beck, E. (1992). VPg gene amplification correlates with infective particle formation in foot-and-mouth disease virus. J. Virol. *66*, 2251-2260.

Fernandez-Miragall, O., and Martinez-Salas, E. (2003). Structural organization of a viral IRES depends on the integrity of the GNRA motif. RNA *9*, 1333-1344.

Fernandez-Miragall, O., Ramos, R., Ramajo, J., and Martinez-Salas, E. (2006). Evidence of reciprocal tertiary interactions between conserved motifs involved in organizing RNA structure essential for internal initiation of translation. RNA *12*, 223-234.

Ferrer-Orta, C., Arias, A., Agudo, R., Perez-Luque, R., Escarmis, C., Domingo, E., and Verdaguer, N. (2006a). The structure of a protein primer-polymerase complex in the initiation of genome replication. EMBO J. *25*, 880-888.

Ferrer-Orta, C., Arias, A., Escarmis, C., and Verdaguer, N. (2006b). A comparison of viral RNA-dependent RNA polymerases. Curr. Opin. Struct. Biol. *16*, 27-34.

Ferrer-Orta, C. and Fita, I. (2004). Structure of foot-and-mouth disease virus particles, In Foot-and-mouth disease. Current perspectives, F. Sobrino, and E. Domingo, eds. (Norfolk, UK: Horizon Bioscience), pp. 77-92.

Filomatori, C.V., Lodeiro, M.F., Alvarez, D.E., Samsa, M.M., Pietrasanta, L., and Gamarnik, A.V. (2006). A 5' RNA element promotes dengue virus RNA synthesis on a circular genome. Genes Dev. *20*, 2238-2249.

Fox, G., Parry, N.R., Barnett, P.V., McGinn, B., Rowlands, D.J., and Brown, F. (1989). The cell attachment site on foot-and-mouth disease virus includes the amino acid sequence RGD (arginine-glycine-aspartic acid). J. Gen. Virol. *70 (Pt 3)*, 625-637.

Francis, M.J., Hastings, G.Z., Syred, A.D., McGinn, B., Brown, F., and Rowlands, D.J. (1987). Non-responsiveness to a foot-and-mouth disease virus peptide overcome by addition of foreign helper T-cell determinants. Nature *330*, 168-170.

Fry, E.E., Lea, S.M., Jackson, T., Newman, J.W., Ellard, F.M., Blakemore, W.E., Abu-Ghazaleh, R., Samuel, A., King, A.M., and Stuart, D.I. (1999). The structure and function of a foot-and-mouth disease virus-oligosaccharide receptor complex. EMBO J. *18*, 543-554.

Gamarnik, A.V., and Andino, R. (1997). Two functional complexes formed by KH domain containing proteins with the 5' noncoding region of poliovirus RNA. RNA *3*, 882-892.

Gamarnik, A.V., and Andino, R. (1998). Switch from translation to RNA replication in a positive-stranded RNA virus. Genes Dev. *12*, 2293-2304.

Gamarnik, A.V., Boddeker, N., and Andino, R. (2000). Translation and replication of human rhinovirus type 14 and mengovirus in Xenopus oocytes. J. Virol. *74*, 11983-11987.

Garcia-Briones, M., Rosas, M.F., Gonzalez-Magaldi, M., Martin-Acebes, M.A., Sobrino, F., and Armas-Portela, R. (2006). Differential distribution of non-structural proteins of foot-and-mouth disease virus in BHK-21 cells. Virology *349*, 409-421.

Gerber, K., Wimmer, E., and Paul, A.V. (2001). Biochemical and genetic studies of the initiation of human rhinovirus 2 RNA replication: purification and enzymatic analysis of the RNA-dependent RNA polymerase 3D(pol). J. Virol. *75*, 10969-10978.

Giraudo, A.T., Beck, E., Strebel, K., de Mello, P.A., La Torre, J.L., Scodeller, E.A., and Bergmann, I.E. (1990). Identification of a nucleotide deletion in parts of polypeptide 3A in two independent attenuated aphthovirus strains. Virology *177*, 780-783.

Goodfellow, I., Chaudhry, Y., Richardson, A., Meredith, J., Almond, J.W., Barclay, W., and Evans, D.J. (2000). Identification of a cis-acting replication element within the poliovirus coding region. J. Virol. *74*, 4590-4600.

Gradi, A., Foeger, N., Strong, R., Svitkin, Y.V., Sonenberg, N., Skern, T., and Belsham, G.J. (2004). Cleavage of eukaryotic translation initiation factor 4GII within foot-and-mouth disease virus-infected cells: identification of the L-protease cleavage site in vitro. J. Virol. *78*, 3271-3278.

Grubman, M.J. (2005). Development of novel strategies to control foot-and-mouth disease: marker vaccines and antivirals. Biologicals *33*, 227-234.

Grubman, M.J., Zellner, M., Bablanian, G., Mason, P.W., and Piccone, M.E. (1995). Identification of the active-site residues of the 3C proteinase of foot-and-mouth disease virus. Virology *213*, 581-589.

Guarne, A., Tormo, J., Kirchweger, R., Pfistermueller, D., Fita, I., and Skern, T. (1998). Structure of the foot-and-mouth disease virus leader protease: a papain-like fold adapted for self-processing and eIF4G recognition. EMBO J. *17*, 7469-7479.

Harris, T.J., and Brown, F. (1977). Biochemical analysis of a virulent and an avirulent strain of foot-and-mouth disease virus. J. Gen. Virol. *34*, 87-105.

Herold, J., and Andino, R. (2001). Poliovirus RNA replication requires genome circularization through a protein-protein bridge. Mol. Cell. *7*, 581-591.

Hunt, S.L., Hsuan, J.J., Totty, N., and Jackson, R.J. (1999). unr, a cellular cytoplasmic RNA-binding protein with five cold-shock domains, is required for internal initiation of translation of human rhinovirus RNA. Genes Dev. *13*, 437-448.

Isken, O., Grassmann, C.W., Yu, H., and Behrens, S.E. (2004). Complex signals in the genomic 3' nontranslated region of bovine viral diarrhea virus coordinate translation and replication of the viral RNA. RNA *10*, 1637-1652.

Jackson, T., Clark, S., Berryman, S., Burman, A., Cambier, S., Mu, D., Nishimura, S., and King, A.M. (2004). Integrin alphavbeta8 functions as a receptor for foot-and-mouth disease virus: role of the beta-chain cytodomain in integrin-mediated infection. J. Virol. *78*, 4533-4540.

Jackson, T., Ellard, F.M., Ghazaleh, R.A., Brookes, S.M., Blakemore, W.E., Corteyn, A.H., Stuart, D.I., Newman, J.W., and King, A.M. (1996). Efficient infection of cells in culture by type O foot-and-mouth disease virus requires binding to cell surface heparan sulfate. J. Virol. *70*, 5282-5287.

Jackson, T., Sheppard, D., Denyer, M., Blakemore, W., and King, A.M. (2000). The epithelial integrin alphavbeta6 is a receptor for foot-and-mouth disease virus. J. Virol. *74*, 4949-4956.

Kirchweger, R., Ziegler, E., Lamphear, B. J., Waters, D., Liebig, H. D., Sommergruber, W., Sobrino, F., Hohenadl, C., Blaas, D., Rhoads, R.E., and et al. (1994). Foot-and-mouth disease virus leader proteinase: purification of the Lb form and determination of its cleavage site on eIF-4 gamma. J. Virol. *68*, 5677-5684.

Kitson, J.D., McCahon, D., and Belsham, G.J. (1990). Sequence analysis of monoclonal antibody resistant mutants of type O foot and mouth disease virus: evidence for the involvement of the three surface exposed capsid proteins in four antigenic sites. Virology *179*, 26-34.

Knowles, N.J. and Samuel, A.R. (2003). Molecular epidemiology of foot-and-mouth disease virus. Virus Res. *91*, 65-80.

Knox, C., Moffat, K., Ali, S., Ryan, M., and Wileman, T. (2005). Foot-and-mouth disease virus replication sites form next to the nucleus and close to the Golgi apparatus, but exclude marker proteins associated with host membrane compartments. J. Gen. Virol. *86*, 687-696.

Kolupaeva, V.G., Pestova, T.V., Hellen, C.U., and Shatsky, I.N. (1998). Translation eukaryotic initiation factor 4G recognizes a specific structural element within the internal ribosome entry site of encephalomyocarditis virus RNA. J. Biol. Chem. *273*, 18599-18604.

Kozak, M. (1989). The scanning model for translation: an update. J. Cell. Biol. *108*, 229-241.

Lea, S., Hernandez, J., Blakemore, W., Brocchi, E., Curry, S., Domingo, E., Fry, E., Abu-Ghazaleh, R., King, A., Newman, J., and et al. (1994). The structure and antigenicity of a type C foot-and-mouth disease virus. Structure *2*, 123-139.

Li, W., Ross-Smith, N., Proud, C.G., and Belsham, G.J. (2001). Cleavage of translation initiation factor 4AI (eIF4AI) but not eIF4AII by foot-and-mouth disease virus 3C protease: identification of the eIF4AI cleavage site. FEBS Lett. *507*, 1-5.

Lopez de Quinto, S., Lafuente, E., and Martinez-Salas, E. (2001). IRES interaction with translation initiation factors: functional characterization of novel RNA contacts with eIF3, eIF4B, and eIF4GII. RNA *7*, 1213-1226.

Lopez de Quinto, S., and Martinez-Salas, E. (1997). Conserved structural motifs located in distal loops of aphthovirus internal ribosome entry site domain 3 are required for internal initiation of translation. J. Virol. *71*, 4171-4175.

Lopez de Quinto, S., and Martinez-Salas, E. (1999). Involvement of the aphthovirus RNA region located between the two functional AUGs in start codon selection. Virology *255*, 324-336.

Lopez de Quinto, S., and Martinez-Salas, E. (2000). Interaction of the eIF4G initiation factor with the aphthovirus IRES is essential for internal translation initiation in vivo. RNA *6*, 1380-1392.

Lopez de Quinto, S., Saiz, M., de la Morena, D., Sobrino, F., and Martinez-Salas, E. (2002). IRES-driven translation is stimulated separately by the FMDV 3'-NCR and poly(A) sequences. Nucleic Acids Res. *30*, 4398-4405.

Luz, N., and Beck, E. (1991). Interaction of a cellular 57-kilodalton protein with the internal translation initiation site of foot-and-mouth disease virus. J. Virol. *65*, 6486-6494.

Mackay, D.K. (1998). Differentiating infection from vaccination in foot-and-mouth disease. Vet. Q. *20 Suppl 2*, 2-5.

Martin, L.R., and Palmenberg, A.C. (1996). Tandem mengovirus 5' pseudoknots are linked to viral RNA synthesis, not poly(C)-mediated virulence. J. Virol. *70*, 8182-8186.

Martinez-Salas, E. (1999). Internal ribosome entry site biology and its use in expression vectors. Curr. Opin. Biotechnol. *10*, 458-464.

Martinez-Salas, E., and Domingo, E. (1995). Effect of expression of the aphthovirus protease 3C on viral infection and gene expression. Virology *212*, 111-120.

Martinez-Salas, E., and Fernandez-Miragall, O. (2004). Picornavirus IRES: structure function relationship. Curr. Pharm. Des. *10*, 3757-3767.

Martinez-Salas, E., Lopez de Quinto, S., Ramos, R., and Fernandez-Miragall, O. (2002). IRES elements: features of the RNA structure contributing to their activity. Biochimie *84*, 755-763.

Martinez-Salas, E., Ortin, J., and Domingo, E. (1985). Sequence of the viral replicase gene from foot-and-mouth disease virus C1-Santa Pau (C-S8). Gene *35*, 55-61.

Martinez-Salas, E., Ramos, R., Lafuente, E., and Lopez de Quinto, S. (2001). Functional interactions in internal translation initiation directed by viral and cellular IRES elements. J. Gen. Virol. *82*, 973-984.

Martinez-Salas, E., Regalado, M.P., and Domingo, E. (1996). Identification of an essential region for internal initiation of translation in the aphthovirus internal ribosome entry site and implications for viral evolution. J. Virol. *70*, 992-998.

Martinez-Salas, E., Saiz, J.C., Davila, M., Belsham, G.J., and Domingo, E. (1993). A single nucleotide substitution in the internal ribosome entry site of foot-and-mouth disease virus leads to enhanced cap-independent translation in vivo. J. Virol. *67*, 3748-3755.

Mason, P.W., Bezborodova, S.V., and Henry, T.M. (2002). Identification and characterization of a cis-acting replication element (cre) adjacent to the internal ribosome entry site of foot-and-mouth disease virus. J. Virol. *76*, 9686-9694.

Mason, P.W., Grubman, M.J., and Baxt, B. (2003). Molecular basis of pathogenesis of FMDV. Virus Res. *91*, 9-32.

Mason, P.W., Rieder, E., and Baxt, B. (1994). RGD sequence of foot-and-mouth disease virus is essential for infecting cells via the natural receptor but can be bypassed by an antibody-dependent enhancement pathway. Proc. Natl. Acad. Sci. USA *91*, 1932-1936.

Mateu, M.G., and Verdaguer, N. (2004). Functional and structural aspects of the interaction of foot-and-mouth disease virus with antibodies, In Foot-and-mouth disease. Current perspectives, F. Sobrino and E. Domingo, eds. (Norfolk, UK: Horizon Bioscience), pp. 223-260.

Mateu, M.G. (1995). Antibody recognition of picornaviruses and escape from neutralization: a structural view. Virus Res. *38*, 1-24.

Mateu, M.G., Valero, M.L., Andreu, D., and Domingo, E. (1996). Systematic replacement of amino acid residues within an Arg-Gly-Asp-containing loop of foot-and-mouth disease virus and effect on cell recognition. J. Biol. Chem. *271*, 12814-12819.

McCullough, K.C., and Sobrino, F. (2004). Immunology of foot-and-mouth disease virus, In Foot-and-mouth disease. Current perspectives, F. Sobrino and E. Domingo, eds. (Norfolk, UK: Horizon Bioscience), pp. 173-222.

McInerney, G.M., King, A.M., Ross-Smith, N., and Belsham, G.J. (2000). Replication-competent foot-and-mouth disease virus RNAs lacking capsid coding sequences. J. Gen. Virol. *81*, 1699-1702.

McKnight, K.L., and Lemon, S.M. (1998). The rhinovirus type 14 genome contains an internally located RNA structure that is required for viral replication. RNA *4*, 1569-1584.

Medina, M., Domingo, E., Brangwyn, J.K., and Belsham, G.J. (1993). The two species of the foot-and-mouth disease virus leader protein, expressed individually, exhibit the same activities. Virology *194*, 355-359.

Mellits, K.H., Meredith, J.M., Rohll, J.B., Evans, D.J., and Almond, J.W. (1998). Binding of a cellular factor to the 3' untranslated region of the RNA genomes of entero- and rhinoviruses plays a role in virus replication. J. Gen. Virol. *79*, 1715-1723.

Moffat, K., Howell, G., Knox, C., Belsham, G.J., Monaghan, P., Ryan, M.D., and Wileman, T. (2005). Effects of foot-and-mouth disease virus nonstructural proteins on the structure and function of the early secretory pathway: 2BC but not 3A blocks endoplasmic reticulum-to-Golgi transport. J. Virol. *79*, 4382-4395.

Monaghan, P., Cook, H., Jackson, T., Ryan, M., and Wileman, T. (2004). The ultrastructure of the developing replication site in foot-and-mouth disease virus-infected BHK-38 cells. J. Gen. Virol. *85*, 933-946.

Monaghan, P., Gold, S., Simpson, J., Zhang, Z., Weinreb, P.H., Violette, S.M., Alexandersen, S., and Jackson, T. (2005). The alpha(v)beta6 integrin receptor for Foot-and-mouth disease virus is expressed constitutively on the epithelial cells targeted in cattle. J. Gen. Virol. *86*, 2769-2780.

Nayak, A., Goodfellow, I.G., Woolaway, K.E., Birtley, J., Curry, S., and Belsham, G.J. (2006). Role of RNA structure and RNA binding activity of foot-and-mouth disease virus 3C protein in VPg uridylylation and virus replication. J. Virol. *80*, 9865-9875.

Neff, S., Sa-Carvalho, D., Rieder, E., Mason, P.W., Blystone, S.D., Brown, E.J., and Baxt, B. (1998). Foot-and-mouth disease virus virulent for cattle utilizes the integrin alpha(v)beta3 as its receptor. J. Virol. *72*, 3587-3594.

Nunez, J.I., Baranowski, E., Molina, N., Ruiz-Jarabo, C.M., Sanchez, C., Domingo, E., and Sobrino, F. (2001). A single amino acid substitution in nonstructural protein 3A can mediate adaptation of foot-and-mouth disease virus to the guinea pig. J. Virol. *75*, 3977-3983.

O'Donnell, V., LaRocco, M., Duque, H., and Baxt, B. (2005). Analysis of foot-and-mouth disease virus internalization events in cultured cells. J. Virol. *79*, 8506-8518.

O'Donnell, V.K., Pacheco, J.M., Henry, T.M., and Mason, P.W. (2001). Subcellular distribution of the foot-and-mouth disease virus 3A protein in cells infected with viruses encoding wild-type and bovine-attenuated forms of 3A. Virology *287*, 151-162.

Ohlmann, T., Rau, M., Pain, V.M., and Morley, S.J. (1996). The C-terminal domain of eukaryotic protein synthesis initiation factor (eIF) 4G is sufficient to support cap-independent translation in the absence of eIF4E. EMBO J. *15*, 1371-1382.

Parry, N., Fox, G., Rowlands, D., Brown, F., Fry, E., Acharya, R., Logan, D., and Stuart, D. (1990). Structural and serological evidence for a novel mechanism of antigenic variation in foot-and-mouth disease virus. Nature *347*, 569-572.

Parsley, T.B., Towner, J.S., Blyn, L.B., Ehrenfeld, E., and Semler, B.L. (1997). Poly (rC) binding protein 2 forms a ternary complex with the 5'-terminal sequences of poliovirus RNA and the viral 3CD proteinase. RNA *3*, 1124-1134.

Paul, A.V., Rieder, E., Kim, D.W., van Boom, J.H., and Wimmer, E. (2000). Identification of an RNA hairpin in poliovirus RNA that serves as the primary template in the in vitro uridylylation of VPg. J. Virol. *74*, 10359-10370.

Pelletier, J., and Sonenberg, N. (1988). Internal initiation of translation of eukaryotic mRNA directed by a sequence derived from poliovirus RNA. Nature *334*, 320-325.

Peng, J.M., Liang, S.M., and Liang, C.M. (2004). VP1 of foot-and-mouth disease virus induces apoptosis via the Akt signaling pathway. J. Biol. Chem. *279*, 52168-52174.

Pestova, T.V., Hellen, C.U., and Shatsky, I.N. (1996). Canonical eukaryotic initiation factors determine initiation of translation by internal ribosomal entry. Mol. Cell. Biol. *16*, 6859-6869.

Pfaff, E. (1997). Recognition sites of RGD-dependent integrins In: Integrin Ligand Interactions RG LAndes Co., Austin, TX USA.

Pfaff, E., Mussgay, M., Bohm, H.O., Schulz, G.E., and Schaller, H. (1982). Antibodies against a preselected peptide recognize and neutralize foot and mouth disease virus. EMBO J. *1*, 869-874.

Pfaff, E., Thiel, H.J., Beck, E., Strohmaier, K., and Schaller, H. (1988). Analysis of neutralizing epitopes on foot-and-mouth disease virus. J. Virol. *62*, 2033-2040.

Piccone, M.E., Rieder, E., Mason, P.W., and Grubman, M.J. (1995). The foot-and-mouth disease virus leader proteinase gene is not required for viral replication. J. Virol. *69*, 5376-5382.

Pilipenko, E.V., Pestova, T.V., Kolupaeva, V.G., Khitrina, E.V., Poperechnaya, A.N., Agol, V.I., and Hellen, C.U. (2000). A cell cycle-dependent protein serves as a template-specific translation initiation factor. Genes Dev. *14*, 2028-2045.

Pilipenko, E.V., Viktorova, E.G., Guest, S.T., Agol, V.I., and Roos, R.P. (2001). Cell-specific proteins regulate viral RNA translation and virus-induced disease. EMBO J. *20*, 6899-6908.

Pisarev, A.V., Chard, L.S., Kaku, Y., Johns, H.L., Shatsky, I.N., and Belsham, G.J. (2004). Functional and structural similarities between the internal ribosome entry sites of hepatitis C virus and porcine teschovirus, a picornavirus. J. Virol. *78*, 4487-4497.

Ramos, R., and Martinez-Salas, E. (1999). Long-range RNA interactions between structural domains of the aphthovirus internal ribosome entry site (IRES). RNA *5*, 1374-1383.

Reigadas, S., Pacheco, A., Ramajo, J., de Quinto, S.L., and Martinez-Salas, E. (2005). Specific interference between two unrelated internal ribosome entry site elements impairs translation efficiency. FEBS Lett. *579*, 6803-6808.

Rigden, R.C., Carrasco, C.P., Summerfield, A., and McCullough, K.C. (2002). Macrophage phagocytosis of foot-and-mouth disease virus may create infectious carriers. Immunology *106*, 537-548.

Robertson, M.E., Seamons, R.A., and Belsham, G.J. (1999). A selection system for functional internal ribosome entry site (IRES) elements: analysis of the requirement for a conserved GNRA tetraloop in the encephalomyocarditis virus IRES. RNA *5*, 1167-1179.

Rodriguez Pulido, M., Serrano, P., Saiz, M., and Martinez-Salas, E. (2007). Foot-and-mouth disease virus infection induces proteolytic cleavage of PTB, eIF3a,b, and PABP RNA-binding proteins. Virology, *364*, 466-474.

Rowlands, D. (2004). Foot-and-mouth disease virus peptide vaccines, In Foot-and-mouth disease. Current perspectives, F. Sobrino and E. Domingo, eds. (Norfolk, UK: Horizon Bioscience). pp. 335-354.

Saiz, M., Gomez, S., Martinez-Salas, E., and Sobrino, F. (2001). Deletion or substitution of the aphthovirus 3' NCR abrogates infectivity and virus replication. J. Gen. Virol. *82*, 93-101.

Salt, J.S. (1993). The carrier state in foot and mouth disease--an immunological review. Br. Vet. J. *149*, 207-223.

Sangar, D.V., Newton, S.E., Rowlands, D.J., and Clarke, B.E. (1987). All foot and mouth disease virus serotypes initiate protein synthesis at two separate AUGs. Nucleic Acids Res. *15*, 3305-3315.

Sanz-Parra, A., Sobrino, F., and Ley, V. (1998). Infection with foot-and-mouth disease virus results in a rapid reduction of MHC class I surface expression. J. Gen. Virol. *79*, 433-436.

Saunders, K., and King, A.M. (1982). Guanidine-resistant mutants of aphthovirus induce the synthesis of an altered nonstructural polypeptide, P34. J. Virol. *42*, 389-394.

Serrano, P., Gomez, J. and Martinez-Salas, E. (2007). Characterization of a cyanobacterial RNase P ribozyme recognition motif in the IRES of foot-and-mouth disease virus reveals a unique structural element. RNA *13*, 849-859.

Serrano, P., Pulido, M.R., Saiz, M., and Martinez-Salas, E. (2006). The 3' end of the foot-and-mouth disease virus genome establishes two distinct long-range RNA-RNA interactions with the 5' end region. J. Gen. Virol. *87*, 3013-3022.

Sobrino, F., and Domingo, E. (2001). Foot-and-mouth disease in Europe. EMBO reports *2*, 459-461.

Sobrino, F., Saiz, M., Jimenez-Clavero, M.A., Nunez, J.I., Rosas, M.F., Baranowski, E., and Ley, V. (2001). Foot-and-mouth disease virus: a long known virus, but a current threat. Vet. Res. *32*, 1-30.

Song, Y., Tzima, E., Ochs, K., Bassili, G., Trusheim, H., Linder, M., Preissner, K.T., and Niepmann, M. (2005). Evidence for an RNA chaperone function of polypyrimidine tract-binding protein in picornavirus translation. RNA *11*, 1809-1824.

Stassinopoulos, I.A., and Belsham, G.J. (2001). A novel protein-RNA binding assay: functional interactions of the foot-and-mouth disease virus internal ribosome entry site with cellular proteins. RNA *7*, 114-122.

Strong, R., and Belsham, G.J. (2004). Sequential modification of translation initiation factor eIF4GI by two different foot-and-mouth disease virus proteases within infected baby hamster kidney cells: identification of the 3Cpro cleavage site. J. Gen. Virol. *85*, 2953-2962.

Sutmoller, P., Barteling, S.S., Olascoaga, R.C., and Sumption, K.J. (2003). Control and eradication of foot-and-mouth disease. Virus Res. *91*, 101-144.

Taboga, O., Tami, C., Carrillo, E., Nunez, J.I., Rodriguez, A., Saiz, J.C., Blanco, E., Valero, M.L., Roig, X., Camarero, J.A., *et al.* (1997). A large-scale evaluation of peptide vaccines against foot-and-mouth disease: lack of solid protection in cattle and isolation of escape mutants. J. Virol. *71*, 2606-2614.

Tami, C., Taboga, O., Berinstein, A., Nunez, J.I., Palma, E.L., Domingo, E., Sobrino, F., and Carrillo, E. (2003). Evidence of the coevolution of antigenicity and host cell tropism of foot-and-mouth disease virus in vivo. J. Virol. *77*, 1219-1226.

Thomas, A.A., Woortmeijer, R.J., Puijk, W., and Barteling, S.J. (1988). Antigenic sites on foot-and-mouth disease virus type A10. J. Virol. *62*, 2782-2789.

Thomson, G.R., Vosloo, W., and Bastos, A.D. (2003). Foot and mouth disease in wildlife. Virus Res. *91*, 145-161.

Tiley, L., King, A.M., and Belsham, G.J. (2003). The foot-and-mouth disease virus cis-acting replication element (cre) can be complemented in trans within infected cells. J. Virol. *77*, 2243-2246.

Todd, S., Towner, J.S., and Semler, B.L. (1997). Translation and replication properties of the human rhinovirus genome in vivo and in vitro. Virology *229*, 90-97.

Towner, J.S., Ho, T.V., and Semler, B.L. (1996). Determinants of membrane association for poliovirus protein 3AB. J. Biol. Chem. *271*, 26810-26818.

van Kuppeveld, F.J., Melchers, W.J., Kirkegaard, K., and Doedens, J.R. (1997). Structure-function analysis of coxsackie B3 virus protein 2B. Virology *227*, 111-118.

van Kuppeveld, F.J., van den Hurk, P.J., Zoll, J., Galama, J.M., and Melchers, W.J. (1996). Mutagenesis of the coxsackie B3 virus 2B/2C cleavage site: determinants of processing efficiency and effects on viral replication. J. Virol. *70*, 7632-7640.

Vende, P., Piron, M., Castagne, N., and Poncet, D. (2000). Efficient translation of rotavirus mRNA requires simultaneous interaction of NSP3 with the eukaryotic translation initiation factor eIF4G and the mRNA 3' end. J. Virol. *74*, 7064-7071.

Walter, B.L., Nguyen, J.H., Ehrenfeld, E., and Semler, B.L. (1999). Differential utilization of poly(rC) binding protein 2 in translation directed by picornavirus IRES elements. RNA *5*, 1570-1585.

Whitton, J.L., Cornell, C.T., and Feuer, R. (2005). Host and virus determinants of picornavirus pathogenesis and tropism. Nat. Rev. Microbiol. *3*, 765-776.

Zhao, Q., Pacheco, J.M., and Mason, P.W. (2003). Evaluation of genetically engineered derivatives of a Chinese strain of foot-and-mouth disease virus reveals a novel cell-binding site which functions in cell culture and in animals. J. Virol. *77*, 3269-3280.

Zibert, A., Maass, G., Strebel, K., Falk, M.M., and Beck, E. (1990). Infectious foot-and-mouth disease virus derived from a cloned full-length cDNA. J. Virol. *64*, 2467-2473.

Chapter 2

Molecular Biology of Pestiviruses

Till Rümenapf and Heinz–Jürgen Thiel

ABSTRACT

Pestiviruses account for important diseases in animals such as Classical swine fever (CSF) and Bovine viral diarrhea / Mucosal disease (BVD/MD). According to the current O.I.E. list CSF and BVD/MD are notifiable diseases and eradication programms are administered in many countries worldwide. The molecular biology of pestiviruses shares many similarities and peculiarities with the human hepaciviruses. Genome organisation and translation strategy are highly similar for the members of both genera. Within the last two decades a number of surprising molecular features have been unraveled that correlate to some aspects of the unique pathogenesis of pestiviral diseases. One hallmark of pestiviruses is their unique strategy to establish persistent infection during pregnancy. Persistent infection with pestiviruses often goes unnoticed; for BVDV frequently nonhomologous RNA recombination events lead to the appearance of genetically distinct viruses that are lethal to the host. The current understanding of the complex mechanisms that define pathogenesis of pestivirus induced diseases, epidemiology as well as eradication and control are reviewed in this article.

GENERAL ASPECTS OF PESTIVIRUSES

TAXONOMY

Pestiviruses belong to a large family of positive strand RNA viruses, the *Flaviviridae*. The family currently consists of three genera: *Flavivirus*, *Pestivirus* (from Latin pestis, "plague") and *Hepacivirus*. In addition to these genera, two groups of unassigned viruses, GBV-A and GBV-C, await formal classification within the family (Thiel et al., 2005). The phylogenetic relationships within the virus family are depicted in Fig. 1. Members of the family are listed in Table 1.

Within the *Flaviviridae*, pestiviruses show greater similarity in genome structure and mechanism of initiating translation to the hepaciviruses than to members of the genus *Flavivirus*.

The positive strand RNA genome of all three genera has a similar organization and serves three distinct roles in the viral life cycle, namely as messenger RNA for translation of viral proteins, as template during RNA replication, and as genetic material packaged within viral particles. All viral proteins are generated as part of a single large polyprotein of > 4000 amino acids that is cleaved by a combination of host and viral proteases.

Table 1 – Members of the family *Flaviviridae*

Taxonomic Unit	Examples
Pestivirus	Bovine viral diarrhea virus-1 (BVDV-1), BVDV-2 Classical swine fever virus (CSFV) Border disease virus (BDV) Pestivirus of giraffe (Tentative species)
Flavivirus Tick-borne viruses Mosquito-borne viruses	 Tick-borne encephalitis virus Yellow fever virus West Nile virus
Hepacivirus	Hepatitis C virus (HCV) with six genotypes GB virus B (GBV-B)
Unclassified	GB virus A (GBV-A) GB virus C (GBV-C)

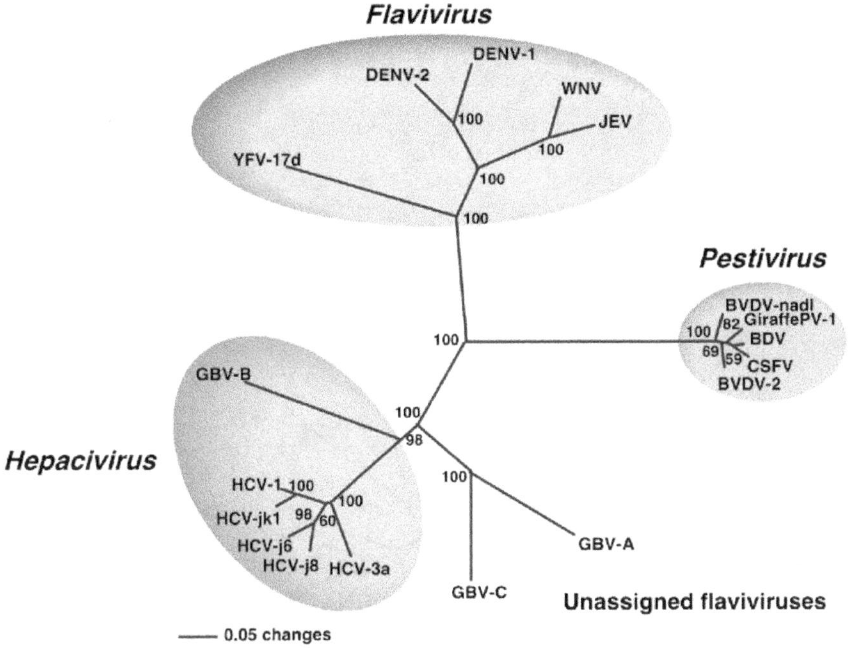

Fig. 1. Phylogenetic relationship of the helicase region of members of the family *Flaviviridae*. Partial gene sequences (~280 aa) from the proposed helicase region were used for the phylogenetic analysis which include representative strains from each genus. CLUSTALX v1.81 was used to create a multiple alignment for the aa sequences which was verified by alignment of the known motifs in the region (e.g., GxGKS/T). An unrooted phylogenetic tree was constructed from the sequence alignment using the distance method Neighbor-joining from the PAUP 4b10 package within a Macintosh environment. The virus names corresponding to the abbreviations can be found in the "List of Species" in each genus and the Genbank accession numbers are AF037405, U18059, M31182, J04358, M87512, M29095, U22303, U22304, NC_002348, AF144617, M62321, D17763, D00944, D10988, X61596, M18370, M12294, NC_002031. Bootstrap replicates of 1000 trees was also examined using PAUP and the percentages are drawn at the branch points of the tree (Courtesy of L. McMullan). Reprinted from Virus Taxonomy, VIIIth Report of ICTV (C.M. Fauquet, M.A. Mayo, J. Maniloff, U. Desselberger, and L.A. Ball, eds), Chapter *Flaviviridae*, p. 997-998 (2005) with permission from Elsevier.

Pestiviruses are animal pathogens of major economic importance for the livestock industry. They include the type member, Bovine viral diarrhea virus (BVDV), as well as Classical swine fever virus (CSFV), and Border disease virus (BVD) of sheep (Table 1). Based primarily on nucleotide and antigenic relatedness, two distinct species of BVDV have been identified, BVDV-1 and BVDV-2 (Thiel et al., 2005). Newly characterized pestiviruses from sheep and also from giraffe and other ruminants indicate additional diversity within this genus (Fig. 2). Accordingly, the giraffe pestivirus (strain H38) has been included as tentative species in the genus *Pestivirus* (Thiel et al., 2005).

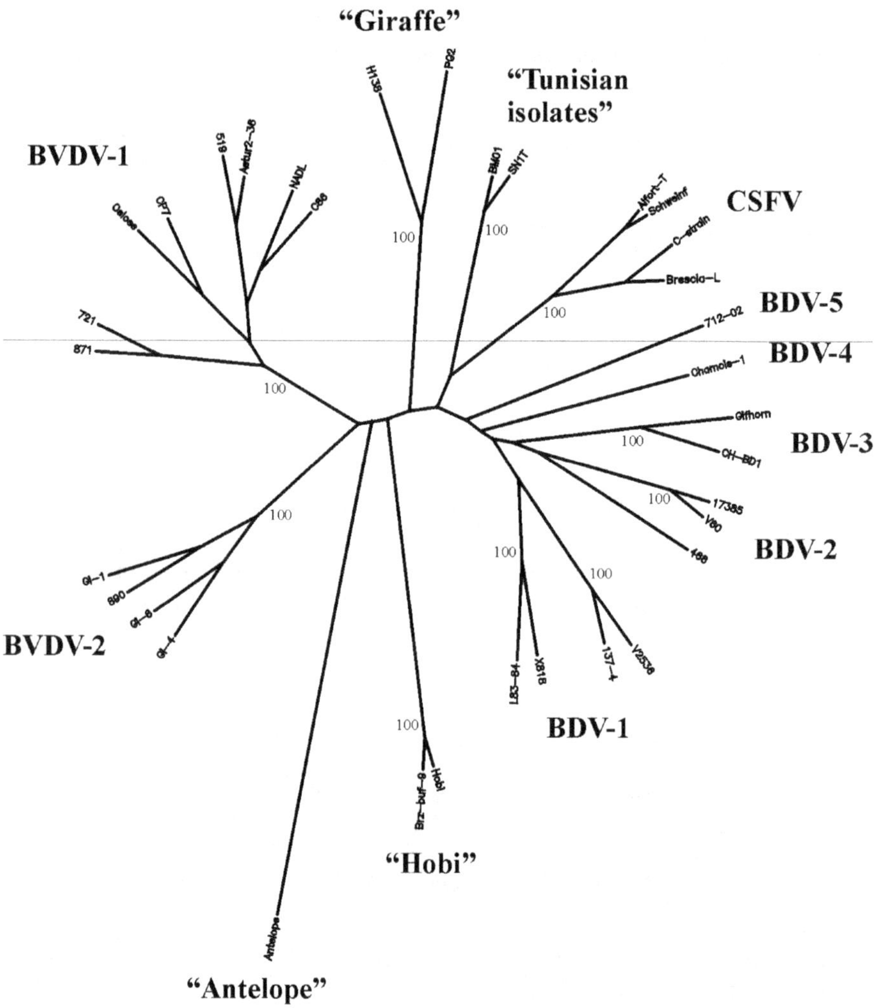

Fig. 2. Dendrogram showing the genetic relationships of pestivirus strains. The phylogenetic tree was constructed from the complete Npro coding sequences. The sequences were obtained from the GenBank database. Multiple sequence alignments were computed by PILEUP included in the GCG software package (Devereux et al., 1984). Distances were calculated by the Kimura 2-parameter method (Kimura, 1980) and used to construct the tree according to the neighbor-joining method (PHYLIP, 3.5c). Branch lengths are proportional to genetic distances. Numbers indicate the percentage of 1000 bootstrap replicates that support each labeled interior branch. This is an unrooted tree. Phylogenetic analysis was performed by P. Becher.

Genetic comparison of pestiviruses has been the subject of several studies. Most frequently partial sequences of the 5' NTR (non translated region) were analyzed resulting in the availability of more than 1000 pestivirus 5' NTR sequences. These can be used for the segregation of pestivirus isolates into the established four species, but are less suited for a further subdivision of the virus species into defined genotypes or subgroups. Segregation of pestiviruses into species as well as genotypes or subgroups can be better accomplished on the basis of N^{pro}, E2 and other coding sequences. As an additional criterion, the antigenic relationship of pestivirus isolates covering most major genotypes has been studied by cross-neutralization (Becher et al., 2003). Available analyses showed that BDV-2, BDV-3, BDV-4 and BDV-5 represent additional groups of pestiviruses and that the isolate PG-2 is closely related to the giraffe pestivirus strain H38 (Fig. 2). The latter two viruses should be members of an approved species of the genus *Pestivirus*. BDV-1, BDV-2, BDV-3, BDV-4 and BDV-5 can be classified either as distinct species or as major genotypes of one species. According to (i) the fact that the BDVs 1-5 are genetically more closely related to each other than to other pestivirus species; (ii) lack of apparent differences concerning the course of disease; and (iii) host range, it is proposed that BDVs 1-5 should be considered as five genotypes of one species (BDV). The criteria outlined above for BDV apply mostly also to BVDVs which are currently divided into two species. Taking into account the proposal for BDV, we suggest that BVDV-1 and BVDV-2 should be two genotypes within one species (Becher et al., 2003). The phylogenetic tree in Fig. 2 depicts the four established pestivirus species together with respective genotypes, subgroups and isolates as well as thus far not classified pestiviruses comprising "Tunisian isolates", Antelope" and "Hobi".

CULTIVATION

Pestiviruses readily adapt to various tissue culture cells that originate from the respective host species. With few exceptions, e.g. rabbit cells, propagation of pestiviruses succeeds in cells from cloven hoofed animals, preferentially from the same animal species that the virus was isolated from. Older reports about the cultivation of CSFV in a number of mammalian cell lines including those of marine mammals (Pirtle and Kniazeff, 1968) have not been confirmed. A number of established cell lines from different tissues exist for the propagation of porcine, ovine and bovine pestiviruses. Most frequently epithelial cell lines from porcine (SK6, PK15 (Moormann and Hulst, 1988; Rümenapf et al., 1989)) or bovine kidney (MDBK (Collett et al., 1988b)) are used but also cell lines from testis, lung, brain or lymphomas are common (Grummer et al., 2006). Titers of up to 10^8 $TCID_{50}$ /ml are observed with pestivirus strains that have a long passage history (e.g. BVDV NADL, CSFV Alfort). BVDV or CSFV isolates that come directly from infected animals grow to lower titers. According to their behaviour in tissue culture, cytopathogenic (cp) and non-cytopathogenic (ncp) biotypes of Pestiviruses can be distinguished .

STRUCTURE AND PHYSICAL PROPERTIES OF THE VIRION

Pestiviruses are enveloped by a lipid membrane and, thus, are readily inactivated by organic solvents, detergents and heat (Rümenapf et al., 1991a). Unlike flaviviruses which are sensitive to low pH, pestiviruses are remarkably resistant to acidic conditions (Liess, 1981). Incubation of BVDV or CSFV at pH 3.0 for 15 minutes did not cause a significant loss of infectivity (Krey et al., 2005). Characterization of the pestivirus particle has proven difficult, mainly because separation of virions from cellular material is insufficient. Neither differential sedimentation nor density gradient centrifugation allow a reliable separation from membranous or vesicular components of the host cell. CSFV and BVDV were reported to sediment at 100 Svedberg units and band at a buoyant density of 1.134 g/ml in sucrose (Laude, 1987) and 1.090 g/ml in Iodixanol (Optiprep) (Maurer et al., 2004). In addition to genome RNA and lipids, four structural proteins have been identified: the capsid protein (Core; C) and three envelope glycoproteins, E^{rns} (for ribonuclease, secreted), E1, and E2 (Rümenapf et al., 1993; Thiel et al., 1991). Characterization of virus particles by electron microscopy revealed spherical enveloped particles, 40 to 60 nm in diameter, with an electron-dense inner core of about 30 nm (Horzinek, 1981). The surface morphology allows no insights into the particle symmetry (Moormann and Hulst, 1988; Weiland et al., 1999).

BINDING AND ENTRY

Entry of pestiviruses is a multistep process involving initial attachment of virions to receptor molecule(s), internalization, and membrane fusion. While recombinant E2 and E^{rns} bind independently to cell surfaces (Bruschke et al., 1997), E2 is the major determinant of cell culture tropism, at least for ruminant pestiviruses. BVD viruses with a chimeric E2 from border disease virus showed a preference for ovine cells (Liang et al., 2003;

Reimann et al., 2004). Adsorption of CSFV E2 competitively inhibits infection with both CSFV and BVDV, whereas inhibition by E^{rns} is seen for CSFV and selected BVDV strains but requires large amounts of E^{rns} (Bruschke et al., 1997). E^{rns} binding thus far has only been reported for E^{rns} molecules which have acquired substitutions with basic residues around amino acid 476 (Ser->Lys). Such a single substitution was shown to significantly increase the binding affinity to heparin sepharose (Hulst et al., 2000). The heparin binding phenotype is selected by propagation in tissue culture cells (SK6) and is lost upon passage in the natural host. There is evidence that E^{rns} is dispensable for infection. CSFV-pseudotyped HIV particles containing E1 and E2 were sufficient to enter porcine cells (Wang et al., 2004). As cellular receptors for BVDV-1 and BVDV-2 several candidate molecules have been claimed (Agnello et al., 1999; Maurer et al., 2004; Minocha et al., 1997; Schelp et al., 2000; Schelp et al., 1995; Xue and Minocha, 1993; Xue and Minocha, 1996; Xue et al., 1997; Zheng et al., 1998). Experiments pointing towards the involvement of the LDL receptor in BVDV entry could not be confimed (Krey et al., 2006b). Only the bovine CD46 molecule has been functionally characterized (Maurer et al., 2004). Antibodies directed against bovine CD46 block infection of bovine cells with most BVDV strains to 70 – 95% (Krey et al., 2006a). CD46 expression leads to a gain of function in that porcine cells become about 100-fold more susceptible to BVDV infection. CD46 is a regulator of complement activation and the ectodomain consists of four "complement control protein repeats" (CCPs). Binding of BVDV to bovine CD46 occurs within a discrete subregion of ten amino acids on antiparallel beta strands within the proximal CCP1 (Krey et al., 2006a). CD46 is an often used cellular receptor for several pathogens. Human CD46 acts as a receptor for several viral pathogens including measles virus, human herpesvirus 6 and certain serotypes of adenovirus (Cattaneo, 2004). The cellular receptor for CSFV, BDV or exotic pestiviruses is so far unrecognized. The viral ligand for CD46 is probably BVDV E2. Several lines of evidence suggest that CD46 is insufficient to mediate infection and that further host cellular molecule(s) are required for pestivirus entry (Hulst and Moormann, 1997; Krey et al., 2006a; Krey et al., 2005; Maurer et al., 2004). After adsorption, BVDV enters its target cells via clathrin-dependent endocytosis and endosomal fusion (Krey et al., 2005; Lecot et al., 2005). During this process the highly acid resistant BVDV particle is modified to respond to low pH prior to fusion (Krey et al., 2005). While the precise mechanism is yet unknown, fusion from without experiments suggest the involvement of disulfide isomerases that are located at the cell surface or in the endosome. Pestivirus glycoproteins are covalently linked by intra- and intermolecular disulfide bonds. In analogy to the fusion process of flaviviruses a pH induced monomerization of E2 was postulated (Krey et al., 2005; Wallin et al., 2004). Interestingly pestivirus E2 contains a conserved CysXXCys motif that resembles active sites of disulfide isomerases (Krey et al., 2005). Whether this enzymatic activity exists in E2 as shown for some retroviruses (Wallin et al., 2004) remains to be established.

STRUCTURE OF THE GENOME

The pestivirus genome consists of a single-stranded RNA of about 12.3 kb in length. Similar to hepaciviruses, pestivirus genomes lack a 5' cap and 3' poly(A) tract (Brock et al., 1992; Moormann and Hulst, 1988). A long ORF of about 4,000 codons is flanked by a 5' NTR of 372 to 385 nucleotides, and a 3' NTR of 185 to 273 nucleotides (Avalos-Ramirez et al., 2001; Brock et al., 1992; Collett et al., 1988a). Two 5' terminal stem-loop structures in the BVDV genome are important for efficient RNA replication (Frolov et al., 1998; Yu et al., 2000). A 5' terminal GUAU sequence is essential for BVDV replication, perhaps via the function of its complement at the 3' end of the minus-strand, as part of a promoter for plus strand synthesis (Frolov et al., 1998). Provided that this tetranucleotide sequence is retained, substitutions and deletions of hairpin Ia and part of Ib do not abolish replication (Becher et al., 2000). Thus, the 5' signals essential for pestivirus genome replication are significantly shorter than for the hepaciviruses. Following the ORF, the 3' NTR consists of a variable region followed by a conserved 3' terminal stem loop preceded by a single-stranded region (Deng and Brock, 1993; Yu et al., 1999). Mutational analyses indicate that the terminal stem loop and the upstream single-stranded region harbour important primary and secondary structural elements that probably function in cis to direct minus-strand initiation, whereas deletions in the variable region are tolerated (Isken et al., 2004; Pankraz et al., 2005; Yu et al., 1999).

BIOSYNTHESIS AND FUNCTION OF VIRAL PROTEINS

Cap-independent translation initiation of the pestivirus genome is mediated by an IRES that bears structural and functional similarity to that of HCV. The minimal pestivirus IRES includes 5' NTR domains II and III, and can be influenced by sequences downstream from the initiator AUG (Chon et al., 1998; Di Marco et al., 2005; Poole et al., 1995; Rijnbrand et al., 1997). As for HCV, the pestivirus IRES binds ribosomal 40S subunits without the need for translation initiation factors eIF4A, eIF4B, and eIF4F (Pestova and Hellen, 1999; Pestova et al., 1998; Sizova et al., 1998). Pestivirus proteins are translated from genomic RNA as a single large polyprotein, which is processed into individual viral proteins (Miller and Purcell, 1990; Purchio et al., 1984). The order of the cleavage products in the pestiviral protein is NH_2-N^{pro}-C-E^{rns}-E1-E2-p7-NS2-NS3-NS4A-NS4B-NS5A-NS5B-COOH (Fig. 3) (Collett et al., 1988b; Collett et al., 1991; Meyers et al., 1992).

Unlike other *Flaviviridae*, the first pestivirus protein encoded in the long ORF is a non-structural protein, N^{pro}, an autoprotease responsible for cleavage at the N^{pro} /C site (Stark et al., 1993; Thiel et al., 1991; Wiskerchen et al., 1991). Processing in the pestivirus structural region appears to be mediated by at least three additional proteases. Host signal peptidase is believed to cleave at the C/E^{rns}, E1/E2, E2/p7, and p7/NS2 sites with incomplete cleavage at the E2/p7 site leading to accumulation of uncleaved NS2-p7 (Elbers et al., 1996; Harada et al., 2000; Rümenapf et al., 1993). As for HCV, signal peptide peptidase mediates further processing of the pestivirus C protein in the transmembrane region to generate the C-terminus of mature C protein (Heimann et al., 2006). The E^{rns}-E1 polyprotein (gp62) is processed slowly, presumably by a cellular protease that has not yet been identified (Rümenapf et al., 1993). The mechanism of NS2-3 cleavage has only recently been elucidated by the identification and characterization of the NS2 autoprotease (Lackner et al., 2004). The cleavage of NS2-3 is essential for viral RNA replication. Nevertheless, processing of NS2-3 is incomplete. The uncleaved form of NS2-3 is apparently required for generation of infectious virions (Agapov et al., 2004). As detailed below, certain cp pestiviruses generate the authentic N-terminus of NS3 via several different mechanisms. The remaining cleavages in the polyprotein that generate NS4A, NS4B, NS5A and NS5B are catalyzed by the pestivirus NS3-4A serine protease (Tautz et al., 1997; Wiskerchen et al., 1991; Xu et al., 1997) (see Fig. 3).

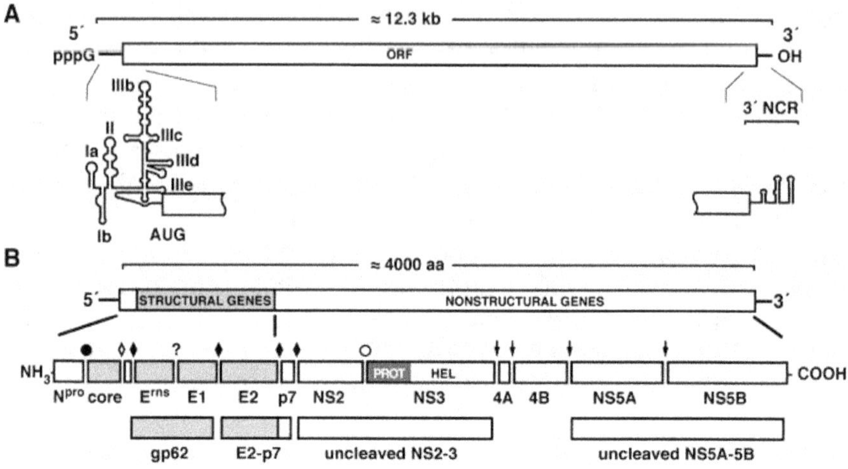

Fig. 3. Pestivirus genome structure and expression. A: Genome structure
B: Polyprotein processing and cleavage products. Symbols identifying proteolytic cleavage sites for
• N^{pro} (autocatalytic cysteine proteinase), ◊ Signal peptide peptidase (host), ♦ Signal peptidase (host),
○ NS2 (autocatalytic cysteine proteinase), ↓ NS3 (serine proteinase), ? unknown proteinase

N^{pro} AUTOPROTEASE

N^{pro} is a non-structural autoprotease that cleaves at a conserved site between Cys-168 and Ser-169 of the polyprotein (Stark et al., 1993; Wiskerchen et al., 1991). N^{pro} was initially described as p20 and was suggested to represent the structural capsid protein. In analogy to the biosynthesis of alphavirus capsid protein it was speculated that p20 releases itself from the nascent polyprotein autoproteolytically (Wiskerchen et al., 1991). Later it was shown that p20 is a nonstructural protease that generates the N-terminus of the structural Core protein (Stark et al., 1993). Residues Glu-22, His-49, and Cys-69 were identified as essential for protease activity by site directed mutagenesis. This arrangement is reminiscent of subtilisin-like cysteine proteinases (Rümenapf et al., 1998), but significant sequence homologies to other proteinases of this family are absent. N^{pro} is dispensable for autonomous viral replication in engineered and spontaneously derived subgenomic RNAs (Behrens et al., 1998; Moser et al., 1999; Tautz et al., 1999; Tratschin et al., 1998). In polyprotein processing, N^{pro} can be functionally replaced by the cellular ubiquitin gene, which directs cellular hydrolases to cleave at the C-terminus of the ubiquitin monomer (Behrens et al., 1998; Tratschin et al., 1998).

Furthermore, N^{pro} also acts as an inhibitor of interferon induction (Ruggli et al., 2005; Ruggli et al., 2003). It was claimed that N^{pro} blocks IRF-3 transcription in CSFV-infected cells (La Rocca et al., 2005), but recent studies on the action of N^{pro} of BVDV and CSFV have shown a direct interaction of N^{pro} with IRF-3 which leads to proteasomal degradation of IRF-3 (Bauhofer et al., 2007; Hilton et al., 2006). Experiments with BVDV have shown that the protease activity of N^{pro} is not required for interferon antagonism (Gil et al., 2006). In accordance with these cell-based studies, N^{pro}-deletion mutants of CSFV are attenuated *in vivo* and have been proposed as live virus vaccine candidates (Mayer et al., 2004).

PESTIVIRUS STRUCTURAL PROTEINS

Pestiviruses contain four structural proteins, Core, E^{rns}, E1, and E2, and thus exceed the virion complexity of the genera Flavivirus and Hepacivirus (three structural proteins each). Structural proteins are not required for replication of the genome and can be provided *in trans* (Frey et al., 2006; Reimann et al., 2003; Stettler et al., 2002; van Gennip et al., 2002; Widjojoatmodjo et al., 2000).

The capsid (C) protein is a small polypeptide consisting of 86 (CSFV) to 88 (BVDV) amino acids with a net charge of around +11. It lacks hydrophobic domains with the exception of 6 amino acids at the C-terminus (ALLAWA) which are the remnants of the start transfer sequence that directs the following glycoprotein E^{rns} into the endoplasmic reticulum. As mentioned above the N-terminus is generated by N^{pro}. The C-terminus is generated by two host cell proteases, signal peptidase (or signalase) and signal peptide peptidase (SPP). Signalase cleaves in the lumen of the ER and leads to the occurence of an immature form of C protein (16 kDa) that contains the entire transmembrane peptide. This hydrophobic peptide is processed further by signal peptide peptidase, an aspartic protease related to presenilin (Lemberg and Martoglio, 2002). SPP cleavage occurs in the lipid bilayer of the ER membrane and occurs between Ala-255 and Leu-256 of the CSFV polyprotein (Heimann et al., 2006). C protein binds to RNA and is supposed to initiate virus morphogenesis. Further to this, the C protein of CSFV is reported to act as transcriptional regulator of HSP70 and the SV40 early promoter (Liu et al., 1998).

The E^{rns} glycoprotein (44- to 48-kDa, formerly known as E0 or gp44/48) is heavily glycosylated at seven to nine potential N-linked glycosylation sites and forms disulfide-linked homodimers

(Langedijk et al., 2002; Thiel et al., 1991). Besides being associated with released virus via an unusual type of membrane anchor (Fetzer et al., 2005), E^{rns} is also secreted from infected cells (Rümenapf et al., 1993; Weiland et al., 1992; Weiland et al., 1999). The most unusual feature of E^{rns} is a ribonuclease activity. It contains an active site domain with high homology to the ribonuclease T2 family. The RNase T2 family encompasses a variety of secreted endoribonucleases with distribution in all kingdoms (Irie and Ohgi, 2001). T2 RNases are usually larger than 200 amino acids and have no absolute base specificity but a preferred cleavage site. The endonucleolytic activity of E^{rns} cleaves 5' of a uridine residue via a 2',3' cyclophosphate intermediate leaving the phosphate at the 3' end of the 5' cleavage product (Hausmann et al., 2004; Schneider et al., 1993). Using synthetic single stranded substrates a high catalytic activity of 1x 10^7 M^{-1} S^{-1} was determined. There is evidence that E^{rns} may also act as an endoribonuclease after it has modified the substrate by the endonucleolytic cleavage (Hausmann et al., 2004; Schneider et al., 1993). Glycosylation and dimer formation are not required for this activity (Windisch et al., 1996). Although the function of E^{rns} ribonuclease activity is not yet clear, it appears to be important for pathogenesis. Antibodies that inhibit ribonuclease activity tend to neutralize virus infectivity (Windisch et al., 1996), and mutations in E^{rns} that destroy ribonuclease activity give rise to viruses that are attenuated *in vivo* (Meyer et al., 2002; Meyers et al., 1999; von Freyburg et al., 2004). While deletion of a catalytically active His-346 led to fully attenuated CSFV, other substitutions that abrogate RNase activity (His-297 Lys, Trp-300 Gly) led to a much lower degree of attenuation. A direct involvement of E^{rns} in virulence was claimed by the finding that E^{rns} prevents interferon induction by stimulation with heteropolymeric RNA (polyIC) (Iqbal et al., 2004). The effect depends on RNase activity which led to the hypothesis that E^{rns} degrades the heteropolymeric RNA under acidic conditions probably in the lysosome. Recombinant E^{rns} is suspected to induce apoptosis of porcine lymphocytes *in vitro* (Bruschke et al., 1997), an observation that could be linked to the marked leukopenia seen in natural infections (Susa et al., 1992). It remains unclear, whether this effect is due to the ribonuclease activity. Interestingly, the C-terminal domain of E^{rns} was reported to promote an energy independent translocation across the cell membranes (Langedijk, 2002). While this exciting finding would suggest an interaction of E^{rns} with intracellular RNA targets of non infected cells and tissues (Langedijk, 2002), the transport effect yet needs confirmation.

E1 and E2 are integral membrane proteins that contain two to three, and four to six N-linked glycosylation motifs, respectively (Weiland et al., 1990). E1 and E2 are associated as disulfide-linked heterodimers that form slowly in the endoplasmic reticulum (Branza-Nichita et al., 2002; Rümenapf et al., 1993) and E2 is also present in homodimers (Thiel et al., 1991; Weiland et al., 1990). E1 is a hydrophobic protein with probably four membrane spanning domains that account for a close association with the ER membrane. A cytosolic domain has not been identified. Overall, the current knowledge about E1 is very small mainly because of a lack of serological reagents.

Recombinant CSFV E2 can bind to cells and block infection of CSFV and BVDV, suggesting a common E2 receptor/coreceptor for binding and entry of pestiviruses (Hulst and Moormann, 1997). Further to adsorption, it was suggested by computational analysis that the domain necessary for fusion with the endosomal membrane locates within E2 (Garry and Dash, 2003). Although the precise roles of the viral glycoproteins in virus assembly and entry remain to be defined, monoclonal antibodies to E^{rns} (Weiland et al., 1992) or E2 (Donis et al., 1988; Paton et al., 1992a; van Rijn et al., 1994; Weiland et al., 1990; Wensvoort, 1989) can neutralize virus infectivity, and both antigens can elicit protective immunity (Hulst et al., 1993; Rümenapf et al., 1991b; van Zijl et al., 1991). E2 was shown to define the host tropism of ovine pestiviruses. Chimeric BVDV bearing an E2 glycoprotein from border disease virus led to a reduced susceptibility of bovine cells, while ovine cells were efficiently infected (Liang et al., 2003). E2 is retained in the endoplasmic reticulum by means of the C-terminal transmembrane anchor (Kohl et al., 2004).

p7 follows the virion structural proteins. It consists of a central charged region separating hydrophobic termini (Elbers et al., 1996) and is required for production of infectious virus (Harada et al., 2000), but not RNA replication (Behrens et al., 1998). Like the HCV p7 protein, pestivirus p7 is inefficiently cleaved from E2, probably via signal peptidase (Elbers et al., 1996). Uncleaved E2–p7 is not required for replication in cell culture (Harada et al., 2000) and both E2–p7 and p7 appear to remain cell associated. It is not yet clear whether p7 is a structural or a nonstructural protein, although it has not been detected in purified virus (Elbers et al., 1996). Similarly to HCV p7, pestivirus p7 can form ion channels, suggesting that it functions in virus assembly and/or entry (Griffin et al., 2003; Griffin et al., 2004).

PESTIVIRUS NONSTRUCTURAL PROTEINS

NS2 is a newly identified cysteine protease, distantly related to the NS2-3 autoprotease of HCV and GB viruses, which is responsible for processing of NS2-3 (about 125-kDa) (Lackner et al., 2004; Lackner et al., 2006). As discussed below, NS2-3 cleavage is essential for pestivirus RNA replication. The efficiency of NS2-3 cleavage is regulated by a cellular chaperone and can determine viral cytopathogenicity. Similar to other members of the *Flaviviridae*, the NS2-3 region participates in virus assembly (Agapov et al., 2004). As for all members of the *Flaviviridae*, pestivirus NS3 contains an N-terminal serine protease domain (Bazan and Fletterick, 1989; Gorbalenya et al., 1989a; Wiskerchen and Collett, 1991) and a C-terminal RNA helicase (Gorbalenya et al., 1989b). Like HCV, the pestivirus NS3 serine protease requires NS4A as a protein cofactor (Xu et al., 1997). The pestiviral NS3-4A serine protease cleaves between leucine and small uncharged amino acids: Leu |(Ser/Ala/Asn) (Tautz et al., 1997; Xu et al., 1997). Substitutions that eliminate serine protease activity abolish viral RNA replication, confirming its essential role in virus viability (Grassmann et al., 1999; Xu et al., 1997). Interestingly, protease activity is retained when threonine is substituted for the nucleophile serine (Tautz et al., 2000). The NS3 protein of BVDV has been purified and shown to possess RNA helicase (Warrener and Collett, 1995) and RNA-stimulated NTPase (Tamura et al., 1993) activities. Site-directed mutagenesis of the conserved helicase and NTPase motifs inhibited or destroyed these activities and abolished viral replication (Grassmann et al., 1999). The hydrophobic NS4A and NS4B proteins are similar in size, composition, and hydropathic properties to the NS4A and NS4B proteins of the hepaciviruses. NS4A acts as a cofactor for the NS3 serine protease (Xu et al., 1997). Like HCV, the NS4A cofactor activity involves interaction of its central domain with the N-terminal region of NS3 (Tautz et al., 2000). An analysis of recombinant ncpBVDV strains revealed that an insertion of viral sequences between the NS4A and NS4B genes did not interfere with efficient viral replication; thus, the NS4AB precursor has no essential role, at least in cell culture (Gallei et al., 2005). The remaining two proteins, NS5A (about 58-kDa) and NS5B (about 75-kDa), are present as mature cleavage products as well as uncleaved NS5A-5B (Collett et al., 1988b; Collett et al., 1991). Little is known about the function of NS5A. This protein is phosphorylated by a cellular serine/threonine kinase with properties similar to kinases that phosphorylate flavivirus NS5 and hepacivirus NS5A (Reed et al., 1998). Genetic analysis revealed that defects in the NS5A gene could be readily complemented *in trans*, while mutations in other NS genes could not (Grassmann et al., 2001). NS5B contains motifs characteristic of RdRPs (Collett et al., 1988b; Collett et al., 1991). The RNA polymerase activity of recombinant NS5B has been characterized *in vitro* and found to extend template-primed RNAs into double-stranded "copy-back" products (Lai et al., 1999; Lohmann et al., 1999; Steffens et al., 1999; Zhong et al., 1998) or to catalyze *de novo* initiation from short synthetic RNA or DNA templates (Kao et al., 1999; Lai et al., 1999). The structure of BVDV NS5B is similar to HCV NS5B and other RdRPs containing a palm subdomain surrounded by finger and thumb subdomains (Choi et al., 2004). The pestivirus NS5B structure reveals a unique N-terminal region, which suggests a role for GTP in *de novo* initiation, and provides a framework for understanding the molecular mechanisms of small molecule inhibitors of the BVDV RdRP (Baginski et al., 2000; Paeshuyse et al., 2006; Sun et al., 2003).

REPLICATION

The basic aspects of pestivirus RNA replication are similar to those described for HCV. RNA replication is associated with cytoplasmic membranes and requires NS3 through NS5B as well as cellular components. Negative- and positive-strand pestivirus RNAs have been detected from 4 to 6 hours after infection, followed by the asymmetric accumulation of additional minus- and excess plus-strand RNAs (Gong et al., 1998; Gong et al., 1996). Accumulation of genome-length intracellular pestivirus RNAs, which co-migrate with virion RNA, generally follows the time course of infectious virus release. Maximum virus titers are achieved about 12 to 24 hours after infection (Gong et al., 1996; Moormann and Hulst, 1988; Purchio et al., 1984). Double-stranded RF RNAs and partial duplex RI RNAs have been tentatively identified (Gong et al., 1998; Gong et al., 1996; Purchio et al., 1984).

Recently, new insights into the regulation of BVDV RNA replication and virus assembly have emerged. Originally, non-cytopathogenic (ncp) BVDV was thought to express only uncleaved NS2-3 leading to the idea that NS2-3 could serve as a functional RNA replicase component. Consistent with this idea, cytopathogenic (cp) BVDV isolates produce both, uncleaved NS2-3 and NS3. Recent work has shown, however, that NS2-3 cleavage is required for RNA replication and regulated by limiting quantities of the cellular chaperone Jiv (J-domain protein interacting with viral protein) (Lackner et al., 2005). NS2-3 cleavage is nearly complete early in ncpBVDV infection, leading to efficient NS3 production and RNA replication. At later times, when Jiv becomes limiting, NS2-3 autoprocessing becomes inefficient, NS2-3 levels increase and viral RNA synthesis rates decline (Lackner et al., 2004). As for HCV, mutations that abolish BVDV NS2-3 processing and thereby NS3 production block RNA replication. Jiv interacts with NS2 and acts as a cofactor for the autoprotease. The amount of Jiv in the infected cell controls pestiviral replication by limiting the generation of active protease molecules and replication complexes. In addition to viral template RNA and replicase proteins, a group of cellular NFAR proteins are recruited by the replication machinery of BVDV; the NFAR proteins associate specifically with the 5' and 3' termini of the viral RNA genome (Isken et al., 2003). The 3' NCR also contains determinants that ensure efficient termination of translation, which is essential for efficient pestivirus RNA replication (Isken et al., 2004).

As described later, there is good evidence that nonhomologous recombination can occur within pestivirus RNAs, as well as between pestivirus RNAs and host cellular RNAs (reviewed in Meyers and Thiel, 1996). One likely mechanism is via copy-choice template recruitment during minus-strand synthesis, which is consistent with the coding orientation of cellular inserts. Alternatively, the existence of a viral RdRP-independent mechanism of RNA recombination has recently been demonstrated for pestiviruses using a cell culture based RNA recombination system in which homologous and nonhomologous recombination occurred between two overlapping transcripts, each lacking different essential parts of the viral RdRP (Gallei et al., 2004).

ASSEMBLY AND RELEASE OF VIRUS PARTICLES

Other than the features of the involved structural proteins described above, little information is available on the assembly and release of pestiviruses from infected cells. Electron microscopic examination of virus-infected cells (Bielefeldt-Ohmann and Bloch, 1982; Gray and Nettleton, 1987) suggests that pestiviruses mature in intracellular vesicles and are released by exocytosis. Consistent with intracellular budding, pestivirus envelope proteins do not accumulate at the plasma membrane (Greiser-Wilke et al., 1991; Weiland et al., 1999). ER alpha-glucosidase inhibitors block infectious BVDV-1 particle formation (Durantel et al., 2001; Zitzmann et al., 1999). Interestingly, E^{rns} and E2 have been immunolocalized on isolated virus particles by electron microscopy, but E2 was not detected in particles undergoing secretion (or perhaps reattachment) at the cell surface (Weiland et al., 1999). This suggests that E2 is conformationally inaccessible to antibodies prior to maturation. Nonstructural proteins also play a role in pestivirus virion assembly or release. These include uncleaved NS2-3 (Agapov et al., 2004) and the C-terminus of NS5B (Ansari et al., 2004).

DISEASES CAUSED BY PESTIVIRUSES

Pestivirus infections may be subclinical or produce a range of clinical conditions including acute diarrhea, acute hemorrhagic syndrome, acute fatal disease, and a wasting disease (previously reviewed in Thiel et al., 1996). CSFV, typically transmitted oronasally, leads to acute or chronic hemorrhagic syndromes with significant mortality. Ruminant pestiviruses, on the other hand, usually result in inapparent infections or cause mild symptoms in adult animals. A notable exception is BVDV-2, which can be associated with a severe acute hemorrhagic condition in cattle (Corapi et al., 1989; Pellerin et al., 1994; Rebhun et al., 1989). In addition, diaplacental transmission of pestiviruses can cause fetal death, malformation and acute syndromes of the newborn, or it can lead to persistent infection of the offspring in a carrier state. Pestiviruses that cause persistent infection are noncytopathogenic (ncp) in cell culture. Persistently infected cattle may come down with fatal mucosal disease (MD). Interestingly, development of MD correlates with the occurrence of virus variants with a cytopathogenic phenotype (see below).

CLASSICAL SWINE FEVER

According to early descriptions, CSF was a peracute to acute disease with a short incubation period. Animals showed little more than a rise in body temperature of about 41°C before they died 2 to 5 days after exposure to the virus. Mortality varied from 90-100% to as low as 30% (Dunne, 1973). The course of the disease has long since changed. First, a chronic form following acute illness was recognized. Today, the peracute form has disappeared and the acute illness no longer reaches the former extremely high mortality rates. In addition, chronic and "late onset" types of the infection are common (van Oirschot, 1992; Wensvoort and Terpstra, 1985). To some extent host factors determine the outcome of the infection. However, viral factors appear to be decisive, i.e., virulence and dose of infection. With respect to virulence, CSFV strains and isolates represent a continuous gradient ranging from high, moderate, low, to no apparent virulence. Strains of moderate to low virulence have probably emerged in the course of the known history of CSF by natural selection of adapted viruses (Steinhauer and Holland, 1987). The virulence of a particular virus strain or isolate is best determined by the experimental infection of pigs under standardized conditions. Sequence comparisons of related virulent and avirulent CSFV strains revealed a multitude of mutations spread over the entire genome (Mayer et al., 2003). Reverse genetic approaches have shown that mutations concerning N^{pro} (Mayer et al., 2004), E^{rns} (Meyer et al., 2002; Meyers et al., 2007; Meyers et al., 1999), E1 (Risatti et al., 2005b), and E2 (Risatti et al., 2005a; Risatti et al., 2006; Risatti et al., 2007; Van Gennip et al., 2004) can account for attenuation.

Acute CSF is caused by viral variants of high virulence. After an incubation period of 6 to 8 days, pigs develop persistent pyrexia and leukopenia. As a result of central nervous system (CNS) disorders, they become progressively depressed and uncoordinated. Symptoms such as staggering movement of the hind legs, convulsions, grinding of teeth, difficulty in locomotion, and posterior paresis are common. Lacrimation, nasal discharge, and obstipation followed by diarrhea are also characteristic clinical findings. The most pathognomic alteration, however, is the hemorrhagic syndrome. Petechial hemorrhages of the outer skin and mucosae, cyanotic coloration of the abdomen, snout, ears, and medial sides of the extremities are frequent. At post-mortem, petechial bleedings of inner organs, e.g. spleen, kidney, lymph nodes, and gut are prominent. Spleen infarctions are also observed.

Like all pestiviruses, CSFV is immunosuppressive during acute infection. About 10 to 12 days post infection, neutralizing antibodies are demonstrable, and surviving animals mount a long-lasting humoral immunity (Van Oirschot et al., 1983).

Less virulent variants of CSFV produce chronic disease. After infection, animals initially display clinical signs resembling acute CSF. However, symptoms are milder and viremia is less pronounced due to an apparently less efficient multiplication of the virus in its host. Thereafter, the clinical condition of the pigs improves temporarily. During that period the production of neutralizing antibodies may lead to failure of virus isolation from peripheral blood. The simultaneous presence of antibodies and CSFV in the peripheral blood can result in the formation of an immune complex glomerulonephritis (Cheville et al., 1970). Several weeks or up to 3 months later, the disease exacerbates and the animals die. During the last stages of the disease the virus is demonstrable in all organs. Postmortem lesions resemble the acute form of CSF.

The outcome of diaplacental infection of the fetus depends on the time of gestation (van Oirschot and Terpstra, 1977). The age of the gilts and the virulence of the strain are additional factors. After experimental infection of gilts at different stages of gestation with a virus of low virulence, three different groups of piglets were recognized: (a) aborted or stillborn fetuses; (b) piglets with congenital, persistent viremia; and (c) normal piglets non infected with CSFV (Meyer et al., 1980). The results led to the following general conclusion: early infections of gilts (before day 41 of gestation) usually result in abortions and stillbirths, whereas inoculations between days 41 and 85 yield viremic animals. After about day 85 normal, nonviremic animals are produced since the developing immune system clears the virus. Due to the anatomy of the porcine placenta not all fetuses may become infected, i.e. in one litter there may be stillborn and/or mummified piglets, persistently viremic piglets, and noninfected piglets (Frey et al., 1980; Meyer et al., 1981; Meyer et al., 1980). Though they may be clinically normal at birth, persistently infected offspring invariably die. At first they fail to thrive and after weeks or months their condition deteriorates and they ultimately succumb to the infection. Survival times up to 11 months after birth have been observed. This course of infection is termed late onset CSF (van Oirschot, 1992).

BORDER DISEASE

Border disease of sheep and goats received its name because it was first investigated in the Border Counties of England and Wales (Hughes et al., 1959). Infections with BDV resemble in many aspects those with BVDV, especially since congenital infection represents the major problem. Acute postnatal infections produce inapparent or sometimes mild clinical disease with transient depression, pyrexia and leucopenia (Shaw et al., 1967; Vantsis et al., 1979). The affected animals mount an efficient immune response and after about 2 weeks the appearance of neutralizing antibodies termimates the short viremic period and eliminates the virus.

During acute infection of pregnant animals BDV crosses the placenta to invade the fetus. With an overall gestation period of about 150 days immunocompetence of the ovine fetus develops between days 64 and 82 (Fahey and Morris, 1978). Before day 16 of gestation the zygote/embryo is not susceptible to infection. This leaves a "window of susceptibility" of about 60 days for establishment of persistent infection after diaplacental infection. In the uterus the virus causes placentitis, and in goats this is associated with a high incidence of fetal death and abortion (Barlow et al., 1975). In sheep early transplacental infection may cause fetal death resulting in resorption, mummification, abortion, or stillbirth (Barlow, 1972). Lambs surviving diaplacental infection show a wide range of abnormalities, of which the hairy shaker or fuzzy-lamb syndrome is the most prominent. Experimental infections between days 50 and 63 of gestation are most likely the peculiar hairy shaker syndrome characterized by hairy fleece, low birth weight, tonic-clonic tremors and ataxia (Nettleton et al., 1992). Lambs infected after day 80 of gestation are born clinically normal with neutralizing antibodies agianst BDV and no infectious virus in their blood. However, after experimental infections, such animals showed characteristic nodular periarthritis which is thought to be the result of an allergic reaction (Gardiner et al., 1980).

The recently discovered remarkable heterogeneity of BDV with at least five genotypes (see Fig. 2) has not been correlated so far with differences concerning the course of the disease. Interestingly, BVDV can cause in sheep and goats the same symptoms as BDV. Moreover, BVDV and BDV infect swine and can cause disease symptoms (Paton et al., 1992b; Vilcek and Belak, 1996; Walz et al., 2004; Wieringa-Jelsma et al., 2006). There are, however, no confirmed reports about diseases of cattle induced by BDV.

BOVINE VIRAL DIARRHEA / MUCOSAL DISEASE

At first sight, infections of ruminants with pestiviruses are far less dramatic than those of pigs with CSFV. This may explain their relatively late discovery. Olafson et al. (Olafson et al., 1946) described a transmissible disease of cattle with a high morbidity and low mortality. Affected cattle display elevated temperature, diarrhea, and coughing. The condition was termed bovine viral diarrhea (BVD). Seven years later, Ramsey and Chivers (Ramsey and Chivers, 1953) reported a fatal disease of cattle with a low morbidity rate. This disease was characterized by pyrexia, anorexia, lethargy, profuse diarrhea containing fresh or clotted blood, mucopurulent nasal discharge, and dehydration. Death usually occurred within 2 weeks after the onset of clinical signs. At postmortem, extensive ulceration of the gastrointestinal tract was regularly found. The disease was called *mucosal disease* (MD). Years later it was discovered that both diseases were caused by the same virus (Gillespie et al., 1961). However, MD remained an enigma for many years since the disease could not be reproduced experimentally. Today, BVDV is considered a major pathogen of cattle with a world-wide economic impact. Acute postnatal infection with BVDV causes clinical pictures ranging from inapparent infections to various symptoms. Importantly, the major economic damage caused by ruminant pestiviruses in susceptible animals is primarily the result of diaplacental infection (Doherty, 1986; Frey et al., 1987; Ross et al., 1986).

Seroepidemiological surveys indicate that most cattle contract a BVDV infection during their lifetime. However, the prevalence is not correlated with disease, indicating that the vast majority of infections remains unnoticed. This is in accord with the difficulties encountered in reproducing BVD by experimental infection (Pritchard, 1963). A more detailed clinical examination of acute infections may reveal a short febrile period, leucopenia, salivation, nasal discharge, coughing, and/or sometimes diarrhea. There is experimental evidence that mixed infections including BVDV and other pathogens may lead to more pronounced clinical disease (Perdrizet et al., 1987; Pritchard et al., 1989). *In vivo* studies have shown that simultanous infection with BVDV and *Pasteurella hemolytica* or bovine herpesvirus 1, respectively, aggravates respiratory disease (Potgieter et al., 1984a; Potgieter et al., 1984b).

An apparently rare but more severe clinical condition in calves associated with acute BVDV infection was first described in North

America more than 15 years ago (Rebhun et al., 1989). The disease is characterized by transient severe thrombocytopenia and petechial and ecchymotic hemorrhages in all mucosal surfaces and various internal organs. The symptoms resemble those of other hemorrhagic diseases, e.g. CSF, African swine fever and Dengue. Meanwhile, this clinical picture has been observed in several European countries as well. The capacity to induce hemorrhagic disease evidently is an indigenous property of certain non-cp BVDV strains, mostly BVDV-2. Whether these strains have been overlooked in the past or have emerged only recently is not clear (Corapi et al., 1989).

In susceptible pregnant cattle, BVDV is able to invade the placentome, to replicate, and to cross to the fetus (Brownlie et al., 1989; Casaro et al., 1971). The outcome of diaplacental infection depends on the time point of infection during gestation. There is no experimental evidence that either germ line cells or early embryonic stages are susceptible to BVDV infection. In vitro studies indicated that even a damaged or removed zona pellucida does not affect survival of bovine embryos in the presence of infectious BVDV (Bielanski and Hare, 1988). Most likely, the fetus becomes susceptible to infection after transplantation at days 19 to 20 post conception or development of fetal cotyledons around day 30 p.c. The fetal immune system does not identify BVDV as "not self" at any time during the first trimester of gestation. Until around day 125 of gestation BVDV infection can result in persistent infection, embryonic death, resorption, and stillbirth or in a whole array of nonfatal teratogenic effects. Since BVDV replicates in the fetus at different stages of organogenesis, various malformations are observed, e.g. cerebellar hypoplasia, hydranencephaly, dysmyelogenesis, chronic skin lesions, thymus atrophy, and reduced lung maturation (Brownlie, 1991; Casaro et al., 1971; Done et al., 1980; Kahrs, 1973; Trautwein et al., 1986). Many infected calves show significant intrauterine growth retardation ("runt"), while others appear to be clinically normal.

The correlation between time of infection and cerebellar development has been studied in detail. Infections between days 79 and 150 of gestation yield calves with cerebellar degeneration. Severity of malformations increases with age of the fetus at time of infection, i.e., infections after day 100 p.c. produce the most severe cerebellar lesions. In addition, ocular lesions, e.g. cataracts, retinal degeneration, and hypoplasia and neuritis of the optic nerves are found (Brown et al., 1973).

Calves born viremic are immunotolerant to the virus and persistently infected for the rest of their lives. The immunotolerance was shown to be highly specific, since the competence toward different BVDV strains remained intact (Bolin, 1988; Coria and McClurkin, 1978) (Steck et al., 1980). Apart from their risk of developing mucosal disease (MD) (see below) these animals can be predisposed to infection with other microorganisms leading to serious disease, e.g. pneumonia or enteric disease (Barber et al., 1985). Mucosal disease (MD) is a sporadic form of disease in cattle which generally occurs between the ages of 6 months and 2 years. The disease is characterized by low morbidity and high case-fatality (Baker, 1987). The clinical signs comprise pyrexia, anorexia, lethargy, profuse diarrhea containing fresh or clotted blood, mucopurulent nasal discharge, and dehydration. Death usually occurs within 2 weeks after the onset of clinical signs. Extensive ulceration of the gastrointestinal tract is the most prominent lesion (Baker, 1987; Moennig and Plagemann, 1992).

PATHOGENESIS

CLASSICAL SWINE FEVER

Under natural conditions the most frequent route by which CSFV enters the host animal is oronasal. First sites of viral replication are epithelial cells of the tonsillar surface and crypts, and the ileum. The major target cells for the virus are endothelial cells, lymphoreticular cells, macrophages, monocyte- and some bone marrow-derived dendritic cells and certain epithelial cells (Carrasco et al., 2004). Classical swine fever is a hemorrhagic disease in the classical sense. The extent of pathological lesions depends on the virulence of CSFV. In postnatal infection, lesions are generally caused by widespread thrombosis and/or endothelial damage. Gross lesions are hemorrhagic diathesis and petechiation generally present in most organ systems, especially in kidneys, urinary bladder, lymph nodes, followed by spleen, brain, larynx, skin, and mucosal and serous membranes. Skin lesions are characterized by erythema, cyanosis, and petechial hemorrhages. Multifocal infarcts at the margin of the spleen are a highly characteristic but inconsistent feature. In subacute and chronic cases, blood vessel damage and subsequent necrosis in the large intestine leads to "button ulcers" which are considered pathognomonic for CSF. Bronchopneumonia is another regular feature. In a high proportion of fatal cases, histopathology of the brain shows a nonsuppurative encephalitis with severe vasculitis. The bone marrow atrophy observed during CSF has been studied by infection of bone marrow hematopoietic cells (Summerfield et al., 2001b).

It was shown that early after infection with CSFV there is a marked depletion of B and T lymphocytes as well as granulocytes (Summerfield et al., 2000; Summerfield et al., 2001a; Susa et al., 1992). At the onset of disease, viral replication in lymphoid tissues demarcates the germinal centres. The viral genome remains localized to that site as the disease progresses even after morphological disintegration of the follicular structure. A block in B-lymphocyte maturation by infection and destruction of germinal centers is suspected to be a major event in the pathogenesis of acute, lethal swine fever (Susa et al., 1992). Severe thrombocytopenia in the course of CSFV infection is a consistent finding. In experimentally induced, chronic CSF, a severe to complete atrophy of the thymus, depletion of lymphocytes and germinal follicles in the peripheral lymphoid tissues, and renal glomerular alterations indicative of immune complex deposits were described. Splenic changes included neutrophilic leukocyte accumulation, extramedullary hematopoiesis, and plasmocytosis (Cheville and Mengeling, 1969; Cheville et al., 1970).

BOVINE VIRAL DIARRHEA

The natural route of acute postnatal infection is oronasal. Initial virus replication occurs in the mucosa around the site of entrance and is responsible for mucosal ulceration and subsequent salivation or nasal discharge (Baker, 1987). From there the virus enters its target cells in the lymphoid tissue lining the oropharynx, especially the epithelial cells in tonsillar crypts where viral antigen is readily found during acute infection. The virus appears to be able to multiply in all of the major lymphocyte populations as well as in accessory cells (Howard, 1990; Ohmann, 1983). Virus is detectable in peripheral blood of experimentally infected calves 2 to 4 days after oral and nasal infection (Mills and Luginbuhl, 1968). A transient leukopenia and impairment of polymorphonuclear leukocyte function might provide explanation for serious disease that occasionally occurs in conjunction with other microorganisms (Brown et al., 1991; Roth et al., 1981). The leukopenia is characterized by decreases in the absolute number of B lymphocytes, T lymphocytes including both, helper (BoT4+) and cytotoxic/suppressor (BoT8+) cells, and neutrophils. The numbers of null cells and monocytes are not affected (Ellis et al., 1988). The description of a hemorrhagic condition incriminates platelets or their precursor cells in the pathogenesis of BVDV infection. The mechanism responsible for platelet deficiency is not yet understood. Both, immune-mediated and direct cell-virus interactions have been discussed (Corapi et al., 1989). Immunohistological investigations have shown that the virus is present in all parts of persistently viremic animals infected in utero. Up to 30% of peripheral blood mononuclear leukocytes (PBL), notably null cells, contain viral antigen (Bielefeldt Ohmann, 1988a).

Another preferred site of virus persistence is the gastrointestinal tract. In the epithelium of the large intestine only single crypt cells contain viral antigen(s), whereas in the small intestine small groups of crypt cells are antigen-positive. The lamina propria, Peyer's patches, colonic and rectal lymphoid patches display a small number of antigen-positive mononuclear cells. In glands, e.g., the glandulae duodenales (Brunner's) and the parotis, a relatively large number of epithelial cells contains viral antigen. Walls of small arterioles stain in a multifocal pattern when analyzed immunohistochemically (Liebler et al., 1991).

The CNS is another important location for viral persistence. Viral antigen is present almost exclusively in neurons. Predilection sites are cerebral cortex and hippocampus. In other areas of the brain and spinal cord, virus is located in single or small groups of neurons. No evidence for inflammation or cellular alterations was seen (Fernandez et al., 1989; Trautwein et al., 1986). In serum of persistently viremic calves the virus is readily detectable with average titers of 10^3 to 10^4 $TCID_{50}$ per millilitre (Bielefeldt Ohmann, 1988b).

MUCOSAL DISEASE

For many years the sporadic incidence of MD remained enigmatic. Importantly, MD occurs only in cattle which had been infected with a non-cp BVDV during the first trimester of gestation (Liess, 1973; Liess et al., 1974). Such animals acquire immunological tolerance with regard to the particular BVDV strain and develop a persistent infection. A major breakthrough was the observation that in naturally occurring cases of MD both viral biotypes, non-cp BVDV as well as cpBVDV, were consistently found (McClurkin et al.,

1985; McKercher et al., 1968; Wilhelmsen et al., 1991). In contrast to the described antigenic variability of BVDV field isolates the members of such a "pair" of noncp BVDV and cp BVDV are antigenically very closely related. This observation led to the hypothesis that during pathogenesis of MD a cp virus develops from the non-cp virus by mutation (Corapi et al., 1988; Howard, 1990). Moreover, MD was reproduced experimentally by superinfection of persistently infected viremic animals with cp BVDV. The cp virus used for superinfection has to be antigenically closely related to the "endogenous" non-cp BVDV (Bolin et al., 1985b; Brownlie et al., 1984; Moennig et al., 1990). This explains why some researchers failed to reproduce the disease by inoculating persistently viremic calves with cp BVDV (Harkness et al., 1984).

One aspect of BVDV infection was for some time subject of confusion and controversy. It concerned the finding that some BVDV-infected animals suffer from clinical illness for a prolonged period, a condition termed chronic MD. There was evidence that affected animals are also persistently viremic. In the meantime it has been shown that chronic disease might develop when viremic animals are superinfected with a cpBVDV sharing only partial antigenic homology with the "endogenous" non-cp BVDV. The cp virus undergoes recombination with the persisting non-cp virus (Fritzemeier et al., 1997): The genomic region responsible for cytopathogenicity comes from the cp virus and the persisting ncp virus provides the genes encoding the structural proteins and others. Thereby the acquired immunotolerance allows spread of the resulting cpBVDV chimera.

When comparing target cell numbers and locations of noncp BVDV in persistently infected animals with those of cp BVDV in cattle suffering from MD a striking difference was noted: The cpBVDV seemed to be more promiscuous than the noncp biotype, i.e., the number of virus-infected cells in animals with experimental MD was dramatically increased compared to the persistently viremic control calves. The comparison was made possible by marker epitopes on the superinfecting cpBVDV in experimental MD, allowing the proper identification of cells infected by cp BVDV with MAbs (Liebler et al., 1991). The basis for the higher efficiency of cpBVDV to infect cells *in vivo* is not clear.

GENERATION AND CHARACTERIZATION OF CYTOPATHOGENIC BVD VIRUSES

Genetic characterization of a number of ncp-cp pairs led to the remarkable discovery that the vast majority of cpBVDV strains are generated via RNA recombination, although a few cpBVDV stains lack obvious genome rearrangements (Meyers and Thiel, 1996; Pellerin et al., 1995 and references therein). The presence of these genome rearrangements strongly correlates with increased NS3 expression, enhanced RNA replication, and cytopathogenicity in cell culture. Fig. 4 illustrates a few of the remarkable cpBVDV genome alterations that have been discovered.

Common features of some cpBVDV variants concern genome rearrangements or point mutations that activate NS2-3 autoprotease activity, leading to increased production of NS3 and RNA replication. For instance, strain NADL (Fig. 4) contains a fragment of the cellular Jiv gene inserted in NS2 (Becher et al., 1996; Meyers and Thiel, 1996). As described above, Jiv is an essential cofactor of the NS2 protease and overexpression of a critical 90 amino acid Jiv subdomain in the context of the viral genome enhances NS2-3 cleavage, regardless whether the fragment is provided *in cis* or *trans* (Lackner et al., 2004). Interestingly, a much smaller insertion is found in the NS2 gene of cpBVDV strain CP7, which contains a 27-nt duplication from an upstream region of the NS2 gene in an alternate reading frame (Tautz et al., 1996). Like the Jiv fragment, this insertion also leads to increased NS2-3 processing and a virus that is cytopathic in culture (Mendez et al., 1998; Meyers et al., 1996). Other viral insertions at or very close to the same site have been described (Balint et al., 2005). For other cpBVDV isolates, increased NS2-3 autoprotease activity appears to result from point mutations that have accumulated within the NS2 gene (Kümmerer and Meyers, 2000).

Another common rearrangement in cpBVDV isolates involves the insertion of ubiquitin or ubiquitin-like genes immediately upstream of NS3 (Baroth et al., 2000; Meyers et al., 1991; Qi et al., 1998; Tautz et al., 1993), leading to NS2-3 processing by ubiquitin C-terminal hydrolase or related enzymes (e.g., strain Osloss in Fig. 4). In some cases, this may be accompanied by additional genome rearrangements such as duplication of the NS3 and NS4A genes (e.g. strain CP1 in Fig. 4) (Meyers et al., 1992). Such strains will also express uncleaved NS2-3, which is required for infectious virion production, as described above. A related type of insertion includes the light chain 3 genes of cellular microtubule-associated proteins, which is targeted for processing also by a cellular protease (Fricke et al., 2004; Meyers et al., 1998).

A third type of cpBVDV genome rearrangement involves the repositioning of the N^{pro} autoprotease immediately upstream of NS3. In one such cpBVDV strain, Pe515CP (Fig. 4), the N^{pro} gene is duplicated together with NS3 and NS4A genes (Meyers et al., 1992). Alternatively, cpBVDV isolates such as CP 9 (Fig. 4) contain a precise deletion of the C-E^{rns}-E1-E2-p7-NS2 genes resulting in an in-frame fusion of N^{pro} and NS3 (Tautz et al., 1994). Such subgenomic RNAs replicate autonomously, express NS3, and induce a cytopathic effect (Behrens et al., 1998) but require non-cpBVDV helper viruses to provide packaging functions *in trans*. For CSFV, cp subgenomic RNAs have been isolated in which the entire coding sequence upstream of NS3 has been deleted (Meyers and Thiel, 1995).

Based on the common feature of increased NS3 expression by cp pestiviruses, it has been hypothesized that NS3 may be directly cytopathic. Alternatively, cytopathic effects could be a byproduct of enhanced viral RNA accumulation and activation of cellular pathways triggering apoptosis. Nevertheless, selection for non-cp variants in cell culture yielded a non-cp BVDV that still produces NS3 and viral RNA at levels comparable to those of the cp parent (Qu et al., 2001). Further analysis revealed that this variant encodes a point mutation in NS4B that attenuates the cytopathic effect of BVDV. Other non-cp BVDVs have been selected in culture with duplication of 14 codons in NS4B or insertion of viral sequences between NS4A and NS4B (Gallei et al., 2005). Given that NS4B is likely to be involved in membrane reorganization (based on functions of the HCV proteins), and that cpBVDV can cause ER stress leading to apoptosis (Jordan et al., 2002), it seems plausible that cytopathic effects may indirectly result from the overcommitment of cellular membranes to replication complexes.

A larger question is how cytopathic effects on the cellular level are linked to MD in infected animals. Increased cell death could directly contribute to tissue injury and induce inflammation. However, as mentioned above, there is evidence that animals with MD show increased numbers of infected cells, possible due to differences in tropism, which could also contribute to disease (Liebler et al., 1991).

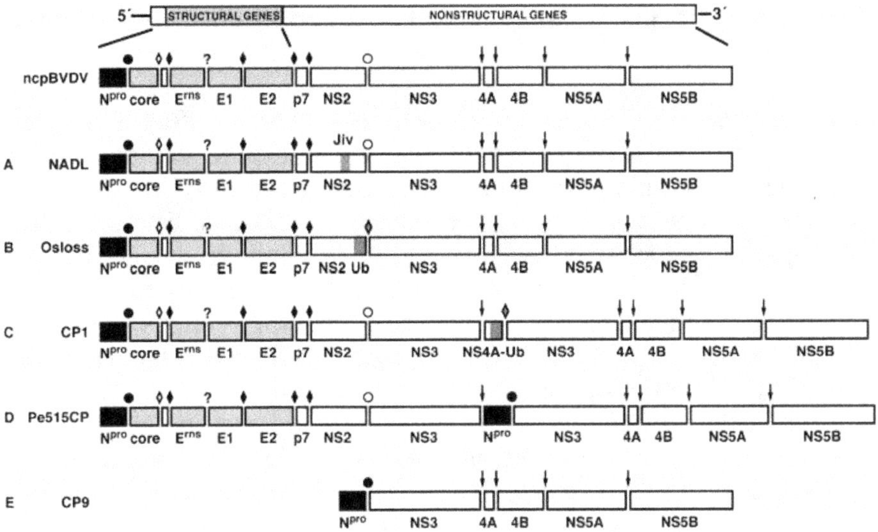

Fig. 4. Genome rearrangements associated with the generation of cytopathogenic bovine viral diarrhea virus (cpBVDV). The **top** diagram indicates the polyprotein of a typical noncytopathogenic BVDV (ncpBVDV) isolate. **Below**, the polyproteins encoded by five different cpBVDV strains generated by RNA recombination are shown: NADL, Osloss, CP1, Pe515CP, and CP9. In-frame insertions of host sequences are present in NADL, Osloss, and CP1. The NS2 autoprotease is responsible for NS3 production in the NADL strain, but the inserted ubiquitin sequences in Osloss and CP1 provide sites for processing by host ubiquitin C-terminal hydrolase. For Pe515CP and the CP9 DI RNA, the N^{pro} autoprotease mediates the cleavage producing the NS3 N terminus. The nomenclature and organization of the cleavage products and the symbols for the normal processing enzymes are defined in Figure 3. Reprinted from Fields Virology, 5th Ed. (David. M. Knipe/Peter N, Howley, eds), Chapter 33: Flaviviridae, p. 1130 (2007) with permission from Lippincott, Williams & Wilkins.

EPIDEMIOLOGY

The economic importance of CSF is closely correlated to the density of the pig population, the technical standard of pig farming, and the enforcement of legislation for the control of infectious diseases of animals. Financial losses directly or indirectly caused by CSF outbreaks may be significant (Terpstra, 1988). One of the major causes of new CSF outbreaks is feeding of swill to pigs, because in meat and meat products, e.g., salami and raw ham, the virus is able to survive for several weeks or months (Savi et al., 1965). In recognition of this danger, swill feeding is officially banned in many countries. In addition, persistently infected piglets excrete large amount of virus, and they are ideal vehicles for CSFV. Transport and trade with these animals contributes significantly to the spread of CSF over long distances (Picard, 1989).

The spread of the disease in a pig herd may take different courses depending on the virulence of the virus. Highly virulent CSFV multiplies very efficiently in its host, and high amounts of the virus are excreted by oral and nasal discharge and feces. Spread of the infection in pig herds is relatively rapid. In contrast, due to the low replication of the virus and the production of neutralizing antibodies, the excretion of low-virulence CSFV is less efficient. Thus, the spread of this type of infection in a herd is relatively slow, and CSF is often diagnosed only 6 to 8 weeks after introduction of the virus (van Oirschot, 1992). Current epidemics in the field are caused by a spectrum of viruses with high to low virulence (Wensvoort and Terpstra, 1985). Migration of rats, airborne transmission, and spraying of manure play a role in the spread of the virus in areas with dense pig populations.

Ruminant pestiviruses are ubiquitous in areas where cattle and sheep are held, and animal populations are seropositive at an average level of 60 to 90% (Harkness et al., 1978; Liess et al., 1983; Steck et al., 1980). The major reservoirs for the maintenance of BVDV and BDV are persistently infected animals. Since some BVDV/BDV strains seem to be well adapted to their hosts, persistently viremic animals without clinical signs may breed and produce several generations of viremic progeny (Liess et al., 1983). In many countries, an average of about 1% of the cattle population is persistently viremic (Bolin et al., 1985a; Houe and Meyling, 1992; Peters et al., 1987). However, on a herd basis this figure may vary widely. In addition, man-made inoculation routes, i.e., artificial insemination, embryo transfer, and vaccines, all of which may contain material accidentally contaminated with BVDV of BDV, play an important role in the spread of these viruses (Liess et al., 1987). The spread of newly introduced BVDV in a susceptible herd is slow, taking weeks to a few months, depending on the size of the herd.

DIAGNOSIS, SURVEILLANCE AND CONTROL OF PESTIVIRUS INFECTIONS

DETECTION OF VIRUS

Pestiviruses can be isolated from buffy coat cells, skin, plasma, serum, or organ suspensions of viremic animals and grown on susceptible cell cultures of the respective host animal, i.e., porcine, bovine, or ovine cells. Because most pestiviruses are of the non-cp biotype, antibodies are a prerequisite for the detection of virus infection in cell culture and identification of viral isolates. Alternatively, the viral genome can be detected by RT-PCR.

The rapid diagnosis of CSFV infection can be based on the demonstration of viral antigen(s) in organ tissue sections using conjugated antibodies for direct immunofluorescence. Suitable materials are tonsils, lymph nodes, spleen and parotid glands. Alternatively, organ suspensions of buffy coat cells can be cultured on susceptible porcine cells with the subsequent detection of viral antigen(s). A direct discrimination of CSFV from ruminant pestiviruses can be achieved using defined MAbs (Hess et al., 1988; Moennig and Plagemann, 1992; Wensvoort et al., 1986; Zhou et al., 1989). Moreover, RT-PCR is suited for detection of virus as well as discrimination from other pestiviruses. With regard to ruminant pestiviruses the demonstration of virus by one of the described methods is of prime importance for the control of the infection, especially since viremic animals are a most efficient reservoir for spreading the virus (Bolin et al., 1985a; Edwards et al., 1987; Howard et al., 1986). The percentage of viremic animals may vary in individual herds. In general, viremic cattle and sheep are seronegative. The conventional isolation of BVDV or BDV from buffy coat cells is a reliable technique. In addition, enzyme immunoassays were developed for the direct demonstration of pestiviral antigen(s) in peripheral blood leukocytes of sheep and cattle as well as skin/ear notch samples (Fenton et al., 1990; Fenton et al., 1991; Moennig and Plagemann, 1992; Shannon et al., 1991). The demonstration of viral antigen or RNA in ear notch samples is currently widely applied for identification of persistently infected animals, also shortly after birth.

Detection of viral nucleic acids has become an increasingly important option for laboratory diagnosis. In particular, the 5' untranslated region has been used for amplification by the reverse transcriptase polymerase chain reaction (RT-PCR) (Boye et al., 1991; De Moerlooze et al., 1993; Hertig et al., 1991; Hooft van Iddekinge et al., 1992; Muyldermans et al., 1993; Wirz et al., 1993). Since the 5' region is highly conserved, amplification can be achieved with all pestiviruses. Subsequent nucleotide sequencing of the respective region allows discrimination between the different pestivirus species (De Moerlooze et al., 1993). For further investigation, additional genomic regions (N^{pro}, E2 etc.) can be analyzed (see also General Aspects, Taxonomy).

SERODIAGNOSIS

The humoral immune response allows detection of antibodies against the structural envelope proteins E2 and E^{rns} as well as against the non-structural protein NS3. Antibodies against E2 can neutralize the virus efficiently. Antibodies against all three proteins are suited to indirectly demonstrate an infection with a pestivirus. Serodiagnostic analyses are of particular importance for the control and eradication of CSF, and for this purpose enzyme immunoassays and neutralization assays can be used. However, cross-reacting antibodies elicited by ruminant pestivirus infections of pigs are usually registered by these tests. As an alternative, enzyme immunoassays have been developed which will detect only antibodies induced by E2 of CSFV. In conjunction with marker vaccines against CSF discriminating assays have been developed.

For BVDV licensed vaccines are based on inactivated wildtype or live attenuated virus. Accordingly, the induced antibodies do not allow discrimination between vaccination and field virus infection. Maternal antibodies will often be present in young animals, in some cases up to an age of nine months. Detection of antibodies is mostly based on ELISA which in contrast to serum neutralization test (SNT) can be run in large numbers and is easier to standardize. The SNT can be used for specific questions like induction of antibodies after vaccination and differentiation of antibody titres against BVDV-1 and BVDV-2. The currently available ELISAs are based on indirect or competitive tests usually with lysate from infected cells as antigen. The indirect tests use secondary antibodies against bovine immunoglobulin to detect BVDV-specific antibodies. The competitive tests employ monoclonal antibodies against NS3. Serum, plasma and milk can be used to detect antibodies against pestiviruses. With regard to milk, sensitivity and specificity of the methods is usually lower.

Pools of milk samples can be used to determine whether antibodies against BVDV are present in a herd.

SURVEILLANCE AND CONTROL

The economic impact of pestivirus infection varies according to their pathogenesis. Classical swine fever is an overt infection with severe lesions aggravated by diaplacental infections, sometimes leading to the birth of viremic piglets spreading the infection. In contrast, ruminant pestiviruses in general do not induce clinical signs during acute infection, and the infection spreads unnoticed. Economic damage is primarily caused by diaplacental infections and their sequelae. Measures to control pestiviruses vary accordingly.

In most industrialized nations, CSF is a notifiable disease, and attempts are made either to eradicate the infection or to vaccinate systematically. Eradication campaigns have been successful in Australia, Canada, U.S. and in most countries of the European Union (Bendixen, 1989; Stahlheim, 1988). Occasional CSF epidemics in wild boar (*sus scrofa ferus*) may interfere with the eradication in domestic pigs. Strategies for oral vaccination of wild boar with attenuated CSF virus are being used within the European Union (Kaden et al., 2000; Kaden et al., 2003).

In contrast ruminant pestiviruses are widespread, and there are only few countries with official guidelines on how to deal with infections with BVDV. Some countries, in particular in Scandinavia, have been extremely successful in reducing the numbers of infected animals. Control efforts concentrate on the prevention of diaplacental infections, e.g. by the elimination of persistently viremic animals as they are virus reservoirs and/or systematic vaccination of seronegative female animals prior to breeding. However, the considerable degree of diversity observed for BVDV sheds some doubt on the prospects of effectively controlling the infection by vaccination (Dubovi, 1992; Shimizu et al., 1989). This applies especially to BVDV-1 where more than eleven subgroups have been described (Fig. 2).

Classical swine fever virus has been serially passaged in rabbits ("Chinese" strain) or in tissue culture cells (Japanese GPE-negative strain, French Thiverval strain). Such approaches led to attenuated viruses that can be used as efficient and safe vaccines. After parenteral inoculation the attenuated virus appears to replicate mainly in tonsils and lymph nodes. Vaccination confers a rapid and long-lasting immunity (Moennig and Plagemann, 1992). However, there are no means of differentiating polyclonal antibodies in sera from pigs that were exposed to either vaccine or field virus. This is considered to be a major impediment for an efficient eradication of CSF and was the main argument for prohibiting vaccination with live virus by the Commission of the European Community.

Because of the ubiquitous presence of BVDV, vaccination still plays an important role world-wide, and its major objective is the prevention of diaplacental infection and the resultant economic losses. Both, inactivated and live virus vaccines are available. Most live vaccine strains have been attenuated using cell culture passages. Pregnant animals must not be vaccinated because live virus can cross the placenta and infect the fetus. As most vaccine viruses are of the cp biotype, vaccination of persistently infected animals with live vaccine virus leads to a superinfection, and MD might be induced (Moennig and Plagemann, 1992).

There is substantial need for improved vaccines against pestiviruses. The major requirements are long-lasting and complete protection, in particular with regard to diaplacental infection. First steps toward novel, efficient, and safe vaccines have been taken. Expression of CSFV glycoproteins in viral vectors which replicate after immunization, i.e., vaccinia virus and pseudorabies virus, leads to a protective immune response in vaccinated animals (Rümenapf et al., 1991b; van Zijl et al., 1991). Interestingly, a construct that lacked most of the CSFV E2 gene failed to induce neutralizing antibodies but induced protective immunity. In addition, a subunit vaccine based on immunization with CSFV glycoprotein E2 expressed in the baculovirus system resulted in immunity and respective vaccines have been licensed. These are marker vaccines which allow serological discrimination of animals which were either vaccinated or infected with field virus (Hulst et al., 1993). The available subunit vaccines can be improved with regard to time span required for induction of protective immunity as well as prevention of diaplacental infection. It is expected that vaccines analogous to the ones described above against CSF will be developed against diseases caused by ruminant pestiviruses. The availability of so called infectious pestivirus clones allows to introduce attenuating mutations into the viral genomes. Such manipulations comprise deletion of genes like N^{pro}, deletion of sequences in the NTRs, introduction of point mutations to destroy enzymatic activities like the one of E^{rns} (Makoschey et al., 2004; Meyers et al., 2007; Meyers et al., 1999; Pankraz et al., 2005; Tratschin et al., 1998). The resulting viruses are being tested for suitability as live vaccines. Moreover, complementing cell lines have been established to provide viral proteins *in trans*, in particular structural proteins (Frey et al., 2006; Reimann et al., 2003; Stettler et al., 2002; van Gennip et al., 2002; Widjojoatmodjo et al., 2000). Transfection of pestiviral RNA with deletion of a structural gene into respective cells leads to viruses which will infect host cells but not generate infectious viruses. It remains to be established whether this approach allows mass production of a vaccine. Aspects like lack of capability to transfer the placenta (live virus vaccine) and possible use as marker vaccines (live and inactivated vaccines) are considered to be of particular importance with respect to pestiviral vaccines.

References

Agapov, E.V., Murray, C.L., Frolov, I., Qu, L., Myers, T.M., and Rice, C.M. (2004). Uncleaved NS2-3 is required for production of infectious bovine viral diarrhea virus. J. Virol. *78*, 2414-2425.

Agnello, V., Abel, G., Elfahal, M., Knight, G.B., and Zhang, Q.X. (1999). Hepatitis C virus and other Flaviviridae viruses enter cells via low density lipoprotein receptor. Proc. Natl. Acad. Sci. USA *96*, 12766-12771.

Ansari, I.H., Chen, L.M., Liang, D., Gil, L.H., Zhong, W., and Donis, R.O. (2004). Involvement of a bovine viral diarrhea virus NS5B locus in virion assembly. J. Virol. *78*, 9612-9623.

Avalos-Ramirez, R., Orlich, M., Thiel, H.J., and Becher, P. (2001). Evidence for the presence of two novel pestivirus species. Virology *286*, 456-465.

Baginski, S.G., Pevear, D.C., Seipel, M., Sun, S.C., Benetatos, C.A., Chunduru, S.K., Rice, C.M., and Collett, M.S. (2000). Mechanism of action of a pestivirus antiviral compound. Proc. Natl. Acad. Sci. USA *97*, 7981-7986.

Baker, J.C. (1987). Bovine viral diarrhea virus: a review. J. Am. Vet. Med. Assoc. *190*, 1449-1458.

Balint, A., Palfi, V., Belak, S., and Baule, C. (2005). Viral sequence insertions and a novel cellular insertion in the NS2 gene of cytopathic isolates of bovine viral diarrhea virus as potential cytopathogenicity markers. Virus Genes *30*, 49-58.

Barber, D.M., Nettleton, P.F., and Herring, J.A. (1985). Disease in a dairy herd associated with the introduction and spread of bovine virus diarrhoea virus. Vet. Rec. *117*, 459-464.

Barlow, R.M. (1972). Experiments in border disease. IV. Pathological changes in ewes. J. Comp. Pathol. *82*, 151-157.
Barlow, R.M., Rennie, J.C., Keir, W.A., Gardiner, A.C., and Vantsis, J.T. (1975). Experiments in Border disease. VII. The disease in goats. J. Comp. Pathol. *85*, 291-297.
Baroth, M., Orlich, M., Thiel, H.J., and Becher, P. (2000). Insertion of cellular NEDD8 coding sequences in a pestivirus. Virology *278*, 456-466.
Bauhofer, O., Summerfield, A., Sakoda, Y., Tratschin, J.D., Hofmann, M.A., and Ruggli, N. (2007). Classical swine fever virus Npro interacts with interferon regulatory factor 3 and induces its proteasomal degradation. J. Virol. *81*, 3087-3096.
Bazan, J.F., and Fletterick, R.J. (1989). Detection of a trypsin-like serine protease domain in flaviviruses and pestiviruses. Virology *171*, 637-639.
Becher, P., Avalos Ramirez, R., Orlich, M., Cedillo Rosales, S., Konig, M., Schweizer, M., Stalder, H., Schirrmeier, H., and Thiel, H.J. (2003). Genetic and antigenic characterization of novel pestivirus genotypes: implications for classification. Virology *311*, 96-104.
Becher, P., Meyers, G., Shannon, A.D., and Thiel, H.J. (1996). Cytopathogenicity of border disease virus is correlated with integration of cellular sequences into the viral genome. J. Virol. *70*, 2992-2998.
Becher, P., Orlich, M., and Thiel, H.J. (2000). Mutations in the 5′ nontranslated region of bovine viral diarrhea virus result in altered growth characteristics. J. Virol. *74*, 7884-7894.
Behrens, S.E., Grassmann, C.W., Thiel, H.J., Meyers, G., and Tautz, N. (1998). Characterization of an autonomous subgenomic pestivirus RNA replicon. J. Virol. *72*, 2364-2372.
Bendixen, H.J. (1989). Incidence and control of classical swine fever, with special reference to the situation in the EEC. Outlook-on-Agriculture, 12-17.
Bielanski, A., and Hare, W.C. (1988). Effect in vitro of bovine viral diarrhea virus on bovine embryos with the zona pellucida intact, damaged and removed. Vet. Res. Commun. *12*, 19-24.
Bielefeldt Ohmann, H. (1988a). BVD virus antigens in tissues of persistently viraemic, clinically normal cattle: implications for the pathogenesis of clinically fatal disease. Acta Vet. Scand. *29*, 77-84.
Bielefeldt Ohmann, H. (1988b). In situ characterization of mononuclear leukocytes in skin and digestive tract of persistently bovine viral diarrhea virus-infected clinically healthy calves and calves with mucosal disease. Vet. Pathol. *25*, 304-309.

Bielefeldt-Ohmann, H., and Bloch, B. (1982). Electron microscopic studies of bovine viral diarrhea virus in tissues of diseased calves and in cell cultures. Arch. Virol. *71*, 57-74.

Bolin, S.R. (1988). Viral and viral protein specificity of antibodies induced in cows persistently infected with noncytopathic bovine viral diarrhea virus after vaccination with cytopathic bovine viral diarrhea virus. Am. J. Vet. Res. *49*, 1040-1044.

Bolin, S.R., McClurkin, A.W., and Coria, M.F. (1985a). Frequency of persistent bovine viral diarrhea virus infection in selected cattle herds. Am. J. Vet. Res. *46*, 2385-2387.

Bolin, S.R., McClurkin, A.W., Cutlip, R.C., and Coria, M.F. (1985b). Severe clinical disease induced in cattle persistently infected with noncytopathic bovine viral diarrhea virus by superinfection with cytopathic bovine viral diarrhea virus. Am. J. Vet. Res. *46*, 573-576.

Boye, M., Kamstrup, S., and Dalsgaard, K. (1991). Specific sequence amplification of bovine virus diarrhea virus (BVDV) and hog cholera virus and sequencing of BVDV nucleic acid. Vet. Microbiol. *29*, 1-13.

Branza-Nichita, N., Lazar, C., Durantel, D., Dwek, R.A., and Zitzmann, N. (2002). Role of disulfide bond formation in the folding and assembly of the envelope glycoproteins of a pestivirus. Biochem. Biophys. Res. Commun. *296*, 470-476.

Brock, K.V., Deng, R., and Riblet, S.M. (1992). Nucleotide sequencing of 5' and 3' termini of bovine viral diarrhea virus by RNA ligation and PCR. J. Virol. Methods *38*, 39-46.

Brown, G.B., Bolin, S.R., Frank, D.E., and Roth, J.A. (1991). Defective function of leukocytes from cattle persistently infected with bovine viral diarrhea virus, and the influence of recombinant cytokines. Am. J. Vet. Res. *52*, 381-387.

Brown, T.T., De Lahunte, A., Scott, F.W., Kahrs, R.F., McEntee, K., and Gillespie, J.H. (1973). Virus induced congenital anomalies of the bovine fetus. II. Histopathology of cerebellar degeneration (hypoplasia) induced by the virus of bovine viral diarrhea-mucosal disease. Cornell Vet. *63*, 561-578.

Brownlie, J. (1991). The pathways for bovine virus diarrhoea virus biotypes in the pathogenesis of disease. Arch. Virol. Suppl. *3*, 79-96.

Brownlie, J., Clarke, M.C., and Howard, C.J. (1984). Experimental production of fatal mucosal disease in cattle. Vet. Rec. *114*, 535-536.

Brownlie, J., Clarke, M.C., and Howard, C.J. (1989). Experimental infection of cattle in early pregnancy with a cytopathic strain of bovine virus diarrhoea virus. Res. Vet. Sci. *46*, 307-311.

Bruschke, C.J., Hulst, M.M., Moormann, R.J., van Rijn, P.A., and van Oirschot, J.T. (1997). Glycoprotein Erns of pestiviruses induces apoptosis in lymphocytes of several species. J. Virol. *71*, 6692-6696.
Carrasco, C.P., Rigden, R.C., Vincent, I.E., Balmelli, C., Ceppi, M., Bauhofer, O., Tache, V., Hjertner, B., McNeilly, F., van Gennip, H.G., *et al.* (2004). Interaction of classical swine fever virus with dendritic cells. J. Gen. Virol. *85*, 1633-1641.
Casaro, A.P., Kendrick, J.W., and Kennedy, P.C. (1971). Response of the bovine fetus to bovine viral diarrhea-mucosal disease virus. Am. J. Vet. Res. *32*, 1543-1562.
Cattaneo, R. (2004). Four viruses, two bacteria, and one receptor: membrane cofactor protein (CD46) as pathogens' magnet. J. Virol. *78*, 4385-4388.
Cheville, N.F., and Mengeling, W.L. (1969). The pathogenesis of chronic hog cholera (swine fever). Histologic, immunofluorescent, and electron microscopic studies. Lab. Invest. *20*, 261-274.
Cheville, N.F., Mengeling, W.L., and Zinober, M.R. (1970). Ultrastructural and immunofluorescent studies of glomerulonephritis in chronic hog cholera. Lab. Invest. *22*, 458-467.
Choi, K.H., Groarke, J.M., Young, D.C., Kuhn, R.J., Smith, J.L., Pevear, D.C., and Rossmann, M.G. (2004). The structure of the RNA-dependent RNA polymerase from bovine viral diarrhea virus establishes the role of GTP in de novo initiation. Proc. Natl. Acad. Sci. USA *101*, 4425-4430.
Chon, S.K., Perez, D.R., and Donis, R.O. (1998). Genetic analysis of the internal ribosome entry segment of bovine viral diarrhea virus. Virology *251*, 370-382.
Collett, M.S., Anderson, D.K., and Retzel, E. (1988a). Comparisons of the pestivirus bovine viral diarrhoea virus with members of the flaviviridae. J. Gen. Virol. *69*, 2637-2643.
Collett, M.S., Larson, R., Belzer, S.K., and Retzel, E. (1988b). Proteins encoded by bovine viral diarrhea virus: the genomic organization of a pestivirus. Virology *165*, 200-208.
Collett, M.S., Wiskerchen, M., Welniak, E., and Belzer, S.K. (1991). Bovine viral diarrhea virus genomic organization. Arch. Virol. Suppl. *3*, 19-27.
Corapi, W.V., Donis, R.O., and Dubovi, E.J. (1988). Monoclonal antibody analyses of cytopathic and noncytopathic viruses from fatal bovine viral diarrhea virus infections. J. Virol. *62*, 2823-2827.
Corapi, W.V., French, T.W., and Dubovi, E.J. (1989). Severe thrombocytopenia in young calves experimentally infected with noncytopathic bovine viral diarrhea virus. J. Virol. *63*, 3934-3943.

Coria, M.F., and McClurkin, A.W. (1978). Specific immune tolerance in an apparently healthy bull persistently infected with bovine viral diarrhea virus. J. Am. Vet. Med. Assoc. *172*, 449-451.
De Moerlooze, L., Lecomte, C., Brown-Shimmer, S., Schmetz, D., Guiot, C., Vandenbergh, D., Allaer, D., Rossius, M., Chappuis, G., Dina, D., and et al. (1993). Nucleotide sequence of the bovine viral diarrhoea virus Osloss strain: comparison with related viruses and identification of specific DNA probes in the 5' untranslated region. J. Gen. Virol. *74*, 1433-1438.
Deng, R., and Brock, K.V. (1993). 5′ and 3′ untranslated regions of pestivirus genome: primary and secondary structure analyses. Nucleic Acids Res. *21*, 1949-1957.
Devereux, J., Haeberli, P., and Smithies, O. (1984). A comprehensive set of sequence analysis programs for the VAX. Nucleic Acids Res. *12*, 387-395.
Di Marco, S., Volpari, C., Tomei, L., Altamura, S., Harper, S., Narjes, F., Koch, U., Rowley, M., De Francesco, R., Migliaccio, G., and Carfí, A. (2005). Interdomain communication in hepatitis C virus polymerase abolished by small molecule inhibitors bound to a novel allosteric site. J. Biol. Chem. *280*, 29765-29770.
Doherty, M.L. (1986). Serious perinatal mortality in cattle associated with bovine virus diarrhoea-mucosal disease (BVD-MD) virus. Irish Vet. J., 189-190.
Done, J.T., Terlecki, S., Richardson, C., Harkness, J.W., Sands, J.J., Patterson, D.S., Sweasey, D., Shaw, I.G., Winkler, C.E., and Duffell, S.J. (1980). Bovine virus diarrhoea-mucosal disease virus: pathogenicity for the fetal calf following maternal infection. Vet. Rec. *106*, 473-479.
Donis, R.O., Corapi, W., and Dubovi, E.J. (1988). Neutralizing monoclonal antibodies to bovine viral diarrhoea virus bind to the 56K to 58K glycoprotein. J. Gen. Virol. *69*, 77-86.
Dubovi, E.J. (1992). Genetic diversity and BVD virus. Comp. Immunol. Microbiol. Infect. Dis. *15*, 155-162.
Dunne, H.W. (1973). Hog cholera (European swine fever). Adv. Vet. Sci. Comp. Med. *17*, 315-359.
Durantel, D., Branza-Nichita, N., Carrouee-Durantel, S., Butters, T.D., Dwek, R.A., and Zitzmann, N. (2001). Study of the mechanism of antiviral action of iminosugar derivatives against bovine viral diarrhea virus. J. Virol. *75*, 8987-8998.
Edwards, S., Drew, T.W., and Bushnell, S.E. (1987). Prevalence of bovine virus diarrhoea virus viraemia. Vet. Rec. *120*, 71.

Elbers, K., Tautz, N., Becher, P., Stoll, D., Rümenapf, T., and Thiel, H.J. (1996). Processing in the pestivirus E2-NS2 region: identification of proteins p7 and E2p7. J. Virol. *70*, 4131-4135.
Ellis, J.A., Davis, W.C., Belden, E.L., and Pratt, D.L. (1988). Flow cytofluorimetric analysis of lymphocyte subset alterations in cattle infected with bovine viral diarrhea virus. Vet. Pathol. *25*, 231-236.
Fahey, K.J., and Morris, B. (1978). Humoral immune responses in foetal sheep. Immunology *35*, 651-661.
Fenton, A., Entrican, G., Herring, J.A., and Nettleton, P.F. (1990). An ELISA for detecting pestivirus antigen in the blood of sheep persistently infected with border disease virus. J. Virol. Methods *27*, 253-260.
Fenton, A., Nettleton, P.F., Entrican, G., Herring, J.A., Malloy, C., Greig, A., and Low, J.C. (1991). Identification of cattle infected with bovine virus diarrhoea virus using a monoclonal antibody capture ELISA. Arch. Virol. Suppl. *3*, 169-174.
Fernandez, A., Hewicker, M., Trautwein, G., Pohlenz, J., and Liess, B. (1989). Viral antigen distribution in the central nervous system of cattle persistently infected with bovine viral diarrhea virus. Vet. Pathol. *26*, 26-32.
Fetzer, C., Tews, B.A., and Meyers, G. (2005). The carboxy-terminal sequence of the pestivirus glycoprotein E(rns) represents an unusual type of membrane anchor. J. Virol. *79*, 11901-11913.
Frey, C.F., Bauhofer, O., Ruggli, N., Summerfield, A., Hofmann, M.A., and Tratschin, J.D. (2006). Classical swine fever virus replicon particles lacking the Erns gene: a potential marker vaccine for intradermal application. Vet. Res. *37*, 655-670.
Frey, H.R., Liess, B., Richter-Reichhelm, H.B., von Benten, K., and Trautwein, G. (1980). Experimental transplacental transmission of hog cholera virus in pigs. I. Virological and serological studies. Zbl. Vet. Med., 154-164.
Frey, H.R., Liess, B., Peters, W., Rommig, L., and Lehmann, B. (1987). Ätiologische Beziehungen zwischen BVD-Virusinfektionen bei Rindern und Fällen mit klinischem Verdacht auf Bovine Virusdiarrhoe. Dtsch. Tierärztl. Wschr., 580-583.
Fricke, J., Voss, C., Thumm, M., and Meyers, G. (2004). Processing of a pestivirus protein by a cellular protease specific for light chain 3 of microtubule-associated proteins. J. Virol. *78*, 5900-5912.
Fritzemeier, J., Haas, L., Liebler, E., Moennig, V., and Greiser-Wilke, I. (1997). The development of early vs. late onset mucosal disease is a consequence of two different pathogenic mechanisms. Arch. Virol. *142*, 1335-1350.

Frolov, I., McBride, M.S., and Rice, C.M. (1998). cis-acting RNA elements required for replication of bovine viral diarrhea virus-hepatitis C virus 5' nontranslated region chimeras. RNA *4*, 1418-1435.

Gallei, A., Orlich, M., Thiel, H.J., and Becher, P. (2005). Noncytopathogenic pestivirus strains generated by nonhomologous RNA recombination: alterations in the NS4A/NS4B coding region. J. Virol. *79*, 14261-14270.

Gallei, A., Pankraz, A., Thiel, H.J., and Becher, P. (2004). RNA recombination in vivo in the absence of viral replication. J. Virol. *78*, 6271-6281.

Gardiner, A.C., Zakarian, B., and Barlow, R.M. (1980). Periarteritis in experimental border disease of sheep. III. Immunopathological observations. J. Comp. Pathol. *90*, 469-474.

Garry, R.F., and Dash, S. (2003). Proteomics computational analyses suggest that hepatitis C virus E1 and pestivirus E2 envelope glycoproteins are truncated class II fusion proteins. Virology *307*, 255-265.

Gil, L.H., Ansari, I.H., Vassilev, V., Liang, D., Lai, V.C., Zhong, W., Hong, Z., Dubovi, E.J., and Donis, R.O. (2006). The amino-terminal domain of bovine viral diarrhea virus Npro protein is necessary for alpha/beta interferon antagonism. J. Virol. *80*, 900-911.

Gillespie, J.H., Coggins, L., Thompson, J., and Baker, J.A. (1961). Comparison by neutralization tests of strains of virus isolated from virus diarrhea and mucosal disease. Cornell Vet. *51*, 155-159.

Gong, Y., Shannon, A., Westaway, E.G., and Gowans, E.J. (1998). The replicative intermediate molecule of bovine viral diarrhoea virus contains multiple nascent strands. Arch. Virol. *143*, 399-404.

Gong, Y., Trowbridge, R., Macnaughton, T.B., Westaway, E.G., Shannon, A.D., and Gowans, E.J. (1996). Characterization of RNA synthesis during a one-step growth curve and of the replication mechanism of bovine viral diarrhoea virus. J. Gen. Virol. *77*, 2729-2736.

Gorbalenya, A.E., Donchenko, A.P., Koonin, E.V., and Blinov, V.M. (1989a). N-terminal domains of putative helicases of flavi- and pestiviruses may be serine proteases. Nucleic Acids Res. *17*, 3889-3897.

Gorbalenya, A.E., Koonin, E.V., Donchenko, A.P., and Blinov, V.M. (1989b). Two related superfamilies of putative helicases involved in replication, recombination, repair and expression of DNA and RNA genomes. Nucleic Acids Res. *17*, 4713-4729.

Grassmann, C.W., Isken, O., and Behrens, S.E. (1999). Assignment of the multifunctional NS3 protein of bovine viral diarrhea virus during RNA replication: an in vivo and in vitro study. J. Virol. *73*, 9196-9205.

Grassmann, C.W., Isken, O., Tautz, N., and Behrens, S.E. (2001). Genetic analysis of the pestivirus nonstructural coding region: defects in the NS5A unit can be complemented in trans. J. Virol. *75*, 7791-7802.

Gray, E.W., and Nettleton, P.F. (1987). The ultrastructure of cell cultures infected with border disease and bovine virus diarrhoea viruses. J. Gen. Virol. *68*, 2339-2346.

Greiser-Wilke, I., Dittmar, K.E., Liess, B., and Moennig, V. (1991). Immunofluorescence studies of biotype-specific expression of bovine viral diarrhoea virus epitopes in infected cells. J. Gen. Virol. *72*, 2015-2019.

Griffin, S.D., Beales, L.P., Clarke, D.S., Worsfold, O., Evans, S.D., Jaeger, J., Harris, M.P., and Rowlands, D.J. (2003). The p7 protein of hepatitis C virus forms an ion channel that is blocked by the antiviral drug, Amantadine. FEBS Lett. *535*, 34-38.

Griffin, S.D., Harvey, R., Clarke, D.S., Barclay, W.S., Harris, M., and Rowlands, D.J. (2004). A conserved basic loop in hepatitis C virus p7 protein is required for amantadine-sensitive ion channel activity in mammalian cells but is dispensable for localization to mitochondria. J. Gen. Virol. *85*, 451-461.

Grummer, B., Fischer, S., Depner, K., Riebe, R., Blome, S., and Greiser-Wilke, I. (2006). Replication of classical swine fever virus strains and isolates in different porcine cell lines. Dtsch. Tierarztl. Wochenschr. *113*, 138-142.

Harada, T., Tautz, N., and Thiel, H.J. (2000). E2-p7 region of the bovine viral diarrhea virus polyprotein: processing and functional studies. J. Virol. *74*, 9498-9506.

Harkness, J.W., Sands, J.J., and Richards, M.S. (1978). Serological studies of mucosal disease virus in England and Wales. Res. Vet. Sci. *24*, 98-103.

Harkness, J.W., Wood, L., and Drew, T. (1984). Mucosal disease in cattle. Vet. Rec. *115*, 283.

Hausmann, Y., Roman-Sosa, G., Thiel, H.J., and Rumenapf, T. (2004). Classical swine fever virus glycoprotein E rns is an endoribonuclease with an unusual base specificity. J. Virol. *78*, 5507-5512.

Heimann, M., Sosa, G.R., Martoglio, B., Thiel, H.J., and Rumenapf, T. (2006). Core protein of pestiviruses is processed at the C terminus by signal peptide peptidase. J. Virol. *80*, 1915-1921.

Hertig, C., Pauli, U., Zanoni, R., and Peterhans, E. (1991). Detection of bovine viral diarrhea (BVD) virus using the polymerase chain reaction. Vet. Microbiol. *26*, 65-76.

Hess, R.G., Coulibaly, C.O., Greiser-Wilke, I., Moennig, V., and Liess, B. (1988). Identification of hog cholera viral isolates by use of monoclonal antibodies to pestiviruses. Vet. Microbiol. *16*, 315-321.

Hilton, L., Moganeradj, K., Zhang, G., Chen, Y.H., Randall, R.E., McCauley, J.W., and Goodbourn, S. (2006). The NPro product of bovine viral diarrhea virus inhibits DNA binding by interferon regulatory factor 3 and targets it for proteasomal degradation. J. Virol. *80*, 11723-11732.

Hooft van Iddekinge, B.J., van Wamel, J.L., van Gennip, H.G., and Moormann, R.J. (1992). Application of the polymerase chain reaction to the detection of bovine viral diarrhea virus infections in cattle. Vet. Microbiol. *30*, 21-34.

Horzinek, M.C. (1981). Non-Arthropod-Borne Togaviruses (London: Academic Press).

Houe, H., and Meyling, A. (1992). Prevalence of bovine virus diarrhoea (BVD) in 19 Danish herds and estimation of incidence of infection in early pregnancy. Prev. Vet. Med. 9-16.

Howard, C.J. (1990). Immunological responses to bovine virus diarrhoea virus infections. Rev. Sci. Tech. *9*, 95-103.

Howard, C.J., Brownlie, J., and Thomas, L.H. (1986). Prevalence of bovine virus diarrhoea virus viraemia in cattle in the UK. Vet. Rec. *119*, 628-629.

Hughes, L.E., Kershaw, G.F., and Shaw, I.G. (1959). "B" or border disease. An undescribed disease of sheep. Vet. Rec., 313-317.

Hulst, M.M., and Moormann, R.J. (1997). Inhibition of pestivirus infection in cell culture by envelope proteins E(rns) and E2 of classical swine fever virus: E(rns) and E2 interact with different receptors. J. Gen. Virol. *78*, 2779-2787.

Hulst, M.M., van Gennip, H.G., and Moormann, R.J. (2000). Passage of classical swine fever virus in cultured swine kidney cells selects virus variants that bind to heparan sulfate due to a single amino acid change in envelope protein E(rns). J. Virol. *74*, 9553-9561.

Hulst, M.M., Westra, D.F., Wensvoort, G., and Moormann, R.J. (1993). Glycoprotein E1 of hog cholera virus expressed in insect cells protects swine from hog cholera. J. Virol. *67*, 5435-5442.

Iqbal, M., Poole, E., Goodbourn, S., and McCauley, J.W. (2004). Role for bovine viral diarrhea virus Erns glycoprotein in the control of activation of beta interferon by double-stranded RNA. J. Virol. *78*, 136-145.

Irie, M., and Ohgi, K. (2001). Ribonuclease T2. Methods Enzymol. *341*, 42-55.

Isken, O., Grassmann, C.W., Sarisky, R.T., Kann, M., Zhang, S., Grosse, F., Kao, P.N., and Behrens, S.E. (2003). Members of the NF90/NFAR protein group are involved in the life cycle of a positive-strand RNA virus. EMBO J. *22*, 5655-5665.

Isken, O., Grassmann, C.W., Yu, H., and Behrens, S.E. (2004). Complex signals in the genomic 3' nontranslated region of bovine viral diarrhea virus coordinate translation and replication of the viral RNA. RNA *10*, 1637-1652.

Jordan, R., Wang, L., Graczyk, T.M., Block, T.M., and Romano, P.R. (2002). Replication of a cytopathic strain of bovine viral diarrhea virus activates PERK and induces endoplasmic reticulum stress-mediated apoptosis of MDBK cells. J. Virol. *76*, 9588-9599.

Kaden, V., Lange, E., Fischer, U., and Strebelow, G. (2000). Oral immunisation of wild boar against classical swine fever: evaluation of the first field study in Germany. Vet. Microbiol. *73*, 239-252.

Kaden, V., Renner, C., Rothe, A., Lange, E., Hanel, A., and Gossger, K. (2003). Evaluation of the oral immunisation of wild boar against classical swine fever in Baden-Wurttemberg. Berl. Munch. Tierarztl. Wochenschr. *116*, 362-367.

Kahrs, R.F. (1973). Effects of bovine viral diarrhea on the developing fetus. J. Am. Vet. Med. Assoc. *163*, 877-878.

Kao, C.C., Del Vecchio, A.M., and Zhong, W. (1999). De novo initiation of RNA synthesis by a recombinant flaviviridae RNA-dependent RNA polymerase. Virology *253*, 1-7.

Kimura, M. (1980). A simple method for estimating evolutionary rates of base substitutions through comparative studies of nucleotide sequences. J. Mol. Evol. *16*, 111-120.

Kohl, W., Zimmer, G., Greiser-Wilke, I., Haas, L., Moennig, V., and Herrler, G. (2004). The surface glycoprotein E2 of bovine viral diarrhoea virus contains an intracellular localization signal. J. Gen. Virol. *85*, 1101-1111.

Krey, T., Himmelreich, A., Heimann, M., Menge, C., Thiel, H.J., Maurer, K., and Rumenapf, T. (2006a). Function of Bovine CD46 as a Cellular Receptor for Bovine Viral Diarrhea Virus Is Determined by Complement Control Protein 1. J. Virol. *80*, 3912-3922.

Krey, T., Moussay, E., Thiel, H.J., and Rumenapf, T. (2006b). Role of the low-density lipoprotein receptor in entry of bovine viral diarrhea virus. J. Virol. *80*, 10862-10867.

Krey, T., Thiel, H.J., and Rumenapf, T. (2005). Acid-resistant bovine pestivirus requires activation for pH-triggered fusion during entry. J. Virol. *79*, 4191-4200.
Kümmerer, B.M., and Meyers, G. (2000). Correlation between point mutations in NS2 and the viability and cytopathogenicity of Bovine viral diarrhea virus strain Oregon analyzed with an infectious cDNA clone. J. Virol. *74*, 390-400.
La Rocca, S.A., Herbert, R.J., Crooke, H., Drew, T.W., Wileman, T.E., and Powell, P.P. (2005). Loss of interferon regulatory factor 3 in cells infected with classical swine fever virus involves the N-terminal protease, Npro. J. Virol. *79*, 7239-7247.
Lackner, T., Muller, A., Konig, M., Thiel, H.J., and Tautz, N. (2005). Persistence of bovine viral diarrhea virus is determined by a cellular cofactor of a viral autoprotease. J. Virol. *79*, 9746-9755.
Lackner, T., Muller, A., Pankraz, A., Becher, P., Thiel, H.J., Gorbalenya, A.E., and Tautz, N. (2004). Temporal modulation of an autoprotease is crucial for replication and pathogenicity of an RNA virus. J. Virol. *78*, 10765-10775.
Lackner, T., Thiel, H.J., and Tautz, N. (2006). Dissection of a viral autoprotease elucidates a function of a cellular chaperone in proteolysis. Proc. Natl. Acad. Sci. USA *103*, 1510-1515.
Lai, V.C., Kao, C.C., Ferrari, E., Park, J., Uss, A.S., Wright-Minogue, J., Hong, Z., and Lau, J. Y. (1999). Mutational analysis of bovine viral diarrhea virus RNA-dependent RNA polymerase. J. Virol. *73*, 10129-10136.
Langedijk, J.P. (2002). Translocation activity of C-terminal domain of pestivirus Erns and ribotoxin L3 loop. J. Biol. Chem. *277*, 5308-5314.
Langedijk, J.P., van Veelen, P.A., Schaaper, W.M., de Ru, A.H., Meloen, R.H., and Hulst, M.M. (2002). A structural model of pestivirus E(rns) based on disulfide bond connectivity and homology modeling reveals an extremely rare vicinal disulfide. J. Virol. *76*, 10383-10392.
Laude, H. (1987). Hog cholera virus: art and facts. Ann. Rech. Vet. *18*, 127-138.
Lecot, S., Belouzard, S., Dubuisson, J., and Rouille, Y. (2005). Bovine viral diarrhea virus entry is dependent on clathrin-mediated endocytosis. J. Virol. *79*, 10826-10829.
Lemberg, M.K., and Martoglio, B. (2002). Requirements for signal peptide peptidase-catalyzed intramembrane proteolysis. Mol. Cell *10*, 735-744.
Liang, D., Sainz, I.F., Ansari, I.H., Gil, L.H., Vassilev, V., and Donis, R.O. (2003). The envelope glycoprotein E2 is a determinant of cell

culture tropism in ruminant pestiviruses. J. Gen. Virol. *84*, 1269-1274.

Liebler, E.M., Waschbusch, J., Pohlenz, J.F., Moennig, V., and Liess, B. (1991). Distribution of antigen of noncytopathogenic and cytopathogenic bovine virus diarrhea virus biotypes in the intestinal tract of calves following experimental production of mucosal disease. Arch. Virol. Suppl. *3*, 109-124.

Liess, B. (1973). Pathogenese virusbedingter Darminfektionen der Tiere. Dtsch. Tierärztl. Wschr., 341-368.

Liess, B. (1981). Hog Cholera, In A world geography of epidemiology and control (London: Academic Press), pp. 627-650.

Liess, B., Frey, H.R., Grambow, H., and Stahl, C. (1987). Embryotransfer und BVD-Virusinfektion bei Rindern. Dtsch. Tierärztl. Wschr., 506-508.

Liess, B., Frey, H.R., Kittsteiner, H., Baumann, F., and Neumann, W. (1974). Beobachtungen und Untersuchungen über die "mucosal Disease" des Rindes - einer immunologisch erklärbaren Spätform der BVD-MD-Virusinfektion mit Kriterien einer "slow virus infection"? Dtsch. Tierärztl. Wschr., 481-487.

Liess, B., Frey, H.R., Orban, S., and Hafez, S.M. (1983). Bovine Virusdiarrhoe (BVD) - "Mucosal Disease". Persistente BVD-Feldvirusinfektionen bei serologisch selektierten Rindern. Dtsch. Tierärztl. Wschr., 249-296, 261-266.

Lindenbach, B.D., Thiel, H.J., and Rice, C.M. (2007). Flaviviridae: The viruses and their replication, In Fields Virology, D.M. Knipe, and P.M. Howley, eds. (Philadelphia: Lippincott-Raven Publishers), pp. 1101-1152.

Liu, J.J., Wong, M.L., and Chang, T.J. (1998). The recombinant nucleocapsid protein of classical swine fever virus can act as a transcriptional regulator. Virus Res. *53*, 75-80.

Lohmann, V., Overton, H., and Bartenschlager, R. (1999). Selective stimulation of hepatitis C virus and pestivirus NS5B RNA polymerase activity by GTP. J. Biol. Chem. *274*, 10807-10815.

Makoschey, B., Becher, P., Janssen, M.G., Orlich, M., Thiel, H.J., and Lutticken, D. (2004). Bovine viral diarrhea virus with deletions in the 5'-nontranslated region: reduction of replication in calves and induction of protective immunity. Vaccine *22*, 3285-3294.

Maurer, K., Krey, T., Moennig, V., Thiel, H.J., and Rumenapf, T. (2004). CD46 is a cellular receptor for bovine viral diarrhea virus. J. Virol. *78*, 1792-1799.

Mayer, D., Hofmann, M.A., and Tratschin, J.D. (2004). Attenuation of classical swine fever virus by deletion of the viral N(pro) gene. Vaccine *22*, 317-328.

Mayer, D., Thayer, T.M., Hofmann, M.A., and Tratschin, J.D. (2003). Establishment and characterisation of two cDNA-derived strains of classical swine fever virus, one highly virulent and one avirulent. Virus Res. *98*, 105-116.

McClurkin, A.W., Bolin, S.R., and Coria, M.F. (1985). Isolation of cytopathic and noncytopathic bovine viral diarrhea virus from the spleen of cattle acutely and chronically affected with bovine viral diarrhea. J. Am. Vet. Med. Assoc. *186*, 568-569.

McKercher, D.G., Saito, J.K., Crenshaw, G.L., and Bushnell, R.B. (1968). Complications in cattle following vaccination with a combined bovine viral diarrhea-infectious bovine rhinotracheitis vaccine. J. Am. Vet. Med. Assoc. *152*, 1621-1624.

Mendez, E., Ruggli, N., Collett, M.S., and Rice, C.M. (1998). Infectious bovine viral diarrhea virus (strain NADL) RNA from stable cDNA clones: a cellular insert determines NS3 production and viral cytopathogenicity. J. Virol. *72*, 4737-4745.

Meyer, C., Von Freyburg, M., Elbers, K., and Meyers, G. (2002). Recovery of virulent and RNase-negative attenuated type 2 bovine viral diarrhea viruses from infectious cDNA clones. J. Virol. *76*, 8494-8503.

Meyer, H., Liess, B., Frey, H.R., Hermanns, W., and Trautwein, G. (1981). Experimental transplacental transmission of hog cholera virus in pigs. IV. Virological and serological studies in newborn piglets. Zentralbl. Veterinarmed. B *28*, 659-668.

Meyer, H., Liess, B., Hermanns, W., and Trautwein, G. (1980). Experimentelle, diaplazentare Infektion von Schweinefeten mit dem Virus der Europäischen Schweinepest (ESP). Virologische und serologische Untersuchungen in der postnatalen Phase. Fortschritte der Veterinärmedizin. Suppl. Zbl. Vet. Med., 140-144.

Meyers, G., Ege, A., Fetzer, C., von Freyburg, M., Elbers, K., Carr, V., Prentice, H., Charleston, B., and Schurmann, E.M. (2007). Bovine viral diarrhea virus: prevention of persistent fetal infection by a combination of two mutations affecting Erns RNase and Npro protease. J. Virol. *81*, 3327-3338.

Meyers, G., Saalmuller, A., and Buttner, M. (1999). Mutations abrogating the RNase activity in glycoprotein E(rns) of the pestivirus classical swine fever virus lead to virus attenuation. J. Virol. *73*, 10224-10235.

Meyers, G., Stoll, D., and Gunn, M. (1998). Insertion of a sequence encoding light chain 3 of microtubule-associated proteins 1A and 1B in a pestivirus genome: connection with virus cytopathogenicity and induction of lethal disease in cattle. J. Virol. *72*, 4139-4148.

Meyers, G., Tautz, N., Becher, P., Thiel, H.J., and Kümmerer, B.M. (1996). Recovery of cytopathogenic and noncytopathogenic bovine viral diarrhea viruses from cDNA constructs. J. Virol. *70*, 8606-8613.

Meyers, G., Tautz, N., Dubovi, E.J., and Thiel, H.J. (1991). Viral cytopathogenicity correlated with integration of ubiquitin-coding sequences. Virology *180*, 602-616.

Meyers, G., Tautz, N., Stark, R., Brownlie, J., Dubovi, E.J., Collett, M.S., and Thiel, H.J. (1992). Rearrangement of viral sequences in cytopathogenic pestiviruses. Virology *191*, 368-386.

Meyers, G., and Thiel, H.J. (1995). Cytopathogenicity of classical swine fever virus caused by defective interfering particles. J. Virol. *69*, 3683-3689.

Meyers, G., and Thiel, H.J. (1996). Molecular characterization of pestiviruses. Adv. Virus Res. *47*, 53-118.

Miller, R.H., and Purcell, R.H. (1990). Hepatitis C virus shares amino acid sequence similarity with pestiviruses and flaviviruses as well as members of two plant virus supergroups. Proc. Natl. Acad. Sci. USA *87*, 2057-2061.

Mills, J.H., and Luginbuhl, R.E. (1968). Distribution and persistence of mucosal disease virus in experimentally exposed calves. Am. J. Vet. Res. *29*, 1367-1375.

Minocha, H.C., Xue, W., and Reddy, J.R. (1997). A 50 kDa membrane protein from bovine kidney cells is a putative receptor for bovine viral diarrhea virus (BVDV). Adv. Exp. Med. Biol. *412*, 145-148.

Moennig, V., Frey, H.R., Liebler, E., Pohlenz, J., and Liess, B. (1990). Reproduction of mucosal disease with cytopathogenic bovine viral diarrhoea virus selected in vitro. Vet. Rec. *127*, 200-203.

Moennig, V., and Plagemann, P.G. (1992). The pestiviruses. Adv. Virus Res. *41*, 53-98.

Moormann, R.J., and Hulst, M.M. (1988). Hog cholera virus: identification and characterization of the viral RNA and the virus-specific RNA synthesized in infected swine kidney cells. Virus Res. *11*, 281-291.

Moser, C., Stettler, P., Tratschin, J.D., and Hofmann, M.A. (1999). Cytopathogenic and noncytopathogenic RNA replicons of classical swine fever virus. J. Virol. *73*, 7787-7794.

Muyldermans, G., San Gabriel, M.C., Caij, A., De Smet, A., and Hamers, R. (1993). Polymerase chain reaction-mediated cloning and in vitro translation of the genes coding for the structural proteins of hog cholera virus. Arch. Virol. *132*, 429-435.

Nettleton, P.F., Gilmour, J.S., Herring, J.A., and Sinclair, J.A. (1992). The production and survival of lambs persistently infected with border disease virus. Comp. Immunol. Microbiol. Infect. Dis. *15*, 179-188.

Ohmann, H.B. (1983). Pathogenesis of bovine viral diarrhoea-mucosal disease: distribution and significance of BVDV antigen in diseased calves. Res. Vet. Sci. *34*, 5-10.

Olafson, P., MacCallum, A.D., and Fox, F.H. (1946). An apparently new transmissible disease of cattle. Cornell Vet., 205-213.

Paeshuyse, J., Leyssen, P., Mabery, E., Boddeker, N., Vrancken, R., Froeyen, M., Ansari, I. H., Dutartre, H., Rozenski, J., Gil, L.H., *et al.* (2006). A Novel, Highly Selective Inhibitor of Pestivirus Replication That Targets the Viral RNA-Dependent RNA Polymerase. J. Virol. *80*, 149-160.

Pankraz, A., Thiel, H.J., and Becher, P. (2005). Essential and nonessential elements in the 3' nontranslated region of Bovine viral diarrhea virus. J. Virol. *79*, 9119-9127.

Paton, D.J., Lowings, J.P., and Barrett, A.D. (1992a). Epitope mapping of the gp53 envelope protein of bovine viral diarrhea virus. Virology *190*, 763-772.

Paton, D.J., Simpson, V., and Done, S.H. (1992b). Infection of pigs and cattle with bovine viral diarrhoea virus on a farm in England. Vet. Rec. *131*, 185-188.

Pellerin, C., Moir, S., Lecomte, J., and Tijssen, P. (1995). Comparison of the p125 coding region of bovine viral diarrhea viruses. Vet. Microbiol. *45*, 45-57.

Pellerin, C., van den Hurk, J., Lecomte, J., and Tussen, P. (1994). Identification of a new group of bovine viral diarrhea virus strains associated with severe outbreaks and high mortalities. Virology *203*, 260-268.

Perdrizet, J.A., Rebhun, W.C., Dubovi, E.J., and Donis, R.O. (1987). Bovine virus diarrhea--clinical syndromes in dairy herds. Cornell Vet. *77*, 46-74.

Pestova, T.V., and Hellen, C.U. (1999). Internal initiation of translation of bovine viral diarrhea virus RNA. Virology *258*, 249-256.

Pestova, T.V., Shatsky, I.N., Fletcher, S.P., Jackson, R.J., and Hellen, C.U. (1998). A prokaryotic-like mode of cytoplasmic eukaryotic ribosome binding to the initiation codon during internal translation initiation of hepatitis C and classical swine fever virus RNAs. Genes Dev. *12*, 67-83.

Peters, W., Liess, B., Frey, H.R., and Trautwein, G. (1987). Incidence and impact of persistent infections with BVD virus in the field. In Pestivirus infections in ruminants, J.W. Harkness, ed. (European Communities), pp. 133-145.

Picard, M. (1989). Classical and African swine fever in France and Europe in 1988. Epidemiologie et Santé Animale, 27-32.

Pirtle, E.C., and Kniazeff, A.J. (1968). Susceptibility of cultured mammalian cells to infection with virulent and modified hog cholera viruses. Am. J. Vet. Res. *29*, 1033 - 1040.

Poole, T.L., Wang, C., Popp, R.A., Potgieter, L.N., Siddiqui, A., and Collett, M.S. (1995). Pestivirus translation initiation occurs by internal ribosome entry. Virology *206*, 750-754.

Potgieter, L.N., McCracken, M.D., Hopkins, F.M., and Walker, R.D. (1984a). Effect of bovine viral diarrhea virus infection on the distribution of infectious bovine rhinotracheitis virus in calves. Am. J. Vet. Res. *45*, 687-690.

Potgieter, L.N., McCracken, M.D., Hopkins, F.M., Walker, R.D., and Guy, J.S. (1984b). Experimental production of bovine respiratory tract disease with bovine viral diarrhea virus. Am. J. Vet. Res. *45*, 1582-1585.

Pritchard, G.C., Borland, E.D., Wood, L., and Pritchard, D.G. (1989). Severe disease in a dairy herd associated with acute infection with bovine virus diarrhoea virus, Leptospira hardjo and Coxiella burnetii. Vet. Rec. *124*, 625-629.

Pritchard, W.R. (1963). The Bovine Viral Diarrhea-Mucosal Disease Complex. Adv. Vet. Sci. Comp. Med. *41*, 1-47.

Purchio, A.F., Larson, R., Torborg, L.L., and Collett, M.S. (1984). Cell-free translation of bovine viral diarrhea virus RNA. J. Virol. *52*, 973-975.

Qi, F., Ridpath, J.F., and Berry, E.S. (1998). Insertion of a bovine SMT3B gene in NS4B and duplication of NS3 in a bovine viral diarrhea virus genome correlate with the cytopathogenicity of the virus. Virus Res. *57*, 1-9.

Qu, L., McMullan, L.K., and Rice, C.M. (2001). Isolation and characterization of noncytopathic pestivirus mutants reveals a role for nonstructural protein NS4B in viral cytopathogenicity. J. Virol. *75*, 10651-10662.

Ramsey, F.K., and Chivers, W.H. (1953). Mucosal disease of cattle. North Am. Vet., 629-633.

Rebhun, W.C., French, T.W., Perdrizet, J.A., Dubovi, E.J., Dill, S.G., and Karcher, L.F. (1989). Thrombocytopenia associated with acute bovine virus diarrhea infection in cattle. J. Vet. Intern. Med. *3*, 42-46.

Reed, K.E., Gorbalenya, A.E., and Rice, C.M. (1998). The NS5A/NS5 proteins of viruses from three genera of the family

Flaviviridae are phosphorylated by associated serine/threonine kinases. J. Virol. *72*, 6199-6206.
Reimann, I., Depner, K., Trapp, S., and Beer, M. (2004). An avirulent chimeric pestivirus with altered cell tropism protects pigs against lethal infection with classical swine fever virus. Virology *322*, 143-157.
Reimann, I., Meyers, G., and Beer, M. (2003). Trans-complementation of autonomously replicating Bovine viral diarrhea virus replicons with deletions in the E2 coding region. Virology *307*, 213-227.
Rijnbrand, R., van der Straaten, T., van Rijn, P.A., Spaan, W.J., and Bredenbeek, P.J. (1997). Internal entry of ribosomes is directed by the 5′ noncoding region of classical swine fever virus and is dependent on the presence of an RNA pseudoknot upstream of the initiation codon. J. Virol. *71*, 451-457.
Risatti, G.R., Borca, M.V., Kutish, G.F., Lu, Z., Holinka, L.G., French, R.A., Tulman, E.R., and Rock, D.L. (2005a). The E2 glycoprotein of classical swine fever virus is a virulence determinant in swine. J. Virol. *79*, 3787-3796.
Risatti, G.R., Holinka, L.G., Carrillo, C., Kutish, G.F., Lu, Z., Tulman, E.R., Sainz, I.F., and Borca, M.V. (2006). Identification of a novel virulence determinant within the E2 structural glycoprotein of classical swine fever virus. Virology *355*, 94-101.
Risatti, G.R., Holinka, L.G., Fernandez Sainz, I., Carrillo, C., Lu, Z., and Borca, M.V. (2007). N-linked glycosylation status of classical swine fever virus strain Brescia E2 glycoprotein influences virulence in swine. J. Virol. *81*, 924-933.
Risatti, G.R., Holinka, L.G., Lu, Z., Kutish, G.F., Tulman, E.R., French, R.A., Sur, J.H., Rock, D.L., and Borca, M.V. (2005b). Mutation of E1 glycoprotein of classical swine fever virus affects viral virulence in swine. Virology *343*, 116-127.
Ross, C.E., Dubovi, E.J., and Donis, R.O. (1986). Herd problem of abortions and malformed calves attributed to bovine viral diarrhea. J. Am. Vet. Med. Assoc. *188*, 618-619.
Roth, J.A., Kaeberle, M.L., and Griffith, R.W. (1981). Effects of bovine viral diarrhea virus infection on bovine polymorphonuclear leukocyte function. Am. J. Vet. Res. *42*, 244-250.
Ruggli, N., Bird, B.H., Liu, L., Bauhofer, O., Tratschin, J.D., and Hofmann, M.A. (2005). N(pro) of classical swine fever virus is an antagonist of double-stranded RNA-mediated apoptosis and IFN-alpha/beta induction. Virology *340*, 265-276.
Ruggli, N., Tratschin, J.D., Schweizer, M., McCullough, K.C., Hofmann, M.A., and Summerfield, A. (2003). Classical swine fever

virus interferes with cellular antiviral defense: evidence for a novel function of N(pro). J. Virol. *77*, 7645-7654.

Rümenapf, T., Meyers, G., Stark, R., and Thiel, H.J. (1989). Hog cholera virus - characterization of specific antiserum and identification of cDNA clones. Virology *171*, 18-27.

Rümenapf, T., Meyers, G., Stark, R., and Thiel, H.J. (1991a). Molecular characterization of hog cholera virus. Arch. Virol. Suppl. *3*, 7-18.

Rümenapf, T., Stark, R., Heimann, M., and Thiel, H.J. (1998). N-terminal protease of pestiviruses: identification of putative catalytic residues by site-directed mutagenesis. J. Virol. *72*, 2544-2547.

Rümenapf, T., Stark, R., Meyers, G., and Thiel, H.J. (1991b). Structural proteins of hog cholera virus expressed by vaccinia virus: further characterization and induction of protective immunity. J. Virol. *65*, 589-597.

Rümenapf, T., Unger, G., Strauss, J.H., and Thiel, H.J. (1993). Processing of the envelope glycoproteins of pestiviruses. J. Virol. *67*, 3288-3294.

Savi, P., Torlone, V., and Titoli, F. (1965). Recherches de la survie du virus de la peste porcine classique dans certains produits de charcuterie. Bull. Off. int. Epiz., 87-96.

Schelp, C., Greiser-Wilke, I., and Moennig, V. (2000). An actin-binding protein is involved in pestivirus entry into bovine cells. Virus Res. *68*, 1-5.

Schelp, C., Greiser-Wilke, I., Wolf, G., Beer, M., Moennig, V., and Liess, B. (1995). Identification of cell membrane proteins linked to susceptibility to bovine viral diarrhoea virus infection. Arch. Virol. *140*, 1997-2009.

Schneider, R., Unger, G., Stark, R., Schneider-Scherzer, E., and Thiel, H.J. (1993). Identification of a structural glycoprotein of an RNA virus as a ribonuclease. Science *261*, 1169-1171.

Shannon, A.D., Richards, S.G., Kirkland, P.D., and Moyle, A. (1991). An antigen-capture ELISA detects pestivirus antigens in blood and tissues of immunotolerant carrier cattle. J. Virol. Methods *34*, 1-12.

Shaw, I.G., Winkler, C.E., and Terlecki, S. (1967). Experimental reproduction of hypomyelinogenesis congenita of lambs. Vet. Rec. *81*, 115-116.

Shimizu, M., Watanabe, H., Satou, K., and Murakami, S. (1989). Antigenic diversity of bovine viral diarrhea-mucosal disease (BVD-MD) viruses recently isolated from persistently infected cattle and mucosal disease, and serologic survey on bovine sera using antigenically different BVD-MD viruses. Nippon Juigaku Zasshi *51*, 1115-1122.

Sizova, D.V., Kolupaeva, V.G., Pestova, T.V., Shatsky, I.N., and Hellen, C.U. (1998). Specific interaction of eukaryotic translation initiation factor 3 with the 5' nontranslated regions of hepatitis C virus and classical swine fever virus RNAs. J. Virol. *72*, 4775-4782.

Stahlheim, O.H.V. (1988). The hog cholera battle and veterinary professionalism. Agricult. Hist., 116-121.

Stark, R., Meyers, G., Rümenapf, T., and Thiel, H.J. (1993). Processing of pestivirus polyprotein: cleavage site between autoprotease and nucleocapsid protein of classical swine fever virus. J. Virol. *67*, 7088-7095.

Steck, F., Lazary, S., Fey, H., Wandeler, A., Huggler, C., Oppliger, G., Baumberger, H., Kaderli, R., and Martig, J. (1980). Immune responsiveness in cattle fatally affected by bovine virus diarrhea-mucosal disease. Zentralbl. Veterinarmed. B *27*, 429-445.

Steffens, S., Thiel, H.J., and Behrens, S.E. (1999). The RNA-dependent RNA polymerases of different members of the family Flaviviridae exhibit similar properties in vitro. J. Gen. Virol. *80*, 2583-2590.

Steinhauer, D.A., and Holland, J.J. (1987). Rapid evolution of RNA viruses. Annu. Rev. Microbiol., 409-433.

Stettler, P., Devos, R., Moser, C., Tratschin, J.D., and Hofmann, M.A. (2002). Establishment and application of bicistronic classical swine fever virus genomes for foreign gene expression and complementation of E2 deletion mutants. Virus Res. *85*, 173-185.

Summerfield, A., Knoetig, S.M., Tschudin, R., and McCullough, K.C. (2000). Pathogenesis of granulocytopenia and bone marrow atrophy during classical swine fever involves apoptosis and necrosis of uninfected cells. Virology *272*, 50-60.

Summerfield, A., McNeilly, F., Walker, I., Allan, G., Knoetig, S.M., and McCullough, K.C. (2001a). Depletion of CD4(+) and CD8(high+) T-cells before the onset of viraemia during classical swine fever. Vet. Immunol. Immunopathol. *78*, 3-19.

Summerfield, A., Zingle, K., Inumaru, S., and McCullough, K.C. (2001b). Induction of apoptosis in bone marrow neutrophil-lineage cells by classical swine fever virus. J. Gen. Virol. *82*, 1309-1318.

Sun, J.H., Lemm, J.A., O'Boyle, D.R., 2nd, Racela, J., Colonno, R., and Gao, M. (2003). Specific inhibition of bovine viral diarrhea virus replicase. J. Virol. *77*, 6753-6760.

Susa, M., Konig, M., Saalmuller, A., Reddehase, M.J., and Thiel, H.J. (1992). Pathogenesis of classical swine fever: B-lymphocyte deficiency caused by hog cholera virus. J. Virol. *66*, 1171-1175.

Tamura, J.K., Warrener, P., and Collett, M.S. (1993). RNA-stimulated NTPase activity associated with the p80 protein of the pestivirus bovine viral diarrhea virus. Virology *193*, 1-10.

Tautz, N., Elbers, K., Stoll, D., Meyers, G., and Thiel, H.J. (1997). Serine protease of pestiviruses: determination of cleavage sites. J. Virol. *71*, 5415-5422.

Tautz, N., Harada, T., Kaiser, A., Rinck, G., Behrens, S., and Thiel, H.J. (1999). Establishment and characterization of cytopathogenic and noncytopathogenic pestivirus replicons. J. Virol. *73*, 9422-9432.

Tautz, N., Kaiser, A., and Thiel, H.J. (2000). NS3 serine protease of bovine viral diarrhea virus: characterization of active site residues, NS4A cofactor domain, and protease-cofactor interactions. Virology *273*, 351-363.

Tautz, N., Meyers, G., Stark, R., Dubovi, E.J., and Thiel, H.J. (1996). Cytopathogenicity of a pestivirus correlates with a 27-nucleotide insertion. J. Virol. *70*, 7851-7858.

Tautz, N., Meyers, G., and Thiel, H.J. (1993). Processing of poly-ubiquitin in the polyprotein of an RNA virus. Virology *197*, 74-85.

Tautz, N., Thiel, H.J., Dubovi, E.J., and Meyers, G. (1994). Pathogenesis of mucosal disease: a cytopathogenic pestivirus generated by an internal deletion. J. Virol. *68*, 3289-3297.

Terpstra, C. (1988). Epizootiology of hog cholera, In Classical swine fever and related viral infections, B. Liess, ed. (Boston, Dordrecht, Lancaster: Martinus Nijoff), pp. 201-232.

Thiel, H.-J., Collett, M.S., Gould, E.A., Heinz, F.X., Houghton, M., Meyers, G., Purcell, R.H., and Rice, C.M. (2005). Family Flaviviridae, In Virus Taxonomy. VIIIth Report of the International Committee on Taxonomy of Viruses, C.M. Fauquet, M.A. Mayo, J. Maniloff, U. Desselberger, and L.A. Ball, eds. (San Diego, USA: Academic Press), pp. 979-996.

Thiel, H.-J., Plagemann, P.G.W., and Moennig, V. (1996). Pestiviruses, In Fields Virology, B.N. Fields, D.M. Knipe, and P.M. Howley, eds. (New York: Raven Press), pp. 1059-1073.

Thiel, H.J., Stark, R., Weiland, E., Rumenapf, T., and Meyers, G. (1991). Hog cholera virus: molecular composition of virions from a pestivirus. J. Virol. *65*, 4705-4712.

Tratschin, J.D., Moser, C., Ruggli, N., and Hofmann, M.A. (1998). Classical swine fever virus leader proteinase Npro is not required for viral replication in cell culture. J. Virol. *72*, 7681-7684.

Trautwein, G., Hewicker, M., Liess, B., Orban, S., and Grunert, E. (1986). Studies on transplacental transmissibility of a bovine virus diarrhoea (BVD) vaccine virus in cattle. III. Occurrence of central

nervous system malformations in calves born from vaccinated cows. Zentralbl. Veterinarmed. B *33*, 260-268.
van Gennip, H.G., Bouma, A., van Rijn, P.A., Widjojoatmodjo, M.N., and Moormann, R.J. (2002). Experimental non-transmissible marker vaccines for classical swine fever (CSF) by trans-complementation of E(rns) or E2 of CSFV. Vaccine *20*, 1544-1556.
Van Gennip, H.G., Vlot, A.C., Hulst, M.M., De Smit, A.J., and Moormann, R.J. (2004). Determinants of virulence of classical swine fever virus strain Brescia. J. Virol. *78*, 8812-8823.
van Oirschot, J.T. (1992). Hog cholera. In Diseases of swine, Leman et al, eds. (Ames IA: Iowa State University Press) *7*, 274-285.
Van Oirschot, J.T., De Jong, D., and Huffels, N.D. (1983). Effect of infections with swine fever virus on immune functions. II. Lymphocyte response to mitogens and enumeration of lymphocyte subpopulations. Vet. Microbiol. *8*, 81-95.
van Oirschot, J.T., and Terpstra, C. (1977). A congenital persistent swine fever infection. I Clinical and virological observations. II. Immune response to swine fever virus and unrelated antigens. Vet. Microbiol. *2*, 121-142.
van Rijn, P.A., Miedema, G.K., Wensvoort, G., van Gennip, H.G., and Moormann, R.J. (1994). Antigenic structure of envelope glycoprotein E1 of hog cholera virus. J. Virol. *68*, 3934-3942.
van Zijl, M., Wensvoort, G., de Kluyver, E., Hulst, M., van der Gulden, H., Gielkens, A., Berns, A., and Moormann, R. (1991). Live attenuated pseudorabies virus expressing envelope glycoprotein E1 of hog cholera virus protects swine against both pseudorabies and hog cholera. J. Virol. *65*, 2761-2765.
Vantsis, J.T., Linklater, K.A., Rennie, J.C., and Barlow, R.M. (1979). Experimental challenge infection of ewes following a field outbreak of Border disease. J. Comp. Pathol. *89*, 331-339.
Vilcek, S., and Belak, S. (1996). Genetic identification of pestivirus strain Frijters as a border disease virus from pigs. J. Virol. Methods *60*, 103-108.
von Freyburg, M., Ege, A., Saalmuller, A., and Meyers, G. (2004). Comparison of the effects of RNase-negative and wild-type classical swine fever virus on peripheral blood cells of infected pigs. J. Gen. Virol. *85*, 1899-1908.
Wallin, M., Ekstrom, M., and Garoff, H. (2004). Isomerization of the intersubunit disulphide-bond in Env controls retrovirus fusion. EMBO J. *23*, 54-65.

Walz, P.H., Baker, J.C., Mullaney, T.P., and Maes, R.K. (2004). Experimental inoculation of pregnant swine with type 1 bovine viral diarrhoea virus. J. Vet. Med. B *51*, 191-193.

Wang, Z., Nie, Y., Wang, P., Ding, M., and Deng, H. (2004). Characterization of classical swine fever virus entry by using pseudotyped viruses: E1 and E2 are sufficient to mediate viral entry. Virology *330*, 332-341.

Warrener, P., and Collett, M.S. (1995). Pestivirus NS3 (p80) protein possesses RNA helicase activity. J. Virol. *69*, 1720-1726.

Weiland, E., Ahl, R., Stark, R., Weiland, F., and Thiel, H.J. (1992). A second envelope glycoprotein mediates neutralization of a pestivirus, hog cholera virus. J. Virol. *66*, 3677-3682.

Weiland, E., Stark, R., Haas, B., Rumenapf, T., Meyers, G., and Thiel, H.J. (1990). Pestivirus glycoprotein which induces neutralizing antibodies forms part of a disulfide-linked heterodimer. J. Virol. *64*, 3563-3569.

Weiland, F., Weiland, E., Unger, G., Saalmuller, A., and Thiel, H.J. (1999). Localization of pestiviral envelope proteins E(rns) and E2 at the cell surface and on isolated particles. J. Gen. Virol. *80*, 1157-1165.

Wensvoort, G. (1989). Topographical and functional mapping of epitopes on hog cholera virus with monoclonal antibodies. J. Gen. Virol. *70*, 2865-2876.

Wensvoort, G., and Terpstra, C. (1985). [Swine fever: a changing clinical picture]. Tijdschr. Diergeneeskd. *110*, 263-269.

Wensvoort, G., Terpstra, C., Boonstra, J., Bloemraad, M., and Van Zaane, D. (1986). Production of monoclonal antibodies against swine fever virus and their use in laboratory diagnosis. Vet. Microbiol. *12*, 101-108.

Widjojoatmodjo, M.N., van Gennip, H.G., Bouma, A., van Rijn, P.A., and Moormann, R.J. (2000). Classical swine fever virus E(rns) deletion mutants: trans-complementation and potential use as nontransmissible, modified, live-attenuated marker vaccines. J. Virol. *74*, 2973-2980.

Wieringa-Jelsma, T., Quak, S., and Loeffen, W.L. (2006). Limited BVDV transmission and full protection against CSFV transmission in pigs experimentally infected with BVDV type 1b. Vet. Microbiol. *118*, 26-36.

Wilhelmsen, C.L., Bolin, S.R., Ridpath, J.F., Cheville, N.F., and Kluge, J.P. (1991). Lesions and localization of viral antigen in tissues of cattle with experimentally induced or naturally acquired mucosal disease, or with naturally acquired chronic bovine viral diarrhea. Am. J. Vet. Res. *52*, 269-275.

Windisch, J.M., Schneider, R., Stark, R., Weiland, E., Meyers, G., and Thiel, H.J. (1996). RNase of classical swine fever virus: biochemical characterization and inhibition by virus-neutralizing monoclonal antibodies. J. Virol. *70*, 352-358.
Wirz, B., Tratschin, J.D., Muller, H.K., and Mitchell, D.B. (1993). Detection of hog cholera virus and differentiation from other pestiviruses by polymerase chain reaction. J. Clin. Microbiol. *31*, 1148-1154.
Wiskerchen, M., Belzer, S.K., and Collett, M.S. (1991). Pestivirus gene expression: the first protein product of the bovine viral diarrhea virus large open reading frame, p20, possesses proteolytic activity. J. Virol. *65*, 4508-4514.
Wiskerchen, M., and Collett, M.S. (1991). Pestivirus gene expression: protein p80 of bovine viral diarrhea virus is a proteinase involved in polyprotein processing. Virology *184*, 341-350.
Xu, J., Mendez, E., Caron, P.R., Lin, C., Murcko, M.A., Collett, M.S., and Rice, C.M. (1997). Bovine viral diarrhea virus NS3 serine proteinase: polyprotein cleavage sites, cofactor requirements, and molecular model of an enzyme essential for pestivirus replication. J. Virol. *71*, 5312-5322.
Xue, W., and Minocha, H.C. (1993). Identification of the cell surface receptor for bovine viral diarrhoea virus by using anti-idiotypic antibodies. J. Gen. Virol. *74*, 73-79.
Xue, W., and Minocha, H.C. (1996). Identification of bovine viral diarrhea virus receptor in different cell types. Vet. Microbiol. *49*, 67-79.
Xue, W., Zhang, S., and Minocha, H.C. (1997). Characterization of a putative receptor protein for bovine viral diarrhea virus. Vet. Microbiol. *57*, 105-118.
Yu, H., Grassmann, C.W., and Behrens, S.E. (1999). Sequence and structural elements at the 3′ terminus of bovine viral diarrhea virus genomic RNA: functional role during RNA replication. J. Virol. *73*, 3638-3648.
Yu, H., Isken, O., Grassmann, C.W., and Behrens, S.E. (2000). A stem-loop motif formed by the immediate 5′ terminus of the bovine viral diarrhea virus genome modulates translation as well as replication of the viral RNA. J. Virol. *74*, 5825-5835.
Zheng, L., Zhang, S., Xue, W., Kapil, S., and Minocha, H.C. (1998). Expression of a 50 kDa putative receptor for bovine viral diarrhea virus in bovine fetal tissues. Can. J. Vet. Res. *62*, 156-159.
Zhong, W., Gutshall, L.L., and Del Vecchio, A.M. (1998). Identification and characterization of an RNA-dependent RNA

polymerase activity within the nonstructural protein 5B region of bovine viral diarrhea virus. J. Virol. *72*, 9365-9369.

Zhou, Y., Moennig, V., Coulibaly, C.O., Dahle, J., and Liess, B. (1989). Differentiation of hog cholera and bovine virus diarrhoea viruses in pigs using monoclonal antibodies. Zentralbl. Veterinarmed. B *36*, 76-80.

Zitzmann, N., Mehta, A.S., Carrouee, S., Butters, T.D., Platt, F.M., McCauley, J., Blumberg, B.S., Dwek, R.A., and Block, T.M. (1999). Imino sugars inhibit the formation and secretion of bovine viral diarrhea virus, a pestivirus model of hepatitis C virus: implications for the development of broad spectrum anti-hepatitis virus agents. Proc. Natl. Acad. Sci. USA *96*, 11878-11882.

Chapter 3

Arteriviruses

Udeni B. R. Balasuriya and Eric J. Snijder

ABSTRACT

In 1996, the family *Arteriviridae* was included within the order *Nidovirales.* Arteriviruses are small, enveloped, animal viruses with an icosahedral core containing a positive-sense RNA genome. The family includes equine arteritis virus (EAV), porcine reproductive and respiratory syndrome virus (PRRSV), lactate dehydrogenase-elevating virus (LDV) of mice and simian hemorrhagic fever virus (SHFV). Three of these viruses were first discovered and characterized a long time ago (EAV-1953, LDV-1960 and SHFV-1964), whereas PRRSV was first isolated in Europe and in North America in the early 1990s. The arteriviruses are highly species specific, but share many biological and molecular properties, including virion morphology, a unique set of structural proteins, genome organization and replication strategy, and the ability to establish prolonged or true persistent infection in their natural hosts. However, the epidemiology and pathogenesis of the infection caused by each virus is distinct, as are the diseases they cause. Although many questions concerning the basic biology of these viruses remain to be answered, a large amount of data related to the molecular biology of EAV and PRRSV have already been obtained. In this chapter, we will review the basic aspects of arterivirus classification and our current understanding of virion architecture, replication, evolution, molecular epidemiology and genetic variation, pathogenesis, host immune response, and vaccination strategies.

INTRODUCTION

Equine arteritis virus (EAV), the prototype arterivirus, was first isolated from the lung of an aborted fetus following an extensive outbreak of respiratory disease and abortion on a Standardbred breeding farm near Bucyrus, Ohio, USA, in 1953 (Doll et al., 1957a; Doll et al., 1957b). After isolation of the causative virus and description of characteristic vascular lesions, equine viral arteritis (EVA) was identified as an etiologically distinct disease of the horse (Doll et al., 1957a). EAV, the causative agent of EVA, was distinguished from equine influenza and equine rhinopneumonitis (equine herpesviruses 1 and 4) viruses, which cause similar respiratory and reproductive disease syndromes in horses. Although it was not etiologically defined as a separate equine disease until 1953, a clinically indistinguishable syndrome was described in the late eighteenth and early nineteenth centuries as "pinkeye", "infectious or epizootic cellulitis", "influenza erysipelatosa", "Pferdestaupe", "Rotlaufseuche" and "equine influenza" (Plateau, 1888; Clark, 1892). An extensive occurrence of EVA that occurred in Kentucky Thoroughbreds in 1984 generated widespread interest, publicity, and concern (Timoney, 1984; Timoney and McCollum, 1993). Since then, an increasing number of outbreaks have been reported from North America and Europe (Bell et al., 2006).

The murine arterivirus lactate dehydrogenase-elevating virus (LDV) was first isolated in 1960 from tumor-bearing laboratory mice with elevated levels of plasma lactate dehdrogenase enzyme activity, but was later found to be endogenous in house mouse populations (*Mus musculus domesticus*; Riley, 1960; Pope, 1961; Li et al., 2000). LDV causes life long asymptomatic, persistent infections in mice that can only be recognized by elevated levels of plasma lactate dehydrogenase. In most cases, LDV is non-pathogenic for its natural host (Rowson and Mahy, 1975; Plagemann and Moennig, 1992).

The monkey arterivirus simian hemorrhagic fever virus (SHFV) was first isolated in 1964, following devastating outbreaks of hemorrhagic fever in colonies of captive macaque monkeys

(Palmer et al., 1968; Tauraso et al., 1968). The virus is endemic among several genera of African green monkeys (e.g. *Cercopithecus aethiops, Erythrocebus patas, Papio anuibus, and Papio cyanocephalus*) and only causes asymptomatic persistent infections (Gravell et al., 1986). However, SHFV causes severe disease following transmission from asymptomatic African monkeys to highly susceptible Asian macaques (e.g. *M. mulatta, M. Speciosa, M. cynomolgus, M. assamensis*).

The newest member of the genus *Arterivirus* is porcine reproductive and respiratory syndrome virus (PRRSV) that appeared suddenly in the Midwestern United States in 1987 and in Central Europe in 1990. In 1991, it was identified as the etiologic agent of porcine reproductive and respiratory syndrome (PRRS), a serious swine disease, by investigators in The Netherlands and USA (Wensvoort et al., 1991; Collins et al., 1992). Since then the virus has spread rapidly around the world and PRRS has become the most economically important infectious disease of swine.

TAXONOMY AND GENOME ORGANIZATION

Preliminary characterization of EAV particles revealed a number of morphological and physicochemical properties resembling those of alphaviruses (Maess et al., 1970; Magnusson et al., 1970; Horzinek et al., 1971) and flaviviruses (Burki, 1965; 1966; Hyllseth, 1970). However, in two subsequent comparative studies, EAV was found to be distinct (Horzinek, 1973a; 1973b) and was therefore proposed to be classified as a non-arthropod borne togavirus, along with the pestiviruses and rubiviruses, LDV, SHFV, and cell fusion virus (Porterfield et al., 1978). In 1984, EAV, LDV, and SHFV were assigned to a new genus *Arterivirus* within the family *Togaviridae* (Westaway et al., 1985). However, the structural proteins of LDV and EAV were known to differ from those of the togaviruses (Plagemann, 1996). Plagemann and Moening (1992) proposed that LDV, EAV and SHFV be classified in a new group of positive-stranded RNA viruses. Studies in the early nineties showed that the genome organization and replication strategy of EAV strikingly resemble those of corona- and toroviruses, including the production of a 3'-coterminal nested set of subgenomic mRNAs, and that key domains of the arterivirus replicase are homologs of domains found in corona- and torovirus replicases. Therefore, these evolutionarily related virus groups were proposed to form a 'coronavirus-like superfamily' (den Boon et al., 1991; Snijder et al., 1993).

It has been some 15 years since the first genome sequence of an arterivirus (EAV) was reported and since then a wealth of sequence data for all four arteriviruses has been generated. The availability of genome sequences and the advancement of molecular biology fostered many molecular studies, which collectively have defined a common framework for the virus structure and replication strategy of arteriviruses, and have resulted in the development of a genome-based taxonomy. The molecular characterization of EAV (de Vries et al., 1990; den Boon et al., 1991; de Vries et al., 1992), LDV (Godeny et al., 1990; Kuo et al., 1992; Godeny et al., 1993), SHFV (Godeny et al., 1995; Zeng et al., 1995), and the more recently identified PRRSV (also known as Lelystad virus [LV] (Conzelmann et al., 1993; Meulenberg et al., 1993; Meulenberg et al., 1995) has shown that their genome organization and expression strategy are highly conserved. It was also concluded that LDV, EAV, SHFV, and PRRSV are phylogenetically related to a degree that justifies their classification in a new virus family called *Arteriviridae*, containing a single genus *Arterivirus* (Conzelmann et al., 1993; Meulenberg et al., 1993; 1994; Meulenberg et al., 1995). Based on genome organization, replication strategy, and conservation of replicase domains, the International Committee on the Taxonomy of Viruses (ICTV) in 1996 placed the family *Arteriviridae* in the new order *Nidovirales*, along with the family *Coronaviridae* (Cavanagh, 1997). All members of the order *Nidovirales* are enveloped viruses with linear, positive-sense, single-stranded RNA genomes. They share a strikingly similar genome organization and replication strategy, but differ considerably in their genetic complexity and virion architecture. "Nido" is derived from the Latin word *nidus* for nest, which refers to the 3' co-terminal nested set of subgenomic mRNAs that is produced during replication of these viruses in the host cell.

LIFE CYCLE AND GENOME EXPRESSION

Arterivirus tropism is determined, at least in part, by the presence of a receptor on the cell surface, since non-permissive cells can be productively infected after transfection of viral genomic RNA (Meulenberg et al., 1998). Virus entry has been studied in some detail only for PRRSV. By using drugs that affect the pH of intracellular membrane compartments, it was shown that entry most likely occurs via the standard endocytic route (Kreutz and Ackermann, 1996). Sialoadhesin (Vanderheijden et al., 2003), a macrophage-restricted cell-surface protein, has been shown to mediate the internalization of PRRSV by alveolar macrophages, but also heparin sulphate on the cell surface and sialic acid on the virion were implicated in entry (Delputte et al., 2005). Following the poorly defined steps of fusion of viral and endosomal membrane and subsequent uncoating of the genome, arterivirus replication starts with the translation of the genome RNA.

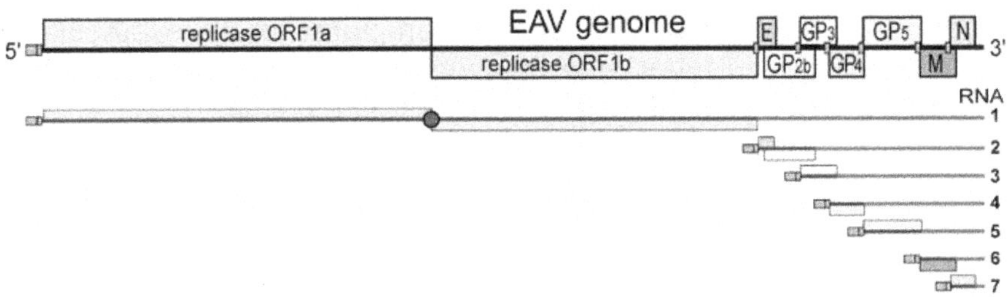

Fig. 1. Arterivirus genome organization and expression are illustrated using the family prototype EAV as an example. The genomic open reading frames are indicated and the names of the corresponding proteins are given. Below the genome, the nested set of mRNAs found in infected cells is depicted, with RNA1 being identical to the viral genome and subgenomic mRNAs 2 to 7 being used to express the structural protein genes located in the 3'-proximal quarter of the genome. With the exception of the bicistronic mRNA2, the subgenomic mRNAs are functionally monocistronic. The box at the 5' end of each mRNA represents the common leader sequence, which is derived from the 5' end of the genome.

The arterivirus genome is a single, positive-sense RNA molecule of ~12-16 kilobases (Fig. 1, den Boon et al., 1991). The genome is polycistronic and contains 9 to 12 open reading frames (ORFs). Its 5'-proximal three-quarters are occupied by two large ORFs that together encode all viral enzyme functions (collectively referred to as "replicase", Fig. 2) required for genome replication and subgenomic RNA production (Molenkamp et al., 2000b). Replicase ORFs 1a and 1b are translated to produce polyproteins pp1a (1727 to 2502 amino acids) and pp1ab (3175 to 3959 amino acids), with the latter being a C-terminally extended version of the former. ORF1b translation depends on a -1 ribosomal frameshift (estimated efficiency 15-20%) just before termination of ORF1a translation (den Boon et al., 1991). The ORF1a/1b overlap region contains signals that promote frameshifting: a so-called "slippery" sequence, the actual ribosomal frameshift site, and a downstream RNA pseudoknot structure.

Following autoproteolytic processing of the replicase pp1a and pp1ab (see below), which yields at least 13 or 14 cleavage products or 'nonstructural proteins' (nsps) and a variety of processing intermediates (Fig. 2, Ziebuhr et al., 2000; van Aken et al., 2006b), the genomic RNA becomes the template for minus strand RNA synthesis. According to the current model for nidovirus RNA synthesis (reviewed by Pasternak et al., 2006; Sawicki et al., 2007), both a full-length minus strand RNA (the template for genome synthesis) and subgenome-length minus strand RNAs are produced (Fig. 3). The latter serve as templates for the synthesis of the subgenomic mRNAs (see below), which in turn are translated to yield the seven viral structural proteins (see below).

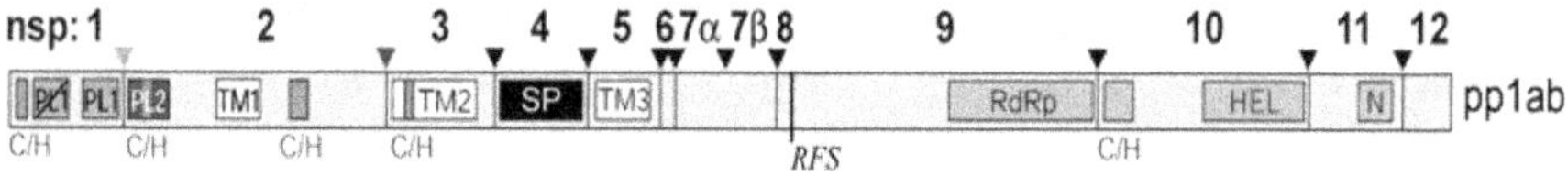

Fig. 2. Arterivirus replicase organization illustrated using the family prototype EAV as an example. Depicted is the replicase polyprotein pp1ab (pp1a is identical to the nsp1-8 part of pp1ab). RFS (ribosomal frameshift) delineates the boundary between amino acids encoded in ORF1a and ORF1b and arrows represent sites in pp1ab that are cleaved by papain-like proteinases (grey) or the main (3CL) proteinase (black). The proteolytic cleavage products (nsps) are numbered and various conserved domains are highlighted. These include domains with conserved Cys and His residues (C/H), putative transmembrane domains (TM), protease domains (PL1, PL2 and SP), the RNA-dependent RNA polymerase domain (RdRp), helicase (HEL), and uridylate-specific endoribonuclease (N). Adapted from Siddell et al., 2005.

Despite the fact that at least two viral proteins (nsp1 and nucleocapsid [N]) are (in part) targeted to the nucleus (Tijms et al., 2002), viral RNA synthesis is thought to occur exclusively in the cytoplasm of the infected cell. The EAV replicase subunits localize to the perinuclear region, where they associate with intracellular membranes that are probably derived from the endoplasmic reticulum (Fig. 4D, van der Meer et al., 1998). Upon infection, host cell membranes are modified into typical vesicular double-membrane structures (double-membrane vesicles [DMVs]) that are thought to carry the viral replication/transcription complex (RTC; Pedersen et al., 1999; Snijder et al., 2001). Expression of EAV nsp2-3 in a heterologous expression system induced strikingly similar structures, indicating that DMV formation does not depend on other viral proteins and/or EAV-specific RNA synthesis (Snijder et al., 2001).

The EAV N protein was found to partially colocalize with the viral RTC (Tijms et al., 2002), suggesting that encapsidation may occur (or begin) at the site of viral RNA synthesis. Arteriviruses acquire their envelope by the budding (Fig. 4B) of nucleocapsids into the smooth endoplasmic reticulum and/or the Golgi complex (Magnusson et al., 1970; Wood et al., 1970; Stueckemann et al., 1982). Most arterivirus envelope proteins (see below) are retained in intracellular membranes, but their specific roles in assembly have not yet been established. After budding, virions accumulate in intracellular vesicles, which are transported to the plasma membrane and release the progeny virus.

ARTERIVIRUS RNA SYNTHESIS

The RNA signals involved in arterivirus genome replication have not been studied in great detail. The coding regions are flanked by 5' and 3' nontranslated regions (NTRs) of 156-221 and 59-117 nucleotides, respectively. However, studies with EAV replicons suggested that approximately 300 nucleotides from both genome termini are likely required for efficient replication, indicating that replication signals extend into the coding sequences (Molenkamp et al., 2000a; Tijms et al., 2001). A comparative analysis confirmed that few - if any - structural elements in the genomic 5' NTR are conserved among arteriviruses (van Den Born et al., 2004). Likewise, the 3' NTR of the arterivirus genome does not contain conserved primary or secondary structure elements, except for a short sequence immediately upstream of the poly (A) tail. In the case of PRRSV, a so-called "kissing interaction" between the loop sequences of RNA hairpins in the 3' NTR and N protein gene was found to be crucial for viral RNA synthesis (Verheije et al., 2002a). Also the 3' end of the EAV genome contains various higher order RNA structures, which are currently under investigation (Beerens and Snijder, 2006). A detailed RNA secondary structure model for the EAV 5' NTR, a domain involved in translation, replication, and transcription (Van Den Born et al., 2004), implicated the so-called "leader TRS hairpin" (LTH) in subgenomic RNA production (see below). The importance of other predicted RNA structures and, for example, their involvement in RNA-protein interactions remain to be investigated.

Following assembly of a functional RTC, the first step in arterivirus RNA synthesis is recognition by the viral enzymes of signals in the 3' end of the genome, where minus strand RNA synthesis is initiated. Whereas genome replication subsequently follows the usual scheme of production of a full-length minus strand RNA that serves as template for the synthesis of new genome RNA, arteriviruses (and other nidoviruses) have attracted attention because of their complex mechanism of subgenomic RNA synthesis. The subgenomic mRNAs (six or more) and the genome of arteriviruses form a 3'-coterminal nested set (Fig. 1). In addition, subgenomic mRNAs contain a common 5' sequence of ~200 nt, the so-called leader sequence, that is identical to the 5' end of the genome RNA (de Vries et al., 1990). Furthermore, as in the case of coronaviruses, arterivirus-infected cells were found to contain a nested set of subgenome-length minus-strand RNAs, the complements of the subgenomic mRNAs (Chen et al., 1994; den Boon et al., 1996).

The production of subgenomic RNAs of arteriviruses and coronaviruses is currently thought to depend on a mechanism of discontinuous minus strand RNA synthesis (Sawicki and Sawicki, 1995). Conserved primary and higher order RNA structures interact with the viral RNA dependent RNA polymerase [RdRp] complex to produce a subgenome-length minus strand template for each subgenomic mRNA. By means of these subgenomic mRNAs, the genes in the 3'-proximal quarter of the genome are rendered accessible for translation. Arterivirus subgenomic mRNAs are produced in non-equimolar, but relatively constant amounts, thus providing a mechanism to regulate the expression of the various structural protein genes. With the exception of the smallest species, the arterivirus subgenomic mRNAs are structurally polycistronic, but generally assumed to be functionally monocistronic. An exception is mRNA2 of EAV, which was proposed to be a functionally bicistronic mRNA from which the partially overlapping genes encoding GP_2 and E are expressed (Snijder et al., 1999). Likewise, mRNA2 of LDV/PRRSV and mRNA4 of SHFV are probably used to translate the same pair of genes, and also mRNA2 of SHFV was proposed to be functionally bicistronic (Godeny et al., 1998).

Conserved short transcription-regulating sequences (TRSs; Fig. 3) are present at the 3' end of the genomic leader sequence ("leader TRS") and at the 5' end of each of the transcription units specifying a subgenomic mRNA "body" ("body TRSs"; reviewed by Pasternak et al., 2006). The observation that the leader TRS in the genome could base-pair with body TRS complements in minus strand RNA (Baric et al., 1983; Spaan et al., 1983), thus somehow facilitating the co-transcriptional joining of subgenomic RNA leader and body, was central to most transcription models proposed over the years.

The detection of subgenome-length minus strands (Sethna et al., 1989; Sawicki and Sawicki, 1990) suggested that this discontinuous step occurs during minus strand RNA synthesis. This concept was subsequently supported by data from biochemical and genetic studies with coronaviruses and arteriviruses. Direct proof for base-pairing between leader TRS and anti-body TRS was obtained from reverse genetics studies using an EAV infectious cDNA clone (van Marle et al., 1999; Pasternak et al., 2001). The RdRp complex is thought to interrupt minus strand RNA synthesis at a body TRS, after which the nascent minus strand is translocated to the 5' end of the genome RNA template in a process that may resemble copy-choice RNA recombination (Brian, 1997; van Marle et al., 1999; Pasternak et al., 2001). Following base-pairing between the body TRS complement in the nascent minus strand and the genomic leader TRS, RNA synthesis is resumed and the subgenome-length minus strand is completed by addition of the complement of the leader sequence.

Obviously, specific interactions between proteins (viral and/or cellular) and RNA sequences (including the leader TRS-presenting hairpin) in both nascent strand and template are likely to control this mechanism. The outcome is that the 3' ends of full-length and subgenome-length minus strand RNAs are identical and that recognition signals in this region may thus be used for the initiation of both genome and subgenomic mRNA synthesis.

THE ARTERIVIRUS REPLICASE

The post-translational processing of the arterivirus replicase pp1a and pp1ab has only been studied extensively for EAV (see Ziebuhr et al., 2000; Siddell, 2005). Eleven cleavage sites have been identified in the EAV replicase, which are cleaved by three different ORF1a-encoded proteinases (Fig. 2; Ziebuhr et al., 2000; van Aken et al., 2006b). Together with the ribosomal frameshift, this yields 13 processing end products (nonstructural protein (nsp) 1 to 12, including an nsp7α and 7β). Of these, the nsp9 to nsp12 subunits are encoded by ORF1b, although nsp9, the viral RdRp, includes a small N-terminal ORF1a-encoded domain, which is identical to nsp8 and is encoded just upstream of the ribosomal frameshift site.

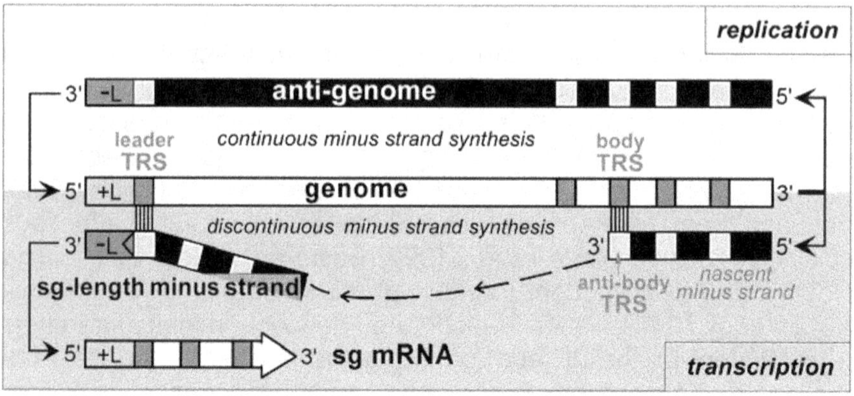

Fig. 3. Model for arterivirus RNA synthesis, using a hypothetical virus producing four subgenomic (sg) mRNAs. Both replication (top panel) and transcription (bottom panel) are depicted. In the "replication mode", the RdRp produced a full-length minus strand RNA (anti-genome) that serves as the template for synthesis of new genome RNA. In the "transcription mode", minus strand RNA synthesis is thought to be discontinuous and regulated by transcription-regulating sequences (TRSs). Body TRSs in the genome are proposed to act as attenuators of minus strand RNA synthesis. Subsequently, the nascent minus strand, with an anti-body TRS at its 3' end, would be redirected to the 5'-proximal region of the genomic template by a base-pairing interaction with the TRS at the 3' end of the leader (+L), which (for EAV) has been shown to reside in a leader TRS hairpin structure. After the addition of the anti-leader (-L) to the nascent minus strands, the subgenome-length minus strands would then serve as templates for subgenomic mRNA synthesis. Adapted from Pasternak et al., 2006.

The three EAV proteinase domains reside in nsp1, nsp2, and nsp4 (for a review, see Snijder and Meulenberg, 1998; Ziebuhr et al., 2000). These enzymes and their corresponding cleavage sites are well conserved in other arteriviruses. EAV nsp1 and nsp2 both contain a cysteine autoproteinase domain that mediates their rapid release from the polyprotein (Snijder et al., 1994). In addition to the homologs of these three EAV proteinases, both LDV and PRRSV possess a fourth nonstructural proteinase (den Boon et al., 1995), which controls the production of an additional N-terminal cleavage product. This additional papain-like proteinase (PCPα) resides in the extreme N-terminus of the replicase and likely is a duplication of the proteinase (PCPβ) present in the C-terminal domain of EAV nsp1. In EAV, PCPα has become inactivated due to the loss of its active site Cys residue (den Boon et al., 1995).

The nsp4 serine proteinase (SP) is the arterivirus main proteinase; it combines the catalytic triad of classical chymotrypsin-like proteinases with the substrate specificity of the 3C-like cysteine proteinases, a subgroup of chymotrypsin-like enzymes named after the picornavirus 3C proteinases (Snijder et al., 1996). Specific residues in the substrate-binding region of the SP are presumed to determine its specificity for cleavage sites containing a (Glu/Gln)-(Gly/Ala/Ser) dipeptide, of which nine have been identified in pp1a/pp1ab (Ziebuhr et al., 2000; van Aken et al., 2006b). EAV nsp4 is the only arterivirus replicase subunit for which an X-ray structure has been obtained (Barrette-Ng et al., 2002). The protein consists of three domains, with domains I and II forming the typical chymotrypsin-like two-β-barrel fold of the SP. The C-terminal domain III was recently found to be dispensable for proteolytic activity and may be involved in fine-tuning replicase polyprotein proteolysis (van Aken et al., 2006a).

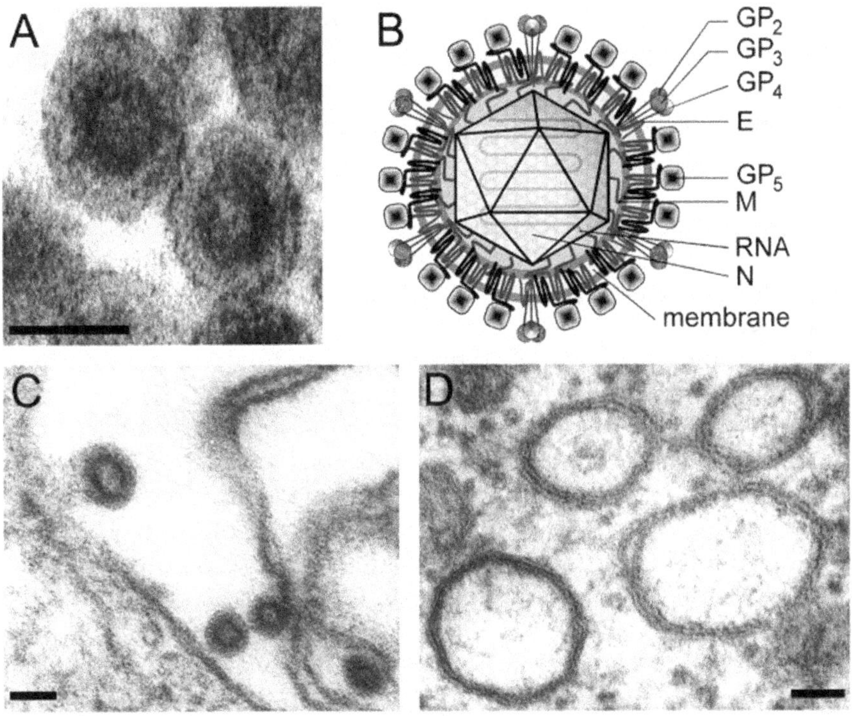

Fig. 4. A. An electron micrograph of extracellular PRRSV particles (bar, 50 nm). B. Schematic representation of an arterivirus particle and its seven structural proteins. C. An electron micrograph of EAV particles budding from smooth membranes in (or close to) the Golgi region of infected baby hamster kidney cells (BHK-21) cells (bar, 50 nm). D. An electron micrograph of typical double membrane vesicles found in the cytoplasm of arterivirus-infected cells which have been implicated in formation of the replication complex and viral RNA synthesis (bar, 50 nm). Panels A, C and D reprinted from Snijder and Meulenberg, 1998. Panel B, reprinted from Snijder et al., 2005.

The functional dissection of the arterivirus nonstructural proteins is still in an early stage. It appears that the ORF1a-encoded proteins are primarily important for regulation of replicase gene expression (by proteolytic processing; see above) and formation of the membrane-anchored 'scaffold' for the replication/transcription complex. The ORF1b-encoded proteins, on the other hand, appear to be more directly involved in viral RNA synthesis. Of the latter, nsp9 (containing the RdRp domain, den Boon et al., 1991) and nsp12 have not been the subject of experimental studies. Nsp11 was predicted to contain a nidovirus-specific endoribonuclease activity (NendoU; Snijder et al., 2003a), a prediction that was experimentally verified for its coronavirus homologue (see Siddell, 2005; Gorbalenya et al., 2006, and references therein). The site-directed mutagenesis of key NendoU residues exerted pleiotropic effects on EAV RNA synthesis, including a complete block in the case of some of the mutants and more moderate effects in others (Posthuma et al., 2006). The exact function of the NendoU domain, a genetic marker of the order Nidovirales (Gorbalenya et al., 2006), remains to be elucidated.

The arterivirus nsp10 contains NTP (nucleotide triphosphate) - binding and superfamily 1 helicase activities that were corroborated in *in vitro* assays (Seybert et al., 2000; Bautista et al., 2002). As in all nidoviruses (Gorbalenya et al., 2006), the helicase is linked to an N-terminal zinc-binding domain that was found to be important for its enzymatic activities (Seybert et al., 2005). EAV reverse genetics studies have demonstrated the general importance of this domain for viral RNA synthesis (van Dinten et al., 2000) and also implicated it in a remarkable transcription-specific defect. A serendipitous mutation at the C-terminal border of the nsp10 zinc-binding domain rendered an EAV full-length cDNA clone non-infectious due to an almost complete block of subgenomic mRNA synthesis (van Dinten et al., 1997). However, this mutant replicated its genomic RNA efficiently, confirming that arterivirus replication and transcription are (at least in part) separate processes. Similar conclusions were drawn from studies on EAV nsp1, which was shown to be fully dispensable for replication but absolutely required for transcription (Tijms et al., 2001). Also, this protein contains an N-terminal (putative) zinc-binding domain, which was implicated in its role in transcription.

VIRION AND STRUCTURAL PROTEINS

Arterivirus particles (Fig. 4A) are formed by a unique set of seven structural proteins (Fig. 4B) that show little or no resemblance to their equivalents from other groups of animal RNA viruses. Major (GP_5, M, and N) and minor (GP_2, GP_3, GP_4, and E) structural proteins are discriminated. The recent analysis of the crystal structure of the C-terminal domain of the PRRSV N protein (Doan and Dokland, 2003) supports the view that arteriviruses have a unique nucleocapsid structure. The surrounding envelope contains two important protein oligomers, a GP_5-M heterodimer and a GP_2-GP_3-GP_4 heterotrimer. Reverse genetics studies with EAV and PRRSV have shown that all seven structural proteins are essential for the production of infectious progeny (Molenkamp et al., 2000b; Wissink et al., 2005). However, it was recently reported that subviral particles, consisting of GP_5, M, N, and RNA, are secreted from cells transfected with knockout mutants lacking one of the minor structural protein genes. The formation of a GP_5-M heterodimer is thought to be a primary determinant of virus budding (Wieringa et al., 2003a; Zevenhoven-Dobbe et al., 2004; Wissink et al., 2005). The transport of GP_5 and M to the Golgi complex appears to depend on their heterodimerization and correlates with the production of infectious virus (Dobbe et al., 2001; Verheije et al., 2002b; Snijder et al., 2003a). After budding, virions accumulate in intracellular vesicles, which are transported to the plasma membrane and release the progeny virus. The generation of aberrant nucleocapsid structures and/or unusual elongated virions was observed for various arteriviruses.

The three major structural proteins GP_5 (and its equivalent GP_7 in SHFV), M, and N are encoded - in this order - by the three most 3'-proximal ORFs in the arterivirus genome. The N protein is only 12-15 kDa and contains many basic residues, in particular in its N-terminal domain that is thought to interact with the genomic RNA during nucleocapsid assembly (Doan and Dokland, 2003). The nonglycosylated M protein (16-20 kDa) is the most conserved structural protein of arteriviruses and resembles the coronavirus M protein in its presumed triple membrane spanning character (de Vries et al., 1992; Meulenberg et al., 1993; Faaberg et al., 1995), exposing just a short 10-18 residue ectodomain at the virion surface. The M protein accumulates in the endoplasmic reticulum, where its heterodimerization with GP_5 is thought to be driven by the formation of a disulfide bridge between a conserved Cys residue in the M ectodomain and a Cys residue in the GP_5 ectodomain (de

Vries et al., 1995; Faaberg et al., 1995; Snijder et al., 2003b). Despite considerable differences in primary structure, the major glycoproteins of arteriviruses (GP_5/GP_7) share common structural features. They contain an N-terminal signal sequence that is assumed to be cleaved from a short ectodomain. The central hydrophobic region is assumed to span the membrane three times and to be followed by a cytoplasmic domain of 50-75 amino acids. In the case of LDV and PRRSV, the ectodomain is only about 30 residues long and contains one to three N-linked glycans (Meulenberg et al., 1993; Faaberg and Plagemann, 1995). The ectodomain of EAV GP_5 is 95 residues long and usually possesses a single N-linked polylactosamine side chain (de Vries et al., 1992), although in some strains an additional N-linked glycan is present. Also in the case of LDV, GP_5 glycosylation occurs by the addition of variable numbers of lactosamine repeats (Li et al., 1998). Non-neuropathogenic and neuropathogenic strains of LDV typically contain three and one polylactosaminoglycan chain(s), respectively.

Minor glycoprotein GP_2 (encoded by ORF2a [PRRSV] or ORF2b [EAV/LDV]) is a conventional class I integral membrane protein with an N-terminal signal peptide, a C-terminal transmembrane segment, and 1 to 4 potential N-glycosylation sites. In EAV-infected cells, it occurs in a variety of monomeric conformations and in complex with other minor GPs (Wieringa et al., 2003a; Wieringa et al., 2003b; Wieringa et al., 2004). Complex formation is a prerequisite for incorporation into virions (Wieringa et al., 2004; Wissink et al., 2005). GP_3 is a heavily glycosylated integral membrane protein with an uncleaved N-terminal signal sequence and a hydrophobic C-terminal domain. The EAV protein is inserted into the lipid bilayer using either or both of its hydrophobic termini and has no detectable endodomain (Hedges et al., 1999a; Wieringa et al., 2002). Like GP_2 and GP_4, EAV GP_3 localizes to the endoplasmic reticulum, both in infected cells and in expression systems (Wieringa et al., 2002). Data obtained for EAV, showed that covalently linked GP_2-GP_4 heterodimers associate with GP_3 to produce the heterotrimer that is required for incorporation of all three minor GPs into virions. Following virus release, GP_3 becomes disulfide-linked to the GP_2-GP_4 heterodimers and this post-assembly maturation event yields the complex of three covalently bound minor GPs. As a result, EAV particles contain both GP_2-GP_4 heterodimers and GP_2-GP_3-GP_4 heterotrimers (Wieringa et al., 2003a; Wieringa et al., 2004). The higher order structure of the small E protein is unknown, but this protein is essential for virus infectivity (Molenkamp et al., 2000b) and may in fact be associated with the minor GP trimer (Wieringa et al., 2004).

Neutralization properties of arteriviruses have been reviewed recently by Balasuriya and MacLachlan (2004). As has conclusively been shown for EAV, GP_5 is the principal target of neutralizing antibodies to PRRSV (both North American [NA] and European [EU]) and LDV. Interestingly, however, MAbs that bind GP_4 have also been shown to neutralize prototype EU PRRSV (Lelystad virus [LV]), but they did not react with the GP_4 protein of the NA prototype strain of PRRSV (VR2332; Meulenberg et al., 1997b; Weiland et al., 1999; Weiland et al., 2000). For both EAV and PRRSV, it has been demonstrated that GP_5-M heterodimer formation is critical for the expression of the neutralization epitopes in their authentic form (Balasuriya et al., 2000; Balasuriya et al., 2002; Jiang et al., 2006). The M protein likely acts as an essential scaffold on which the GP_5 protein folds to form the epitopes that induce neutralizing antibodies in horses and pigs.

EVOLUTION

The common ancestry of major parts of their replicases and the similar genome organization suggested that arteriviruses (family *Arteriviridae*), roniviruses (family *Roniviridae),* and corona- and toroviruses (family *Coronaviridae*) have evolved from a common ancestor. Apparently, during evolution different nidovirus subgroups have acquired unrelated sets of structural protein genes, likely by heterologous RNA recombination (Gorbalenya et al., 2006). There is a relatively large evolutionary distance between EAV and the other three arteriviruses (Godeny et al., 1993; Meulenberg et al., 1997a; Smith et al., 1997). Phylogenetic studies indicate that LDV and PRRSV are more closely related to each other than to EAV and that SHFV occupies an intermediate position but is more closely related to LDV and PRRSV than it is to EAV. A comprehensive review of Nidovirus evolution has been recently published (Gorbalenya et al., 2006).

Recent studies have addressed the origin of the newly emerged porcine arterivirus PRRSV in more detail (Suarez et al., 1996; Forsberg et al., 2001; Forsberg et al., 2002; Plagemann, 2003; Forsberg, 2005; Hanada et al., 2005). Based on retrospective serological surveys, PRRSV emerged in Canada in 1979, but the first of a series of possible outbreaks was reported in North America in the mid-1980s and in Western Europe (Germany and the Netherlands) in 1990. Although PRRSV appeared in North America and Europe within a narrow time frame, the two isolates were genetically distinct as there was less than 60% amino acid identity between these two types (Suarez et al., 1996; Forsberg et al., 2002). Therefore, the first EU isolate from the Netherlands (LV) and the first North American (NA) isolate (VR2332) are now considered as the prototype strains of EU PRRSV genotype (genotype I) and the NA PRRSV genotype (genotype II), respectively.

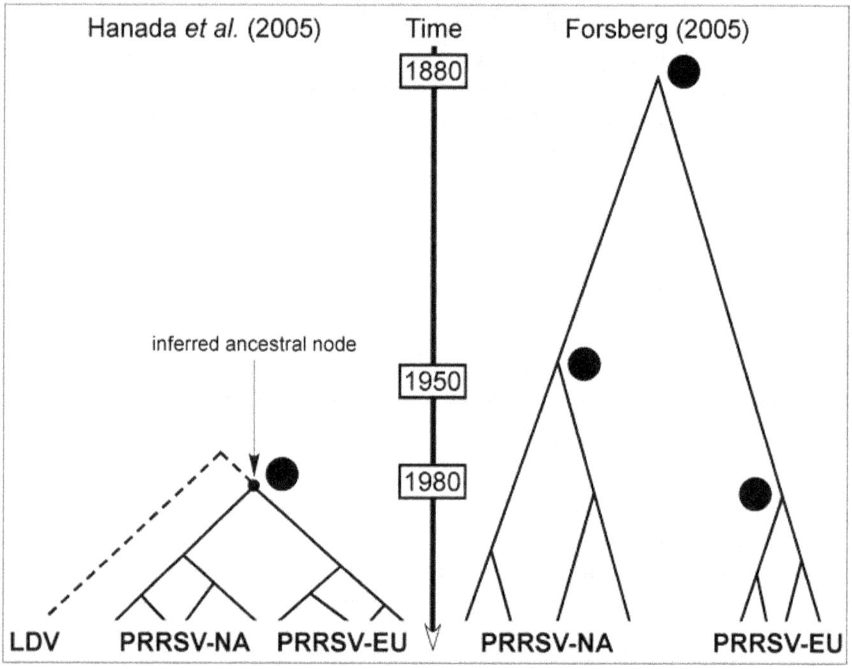

Fig. 5. Schematic phylogeny of North American (NA) and European (EU) genotypes of PRRSV. The left panel shows the subtype divergence time (global most recent common ancestor [MRCA]) as estimated by Hanada et al, 2005 by using the closest relative, LDV, to infer the nucleotide sequence of the subtype ancestor. The right panel shows the divergence time of subtypes (global MRCA) and the lineages within subtypes (local MRCAs) as estimated by Forsberg, 2005 and Forsberg et al., 2001. Reproduced with permission from Forsberg, 2005.

There is no coherent hypothesis to explain the co-temporal emergence of the two distinct PRRSV genotypes in the two continents. Based on the evolutionary history of PRRSV and speculations on its epidemiological background, Hanada *et al.*, (2005) hypothesized that the virus was transmitted from an unknown reservoir host to pigs around 1980. Following this species jump, the virus would have evolved rapidly into two distinct genotypes (Fig. 5). The divergence time and evolution rate of PRRSV were estimated using molecular evolutionary analysis and suggested that PRRSV had evolved at a higher rate (10^{-2}/site/year) compared to other RNA viruses (10^{-3} to 10^{-5}/site/year). This placed the most recent common ancestor (MRCA) of all PRRSV isolates around 1980 and revealed that no significant subtype divergence had occurred before emergence in pigs.

However, two other genetic studies support the idea that the two PRRSV genotypes diverged in separate host reservoirs prior to their emergence in pigs (Stadejek et al., 2002; Forsberg, 2005). In the first study, a large number of ORF3 sequences were used to establish that the MRCA of Western EU PRRSV isolates existed in 1980, which is approximately prior to the emergence of PRRSV in domestic pigs in Western Europe. The second study showed that Eastern European pigs harbored highly diverse and distinct EU PRRSV lineages before the virus emerged in Western Europe. Furthermore, recent genetic analysis by Forsberg (Forsberg, 2005) strongly supported the alternative hypothesis that PRRSV may have made two independent species jumps in North America and Europe, prior to its emergence as a new pathogen of swine (Fig. 5). This study also suggested that the MRCA of all EU and NA PRRSV genotypes existed already around 1880, 100 years before the date predicted by Hanada et al. (Hanada et al., 2005). This second hypothesis is based on the markedly different genetic composition that is associated with distinct geographical origins of these two PRRSV genotypes and on their close relatedness to the mouse arterivirus, LDV. It is also in line with speculations by Plagemann (Plagemann, 2003) that a mutant variant of LDV infected European wild boars and was subsequently imported into North America in the early 1900s. Then the virus evolved independently in wild hog populations on the two continents before entering the domestic population. Based on ORF5 phylogeny he suggested that the first transmission of LDV from wild mice to wild boars could have occurred in Lithuania (first species jump) and subsequently the virus may have been introduced into the wild boar population in the Eastern part of Germany (Sachsen-Anhalt), where the first European seropositive pigs were detected in 1988 and 1989. Presumably virus present in wild boars in this German region jumped into domestic pigs (second species jump), resulting in the first outbreak of PRRS detected in North Western Germany in 1990.

There is, however, still no plausible explanation for the co-temporal emergence of the two PRRSV genotypes on two continents.

Recombination

Similar to coronaviruses and other positive-stranded RNA viruses, arteriviruses also evolve by genetic recombination. Two probable natural recombination events may have led to the partial duplication of ORF1a of PRRSV/LDV and the insertion of three ORFs (4, 5, and 6) in the SHFV genome (Godeny et al., 1998; Yuan et al., 1999). The adjacent PCPα and PCPβ domains in the nsp1 region likely resulted from duplication of this protease domain (den Boon et al., 1995). The proteins encoded by SHFV ORFs 4, 5, and 6 display similarity with proteins encoded by ORFs 2a, 2b, and 3 suggesting that these genes may have been either duplicated by an SHFV RNA recombination event or derived from another, distantly related unidentified arterivirus by heterologous RNA recombination (Godeny et al., 1998). Recently, natural recombination between EU PRRSV field isolates has been reported (Forsberg et al., 2002). There is also evidence for homologous genetic recombination between EAV field strains (Chirnside et al., 1994).

Genetic recombination of arteriviruses has also been observed under experimental conditions. Upon co-inoculation of mice with neuropathogenic and non-neuropathogenic LDV strains (LDV-P and LDV-C, respectively) recombination in the 1276-nt segment covering ORFs 5-7 was found to have a frequency as high as 5%. However, during natural infections genetic recombination may not play a significant role in the evolution of LDV in mice (Li et al., 1999). High frequency RNA recombination between genetically closely related PRRSV sequences has been reported (van Vugt et al., 2001). The inter- and intra-type RNA recombination of PRRSV has been studied in MARC-145 cells following co-infection with EU and NA PRRSV strains or co-infection with closely related EU strains. Interestingly, no RNA recombination was detected between the EU and NA types, but there was 0.1 to 2.5% RNA recombination between EU isolates of PRRSV. This study predicted RNA recombination between EU and NA types to be at least 10,000 times less likely to occur than RNA recombination between diverse EU isolates. In summary, homologous RNA recombination in arteriviruses may allow them to adapt to changing circumstances and to escape selection pressures imposed by the host environment.

Quasispecies

The low fidelity of viral RdRps and lack of a proofreading mechanism lead to high mutation rates in RNA viruses (10^{-3} to 10^{-5} substitutions per nucleotide copied) and as a consequence RNA virus populations exist as heterogeneous mutant swarms or 'quasispecies' (Domingo et al., 1996). These are a repository of genotypic and phenotypic variants with the potential to become dominant in response to selective pressures, such as neutralizing antibodies and antiviral therapy. This evolutionary plasticity accounts for the ability of RNA viruses to evade the host immune response, establish persistent infections, and generate viral variants with novel virulence phenotype and clinical manifestations.

Arterivirus quasispecies have been documented (Balasuriya et al., 1999a; Hedges et al., 1999b; Rowland et al., 1999). Evaluation of sequential semen samples from long-term carrier stallions have demonstrated that EAV behaves as a quasispecies in the reproductive tract of the carrier stallion and that novel genotypic and phenotypic variants arise during persistent infection. In the course of persistent infection, EAV evolves at a rate of approximately 1% nucleotide substitution per 2822 nt (ORFs 2-7) per year, leading to emergence of novel genotypic and phenotypic variants (Hedges et al., 1999b). In contrast, evaluation of the molecular evolution of EAV derived from an infectious cDNA clone demonstrated that the virus was genetically stable during acute, transient infection, with conservation of the master sequence, but that there was increasing quasispecies heterogeneity of the virus population present in the tissues of individual horses (Balasuriya et al., 1999b). Similarly, the genetic stability and conservation of the master sequence of viruses that circulate during an outbreak has been reported (Balasuriya et al., 1999a). Therefore, selective pressures exerted during the course of persistent infection of the reproductive tract of carrier stallions clearly can be responsible for divergence and emergence of phenotypically novel EAV variants, and likely compensate for the relatively limited diversity of EAV that is generated during EVA outbreaks.

Rowland et al., (1999) reported evidence for quasispecies evolution and emergence of novel genetic variants during *in utero* infection of pigs with the VR2332 prototype of NA PRRSV. Evolution of PRRSV during sequential passage in pigs demonstrated that multiple variants arose concurrently in every pig that was sampled over the 367-day observation period and that the virus evolved continuously in infected pigs, with different genes undergoing various degrees of change. Sequence analysis showed ORFs 1b and 7 to be highly conserved, in contrast to ORF5 that was highly variable during sequential passaging. Furthermore, quasispecies evolution of PRRSV was demonstrated during natural infections (Goldberg et al., 2003). Sequence and phylogenetic analysis showed genetic variation between farms, among animals within farms, and within individual animals. Every animal on the farm harbored a genetically distinct virus variant suggesting emergence of variants with unique master sequences. This clearly suggested that PRRSV evolved by generating viral quasispecies during natural infection. Also in LDV the existence of quasispecies has been reported (Plagemann et al., 2001). LDV-P and LDV-vx variants were found to coexist in original isolates from transplantable tumors when cloned by repeated endpoint dilution in mice. Similarly, LDV-C and LDV-v were found to coexist with LDV-P and/or LDV-vx in neuropathogenic LDV populations in infected mice and could be separated by endpoint dilution in mice.

GENETIC VARIATION AND MOLECULAR EPIDEMIOLOGY

The NA and EU isolates of PRRSV share morphological and structural similarities and cause similar disease in infected pigs, but they display significant molecular and antigenic diversity. The PRRSV isolates from the two continents share only 55-80% nucleotide identity in the different viral genes (Suarez et al., 1996). The VR-2332 and LV prototype viruses differ significantly in the 5' leader sequence and regions of ORF1a (Nelsen et al., 1999).

An increasing number of reports indicate that PRRSV isolates are antigenically, genetically, and pathologically very heterogeneous (Shen et al., 2000; Fang et al., 2004; Ropp et al., 2004; Han et al., 2006; Stadejek et al., 2006b). Comparative nucleotide and amino acid sequence analyses have demonstrated significant genetic and antigenic diversity among NA field isolates of PRRSV. Until recently EU genotype viruses were thought to be genetically less heterogeneous than the NA isolates of PRRSV. However, analysis of ORF5 and 7 sequences from isolates from Europe, Italy, and Denmark revealed marked genetic diversity and geographic patterns (Forsberg et al., 2002; Forsberg, 2005; Pesch et al., 2005). In this analysis, three distinct clades of PRRSV field isolates were identified (Lelystad-like [Western EU isolates], Danish and Italian-like) and the isolates from Italy were genetically more heterogeneous than the remainder of the Western European isolates, suggesting that the genetic diversity of EU PRRSV is skewed geographically. Recently, Stadejek et al., (2006) showed that PRRSV strains from Eastern Europe (east of Poland) were significantly different from the EU genotype sequences on the west side of the Poland/Belarus border (Stadejek et al., 2006a; Stadejek et al., 2006b). Phylogenetic analysis grouped viruses from west of the Poland/Belarus border to a single cluster identified as subtype 1. Isolates from Eastern Europe clustered into three different clades identified as subtypes 2, 3 and 4.

Until recently, NA genotypes were detected only in North and Central America and Asia, while EU genotypes were restricted to Europe. Now, NA-type PRRSV has been introduced into Europe through the use of a live vaccine (Botner et al., 1997; Gao et al., 2004; Indik et al., 2005), and EU-type virus (Lelystad-like virus) has been isolated in Canada following importation of swine from Europe (Dewey et al., 2000). Recently, EU-like PRRSV has been discovered in American swine herds (Zeman et al., 1993; Ropp et al., 2004). Sequence, phylogenetic and antigenic analysis have shown that the EU-like viruses isolated in the USA are more closely related to LV and PRRSV isolates from the United Kingdom. Interestingly, EU-like PRRSV viruses isolated in the USA have a 51-nucleotide deletion in ORF1a within the region encoding nsp2 (17-amino acid deletion between residues 734 and 750), as well as several amino acid differences (Fang et al., 2004). In addition, various deletions and an insertion in nsp2 have been identified among USA field isolates and in a vaccine strain of PRRSV (Shen et al., 2000). Also a significant amount of genetic diversity, in terms of gene length and sequence variation, has been reported among PRRSV isolates (Nelsen et al., 1999; Shen et al., 2000; Tan et al., 2001; Han, 2006). It has been shown that the 5' and 3' NTRs of EU isolates are considerably longer than those of NA isolates. However, with a few exceptions, these two regions are conserved among the NA isolates. Significant nucleotide and amino acid sequence differences were reported among isolates within the same genotype, particularly in the ORF1a regions encoding nsp1β and nsp2 and the structural protein genes for GP_3 and GP_5 (ORF3 and ORF5, respectively; (Wootton et al., 2000; Han, 2006). ORF5, encoding the major envelope glycoprotein, displayed the highest genetic variability among PRRSVs from around the world and has therefore been the target gene for molecular epidemiological studies.

In contrast to PRRSV, EAV field isolates from North America and Europe are closely related and differ only by 85% at the nucleotide level. Phylogenetic analyses based on ORF5 cluster EAV field strains into two distinct clades, reflecting their North American and European origin (Balasuriya et al., 1995; Stadejek et al., 1999). Data from individual EU and NA isolates indicates movement of viruses between the two continents due to the transport of carrier stallions and/or their semen. The major nucleotide and amino acid differences between NA and EU EAV strains occur in the nsp2-encoding part of ORF1a, as well as in ORFs 3 and 5, which encode the GP_3 and GP_5 envelope proteins, respectively (Balasuriya and MacLachlan, 2004, and references therein). Comparative nucleotide and amino acid sequence analysis of twenty EAV strains clearly demonstrated that there is considerable variation in nsp2, with aa substitutions being concentrated in the central segment (aa 388-496) of the protein. The biological significance of this extensive variation in nsp2 of EAV field strains is unknown.

There is a paucity of data on the genetic diversity of LDV and SHFV. Comparison of amino acid sequences of M and N proteins of two LDV strains revealed that they are more conserved than the polypeptides encoded by ORFs 1 to 5.

CELL TROPISM, EPIDEMIOLOGY, CLINICAL FEATURES AND PATHOGENESIS

The epidemiology, clinical disease, pathogenesis and mechanism of persistent infection of EAV, PRRSV, LDV and SHFV differ significantly. All four arteriviruses primarily target macrophages and/or endothelial cells but the clinical outcome in their natural hosts is unique for each agent.

EAV

EAV can infect horses, donkeys, mules, and zebras. The virus replicates in primary cultures of equine pulmonary artery endothelial, horse kidney, rabbit kidney, and hamster kidney cells, but also in cell lines such as BHK-21, RK-13, African green monkey kidney cells (Vero), rhesus monkey kidney (LLC-MK2), equine dermis (NBL-6), hamster lung (HmLu), SV-40 transformed equine ovary, and canine hepatitis virus-transformed hamster tumor cells

(HS and HT-7). High passage (passage 399) RK-13 cells are more susceptible to EAV infection than the low passage (passage 199) RK-13 cells.

EVA is a respiratory and reproductive disease in horses that occurs throughout the world (Timoney and McCollum, 1993; Glaser et al., 1996; Bell et al., 2006). While there is only one known EAV serotype, field strains differ in their virulence and neutralization phenotype. The clinical signs displayed by EAV-infected horses depend on a variety of factors including the age and physical condition of the animal, challenge dose and route of infection, strain of virus, and environmental conditions. With the sole exception of the experimentally derived highly horse-adapted virulent Bucyrus strain, field isolates of EAV very rarely cause fatal infection in older horses (McCollum and Timoney, 1998). The vast majority of EAV infections are inapparent (or subclinical), but acutely infected animals may develop a wide range of clinical signs including pyrexia, depression, anorexia, dependent edema (scrotum, ventral trunk, and limbs), stiffness of gait, conjunctivitis, lacrimation and swelling around the eyes (periorbital and supraorbital edema), respiratory distress, urticaria, and leukopenia. The incubation period of 3 to 14 days (usually 6 to 8 days following venereal exposure) is followed by pyrexia of up to 41°C that may persist for 2 to 9 days. The virus causes abortion in pregnant mares with abortion rates during natural outbreaks varying from 10 to 71%, depending on the virus strain. EAV-induced abortions can occur at any time between 3 and 10 months of gestation. Infection can cause a severe fulminating interstitial pneumonia in neonatal foals, and a progressive pneumoenteric syndrome in 1 to 3 months old foals. A variable proportion of acutely infected stallions (10 to 70%) become persistently infected and shed the virus in semen. There is no evidence of EAV persisting in the mares, geldings or foals. The virus persists mainly in the ampulla of the male reproductive tract and the establishment and maintenance of the carrier state in the stallion is testosterone-dependent (see review by Timoney and McCollum, 1993).

The sero-prevalence of EAV infection varies between countries and horses of different breeds and age in the same country (Bell et al., 2006). In the United States, some 70 to 90% of adult Standardbred horses are seropositive for EAV, as compared to only 1-3% of the Thoroughbred population. Similarly, a high percentage of EU Warmblood horses are seropositive. Sero-prevalence increases with age, indicating that horses may be repeatedly exposed to the virus as they age. Persistently infected carrier stallions function as the natural reservoir of EAV and disseminate the virus to susceptible mares at breeding. The two principal modes of transmission are horizontal transmission, by dispersal of infectious respiratory tract secretion from acutely infected horses, and venereal transmission during natural or artificial insemination with infective semen from persistently infected stallions. Persistently infected carrier stallions are the essential reservoir responsible for perpetuation and maintenance of EAV in equine populations. Between 85 and 100% of seronegative mares bred to long-term carrier stallions become infected with the virus and seroconvert within 28 days. Mares that become infected following natural or artificial breeding then can readily transmit the virus by nasal aerosol to susceptible cohorts in close proximity. EAV also can be transmitted through indirect contact with fomites or personnel. Congenital infection results from transplacental transmission (vertical transmission) of virus when pregnant mares are infected late in gestation.

The pathogenesis of EVA has been studied by both the experimental inoculation of horses with laboratory and field strains of different virulence, and by careful evaluation of natural outbreaks (MacLachlan et al., 1996; McCollum and Timoney, 1998). The clinical outcome and pathogenesis of EVA have differed markedly between individual experiments, which likely reflects the inherent differences in the route of infection, virus dose, strain of virus, and quality of the specimens used for virus isolation. Following respiratory infection, initial multiplication of the virus takes place in endothelial cells and pulmonary macrophages in the lung, and the virus soon appears in the regional lymph nodes, especially the bronchial nodes (in two days). Within three days, the virus is present in virtually all organs and tissues (viremia), where it replicates primarily in macrophages and endothelial cells, but it may also replicate in other selected cell types such as epithelial cells, mesothelium, and smooth muscle cells of small blood vessels and myometrium. Virus can be isolated from the nasopharynx for about 2-14 days after infection and from buffy coat for 2-19 days. Virus typically is isolated from serum or plasma for 7-9 days, and the disappearance of virus from serum coincides with the development of virus-specific neutralizing antibodies. Apart from occasional instances where EAV has been isolated from buffy coat cells for several months after infection, and from the reproductive tract of colts (>6 months),

EAV is not generally isolated beyond 28 days after infection except from the semen of carrier stallions. EAV infection of pregnant mares can result in the abortion of fetuses that are usually partially autolyzed at the time of expulsion. Although it is very infrequently observed, the aborted fetus may exhibit interlobular pulmonary edema, pleural and pericardial effusion, and petechial and ecchymotic hemorrhages on the serosal and mucosal surfaces of the small intestine. Neonatal foals can develop a very severe, acute interstitial pneumonia.

The characteristic histological feature of EVA is a severe necrotizing panvasculitis of small blood vessels (MacLachlan et al., 1996; Del Piero, 2000). Affected arteries show foci of intimal, subintimal and medial necrosis, with edema and infiltration of lymphocytes and neutrophils. Prominent vascular lesions are also seen in the placenta, brain, liver, and spleen of aborted fetuses. Lungs of affected neonatal foals have severe interstitial pneumonia. The pathogenesis of EVA is not clearly defined. The characteristic vascular lesions of EVA do not appear to be the result of immune-mediated injury. Vascular injury in EVA is likely the result of direct virus-mediated injury to the lining (endothelium) and walls (media) of affected vessels. EAV infects and replicates in endothelial cells and causes extensive damage to the endothelium and the subjacent internal elastic lamina, which allows access to the media of affected vessels. Increased vascular permeability and leukocyte infiltration resulting from generation of chemotactic factors leads to hemorrhage and edema around these vessels. In addition to endothelial cells (EC), EAV also replicates well in macrophages (MΦ) in infected horses and it recently has been convincingly shown that EAV infection of cultured equine EC and MΦ leads to their activation with increased transcription of genes encoding proinflammatory mediators including IL-1β, IL-6, IL-8 and TNF-α (Moore et al., 2003). Furthermore, virulent and avirulent strains of EAV induced different quantities of TNF-α and other proinflammatory cytokines from both infected ECs and MΦs. These studies strongly suggest that cytokine mediators that are produced by ECs and MΦs have a central role in the pathogenesis of EVA. However, the relative roles and importance of direct virus-induced injury versus virus-induced proinflammatory mediators have not yet been defined.

Recent studies suggest that abortion following EAV infection of pregnant mares is the result of a lethal fetal infection rather than myometritis and/or placental damage that impairs progesterone synthesis leading to fetal expulsion (MacLachlan et al., 1996). The tissues of aborted fetuses contain higher titers of virus than do those of the mares from which they abort, indicating that there is substantial virus replication in the fetus itself. The stress that results from fetal infection would be expected to activate the fetal hypothalamic-pituitary axis, and to so induce abortion.

PRRSV

PRRSV primarily infects domestic and feral pigs, but recently, chickens and ducks that were exposed to PRRSV in drinking water shed the virus in their feces, suggesting that they are susceptible to infection with the virus (Zimmerman J., 2006). NA PRRSV (type II) isolates replicate in porcine alveolar macrophages (PAM), CRL-11171 and African green monkey cell line (MA-104), and derivatives thereof (CL2621 or MARC-145). Most, if not all, EU PRRSV (type I) isolates replicate best or exclusively in PAMs (Wensvoort et al., 1991). However, EU PRRSV isolates have been adapted to grow in the CL2621 cell line. Vaccine strains of PRRSV replicate 100- to 1000-times more efficiently in derivatives of monkey kidney cell lines than in PAMs.

Clinical signs of PRRS are extremely variable and influenced by the strain of virus, immune status of the herd, and management practices (reviewed in Zimmerman J., 2006 and references there in). Low-virulence PRRSV strains may result in widespread infection of swine with minimal occurrence of disease, whereas highly virulent strains can cause severe clinical disease in susceptible herds (Nodelijk, 2002; Nodelijk et al., 2003). All ages of pigs are susceptible to infection in immunologically naive herds. Acute PRRSV infection of susceptible pigs is characterized by anorexia, fever (39-41°C), dyspnea, and lethargy. Affected swine are lymphopenic and exhibit transient cutaneous hyperemia or cyanosis of extremities that is most visible on the ears, snouts, mammary glands, and vulvas. Transplacental transmission of PRRSV occurs most efficiently in the third trimester of pregnancy (usually after 100 days of gestation) and the abortion rate in affected sows can range from 10 to 50%. Sow mortality is considerably lower. Affected litters contain a variable mixture of normal pigs, weak small pigs, stillborn pigs, and partially or completely mummified fetuses (so-called SMEDI; stillbirth, mummification, embryonic death, and infertility). Infected sows also can exhibit nervous signs such as ataxia and circling. PRRSV-infected boars may continue to shed virus in their semen for prolonged periods of time.

PRRSV appears to be endemic in virtually all swine-producing countries of the world (Cho and Dee, 2006). The rapid global spread may be due to the national and international movement of infected pigs and their germplasm (semen), as well as the use of modified live virus vaccines for the control of PRRS (Prieto and Castro, 2005). Transmission of PRRSV usually occurs by close contact between infected and uninfected animals. However, a possible long distance aerosol transmission between farms remains highly controversial (Cho et al., 2007). It has been shown that a virulent strain of PRRSV is shed with high frequency and at high concentration, and is more rapidly transmitted than low virulent strains. Swine are susceptible by a number of exposure routes, including oral, intranasal, vaginal, intramuscular, and intraperitoneal infection. PRRSV is shed in respiratory tract secretions, saliva, semen, mammary secretions, urine, and feces of infected animals. Susceptible pigs are naturally infected by inhalation of infectious aerosols or ingestion of contaminated food. Congenital infection results from transplacental transmission (vertical transmission) of virus. Transmission of PRRSV to females has been demonstrated during breeding with semen from persistently infected carrier boars. The virus can also be transmitted through indirect contact with fomites or personnel, and some pigs harbor the virus in their tonsils long after it has been cleared from other tissues.

PRRSV replicates in macrophages and dendritic cells, especially those in the lungs and lymphoid tissues. Viremia occurs soon after infection and can last for 1 to 2 weeks in mature animals and 8 weeks in young pigs. Gross lesions are usually observed in only a few organ systems (e.g. respiratory and lymphoid). Microscopic lesions include diffuse interstitial pneumonia, myocarditis, vasculitis, and encephalitis. Lymphoid tissues exhibit lymphoid hyperplasia and follicular necrosis with mixed inflammatory cell infiltration. Clinical outbreaks of PRRSV can be complicated by bacterial pneumonia, septicemia, or enteritis.

LDV

LDV primarily replicates in primary cultures of mouse macrophages from 1- to 2 week-old mice and mouse macrophage cell lines, but not in other cell lines (Plagemann, 1996). The virus can cause a life-long persistent infection in inbred laboratory strains of mice, although infection of laboratory mice now is very uncommon (Rowson and Mahy, 1975; Plagemann, 1996; Plagemann, 2003; Coutelier, 2006). The virus has also been isolated from wild house mice (*Mus musculus domesticus*) in several countries, although the worldwide incidence is not known (Rowson and Mahy, 1975; Li et al., 2000). Attempts to infect *Peromyscus* mice, rats, guinea pigs and rabbits with LDV have been unsuccessful. LDV replicates in permissive macrophages in the spleen, lymph nodes, thymus, and liver of infected mice. The subpopulation of macrophages that is permissive to infection also is responsible for the normal clearance of the lactate dehydrogenase (LDH) enzyme from circulation. The continuous destruction of these macrophages by LDV leads to elevated levels of LDH in blood, thus explaining the name of the virus (Riley et al., 1960; Riley, 1974). The persistent infection that characterizes LDV infection of mice is maintained by replication of LDV in new permissive macrophages, which are continuously regenerated from apparently non-permissive precursor cells. Other than the elevated LDH level and subtle changes in host immunity, persistent LDV infection generally is asymptomatic. However, infection of mice of certain strains (C58, AKR, PL/J and C3H/FgBoy) by neurovirulent strains of LDV can lead to fatal age-dependent poliomyelitis (Chen et al., 1998; Chen et al., 2000).

Anti-LDV antibodies begin to appear 4 to 5 days post infection and peak at 3 to 4 weeks (Plagemann, 1996; van den Broek et al., 1997; Plagemann, 2003). Antibodies that neutralize LDV appear in mice only at 4 weeks post infection and are directed against the GP_5 protein of the virus. The replication of LDV in macrophages allows it to avoid host defense mechanisms, although the precise mechanism of immune invasion and persistent infection is still unclear.

LDV transmission between mice is relatively inefficient, despite the life-long viremia that occurs in infected mice and the secretion of the virus in urine, feces and saliva. LDV transmission from mother to offspring through the placenta or via breast milk is highly efficient if the mother is immunologically naive. However, transmission via these routes from persistently infected mothers is rare because anti-LDV antibodies block transplacental transmission of the virus and its release into milk. Generally horizontal transmission of LDV between mice is restricted by mucosal barriers although horizontal transmission of LDV does occur amongst laboratory mice that fight and bite one another. Sexual transmission of LDV among mice has not been demonstrated.

SHFV

SHFV infects and replicates in primary cultures of peritoneal macrophages from rhesus and patas monkeys and in MA-104 cell cultures. The virus was first isolated from rhesus macaques suffering from hemorrhagic fever, in primate research centers in the Soviet Union and the Unites States (Tauraso et al., 1968; Renquist, 1990; Plagemann, 2003). African green monkeys can be persistently infected with SHFV in the wild, but do not exhibit any clinical signs of disease. Accidental transmission of SHFV from African green monkeys to Asian macaque monkeys results in a generally fatal hemorrhagic fever. Clinical signs in macaques include anorexia, fever, cyanosis, skin petechia and hemorrhages, nosebleeds, facial edema, bloody diarrhea, dehydration, adipsia, and proteinuria. Death generally occurs 5-25 days after the onset of clinical signs, and case fatality rates approaches 100% in macaques. The high lethality may be due to an extreme sensitivity of their macrophages to cytocidal infection by SHFV. Infection of captive patas monkeys with SHFV from persistently infected patas monkeys results in persistent infection without any clinical signs. However, infection of captive patas monkeys with SHFV from diseased macaques results in transient mild disease suggesting selection of more virulent variants during epizootics in macaque monkeys. The humoral immune response against SHFV varies with the species of monkey and the infecting virus strain. SHFV induces neutralizing antibodies in patas monkeys 7 days after experimental infection. However, only low levels of anti-SHFV antibodies are found in many persistently infected patas monkeys.

The prevalence and incidence of SHFV infection amongst monkeys in endemic areas of Africa are not known, but the incidence of persistent subclinical infections in wild patas monkeys appears to be high. The method of SHFV transmission amongst these monkeys in the wild is unclear. Infection most likely occurs through wounds and biting, but sexual transmission has not been ruled out. SHFV is not transmitted transplacentally from persistently infected mothers to their offspring. Several epizootics of SHF in captive macaque colonies originated from accidental mechanical transmission of virus from asymptomatic, persistently infected African monkeys. Once illness becomes apparent in the macaque colony, SHFV spreads rapidly throughout the colony, most likely by direct contact and aerosols.

DIAGNOSIS

EAV

It has to be stressed that EVA is uncommon in horses, and that other, more important respiratory and non-respiratory equine infections clinically resemble EVA. Confirmation of diagnosis of EVA currently is based on virus detection and/or serological demonstration of elevated neutralizing antibody titers (4-fold or greater) in paired serum samples taken at a 14-28 day interval. EAV can be isolated from nasal swabs or anticoagulated blood collected from horses with signs of EVA, or the tissues (including placenta) of aborted equine fetuses. Putative carrier stallions are first identified by serology as they always are seropositive, and persistent infection is confirmed by virus isolation from semen in cell culture, by test-breeding using seronegative mares (and monitoring these for seroconversion to EAV after breeding), or by RT-PCR (traditional or real-time) to identify viral nucleic acid in semen. Histopathological examination coupled with immunohistochemical staining also is useful for diagnosis of abortion in particular, as are RT-PCR assays.

PRRSV

Diagnosis of PRRSV may be complicated by the fact that many infections are inapparent, but PRRS should be considered when there are clinical signs of respiratory disease associated with reproductive failure in a herd. PRRSV antibodies are detected by using a variety of serological assays, including ELISA, immunofluorescence assay, immunoperoxidase monolayer assay, and serum virus neutralization assay. Serological examination of acute and convalescent pig sera may provide evidence of seroconversion. Virus isolation from clinical specimens such as bronchoalveolar lavage fluid, lung, lymph node, buffy coat, and serum can be done in PAMs, MA104 and its derivatives, and CRL-11171 cell cultures. Different PRRSV strains or isolates vary in their ability to replicate in different cell types. Various PCR assays have been developed to detect viral nucleic acid in blood, semen and other clinical specimens.

LDV

LDV can only be quantitated by an end-point dilution assay in mice, which is based on the increase in plasma LDH activity that accompanies LDV infection in mice. The presence of LDV in mice or in other materials is readily detected by injecting plasma, tissue homogenates or other materials (e.g. transplantable mouse tumors) into groups of two to three mice and assaying their plasma LDH activity 4 to 5 days later. Transplantable mouse tumors can be readily freed of LDV by a 3-week *in vitro* propagation or passage in a host animal other than mice.

SHFV

Persistently infected monkeys can be identified by the presence of SHFV that replicates in primary cultures of peritoneal macrophages from rhesus and patas monkeys. The most sensitive method for detection of persistent infection in monkeys is experimental inoculation of macaque monkeys, and there currently are no molecular diagnostic assays for detection of SHFV. Indirect immunofluorescence assay, ELISA, and neutralization assays are available for serological diagnosis. However, these assays are not reliable because of the low level of anti-SHFV antibodies found in many persistently infected monkeys. Accidental transmission of SHFV from persistently infected African monkeys to macaque monkeys in primate centers can be minimized by strict adherence to proper sanitary conditions and animal care practices.

IMMUNE RESPONSE

There are several comprehensive reviews on the immune response to arterivirus infections and therefore, it will be discussed only briefly here (Murtaugh et al., 2002; Balasuriya and MacLachlan, 2004).

EAV

Both natural and experimental infection of horses with either virulent or avirulent strains of EAV results in long lasting immunity against re-infection with all strains of the virus, including the most virulent experimentally derived strains (Doll et al., 1968; McCollum, 1970; McCollum, 1986). There is only one known serotype of EAV and all field strains evaluated thus far are neutralized by polyclonal antiserum raised against the prototype Bucyrus strain (EAV ATCC VR-796). The humoral immune response is characterized by the development of both complement-fixing and virus-specific neutralizing antibodies (McCollum., 1970; Fukunaga and McCollum,

1977). Complement-fixing antibodies develop between 1 to 2 weeks after infection, peak after 2 to 3 weeks, and steadily decline to disappear by 8 months post infection, whereas neutralizing antibodies are detected within 1 to 2 weeks following exposure. However, virus does persist in the male reproductive tract of carrier stallions for long periods despite the presence of high titers of virus-specific neutralizing antibodies in their serum.

The cellular immune response to EAV has not been investigated in detail. Castillo-Olivares et al. (2003) recently described methods to assess EAV-specific CTL responses using peripheral blood mononuclear cells (PBMC) from convalescent EAV-infected (experimental) ponies. The data showed that cytotoxicity induced by EAV-stimulated PBMC was virus-specific, genetically-restricted, and mediated by CD8+ T cells, and that EAV-specific CTL precursors persist for at least 1 year after infection. Further studies are required to accurately identify the specific viral protein(s) that is targeted by the CTL response of EAV-infected horses.

PRRSV

Pigs infected with PRRSV produce a variety of virus-specific antibodies. Virus-specific IgM appears by 5-7 days after infection and IgG by 7-14 days (Murtaugh *et al.,* 2002, and references therein). ELISA antibody titers peak by 5-6 weeks and persist thereafter (Horter et al., 2002). However, PRRSV neutralizing antibodies appear slowly, usually between 4-5 weeks post infection, and do not peak until after approximately 10 weeks. The appearance of neutralizing antibodies is associated with the clearance of virus from the lungs of infected pigs. However, the virus can persist for up to 150 days in some pigs and these animals continue to shed virus (Allende et al., 2000). The humoral immune response to PRRSV varies significantly between individual pigs, and there are reports suggesting that protective immunity is strain specific with some degree of heterologous protection against other strains of PRRSV (Nelson et al., 1994; Lager et al., 1999). The central role of antibodies in protecting swine against re-infection with PRRSV, however, is clearly evident as shown in elegant studies by Osorio et al., 2002. They showed that passive transfer of PRRSV antibodies fully protected pregnant sows against challenge with virulent PRRSV. Piglets born to immune sows acquire anti-PRRSV antibodies by ingestion of colostrum, and these maternal antibodies persist in piglets up to 6-8 weeks of age.

PRRSV-specific T-cell proliferative responses (both CD4+ and CD8+) in infected pigs occur between 4 and 12 weeks after infection, which suggests that the M protein may be an important target of the CMI response (Bautista and Molitor, 1997; Bautista et al., 1999).

LDV

Mice develop anti-LDV antibodies 4-5 days after infection leading to a peak at 3-4 weeks, as measured by immunofluorescence assay (Plagemann, 1996). However, neutralizing antibodies to LDV do not appear until approximately 4 weeks after infection and they only poorly neutralize the virus. Both neutralizing and non-neutralizing antibodies are directed against GP_5, the major envelope glycoprotein of the virus. Infectious LDV persists in circulating immune complexes throughout persistent infection, and replication of LDV in permissive macrophages, which are continuously regenerated from apparently non-permissive precursors, provides a continuous source of virus. The replication of LDV in macrophages also allows it to avoid host defense mechanisms, although the precise mechanism of immune invasion and persistent infection is still unclear.

SHFV

The immune response to SHFV has not been studied in detail but monkeys produce complement-fixing, ELISA-positive, and neutralizing antibodies to the virus following infection (Madden et al., 1978; Godeny, 2002). Neutralizing antibodies against one strain of SHFV do not completely neutralize other strains, suggesting that there is significant variation in neutralization determinants of individual strains. However, four strains (LVR, P-180, P-248 and P-741) that have been studied were found to be antigenically related to each other based on ELISA antibodies. It has been documented that the virulent (LVR, P-180) and avirulent (P-248, P-741) strains of SHFV induce different levels of humoral antibody response in patas monkeys. The more virulent LVR strain induced antibodies within 7 days after infection and appearance of these antibodies was associated with the clearance of the virus. In contrast, the two low virulent strains induce minimal or no antibody response with persistent low level of viremia.

VACCINATION AND TREATMENT STRATEGIES

EAV

Modified live attenuated and inactivated virus vaccines are available for prevention of EAV infection in horses. In addition, experimental vaccines using recombinant DNA technology have been described, like a subunit vaccine, Venezuelan equine encephalitis [VEE] replicon particle-based vaccine, infectious clone-based live marker vaccine, and DNA vaccine (reviewed in Balasuriya and MacLachlan, 2004). However, vaccines based on recombinant DNA technology are not yet considered appropriate for use in the field by the equine industry.

Colts should be vaccinated with the attenuated modified live virus (MLV) vaccine after they lose colostral passively acquired immunity, but before they become pubertal. Outbreaks of EVA can be prevented by the identification of persistently infected stallions and the institution of management practices to prevent the introduction of EAV-infected horses. Carrier stallions should be kept physically isolated and bred only to mares that are seropositive from previous natural exposure or vaccination. Mares should be kept isolated from other seronegative horses for a 3 weeks period after being bred by carrier stallions.

Currently there are no fully proven means available of eliminating the carrier state in stallions persistently infected with EAV other than surgical castration. There is no specific treatment for horses infected with EAV. However, there are preliminary data to support that GnRH (Gonadotropin-releasing hormone) vaccines or antagonists can temporarily limit the shedding of the virus in the semen of carrier stallions (Fortier, 2002).

PRRSV

Both MLV and killed vaccines are available for prevention of PRRSV infection of swine. However, these vaccines only provide protection against closely related PRRSV strains and not against genetically diverse field strains. MLV vaccines induce long-lasting protection as compared to killed vaccines, but do not completely prevent re-infection with wild-type virus and subsequent virus transmission. There have been reports of under attenuation and reversion of MLV vaccine strains to more virulent types that can be spread from vaccinated to unvaccinated swine. The shedding and spreading of virus between vaccinated and non-vaccinated pigs have been reported following vaccination with the MLV vaccine derived from the NA strain of PRRSV under experimental and natural conditions (Botner et al., 1997; Madsen et al., 1998). Furthermore, shedding of MLV vaccine strains of PRRSV in semen of vaccinated boars and subsequent spread to naïve breeding herd has also been demonstrated (Nielsen et al., 1997). Little knowledge is available on the effectiveness of inactivated vaccines; however, modified live vaccines, which predominate, have not fully met the expectations and needs of the industry. Several experimental PRRSV vaccines using recombinant DNA technology have been described, but none have been approved for commercial use in the USA. Alternatives to vaccination include a variety of strategies that seek to eliminate the virus, including whole herd depopulation-repopulation, test and removal, herd closure, and partial depopulation. There is no effective treatment for PRRSV infection in pigs.

LDV AND SHFV

There are no vaccines against LDV or SHFV infection.

FUTURE PERSPECTIVES

Over the past 15 years, the molecular characterization of four arterivirus genomes and the viral life cycle has defined, in basic terms, the protein and RNA molecules of this virus family. Unraveling their role in the viral life cycle is one of the important challenges for the immediate future. For example, understanding the role of the seven structural proteins in virus assembly, receptor binding, and entry promises to be a complex issue. Progress has been made towards understanding host cell attachment and entry of one arterivirus (PRRSV), but many issues remain to be resolved, in particular concerning the role of specific molecules on virion and cell surface. In arterivirus replicase research, the combination of structural genomics, biochemistry, and (reverse) genetics promises will aid the characterization of additional replicase domains. When considering the replicase as a whole, the regulatory role of proteolysis, the assembly, structure, and function of the membrane-bound replication/transcription complex, and other virus-host interactions are largely unexplored fields. The systems for EAV and PRRSV reverse genetics that have become available are of great value for molecular biological research. In addition, they are starting to be employed in studies into arterivirus pathogenesis and immunity, and also vaccine development may benefit from this technology, as well as from newly developed strategies to produce "defective" arteriviruses (e.g. lacking one of the minor GP genes) in complementing cell lines. An intriguing final question is whether arteriviruses will emerge or be identified in other host species, although such viruses may be difficult to find if these hosts develop asymptomatic infections. On the basis of the plasticity of their genomes and the fact that they can be transmitted by aerosols, arteriviruses may have ample opportunity to attempt to cross species barriers, and - in fact - may already have done so.

ACKNOWLEDGMENTS

For pleasant contacts and collaborations, and helpful discussions through the years, we would like to thank our past and present colleagues at the University of California at Davis, Maxwell H. Gluck Equine Research Center, Leiden University Medical Center, and many colleagues from the arterivirus field. In particular, we would like to acknowledge the contributions of Alexander Gorbalenya, James MacLachlan, William McCollum, Willy Spaan, Peter Timoney, Fred Wassenaar, and Jessika Zevenhoven-Dobbe. Supported in part by grants from Grayson-Jockey Club Research Foundation, Inc. and American Quarter Horse Foundation to Udeni B. R. Balasuriya and grants from the Council for Chemical Sciences of the Netherlands Organization for Scientific Research and the European Union to Eric J. Snijder.

REFERENCES

Allende, R., Laegreid, W.W., Kutish, G.F., Galeota, J.A., Wills, R.W., and Osorio, F.A. (2000). Porcine reproductive and respiratory syndrome virus: description of persistence in individual pigs upon experimental infection. J. Virol. *74*, 10834-10837.

Balasuriya, U.B., Hedges, J.F., Nadler, S.A., McCollum, W.H., Timoney, P.J., and MacLachlan, N.J. (1999a). Genetic stability of equine arteritis virus during horizontal and vertical transmission in an outbreak of equine viral arteritis. J. Gen. Virol. *80*, 1949-1958.

Balasuriya, U.B., Heidner, H.W., Davis, N.L., Wagner, H.M., Hullinger, P.J., Hedges, J.F., Williams, J.C., Johnston, R.E., David Wilson, W., Liu, I.K., and James MacLachlan, N. (2002). Alphavirus replicon particles expressing the two major envelope proteins of equine arteritis virus induce high level protection against challenge with virulent virus in vaccinated horses. Vaccine *20*, 1609-1617.

Balasuriya, U.B., Heidner, H.W., Hedges, J.F., Williams, J.C., Davis, N.L., Johnston, R.E., and MacLachlan, N.J. (2000). Expression of the two major envelope proteins of equine arteritis virus as a heterodimer is necessary for induction of neutralizing antibodies in mice immunized with recombinant Venezuelan equine encephalitis virus replicon particles. J. Virol. *74*, 10623-10630.

Balasuriya, U.B., and MacLachlan, N.J. (2004). The immune response to equine arteritis virus: potential lessons for other arteriviruses. Vet. Immunol. Immunopathol. *102*, 107-129.

Balasuriya, U.B., Snijder, E.J., van Dinten, L.C., Heidner, H.W., Wilson, W.D., Hedges, J.F., Hullinger, P.J., and MacLachlan, N.J. (1999b). Equine arteritis virus derived from an infectious cDNA clone is attenuated and genetically stable in infected stallions. Virology *260*, 201-208.

Balasuriya, U.B., Timoney, P.J., McCollum, W.H., and MacLachlan, N.J. (1995). Phylogenetic analysis of open reading frame 5 of field isolates of equine arteritis virus and identification of conserved and nonconserved regions in the GL envelope glycoprotein. Virology *214*, 690-697.

Baric, R.S., Stohlman, S.A., and Lai, M.M. (1983). Characterization of replicative intermediate RNA of mouse hepatitis virus: presence of leader RNA sequences on nascent chains. J. Virol. *48*, 633-640.

Barrette-Ng, I.H., Ng, K.K., Mark, B.L., Van Aken, D., Cherney, M.M., Garen, C., Kolodenko, Y., Gorbalenya, A.E., Snijder, E.J., and James, M.N. (2002). Structure of arterivirus nsp4. The smallest chymotrypsin-like proteinase with an alpha/beta C-terminal extension and alternate conformations of the oxyanion hole. J. Biol. Chem. *277*, 39960-39966.

Bautista, E.M., Faaberg, K.S., Mickelson, D., and McGruder, E.D. (2002). Functional properties of the predicted helicase of porcine reproductive and respiratory syndrome virus. Virology *298*, 258-270.

Bautista, E.M., and Molitor, T.W. (1997). Cell-mediated immunity to porcine reproductive and respiratory syndrome virus in swine. Viral Immunol. *10*, 83-94.

Bautista, E.M., Suarez, P., and Molitor, T.W. (1999). T cell responses to the structural polypeptides of porcine reproductive and respiratory syndrome virus. Arch. Virol. *144*, 117-134.

Beerens, N., and Snijder, E.J. (2006). RNA signals in the 3' terminus of the genome of Equine arteritis virus are required for viral RNA synthesis. J. Gen. Virol. *87*, 1977-1983.

Bell, S.A., Balasuriya, U.B.R., MacLachlan, N.J. (2006). Equine viral arteritis. Clin. Tech. Equine Pract. *5*, 233-238.

Botner, A., Strandbygaard, B., Sorensen, K.J., Have, P., Madsen, K.G., Madsen, E.S., and Alexandersen, S. (1997). Appearance of acute PRRS-like symptoms in sow herds after vaccination with a modified live PRRS vaccine. Vet. Rec. *141*, 497-499.

Brian, D.A., Spann, W.J.M. (1997). Recombination and coronavirus defective interfering RNAs. Semin. Virol. *8*, 101-111.
Burki, F. (1965). Properties of the equine arteritis virus. Pathol. Microbiol. (Basel) *28*, 939-949.
Burki, F. (1966). Further properties of equine arteritis virus. Arch. Gesamte Virusforsch. *19*, 123-129.
Castillo-Olivares, J., Tearle, J.P., Montesso, F., Westcott, D., Kydd, J.H., Davis-Poynter, N.J., and Hannant, D. (2003). Detection of equine arteritis virus (EAV)-specific cytotoxic CD8+ T lymphocyte precursors from EAV-infected ponies. J. Gen. Virol. *84*, 2745-2753.
Cavanagh, D. (1997). Nidovirales: a new order comprising Coronaviridae and Arteriviridae. Arch. Virol. *142*, 629-633.
Chen, Z., Faaberg, K.S., and Plagemann, P.G. (1994). Determination of the 5' end of the lactate dehydrogenase-elevating virus genome by two independent approaches. J. Gen. Virol. *75*, 925-930.
Chen, Z., Li, K., and Plagemann, P.G. (2000). Neuropathogenicity and sensitivity to antibody neutralization of lactate dehydrogenase-elevating virus are determined by polylactosaminoglycan chains on the primary envelope glycoprotein. Virology *266*, 88-98.
Chen, Z., Li, K., Rowland, R.R., Anderson, G.W., and Plagemann, P.G. (1998). Lactate dehydrogenase-elevating virus variants: cosegregation of neuropathogenicity and impaired capability for high viremic persistent infection. J. Neurovirol. *4*, 560-568.
Chirnside, E.D., Wearing, C.M., Binns, M.M., and Mumford, J.A. (1994). Comparison of M and N gene sequences distinguishes variation amongst equine arteritis virus isolates. J. Gen. Virol. *75*, 1491-1497.
Cho, J.G., and Dee, S.A. (2006). Porcine reproductive and respiratory syndrome virus. Theriogenology *66*, 655-662.
Cho, J.G., Deen, J., and Dee, S.A. (2007). Influence of isolate pathogenicity on the aerosol transmission of Porcine reproductive and respiratory syndrome virus. Can. J. Vet. Res. *71*, 23-27.
Clark, I. (1892). Transmission of pink-eye from apparently healthy stallions to mares. J. Comp. Pathol. *5*, 261-264.
Collins, J.E., Benfield, D.A., Christianson, W.T., Harris, L., Hennings, J.C., Shaw, D.P., Goyal, S.M., McCullough, S., Morrison, R.B., Joo, H.S., et al. (1992). Isolation of swine infertility and respiratory syndrome virus (isolate ATCC VR-2332) in North America and experimental reproduction of the disease in gnotobiotic pigs. J. Vet. Diagn. Invest. *4*, 117-126.
Conzelmann, K.K., Visser, N., Van Woensel, P., and Thiel, H.J. (1993). Molecular characterization of porcine reproductive and

respiratory syndrome virus, a member of the arterivirus group. Virology *193*, 329-339.
Coutelier, J.-P., Brinton, M.A. (2006). Lactate dehydrogenase-elevating virus. In The Mouse in Biomedical Research, J.G. Fox, C. Newcomer, A. Smith, S. Barthold, F. Quimby, and M. Davisson, eds. (San Diego: Elsevier), pp. 215-234.
de Vries, A.A., Chirnside, E.D., Bredenbeek, P.J., Gravestein, L.A., Horzinek, M.C., and Spaan, W.J. (1990). All subgenomic mRNAs of equine arteritis virus contain a common leader sequence. Nucleic Acids Res. *18*, 3241-3247.
de Vries, A.A., Chirnside, E.D., Horzinek, M.C., and Rottier, P.J. (1992). Structural proteins of equine arteritis virus. J. Virol. *66*, 6294-6303.
de Vries, A.A., Post, S.M., Raamsman, M.J., Horzinek, M.C., and Rottier, P.J. (1995). The two major envelope proteins of equine arteritis virus associate into disulfide-linked heterodimers. J. Virol. *69*, 4668-4674.
Del Piero, F. (2000). Equine viral arteritis. Vet. Pathol. *37*, 287-296.
Delputte, P.L., Costers, S., and Nauwynck, H.J. (2005). Analysis of porcine reproductive and respiratory syndrome virus attachment and internalization: distinctive roles for heparan sulphate and sialoadhesin. J. Gen. Virol. *86*, 1441-1445.
den Boon, J.A., Faaberg, K.S., Meulenberg, J.J., Wassenaar, A.L., Plagemann, P.G., Gorbalenya, A.E., and Snijder, E.J. (1995). Processing and evolution of the N-terminal region of the arterivirus replicase ORF1a protein: identification of two papainlike cysteine proteases. J. Virol. *69*, 4500-4505.
den Boon, J.A., Kleijnen, M.F., Spaan, W.J., and Snijder, E.J. (1996). Equine arteritis virus subgenomic mRNA synthesis: analysis of leader-body junctions and replicative-form RNAs. J. Virol. *70*, 4291-4298.
den Boon, J.A., Snijder, E.J., Chirnside, E.D., de Vries, A.A., Horzinek, M.C., and Spaan, W.J. (1991). Equine arteritis virus is not a togavirus but belongs to the coronavirus-like superfamily. J. Virol. *65*, 2910-2920.
Dewey, C., Charbonneau, G., Carman, S., Hamel, A., Nayar, G., Friendship, R., Eernisse, K., and Swenson, S. (2000). Lelystad-like strain of porcine reproductive and respiratory syndrome virus (PRRSV) identified in Canadian swine. Can. Vet. J. *41*, 493-494.
Doan, D.N., and Dokland, T. (2003). Structure of the nucleocapsid protein of porcine reproductive and respiratory syndrome virus. Structure *11*, 1445-1451.

Dobbe, J.C., van der Meer, Y., Spaan, W.J., and Snijder, E.J. (2001). Construction of chimeric arteriviruses reveals that the ectodomain of the major glycoprotein is not the main determinant of equine arteritis virus tropism in cell culture. Virology *288*, 283-294.

Doll, E.R., Bryans, J.T., McCollum, W.H., and Crowe, M.E. (1957a). Isolation of a filterable agent causing arteritis of horses and abortion by mares; its differentiation from the equine abortion (influenza) virus. Cornell Vet. *47*, 3-41.

Doll, E.R., Bryans, J.T., Wilson, J.C., and McCollum, W.H. (1968). Immunization against equine viral arteritis using modified live virus propagated in cell cultures of rabbit kidney. Cornell Vet. *48*, 497-524.

Doll, E.R., Knappenberger, R.E., and Bryans, J.T. (1957b). An outbreak of abortion caused by the equine arteritis virus. Cornell Vet. *47*, 69-75.

Domingo, E., Escarmis, C., Sevilla, N., Moya, A., Elena, S.F., Quer, J., Novella, I.S., and Holland, J.J. (1996). Basic concepts in RNA virus evolution. FASEB J. *10*, 859-864.

Faaberg, K.S., Even, C., Palmer, G.A., and Plagemann, P.G. (1995). Disulfide bonds between two envelope proteins of lactate dehydrogenase-elevating virus are essential for viral infectivity. J. Virol. *69*, 613-617.

Faaberg, K.S., and Plagemann, P.G. (1995). The envelope proteins of lactate dehydrogenase-elevating virus and their membrane topography. Virology *212*, 512-525.

Fang, Y., Kim, D.Y., Ropp, S., Steen, P., Christopher-Hennings, J., Nelson, E.A., and Rowland, R.R. (2004). Heterogeneity in Nsp2 of European-like porcine reproductive and respiratory syndrome viruses isolated in the United States. Virus Res. *100*, 229-235.

Forsberg, R. (2005). Divergence time of porcine reproductive and respiratory syndrome virus subtypes. Mol. Biol. Evol. *22*, 2131-2134.

Forsberg, R., Oleksiewicz, M.B., Petersen, A.M., Hein, J., Botner, A., and Storgaard, T. (2001). A molecular clock dates the common ancestor of European-type porcine reproductive and respiratory syndrome virus at more than 10 years before the emergence of disease. Virology *289*, 174-179.

Forsberg, R., Storgaard, T., Nielsen, H.S., Oleksiewicz, M.B., Cordioli, P., Sala, G., Hein, J., and Botner, A. (2002). The genetic diversity of European type PRRSV is similar to that of the North American type but is geographically skewed within Europe. Virology *299*, 38-47.

Fortier, G., Vidament, M., DeCraene, F., Ferry, B., Daels, P.F. (2002). The effect of GnRH antagonist on testosterone secretion, spermatogenesis and viral excretion in EVA-virus excreting stallions. Theriogenology *58*, 425-427.

Fukunaga, Y., and McCollum, W.H. (1977). Complement-fixation reactions in equine viral arteritis. Am. J. Vet. Res. *38*, 2043-2046.

Gao, Z.Q., Guo, X., and Yang, H.C. (2004). Genomic characterization of two Chinese isolates of porcine respiratory and reproductive syndrome virus. Arch. Virol. *149*, 1341-1351.

Glaser, A.L., de Vries, A.A., Rottier, P.J., Horzinek, M.C., and Colenbrander, B. (1996). Equine arteritis virus: a review of clinical features and management aspects. Vet. Q. *18*, 95-99.

Godeny, E.K. (2002). Enzyme-linked immunosorbent assay for detection of antibodies against simian hemorrhagic fever virus. Comp. Med. *52*, 229-232.

Godeny, E.K., Chen, L., Kumar, S.N., Methven, S.L., Koonin, E.V., and Brinton, M.A. (1993). Complete genomic sequence and phylogenetic analysis of the lactate dehydrogenase-elevating virus (LDV). Virology *194*, 585-596.

Godeny, E.K., de Vries, A.A., Wang, X.C., Smith, S.L., and de Groot, R.J. (1998). Identification of the leader-body junctions for the viral subgenomic mRNAs and organization of the simian hemorrhagic fever virus genome: evidence for gene duplication during arterivirus evolution. J. Virol. *72*, 862-867.

Godeny, E.K., Speicher, D.W., and Brinton, M.A. (1990). Map location of lactate dehydrogenase-elevating virus (LDV) capsid protein (Vp1) gene. Virology *177*, 768-771.

Godeny, E.K., Zeng, L., Smith, S.L., and Brinton, M.A. (1995). Molecular characterization of the 3' terminus of the simian hemorrhagic fever virus genome. J. Virol. *69*, 2679-2683.

Goldberg, T.L., Lowe, J.F., Milburn, S.M., and Firkins, L.D. (2003). Quasispecies variation of porcine reproductive and respiratory syndrome virus during natural infection. Virology *317*, 197-207.

Gorbalenya, A.E., Enjuanes, L., Ziebuhr, J., and Snijder, E.J. (2006). Nidovirales: evolving the largest RNA virus genome. Virus Res. *117*, 17-37.

Gravell, M., London, W.T., Leon, M.E., Palmer, A.E., and Hamilton, R.S. (1986). Differences among isolates of simian hemorrhagic fever (SHF) virus. Proc. Soc. Exp. Biol. Med. *181*, 112-119.

Han, J., Wang, Y., and Faaberg, K.S. (2006). Complete genome analysis of RFLP 184 isolates of porcine reproductive and respiratory syndrome virus. Virus Res. *122*, 175-182.

Hanada, K., Suzuki, Y., Nakane, T., Hirose, O., and Gojobori, T. (2005). The origin and evolution of porcine reproductive and respiratory syndrome viruses. Mol. Biol. Evol. *22*, 1024-1031.

Hedges, J.F., Balasuriya, U.B., and MacLachlan, N.J. (1999a). The open reading frame 3 of equine arteritis virus encodes an immunogenic glycosylated, integral membrane protein. Virology *264*, 92-98.

Hedges, J.F., Balasuriya, U.B., Timoney, P.J., McCollum, W.H., and MacLachlan, N.J. (1999b). Genetic divergence with emergence of novel phenotypic variants of equine arteritis virus during persistent infection of stallions. J. Virol. *73*, 3672-3681.

Horter, D.C., Pogranichniy, R.M., Chang, C.C., Evans, R.B., Yoon, K.J., and Zimmerman, J.J. (2002). Characterization of the carrier state in porcine reproductive and respiratory syndrome virus infection. Vet. Microbiol. *86*, 213-228.

Horzinek, M., Maess, J., and Laufs, R. (1971). Studies on the substructure of togaviruses. II. Analysis of equine arteritis, rubella, bovine viral diarrhea, and hog cholera viruses. Arch. Gesamte Virusforsch. *33*, 306-318.

Horzinek, M.C. (1973a). Comparative aspects of togaviruses. J. Gen. Virol. *20*, Suppl:87-103.

Horzinek, M.C. (1973b). The structure of togaviruses. Prog. Med. Virol. *16*, 109-156.

Hyllseth, B. (1970). Buoyant density studies on equine arteritis virus. Arch. Gesamte Virusforsch. *30*, 97-104.

Indik, S., Schmoll, F., Sipos, W., and Klein, D. (2005). Genetic variability of PRRS virus in Austria: consequences for molecular diagnostics and viral quantification. Vet. Microbiol. *107*, 171-178.

Jiang, Y., Fang, L., Xiao, S., Zhang, H., Pan, Y., Luo, R., Li, B., and Chen, H. (2006). Immunogenicity and protective efficacy of recombinant pseudorabies virus expressing the two major membrane-associated proteins of porcine reproductive and respiratory syndrome virus. Vaccine 25, 547-560.

Kreutz, L.C., and Ackermann, M.R. (1996). Porcine reproductive and respiratory syndrome virus enters cells through a low pH-dependent endocytic pathway. Virus Res. *42*, 137-147.

Kuo, L., Chen, Z., Rowland, R.R., Faaberg, K.S., and Plagemann, P.G. (1992). Lactate dehydrogenase-elevating virus (LDV): subgenomic mRNAs, mRNA leader and comparison of 3'-terminal sequences of two LDV isolates. Virus Res. *23*, 55-72.

Lager, K.M., Mengeling, W.L., and Brockmeier, S.L. (1999). Evaluation of protective immunity in gilts inoculated with the NADC-8 isolate of porcine reproductive and respiratory syndrome virus

(PRRSV) and challenge-exposed with an antigenically distinct PRRSV isolate. Am. J. Vet. Res. *60*, 1022-1027.

Li, K., Chen, Z., and Plagemann, P. (1998). The neutralization epitope of lactate dehydrogenase-elevating virus is located on the short ectodomain of the primary envelope glycoprotein. Virology *242*, 239-245.

Li, K., Chen, Z., and Plagemann, P. (1999). High-frequency homologous genetic recombination of an arterivirus, lactate dehydrogenase-elevating virus, in mice and evolution of neuropathogenic variants. Virology *258*, 73-83.

Li, K., Schuler, T., Chen, Z., Glass, G.E., Childs, J.E., and Plagemann, P.G. (2000). Isolation of lactate dehydrogenase-elevating viruses from wild house mice and their biological and molecular characterization. Virus Res. *67*, 153-162.

MacLachlan, N.J., Balasuriya, U.B., Rossitto, P.V., Hullinger, P.A., Patton, J.F., and Wilson, W.D. (1996). Fatal experimental equine arteritis virus infection of a pregnant mare: immunohistochemical staining of viral antigens. J. Vet. Diagn. Invest. *8*, 367-374.

Madden, D.L., Fuccillo, D.A., Dorosz, J.A., London, W.T., Palmer, A.E., and Castellano, G.A. (1978). Antigenic relationship of two strains of simian hemorrhagic fever virus. Lab. Anim. Sci. *28*, 422-427.

Madsen, K.G., Hansen, C.M., Madsen, E.S., Strandbygaard, B., Botner, A., and Sorensen, K.J. (1998). Sequence analysis of porcine reproductive and respiratory syndrome virus of the American type collected from Danish swine herds. Arch. Virol. *143*, 1683-1700.

Maess, J., Reczko, E., and Bohm, H.O. (1970). Equine arteritis virus: multiplication in BHK 21-cells buoyant density and electron microscopical demonstration. Arch. Gesamte Virusforsch. *30*, 47-58.

Magnusson, P., Hyllseth, B., and Marusyk, H. (1970). Morphological studies on equine arteritis virus. Arch. Gesamte Virusforsch. *30*, 105-112.

McCollum, W.H. (1970). Vaccination for equine viral arteritis, Proceedings of the second international conference (equine infectious diseases) (Paris: S. Karger, Basel), pp. 143-151.

McCollum W.H., and Timoney, P.J. (1998). Experimental observations on the virulence of isolates of equine arteritis virus, Proceedings of the eighth international conference (equine infectious diseases VIII) (Dubai: R & W Publications (Newmarket) Limited), pp. 558-559.

McCollum, W.H. (1986). Responses of horses vaccinated with avirulent modified-live equine arteritis virus propagated in the E. Derm. (NBL-6) cell line to nasal inoculation with virulent virus. Am. J. Vet. Res. *47*, 1931-1934.

Meulenberg, J.J., Bos-de Ruijter, J.N., van de Graaf, R., Wensvoort, G., and Moormann, R.J. (1998). Infectious transcripts from cloned genome-length cDNA of porcine reproductive and respiratory syndrome virus. J. Virol. *72*, 380-387.

Meulenberg, J.J., Hulst, M.M., de Meijer, E.J., Moonen, P.L., den Besten, A., de Kluyver, E.P., Wensvoort, G., and Moormann, R.J. (1993). Lelystad virus, the causative agent of porcine epidemic abortion and respiratory syndrome (PEARS), is related to LDV and EAV. Virology *192*, 62-72.

Meulenberg, J.J., Hulst, M.M., de Meijer, E.J., Moonen, P.L., den Besten, A., de Kluyver, E.P., Wensvoort, G., and Moormann, R.J. (1994). Lelystad virus belongs to a new virus family, comprising lactate dehydrogenase-elevating virus, equine arteritis virus, and simian hemorrhagic fever virus. Arch. Virol. Suppl. *9*, 441-448.

Meulenberg, J.J., Petersen-den Besten, A., De Kluyver, E.P., Moormann, R.J., Schaaper, W.M., and Wensvoort, G. (1995). Characterization of proteins encoded by ORFs 2 to 7 of Lelystad virus. Virology *206*, 155-163.

Meulenberg, J.J., Petersen den Besten, A., de Kluyver, E., van Nieuwstadt, A., Wensvoort, G., and Moormann, R.J. (1997a). Molecular characterization of Lelystad virus. Vet. Microbiol. *55*, 197-202.

Meulenberg, J.J., van Nieuwstadt, A.P., van Essen-Zandbergen, A., and Langeveld, J.P. (1997b). Posttranslational processing and identification of a neutralization domain of the GP4 protein encoded by ORF4 of Lelystad virus. J. Virol. *71*, 6061-6067.

Molenkamp, R., Rozier, B.C., Greve, S., Spaan, W.J., and Snijder, E.J. (2000a). Isolation and characterization of an arterivirus defective interfering RNA genome. J. Virol. *74*, 3156-3165.

Molenkamp, R., van Tol, H., Rozier, B.C., van der Meer, Y., Spaan, W.J., and Snijder, E.J. (2000b). The arterivirus replicase is the only viral protein required for genome replication and subgenomic mRNA transcription. J. Gen. Virol. *81*, 2491-2496.

Moore, B.D., Balasuriya, U.B., Watson, J.L., Bosio, C.M., MacKay, R.J., and MacLachlan, N.J. (2003). Virulent and avirulent strains of equine arteritis virus induce different quantities of TNF-alpha and other proinflammatory cytokines in alveolar and blood-derived equine macrophages. Virology *314*, 662-670.

Murtaugh, M.P., Xiao, Z., and Zuckermann, F. (2002). Immunological responses of swine to porcine reproductive and respiratory syndrome virus infection. Viral Immunol. *15*, 533-547.

Nelsen, C.J., Murtaugh, M.P., and Faaberg, K.S. (1999). Porcine reproductive and respiratory syndrome virus comparison: divergent evolution on two continents. J. Virol. *73*, 270-280.

Nelson, E.A., Christopher-Hennings, J., and Benfield, D.A. (1994). Serum immune responses to the proteins of porcine reproductive and respiratory syndrome (PRRS) virus. J. Vet. Diagn. Invest. *6*, 410-415.

Nielsen, T.L., Nielsen, J., Have, P., Baekbo, P., Hoff-Jorgensen, R., and Botner, A. (1997). Examination of virus shedding in semen from vaccinated and from previously infected boars after experimental challenge with porcine reproductive and respiratory syndrome virus. Vet. Microbiol. *54*, 101-112.

Nodelijk, G. (2002). Porcine reproductive and respiratory syndrome (PRRS) with special reference to clinical aspects and diagnosis. A review. Vet. Q. *24*, 95-100.

Nodelijk, G., Nielen, M., De Jong, M.C., and Verheijden, J.H. (2003). A review of porcine reproductive and respiratory syndrome virus in Dutch breeding herds: population dynamics and clinical relevance. Prev. Vet. Med. *60*, 37-52.

Osorio, F.A., Galeota, J.A., Nelson, E., Brodersen, B., Doster, A., Wills, R., Zuckermann, F., and Laegreid, W.W. (2002). Passive transfer of virus-specific antibodies confers protection against reproductive failure induced by a virulent strain of porcine reproductive and respiratory syndrome virus and establishes sterilizing immunity. Virology *302*, 9-20.

Palmer, A.E., Allen, A.M., Tauraso, N.M., and Shelokov, A. (1968). Simian hemorrhagic fever. I. Clinical and epizootiologic aspects of an outbreak among quarantined monkeys. Am. J. Trop. Med. Hyg. *17*, 404-412.

Pasternak, A.O., Spaan, W.J., and Snijder, E.J. (2006). Nidovirus transcription: how to make sense...? J. Gen. Virol. *87*, 1403-1421.

Pasternak, A.O., van den Born, E., Spaan, W.J., and Snijder, E.J. (2001). Sequence requirements for RNA strand transfer during nidovirus discontinuous subgenomic RNA synthesis. EMBO J. *20*, 7220-7228.

Pedersen, K.W., van der Meer, Y., Roos, N., and Snijder, E.J. (1999). Open reading frame 1a-encoded subunits of the arterivirus replicase induce endoplasmic reticulum-derived double-membrane vesicles which carry the viral replication complex. J. Virol. *73*, 2016-2026.

Pesch, S., Meyer, C., and Ohlinger, V.F. (2005). New insights into the genetic diversity of European porcine reproductive and respiratory syndrome virus (PRRSV). Vet. Microbiol. *107*, 31-48.

Plagemann, G.W. (1996). Lactate dehydrogenase-elevating virus and related viruses In Fields Virology Third Edition, B.N. Fields, D.M. Knipe, and P.M. Howley, eds. (New York: Lippincott-Reven), pp. 1105-1120.

Plagemann, P.G. (2003). Porcine reproductive and respiratory syndrome virus: origin hypothesis. Emerg. Infect. Dis. *9*, 903-908.

Plagemann, P.G., Chen, Z., and Li, K. (2001). Replication competition between lactate dehydrogenase-elevating virus quasispecies in mice. Implications for quasispecies selection and evolution. Arch. Virol. *146*, 1283-1296.

Plagemann, P.G., and Moennig, V. (1992). Lactate dehydrogenase-elevating virus, equine arteritis virus, and simian hemorrhagic fever virus: a new group of positive-strand RNA viruses. Adv. Virus Res. *41*, 99-192.

Plateau, E., Jacquet, A. (1888). The propagation of influenza from stallion to mares. J. Comp. Pathol. *1*, 37-38.

Pope, J.H. (1961). Studies of a virus isolated from a wild house mouse, *Mus musculus*, and producing spleenomegaly and lymphnode enlargemnt in mice. Aust. J. Exp. Biol. Med. Sci. *39*, 521-536.

Porterfield, J.S., Casals, J., Chumakov, M.P., Gaidamovich, S.Y., Hannoun, C., Holmes, I.H., Horzinek, M.C., Mussgay, M., Oker-Blom, N., Russell, P.K., and Trent, D.W. (1978). Togaviridae. Intervirology *9*, 129-148.

Posthuma, C.C., Nedialkova, D.D., Zevenhoven-Dobbe, J.C., Blokhuis, J.H., Gorbalenya, A.E., and Snijder, E.J. (2006). Site-directed mutagenesis of the Nidovirus replicative endoribonuclease NendoU exerts pleiotropic effects on the arterivirus life cycle. J. Virol. *80*, 1653-1661.

Prieto, C., and Castro, J.M. (2005). Porcine reproductive and respiratory syndrome virus infection in the boar: a review. Theriogenology *63*, 1-16.

Renquist, D. (1990). Outbreak of simian hemorrhagic fever. J. Med. Primatol. *19*, 77-79.

Riley, V. (1974). Persistence and other characteristics of the lactate dehydrogenase-elevating virus (LDH-virus). Prog. Med. Virol. *18*, 198-213.

Riley, V., Lilly, F., Euerto, E., and Bardell, D. (1960). Transmissible agent associated with 26 types of experimental mouse neoplasms. Science *132*, 545-547.

Ropp, S.L., Wees, C.E., Fang, Y., Nelson, E.A., Rossow, K.D., Bien, M., Arndt, B., Preszler, S., Steen, P., Christopher-Hennings, J., *et al.* (2004). Characterization of emerging European-like porcine reproductive and respiratory syndrome virus isolates in the United States. J. Virol. *78*, 3684-3703.
Rowland, R.R., Steffen, M., Ackerman, T., and Benfield, D.A. (1999). The evolution of porcine reproductive and respiratory syndrome virus: quasispecies and emergence of a virus subpopulation during infection of pigs with VR-2332. Virology *259*, 262-266.
Rowson, K.E., and Mahy, B.W. (1975). Lactic dehydrogenase virus. Virol. Monogr. 1-121.
Sawicki, S.G., and Sawicki, D.L. (1990). Coronavirus transcription: subgenomic mouse hepatitis virus replicative intermediates function in RNA synthesis. J. Virol. *64*, 1050-1056.
Sawicki, S.G., and Sawicki, D.L. (1995). Coronaviruses use discontinuous extension for synthesis of subgenome-length negative strands. Adv. Exp. Med. Biol. *380*, 499-506.
Sawicki, S.G., Sawicki, D.L., and Siddell, S.G. (2007). A contemporary view of coronavirus transcription. J. Virol. *81*, 20-29.
Sethna, P.B., Hung, S.L., and Brian, D.A. (1989). Coronavirus subgenomic minus-strand RNAs and the potential for mRNA replicons. Proc. Natl. Acad. Sci. USA *86*, 5626-5630.
Seybert, A., Posthuma, C.C., van Dinten, L.C., Snijder, E.J., Gorbalenya, A.E., and Ziebuhr, J. (2005). A complex zinc finger controls the enzymatic activities of nidovirus helicases. J. Virol. *79*, 696-704.
Seybert, A., van Dinten, L.C., Snijder, E.J., and Ziebuhr, J. (2000). Biochemical characterization of the equine arteritis virus helicase suggests a close functional relationship between arterivirus and coronavirus helicases. J. Virol. *74*, 9586-9593.
Shen, S., Kwang, J., Liu, W., and Liu, D.X. (2000). Determination of the complete nucleotide sequence of a vaccine strain of porcine reproductive and respiratory syndrome virus and identification of the Nsp2 gene with a unique insertion. Arch. Virol. *145*, 871-883.
Siddell, S.G., Ziebuhr, J., Snijder, E.J. (2005). Coronaviruses, toroviruses and arteriviruses, In Topley and Wilson's microbiology and microbial infections; Virology volume, B.W. Mahy, V. ter Meulen, eds. (London: Hodder Arnold), pp. 823-856.
Smith, S.L., Wang, X., and Godeny, E.K. (1997). Sequence of the 3' end of the simian hemorrhagic fever virus genome. Gene *191*, 205-210.

Snijder, E.J., Bredenbeek, P.J., Dobbe, J.C., Thiel, V., Ziebuhr, J., Poon, L.L., Guan, Y., Rozanov, M., Spaan, W.J., and Gorbalenya, A.E. (2003a). Unique and conserved features of genome and proteome of SARS-coronavirus, an early split-off from the coronavirus group 2 lineage. J. Mol. Biol. *331*, 991-1004.

Snijder, E.J., Dobbe, J.C., and Spaan, W.J. (2003b). Heterodimerization of the two major envelope proteins is essential for arterivirus infectivity. J. Virol. *77*, 97-104.

Snijder, E.J., Horzinek, M.C., and Spaan, W.J. (1993). The coronavirus-like superfamily. Adv. Exp. Med. Biol. *342*, 235-244.

Snijder, E.J., and Meulenberg, J.J. (1998). The molecular biology of arteriviruses. J. Gen. Virol. *79*, 961-979.

Snijder, E.J., Siddell, S.G., Gorbalenya, A.E. (2005). The Order Nidovirales, In Topley and Wilson's microbiology and microbial infections; Virology volume, W. Mahy, V. ter Meulen, eds. (London: Hodder Arnold), pp. 390-404.

Snijder, E.J., van Tol, H., Pedersen, K.W., Raamsman, M.J., and de Vries, A.A. (1999). Identification of a novel structural protein of arteriviruses. J. Virol. *73*, 6335-6345.

Snijder, E.J., van Tol, H., Roos, N., and Pedersen, K.W. (2001). Non-structural proteins 2 and 3 interact to modify host cell membranes during the formation of the arterivirus replication complex. J. Gen. Virol. *82*, 985-994.

Snijder, E.J., Wassenaar, A.L., and Spaan, W.J. (1994). Proteolytic processing of the replicase ORF1a protein of equine arteritis virus. J. Virol. *68*, 5755-5764.

Snijder, E.J., Wassenaar, A.L., van Dinten, L.C., Spaan, W.J., and Gorbalenya, A.E. (1996). The arterivirus nsp4 protease is the prototype of a novel group of chymotrypsin-like enzymes, the 3C-like serine proteases. J. Biol. Chem. *271*, 4864-4871.

Spaan, W., Delius, H., Skinner, M., Armstrong, J., Rottier, P., Smeekens, S., van der Zeijst, B.A., and Siddell, S.G. (1983). Coronavirus mRNA synthesis involves fusion of non-contiguous sequences. EMBO J. *2*, 1839-1844.

Stadejek, T., Bjorklund, H., Bascunana, C.R., Ciabatti, I.M., Scicluna, M.T., Amaddeo, D., McCollum, W.H., Autorino, G.L., Timoney, P.J., Paton, D.J., *et al.* (1999). Genetic diversity of equine arteritis virus. J. Gen. Virol. *80*, 691-699.

Stadejek, T., Mittelholzer, C., Oleksiewicz, M.B., Paweska, J., and Belak, S. (2006a). Highly diverse type of equine arteritis virus (EAV) from the semen of a South African donkey: short communication. Acta. Vet. Hung. *54*, 263-270.

Stadejek, T., Oleksiewicz, M.B., Potapchuk, D., and Podgorska, K. (2006b). Porcine reproductive and respiratory syndrome virus strains of exceptional diversity in eastern Europe support the definition of new genetic subtypes. J. Gen. Virol. *87*, 1835-1841.
Stadejek, T., Stankevicius, A., Storgaard, T., Oleksiewicz, M.B., Belak, S., Drew, T.W., and Pejsak, Z. (2002). Identification of radically different variants of porcine reproductive and respiratory syndrome virus in Eastern Europe: towards a common ancestor for European and American viruses. J. Gen. Virol. *83*, 1861-1873.
Stueckemann, J.A., Holth, M., Swart, W.J., Kowalchyk, K., Smith, M.S., Wolstenholme, A.J., Cafruny, W.A., and Plagemann, P.G. (1982). Replication of lactate dehydrogenase-elevating virus in macrophages. 2. Mechanism of persistent infection in mice and cell culture. J. Gen. Virol. *59*, 263-272.
Suarez, P., Zardoya, R., Martin, M.J., Prieto, C., Dopazo, J., Solana, A., and Castro, J.M. (1996). Phylogenetic relationships of European strains of porcine reproductive and respiratory syndrome virus (PRRSV) inferred from DNA sequences of putative ORF-5 and ORF-7 genes. Virus Res. *42*, 159-165.
Tan, C., Chang, L., Shen, S., Liu, D.X., and Kwang, J. (2001). Comparison of the 5' leader sequences of North American isolates of reference and field strains of porcine reproductive and respiratory syndrome virus (PRRSV). Virus Genes *22*, 209-217.
Tauraso, N.M., Shelokov, A., Palmer, A.E., and Allen, A.M. (1968). Simian hemorrhagic fever. 3. Isolation and characterization of a viral agent. Am. J. Trop. Med. Hyg. *17*, 422-431.
Tijms, M.A., van der Meer, Y., and Snijder, E.J. (2002). Nuclear localization of non-structural protein 1 and nucleocapsid protein of equine arteritis virus. J. Gen. Virol. *83*, 795-800.
Tijms, M.A., van Dinten, L.C., Gorbalenya, A.E., and Snijder, E.J. (2001). A zinc finger-containing papain-like protease couples subgenomic mRNA synthesis to genome translation in a positive-stranded RNA virus. Proc. Natl. Acad. Sci. USA *98*, 1889-1894.
Timoney, P.J. (1984). Clinical, virological, and epidemiological features of the 1984 outbreak of equine viral arteritis in the Thoroughbred population in Kentucky, USA, Paper presented at: Proceedings of the Grayson Foundation International Conference of Thoroughbred Breeders Organizations (Dromoland Castle, Ireland), pp. 24-33.
Timoney, P.J., and McCollum, W.H. (1993). Equine viral arteritis. Vet. Clin. North Am. Equine Pract. *9*, 295-309.
van Aken, D., Benckhuijsen, W.E., Drijfhout, J.W., Wassenaar, A.L., Gorbalenya, A.E., and Snijder, E.J. (2006a). Expression,

purification, and in vitro activity of an arterivirus main proteinase. Virus Res. *120*, 97-106.

van Aken, D., Zevenhoven-Dobbe, J., Gorbalenya, A.E., and Snijder, E.J. (2006b). Proteolytic maturation of replicase polyprotein pp1a by the nsp4 main proteinase is essential for equine arteritis virus replication and includes internal cleavage of nsp7. J. Gen. Virol. *87*, 3473-3482.

Van Den Born, E., Gultyaev, A.P., and Snijder, E.J. (2004). Secondary structure and function of the 5'-proximal region of the equine arteritis virus RNA genome. RNA *10*, 424-437.

van den Broek, M.F., Sporri, R., Even, C., Plagemann, P.G., Hanseler, E., Hengartner, H., and Zinkernagel, R.M. (1997). Lactate dehydrogenase-elevating virus (LDV): lifelong coexistence of virus and LDV-specific immunity. J. Immunol. *159*, 1585-1588.

van der Meer, Y., van Tol, H., Locker, J.K., and Snijder, E.J. (1998). ORF1a-encoded replicase subunits are involved in the membrane association of the arterivirus replication complex. J. Virol. *72*, 6689-6698.

van Dinten, L.C., den Boon, J.A., Wassenaar, A.L., Spaan, W.J., and Snijder, E.J. (1997). An infectious arterivirus cDNA clone: identification of a replicase point mutation that abolishes discontinuous mRNA transcription. Proc. Natl. Acad. Sci. USA *94*, 991-996.

van Dinten, L.C., van Tol, H., Gorbalenya, A.E., and Snijder, E.J. (2000). The predicted metal-binding region of the arterivirus helicase protein is involved in subgenomic mRNA synthesis, genome replication, and virion biogenesis. J. Virol. *74*, 5213-5223.

van Marle, G., Dobbe, J.C., Gultyaev, A.P., Luytjes, W., Spaan, W.J., and Snijder, E.J. (1999). Arterivirus discontinuous mRNA transcription is guided by base pairing between sense and antisense transcription-regulating sequences. Proc. Natl. Acad. Sci. USA *96*, 12056-12061.

van Vugt, J.J., Storgaard, T., Oleksiewicz, M.B., and Botner, A. (2001). High frequency RNA recombination in porcine reproductive and respiratory syndrome virus occurs preferentially between parental sequences with high similarity. J. Gen. Virol. *82*, 2615-2620.

Vanderheijden, N., Delputte, P.L., Favoreel, H.W., Vandekerckhove, J., Van Damme, J., van Woensel, P.A., and Nauwynck, H.J. (2003). Involvement of sialoadhesin in entry of porcine reproductive and respiratory syndrome virus into porcine alveolar macrophages. J. Virol. *77*, 8207-8215.

Verheije, M.H., Olsthoorn, R.C., Kroese, M.V., Rottier, P.J., and Meulenberg, J.J. (2002a). Kissing interaction between 3' noncoding and coding sequences is essential for porcine arterivirus RNA replication. J. Virol. *76*, 1521-1526.

Verheije, M.H., Welting, T.J., Jansen, H.T., Rottier, P.J., and Meulenberg, J.J. (2002b). Chimeric arteriviruses generated by swapping of the M protein ectodomain rule out a role of this domain in viral targeting. Virology *303*, 364-373.

Weiland, E., Bolz, S., Weiland, F., Herbst, W., Raamsman, M.J., Rottier, P.J., and De Vries, A.A. (2000). Monoclonal antibodies directed against conserved epitopes on the nucleocapsid protein and the major envelope glycoprotein of equine arteritis virus. J. Clin. Microbiol. *38*, 2065-2075.

Weiland, E., Wieczorek-Krohmer, M., Kohl, D., Conzelmann, K.K., and Weiland, F. (1999). Monoclonal antibodies to the GP5 of porcine reproductive and respiratory syndrome virus are more effective in virus neutralization than monoclonal antibodies to the GP4. Vet. Microbiol. *66*, 171-186.

Wensvoort, G., Terpstra, C., Pol, J.M., ter Laak, E.A., Bloemraad, M., de Kluyver, E.P., Kragten, C., van Buiten, L., den Besten, A., Wagenaar, F., and et al. (1991). Mystery swine disease in The Netherlands: the isolation of Lelystad virus. Vet. Q. *13*, 121-130.

Westaway, E.G., Brinton, M.A., Gaidamovich, S., Horzinek, M.C., Igarashi, A., Kaariainen, L., Lvov, D.K., Porterfield, J.S., Russell, P.K., and Trent, D.W. (1985). Togaviridae. Intervirology *24*, 125-139.

Wieringa, R., De Vries, A.A., Post, S.M., and Rottier, P.J. (2003a). Intra- and intermolecular disulfide bonds of the GP2b glycoprotein of equine arteritis virus: relevance for virus assembly and infectivity. J. Virol. *77*, 12996-13004.

Wieringa, R., de Vries, A.A., Raamsman, M.J., and Rottier, P.J. (2002). Characterization of two new structural glycoproteins, GP(3) and GP(4), of equine arteritis virus. J. Virol. *76*, 10829-10840.

Wieringa, R., de Vries, A.A., and Rottier, P.J. (2003b). Formation of disulfide-linked complexes between the three minor envelope glycoproteins (GP2b, GP3, and GP4) of equine arteritis virus. J. Virol. *77*, 6216-6226.

Wieringa, R., de Vries, A.A., van der Meulen, J., Godeke, G.J., Onderwater, J.J., van Tol, H., Koerten, H.K., Mommaas, A.M., Snijder, E.J., and Rottier, P.J. (2004). Structural protein requirements in equine arteritis virus assembly. J. Virol. *78*, 13019-13027.

Wissink, E.H., Kroese, M.V., van Wijk, H.A., Rijsewijk, F.A., Meulenberg, J.J., and Rottier, P.J. (2005). Envelope protein requirements for the assembly of infectious virions of porcine reproductive and respiratory syndrome virus. J. Virol. *79*, 12495-12506.

Wood, O., Tauraso, N., and Liebhaber, H. (1970). Electron microscopic study of tissue cultures infected with simian haemorrhagic fever virus. J. Gen. Virol. *7*, 129-136.

Wootton, S., Yoo, D., and Rogan, D. (2000). Full-length sequence of a Canadian porcine reproductive and respiratory syndrome virus (PRRSV) isolate. Arch. Virol. *145*, 2297-2323.

Yuan, S., Nelsen, C.J., Murtaugh, M.P., Schmitt, B.J., and Faaberg, K.S. (1999). Recombination between North American strains of porcine reproductive and respiratory syndrome virus. Virus Res. *61*, 87-98.

Zeman, D., Neiger, R., Yaeger, M., Nelson, E., Benfield, D., Leslie-Steen, P., Thomson, J., Miskimins, D., Daly, R., and Minehart, M. (1993). Laboratory investigation of PRRS virus infection in three swine herds. J. Vet. Diagn. Invest. *5*, 522-528.

Zeng, L., Godeny, E.K., Methven, S.L., and Brinton, M.A. (1995). Analysis of simian hemorrhagic fever virus (SHFV) subgenomic RNAs, junction sequences, and 5' leader. Virology *207*, 543-548.

Zevenhoven-Dobbe, J.C., Greve, S., van Tol, H., Spaan, W.J., and Snijder, E.J. (2004). Rescue of disabled infectious single-cycle (DISC) equine arteritis virus by using complementing cell lines that express minor structural glycoproteins. J. Gen. Virol. *85*, 3709-3714.

Ziebuhr, J., Snijder, E.J., and Gorbalenya, A.E. (2000). Virus-encoded proteinases and proteolytic processing in the Nidovirales. J. Gen. Virol. *81*, 853-879.

Zimmerman J., O.F., Benfield D., Murtaugh M., Stevenson G., Torremorell M. (2006). PRRS virus (porcine arterivirus), In Diseases of Swine, Z.J. Straw, S. D'Allaire, D.J. Taylor, eds. (Ames, Iowa: Blackwell Publishing Company), pp. 387-417.

Chapter 4

Coronavirus Replication and Interaction with Host

Luis Enjuanes, Isabel Sola, Sonia Zúñiga and Fernando Almazán

ABSTRACT

Coronavirus (CoV) genome replication takes place in the cytoplasm in a membrane-protected microenvironment, and starts with the translation of the genome to produce the viral replicase. CoV transcription involves a discontinuous RNA synthesis (template switch) during the extension of a negative copy of the subgenomic mRNAs. The requirement for basepairing during transcription has been formally demonstrated in arteriviruses and CoVs. CoV N protein is required for coronavirus RNA synthesis, and has RNA chaperone activity that may be involved in template switch. Both viral and cellular proteins are required for replication and transcription, and the role of selected proteins will be addressed. CoVs initiate translation by cap-dependent and cap-independent mechanisms.

Cell macromolecular synthesis may be controlled after CoV infection by locating some virus proteins in the host cell nucleus. Infection by different coronaviruses cause in the host alteration in the transcription and translation patterns, in the cell cycle, the cytoskeleton, apoptosis and coagulation pathways, inflammation, and immune and stress responses. The balance between the genes up- and down-regulated could explain the pathogenesis caused by these viruses. Specific aspects of CoV-host interaction are reviewed in this chapter.

Coronavirus expression systems based on single genome vectors constructed by targeted recombination, or by using infectious cDNAs, have been developed. The possibility of expressing different genes under the control of transcription regulating sequences (TRSs) with programmable strength, and engineering tissue and species tropism indicates that CoV vectors are very flexible. CoV based vectors have emerged with high potential for vaccine development and, possibly, for gene therapy.

INTRODUCTION

RNA virus genomes vary between 3 and 32 kb, whereas DNA genome varies as much as three orders of magnitude (Gorbalenya et al., 2006). There are four classes of RNA viruses, depending on whether they are double or single stranded and, in this case, whether they have positive or negative polarity, plus the retroviruses. Among these viruses, only the RNA viruses with single-stranded genomes of positive (mRNA) polarity (ssRNA+) have a genome in the upper range of the RNA genomes. Within the *Nidovirales* order, there are three virus families, the *Coronaviridae* comprising the coronaviruses and toroviruses, plus the *Roniviridae* and *Arteriviridae*. The two first virus families have genome sizes of 26-32 kb. Most of our knowledge on Nidovirales is derived from the coronaviruses and arteriviruses (Enjuanes, 2005; Enjuanes et al., 2006; Masters, 2006; Pasternak, Spaan, and Snijder, 2006; Siddell, Ziebuhr, and Snijder, 2005; Ziebuhr, 2005).

The large progress on coronaviruses has been possible by the development of technologies for reverse genetics. Reverse genetics for CoVs was first achieved by targeted recombination (Masters, 1999). Soon after, the first CoV infectious cDNA clones were constructed for transmissible gastroenteritis CoV (TGEV) (Almazán et al., 2000; Yount, Curtis, and Baric, 2000). These achievements were followed by the development of infectious cDNA clones for human coronavirus (HCoV) 229E (Thiel et al., 2001a) and HCoV-OC43 (St-Jean et al., 2006), severe and acute respiratory syndrome coronavirus (SARS-CoV) (Almazan et al.,

2006; Yount et al., 2003), mouse hepatitis virus (MHV) (Yount et al., 2002), and for avian coronavirus (Casais et al., 2001; Youn, Leibowitz, and Collisson, 2005).

Coronavirus infection starts by binding to a receptor on the cell surface, after which fusion of the viral and cellular membrane is mediated by one of the major surface glycoproteins. This fusion event (either at the plasma or endosomal membrane) releases the nucleocapsid into the host cell cytoplasm. After genome uncoating, translation of two replicase open reading frames (ORFs) is initiated by host ribosomes, to yield large polyprotein precursors that undergo autoproteolysis to eventually produce a membrane-bound replicase complex. This complex mediates the synthesis of the genome RNA and a nested set of subgenomic RNAs. New virus particles are assembled by the association of the new genomes with the cytoplasmic nucleocapsid protein and the subsequent envelopment of the nucleocapsid structure. The newly formed virions then leave the cell by following the exocytic pathway towards the plasma membrane.

In this chapter, we will discuss the molecular mechanism of coronavirus replication, transcription, and translation, the effect of coronavirus infection on host cells, and the use of coronaviruses as vectors for vaccine development. Special attention will be provided to the technical approaches applied in these studies.

PATHOGENESIS INDUCED BY CORONA-VIRUSES

CoVs are classified in three groups attending to genetic analysis (Gonzalez et al., 2003; Spaan et al., 2005). CoVs are associated mainly with respiratory, enteric, hepatic and central nervous system diseases. In humans and fowl, CoVs primarily cause upper respiratory tract infections, while porcine and bovine CoVs establish enteric infections that result in severe economic losses (USDA 2002). HCoVs have also been associated with infant necrotizing enterocolitis (Resta et al., 1985) and are tentative candidates for multiple sclerosis (Denison, 1999). Coronaviruses have recently shown high impact on human population due to the apparition of the virus causing the severe acute respiratory syndrome (SARS) responsible of around 800 deads (Drosten et al., 2003; Marra et al., 2003; Rota et al., 2003).

Group 1 CoV includes members infecting human, porcine, canine and feline species, closely related in sequence and, in some cases, also antigenically (Sanchez et al., 1990). In group 1 CoVs, epithelial cells, and widely distributed cells such as macrophages are infected. This group includes HCoVs-229E and NL-63, causing mild respiratory infections and bronchiolitis, respectively (Denison, 1999; van der Hoek et al., 2004). In addtition, it includes TGEV, an enteropathogenic CoV that replicates in both villus epithelial cells of the small intestine and in lung cells, a non-enteropathogenic virus related to TGEV, the porcine respiratory CoV (PRCV) (Callebaut et al., 1988; Pensaert, Callebaut, and Vergote, 1986), that replicates to high titers in the respiratory tract (Cox, Hooyberghs, and Pensaert, 1990; Cox et al., 1990), and the porcine epidemic diarrhea virus (PEDV) (Pensaert and De Bouck, 1978). This virus is closely related to TGEV in sequence, but antigenically distinct (Sanchez et al., 1990), and infects the enteric tract of swine but not the lungs. Canine coronavirus (CCoV) usually produces a mild gastroenteritis, although some virus strains cause a more severe and sometimes fatal diarrhea. Feline coronavirus (FCoV) causes a disease involving an antibody-dependent enhancement of infection and immuno-complex induced lesions (Olsen, 1993).

Group 2 CoVs, like murine CoVs, display different tropisms and levels of virulence. The commonly used laboratory strains infect primarily the liver and the brain and thus provide animal models for hepatitis and encephalitis as well as the immune-mediated demyelinating disease that develops late after infection (Perlman, 1998). SARS-CoV has been classified in group 2b CoV (Gorbalenya et al., 2006; Gorbalenya, Snijder, and Spaan, 2004; Lau et al., 2005; Poon et al., 2005; Woo et al., 2006) and causes severe acute respiratory disease, pneumonia, diarrhea, and sometimes death (Peiris et al., 2003). In addition, a new group 2 human CoV causing pneumonia, HKU-1, has been identified (van der Hoek et al., 2004; Woo et al., 2005). Group 3 CoVs, such as infectious bronchitis virus (IBV), cause respiratory disease in chickens infecting the ciliated epithelia of the nose and trachea. In addition, it may infect the alimentary tract, kidneys, and the oviduct, which may contribute to diminish egg production (Cavanagh, 2003).

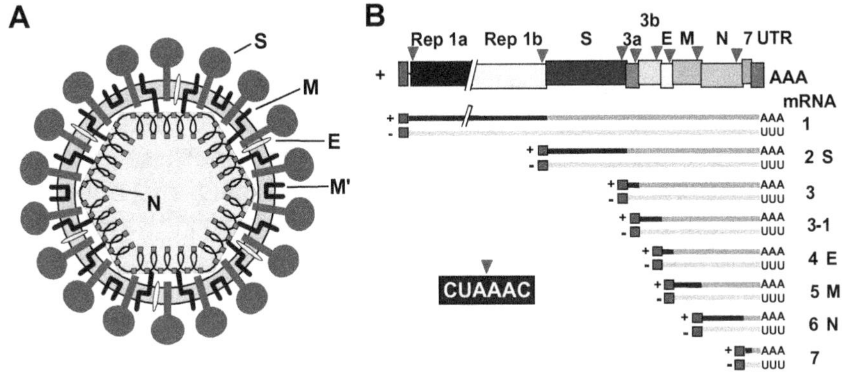

Fig. 1. TGEV structure and genome organization. A. Diagram of CoV structure using TGEV as a prototype. The scheme shows the envelope, the core and the nucleocapsid structure. S, spike protein; M and M', large membrane proteins with the amino-terminus facing the external surface of the virion and the carboxy-terminus towards the inside or the outside face of the virion, respectively; E, small envelope protein; N, nucleoprotein. B. Representation of a prototype TGEV genome and subgenomic RNAs. Beneath the top bar a set of positive and negative-sense mRNA species synthesized in infected cells is shown. Dark and semi-dark thin lines (+), mRNA sequences translated and non-translated into viral proteins, respectively. Light lines (-), RNAs complementary to the different mRNAs. CUAAAC, TGEV core sequence (CS) of the transcription-regulating sequences (TRS). Poly(A) and Poly(U) tails are indicated by AAA or UUU, respectively. Rep 1a and Rep 1b, replicase genes and proteins. 3a, 3b, and 7, non-structural proteins. Other acronyms as in A. Numbers and letters indicate viral genes. Inverted red triangles, CS of TGEV TRSs preceding each gene.

CORONAVIRUS STRUCTURE AND GENOME EXPRESSION

CoVs have the largest known RNA genomes (from 26 to 32 kb) and contain a variable number of functional genes. CoV genes are arranged in the order 5'-Rep1a-Rep1b-S-E-M-N-3' in all family members. All CoVs encode at least four structural proteins, the spike (S), envelope (E), membrane (M), and nucleoprotein (N) (Fig. 1A). In addition, some CoVs, such as MHV, may have another structural protein, hemagglutinin esterase (HE) (Brian, Hogue, and Kienzle, 1995; Kazi et al., 2005). Recently, it has been described that SARS-CoV has at least three additional structural proteins in the envelope (3a, 7a, and 7b) (Huang et al., 2006; Ito et al., 2005; Schaecher, Mackenzie, and Pekosz, 2006).

CoVs are enveloped viruses containing an internal core (Fig. 1A), formed by the genomic RNA coated with the N protein. This nucleoprotein interacts with the M protein carboxy-terminus. M protein is mostly embedded in the virus membrane (Escors et al., 2001). The S protein is responsible for attachment and entry into cells and the major inducer of CoV neutralizing antibodies (Cavanagh, 2003; He et al., 2006; Suñe et al., 1990).

CoVs RNA genomes have positive polarity and use two strategies for protein synthesis. According to one, the synthesis of polyproteins including subunits of the replicative complex, such as the RNA-dependent RNA polymerase (RdRp) and helicase takes place. A second strategy is based on the synthesis of a nested set of subgenomic mRNAs of different size, each encoding a protein. The replicase gene expands two thirds of the genome (about 20 kb) and includes two open reading frames (ORFs) 1a and 1b (Fig. 1B). ORF1a encodes a polyprotein of 450 to 500 kDa, whereas ORF1ab encodes a polyprotein of 750-800 kDa using a –1 ribosomal frameshift signal that directs the translation of ORF1b in a subset of the molecules that were started at the beginning of ORF1a. The subunits of the replicase machinery are produced by proteolytic processing. ORF 1a encodes the 3C-like protease (3Clpro) resembling that of picornavirus 3C proteinases, that is responsible for the processing of the downstream, most conserved part of the replicase polyproteins and is also called main proteinase (Mpro) (Ziebuhr, Snijder, and Gorbalenya, 2000). The CoV replicase also includes a series of domains with enzymatic activities such as adenosine diphosphate-ribose 1''-phosphatase (ADRP), a large RdRp, a nucleoside triphosphate (NTP)-binding 5'-to-3'-helicase domain (HEL), a 3'-to-5' exonuclease, a poly(U)-specific endoribonuclease (NendoU), and a S-adenosyl-methionine-dependent ribose 2'-*O*-methyltransferase (MTR) (Gorbalenya et al., 2006). The acquisition of some of the RNA-processing enzymes may have improved the low fidelity of RNA replication to allow genome expansion and give rise to the CoVs.

CORONAVIRUS REPLICATION

Like all other positive-stranded RNA viruses, CoV genome replication is mediated through the continuous synthesis of a full-length negative-strand RNA, which in turn is the template for the synthesis of progeny virus genomes. This process is mediated by the viral replication complex that apparently includes cellular proteins and up to 16 non-structural viral proteins (nsps) derived from the proteolytic processing of the replicase polyproteins pp1a and pp1ab (Snijder et al., 2005; Ziebuhr, 2005). The replication complex is associated with modified intracellular membranes, which may be important to create a microenvironment suitable for viral RNA synthesis, recruitment of host proteins, and possibly evasion of host defense mechanisms that may be induced by double-stranded RNA intermediates generated during the RNA virus replication. Electron microscopy studies of SARS-CoV and MHV-infected cells have shown that these membrane structures consist of double-membrane vesicles (DMVs) (Gosert et al., 2002) (Snijder et al., 2006). The cellular origin of CoV induced DMVs is under debate and different intracellular compartments, including the Golgi, late endosomal membranes, autophagosomes, and the endoplasmic reticulum have been implicated in DMVs formation (Prentice et al., 2004a; Prentice et al., 2004b; Shi et al., 1999; Snijder et al., 2006; van der Meer et al., 1999). Structures similar to DMVs have been found in the equine arteritis virus (EAV) infection (Pedersen et al., 1999), suggesting that CoV and arterivirus have a common replication strategy.

In addition to the nsps encoded by the replicase gene, the CoV structural N protein has been implicated in viral RNA synthesis by several experimental evidences. Based on the specific binding of N protein to the leader sequence, it has been proposed that N protein could participate in transcription (Baric et al., 1988; Stohlman et al., 1988). At early times after infection, CoV N protein colocalizes with the replication complex in DMVs, the site of CoV RNA synthesis (Bost, Prentice, and Denison, 2001). N protein has also been found to enhance the rescue of infectious virus from cDNA clones generated from different CoVs, such as IBV (Casais et al., 2001), HCoV-229E (Schelle et al., 2005; Thiel et al., 2003), and TGEV using RNA in vitro transcripts (Yount, Curtis, and Baric, 2000). Finally, using TGEV and HCoV-229E derived replicons two groups have shown that only background levels of CoV RNA synthesis are produced in the absence of N protein (Almazan, Galan, and Enjuanes, 2004; Schelle et al., 2005). In these systems, the presence either *in cis* or *in trans* of N protein is required for efficient CoV RNA synthesis (Almazan, Galan, and

Enjuanes, 2004) (Fig. 2). A quantitative analysis of TGEV replicon activity showed an increase of more than 100-fold when N protein was provided in cis, and more than 1000-fold when N protein was provided in trans (Almazan, Galan, and Enjuanes, 2004). Whether the effect of N protein is at the level of replication, transcription or both, remains to be determined. Despite the variable size of N protein from different CoVs, a conserved three-domain organization has been proposed (Parker and Masters, 1990). N protein domain II is more conserved than domains I and III, especially in a serine-rich region that is probably important for the biological function of N protein. Recently, a modular organization common for all CoV N proteins, based on protein disorder, has been proposed (Chang et al., 2006; Zuñiga et al., 2007).

Several active N protein domains have been mapped, such as a RNA binding domain (Laude and Masters, 1995; Nelson, Stohlman, and Tahara, 2000), the oligomerization binding domain (He et al., 2004a; Surjit et al., 2004b), and the M protein binding domain (He et al., 2004b; Hurst et al., 2005), also involved in the N protein oligomerization. N protein activity should be a consequence of its self-interaction and interaction with other viral and cellular proteins. Furthermore, N protein has been described as a phosphoprotein. Although the 15% of the amino acids of CoV N proteins are candidate phosphoacceptor serines and threonines, so far only four phosphorylated serine residues have been identified in the N protein of TGEV (Calvo et al., 2005) and IBV (Chen et al., 2005) by mass spectrometry. SARS-CoV phosphorylation has also been demonstrated, but the precise amino acid residues involved have not been mapped (Surjit et al., 2005). Phosphorylation could cause conformational changes in N protein structure (Chen et al., 2005; Mohandas and Dales, 1991; Stohlman et al., 1983) by the introduction of negative charges in a basic environment affecting N protein RNA binding activity. In fact, the N protein phosphorylation state has been proposed to regulate N protein functions.

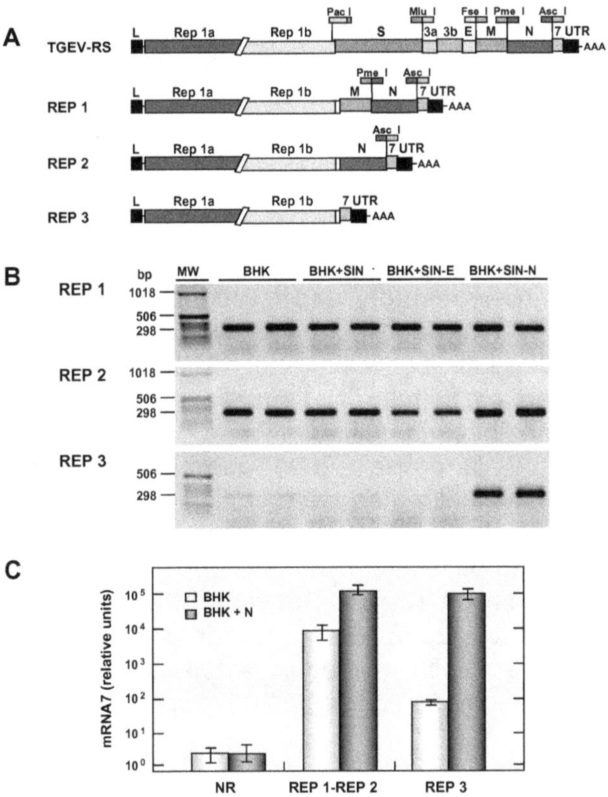

Fig. 2. Effect of N protein on TGEV-derived replicon activity. (A) Genetic structure of TGEV derived replicons. The genetic structure of the TGEV cDNA clone (TGEV-RS) and the replicons generated from this cDNA (REP 1, REP 2, and REP 3) are illustrated. Letters and numbers above the bars indicate the viral genes. L, leader sequence; UTR, untranslated region. Relevant restriction sites are indicated. (B) Functional analysis of TGEV replicons in the presence or absence of N protein. BHK-pAPN (BHK), or BHK-pAPN cells transformed with either the Sindbis virus replicon pSINrep21 alone (BHK + SIN$^-$), or expressing TGEV E protein (BHK + SIN-E), or TGEV N protein (BHK + SIN-N), were transfected with TGEV replicons (REP 1, 2, and 3) using Lipofectamine 2000 (Invitrogen) following the manufacturer's specifications. At 36 hpt total intracellular RNA was extracted and the replicon activity was determined by RT-PCR analysis with specific primers to detect gene 7 mRNA. Duplicate RT-PCR products amplified in parallel were resolved by electrophoresis in 1% agarose gels. MW, molecular weight markers. (C) Real-time RT-PCR quantification of gene 7 mRNA. The amount of mRNA7, expressed as relative units, was determined by real-time RT-PCR with specific oligonucleotides in RNA samples isolated at 36 hpt from BHK-pAPN (BHK) or BHK-pAPN cells expressing N protein (BHK + N) transfected with either a non-replicative cDNA clone (NR), REP 1, REP 2 or REP 3. REP 1-REP 2 indicates mean values from REP 1 and 2, both encoding N gene. Mean values from three experiments are represented, with standard deviations shown as error bars.

In addition to the role of N protein in virus assembly (Escors et al., 2001; Risco et al., 1998) and CoV RNA synthesis, a number of potential activities have been postulated for N protein. It was shown that in addition to its presence in the cytoplasm, CoV N protein has also been localized in the cell nucleus. Up to date, the N protein from CoV group 1 (TGEV), 2 (MHV), and 3 (IBV) (Hiscox et al., 2001; Timani et al., 2005; Wurm et al., 2001; Yan et al., 2003; Yount, Curtis, and Baric, 2000), and also N protein from the arteriviruses porcine respiratory and reproductive syndrome virus (PRRSV) and EAV (Rowland et al., 1999; Tijms, van der Meer, and Snijder, 2002) localize within the nucleolus. However, the association of N protein with the nucleus may be cell dependent as TGEV N protein has been identified in the nucleus of LLC-PK1 and Vero infected cells (Wurm et al., 2001), but not in ST cells (Calvo et al., 2005). The nucleolus has been implicated in many aspects of cell biology that include functions such as ribosomal rRNA synthesis and ribosome biogenesis, gene silencing, senescence, and cell cycle regulation (Carmo-Fonseca, Mendes-Soares, and Campos, 2000; Hiscox, 2002; Olson, Dundr, and Szebeni, 2000). Viruses may interact with nucleolar antigens leading to their redistribution during infection (Hiscox, 2002) and preparing the cell from viral infection.

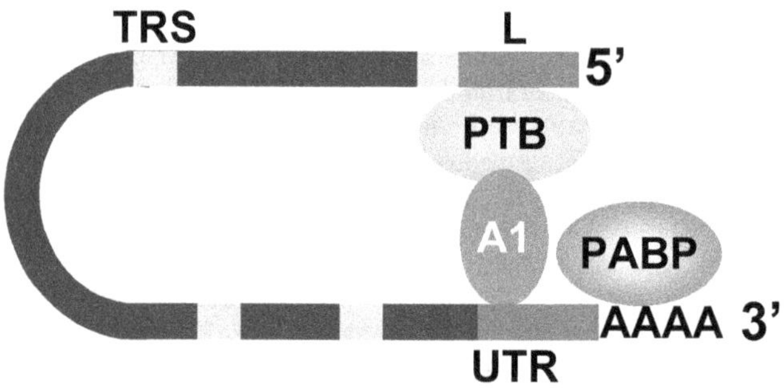

Fig. 3. Model representing the postulated interaction between the 5' and 3' ends of CoV genome. The 5'-3' cross-talk mediated by hnRNP A1 (A1), PTB, and PABP is illustrated. L, leader sequence; TRS, transcription regulating sequence; UTR, untranslate region.

CoV replication employs complex mechanisms that involve both, viral and cellular proteins. A number of host factors have been proposed to participate in CoV RNA synthesis on the basis of their ability to bind in vitro to RNA segments of functional importance. Heterogeneous nuclear ribonucleoprotein (hnRNP) A1 has been identified as a major protein species strongly binding to nt 90-170 and weakly to nt 260-350 from the 3' end of MHV RNA. These binding sites are complementary to the sites on the negative-strand RNA that bind another cellular protein, polypyrimidine tract-binding protein (PTB). Furthermore, hnRNA A1 and PTB, also bind to the complementary strands at the 5' end of MHV RNA and together mediate the in vitro formation of an RNP complex involving the 5' and 3' ends of MHV RNA (Huang and Lai, 2001; Li et al., 1999; Li et al., 1997; Zhang and Lai, 1995) (Fig. 3). The functional relevance of hnRNP A1 in MHV replication has been established by experiments showing that over-expression of hnRNP A1 facilitates MHV replication, whereas dominant-negative mutants of hnRNP A1 reduce replication (Shi et al., 2000). Furthermore, mutations of the hnRNP A1 and PTB binding regions also impaired transcription of subgenomic RNA (Huang and Lai, 1999), suggesting a functional role in replication and transcription for the interaction between hnRNP A1 and PTB. Finally, there are experimental evidences indicating that the poly (A) binding protein (PABP) specifically interacts with the poly (A) of CoV genomes, and that this interaction may affect CoV replication (Huang and Lai, 2001; Liu, Yu, and Leibowitz, 1997; Spagnolo and Hogue, 2000; Yu and Leibowitz, 1995a; Yu and Leibowitz, 1995b).

Although several experimental evidences supports the importance of viral and cellular proteins in CoV replication, our understanding of the role of these proteins is still limited. The recent development of CoV reverse genetic systems enabling site-directed mutagenesis of any nucleotide of the genome (Almazan et al., 2006; Almazán et al., 2000; Casais et al., 2001; Thiel et al., 2001a; Yount, Curtis, and Baric, 2000; Yount et al., 2003; Yount et al., 2002) is expected to greatly increase the understanding of CoV replication, as well as the functional importance of viral and cellular proteins. The recent advances in the RNA interference technology will be a valuable tool in understanding the functional relevance of these cellular proteins. However, the validation of host factors that are critical or essential to normal cellular functions will require the establishment of an *in vitro* replication system, in which reconstitution of CoV RNA synthesis could be achieved from isolated components.

CoV TRANSCRIPTION

CoV transcription, and in general transcription in the Nidovirales order, is an RNA-dependent RNA synthesis including a discontinuous step during the production of subgenomic mRNAs (sgmRNAs) (Lai and Cavanagh, 1997; Sawicki and Sawicki, 1998; Sawicki, Sawicki, and Siddell, 2007). This discontinuous transcription mechanism implies the fusion of the common 5'-terminal leader sequence to the 5' end of the mRNA coding sequence (body), generating a nested set of subgenomic mRNAs that are 5'- and 3'-coterminal with the virus genome. Sequences preceding each gene represent signals for discontinuous transcription of sgmRNAs. These transcription-regulating sequences (TRS) include a conserved core sequence (CS), identical in all genes (CS-B), and the relatively variable 5' and 3' flanking sequences (5' TRS and 3' TRS, respectively) that regulate transcription (Alonso et al., 2002). As the CS sequence is also found at the 3' end of the leader sequence (CS-L), it could basepair with the nascent minus strand complementary to each CS-B (cCS-B). In fact, the requirement for basepairing during transcription has been formally demonstrated in CoVs (Zuñiga et al., 2004) and arteriviruses (Pasternak et al., 2001; van Marle et al., 1999) by experiments in which basepairing between CS-L and the complement of CS-B was engineered in infectious genomic cDNAs. The most frequently used CS of group 1 CoVs (hexamer 5'-CUAAAC-3') and group 2 (heptamer 5'-UCUAAAC-3') share homology, whereas the CS of CoVs belonging to group 3, like that of IBV, has the most divergent sequence (5'-CUUAACAA-3'). Also, arterivirus CSs have a sequence (5'-UCAACU-3') that partially resembles that of IBV.

The constrution of infectious cDNA clones (see above) has been instrumental to study the mechanism of CoV transcription. Therefore, in general it is recommended assembling these cDNA clones before starting the study of the mechanism of virus transcription.

Current experimental data are compatible with a model of discontinuous transcription during negative-strand synthesis (Sawicki, Wang, and Sawicki, 2001; Sawicki and Sawicki, 1990; Sawicki and Sawicki, 1998; Sawicki, Sawicki, and Siddell, 2007). This concept was reinforced by demonstrating in CoVs and arteriviruses that the CS included in the sgmRNA was derived from the CS preceding each gene and not from the CS present at the 3' end of the leader sequence (Pasternak et al., 2001; van Marle et al., 1999; Zuñiga et al., 2004). In both viral systems, the equine arteritis virus (EAV) arterivirus (Pasternak et al., 2001; Pasternak et al., 2003; van Marle et al., 1999) and in TGEV (Zuñiga et al., 2004), it has been demonstrated that transcription depends on the duplex formation between leader CS and the complement of body CS at the 3' end of nascent (-) RNA and, more precisely, to the free energy (ΔG) associated to the formation of the RNA duplex. Sequences immediately flanking the CS in the TRS also decisively contribute to the leader-body RNA duplex formation (Pasternak et al., 2000; Pasternak et al., 2003; Sola et al., 2005). In fact, if the ΔG associated to the formation of this duplex reaches a minimum threshold, template switching would take place leading to the synthesis of sgmRNA. The working model proposed for CoV transcription includes three main steps (Fig. 4): (i) Formation of 5'-3' complexes in the genomic RNA to approximate the leader region to the 3' end of the genome, where the synthesis of negative RNA starts, (ii) base pairing scanning of the (-) nascent RNA by the TRS-L to find sequence domains with a favorable ΔG and, (iii) template switch to the leader TRS. In this model, the TRS-B would act as an attenuation and dissociation signal for the transcription complex during synthesis of the RNA negative strand. Template-switch at the sites of RdRp pausing resembles a high-frequency similarity-assisted copy-choice RNA recombination (Brian and Spaan, 1997; Nagy and Simon, 1997; Pasternak et al., 2000) in which the noncontiguous TRS-B and TRS-L sequences are probably brought into physical proximity by RNA-protein and protein-protein interactions. The nascent negative RNA with the TRS complement at its 3' end (cTRS-B) (donor molecule) would dissociate from the gRNA template at the CS domain to join the TRS-L (acceptor) and resume synthesis of subgenomic RNA. This sgRNA would serve as a template for multiple rounds of sgmRNA synthesis (Sola et al., 2005).

An extensive mutagenesis analysis of the TRS sequences in the TGEV infectious genomic cDNA has confirmed that complementarity between TRS-L and cTRS-B is a crucial regulating factor in sgRNA synthesis (Sola et al., 2005; Zuñiga et al., 2004). The proximity of body TRS to the 3' end of the genome might also regulate sgRNA synthesis. This is the case for all viruses that produce multiple sgRNAs (Miller and Koev, 2000) and, in general, for CoVs and arteriviruses. During infection, RNA species produced from more 3'-proximal TRSs are usually more abundant than the more distal ones, in accordance with the proposed role of the body TRS as an attenuator of minus-strand synthesis (de Vries et al., 1990; Lai, 1997; Penzes et al., 2001). In addition, viral and host components involved in protein-RNA and protein-protein recognition will also be decisive in transcription (Lai, 1998). Differences in affinity of such factors for body TRSs might also regulate transcription in CoVs. A similar mechanism has been described in the DNA-dependent RNA-polymerase I termination system, in which specific proteins bind to termination sequences (Reeder and Lang, 1997).

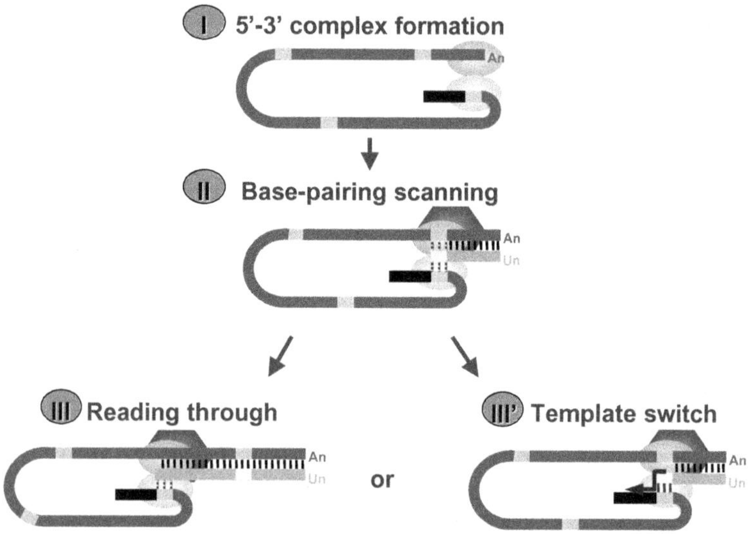

Fig. 4. Three-step working model of coronavirus transcription. (I) 5'-3' complex formation step. Proteins binding the 5' and 3' end TGEV sequences are represented by ellipsoids. Leader sequence is indicated with the black bar preceeding the first CS, CS sequences are indicated with a light grey bar. An, poly(A) tail. (II) Base pairing scanning step. Minus strand RNA is in a lighter color compared with positive strand RNA. The transcription complex is represented by the hexagon. Vertical dotted bars represent the base pairing scanning by the TRS-L sequence in the transcription process. Vertical solid bars indicate complementarity between the genomic (gRNA) and the nascent minus strand. Un, poly(U) tail. (III) The synthesis of the negative strand can continue to make a longer sgRNA (III), or a template switch step can take place (III') as indicated in the text. The thick arrow indicates the switch in the template made by the transcription complex to complete the synthesis of (-) sgRNA by adding the complement of the leader sequence.

The identification of host factors participating in coronavirus transcription is still a field under development. A limited number of cellular proteins have been identified from specific binding assays to *cis*-acting RNA regulatory sequences from mouse hepatitis coronavirus (MHV). These include PTB, binding to 5'-leader sequences (Li et al., 1999) and the sequences complementary to the 3'-UTR (Huang and Lai, 1999), hnRNPA1, binding to the complementary sequences of leader and TRS (Li et al., 1997) and to the 3'-UTR (Huang and Lai, 2001), poly(A)-binding protein (Spagnolo and Hogue, 2000), mitochondrial aconitase, binding to 3'-UTR (Nanda and Leibowitz, 2001), and SYNCRIP binding either the positive or the negative leader sequence (Choi, Mizutani, and Lai, 2004). In a first approach, mutations of the PTB-binding sites inhibited the replication and transcription of MHV genomic and defective-interfering RNA, in direct proportion to the extent of reduction of PTB binding, suggesting that PTB played a role in regulating viral RNA synthesis (Li et al., 1999). Similarly, when the transcription regulating sequences from an MHV DI RNA were mutated, transcription was also inhibited proportionally to the reduction of hnRNP A1 binding (Li et al., 1997; Zhang and Lai, 1995). However, functional studies were essential to confirm the involvement of these proteins on viral RNA synthesis. A primary effect of PTB on viral RNA synthesis was established by *in vivo* assays on cells overexpressing PTB, demonstrating that overexpression had a dominant-negative effect on MHV replication/transcription (Choi, Huang, and Lai, 2002). The effect of hnRNP A1 on MHV transcription was further confirmed in cell lines expressing a dominant-negative mutant of hnRNA A1 (Shi et al., 2000). Although MHV can replicate in a cell line deficient in hnRNP A1 (Shen and Masters, 2001), additional studies showed that other hnRNP A1 related proteins substitute for hnRNP A1 function (Shi, Yu, and Lai, 2003). A disadvantage of overexpression assays is that the overexpressed protein may titrate out other replication components, resulting in the inhibition of replication and transcription. An alternative functional assay to confirm the relevance of the candidate proteins is the RNA interference approach, to down-regulate the protein of interest by small interfering RNAs (siRNAs). Using this technology, it was demonstrated the involvement of SYNCRIP in MHV RNA synthesis (Choi, Mizutani, and Lai, 2004).

It is likely that some of these proteins, such as PTB, hnRNP A1 and SYNCRIP, binding to MHV 5'-UTR of both strands, form a large protein complex to regulate viral RNA synthesis. One of the largest difficulties in the identification of factors affecting replication or transcription is to distinguish between direct and indirect effects. In addition, frequently, factors affecting RNA synthesis also affect translation. Furthermore, viral translation, replication and transcription are processes tightly associated. This situation adds a further level of complication to the identification of host factors involved in a specific viral function (Boguszewska-Chachulska and Haenni, 2005).

Studies on the effect of TRS copy number on CoV transcription were performed mostly in defective-interfering RNA genomes, using more than one contiguous TRS in order to express the same mRNA. The accumulated sgmRNA remained nearly the same for constructs with one, two, or three TRSs, and transcription preferentially occurred in all cases at the 3'-most TRS, the closest from the site at which minus strand synthesis is initiated (Joo and Makino, 1995; Krishnan, Chang, and Brian, 1996; Stirrups et al., 2000; van Marle et al., 1995). Similar conclusion was obtained using arterivirus recombinant virus in which three TRSs were introduced (Pasternak, Spaan, and Snijder, 2004). As a general conclusion, for similar TRSs, transcription progressively favored smaller RNA species, leading to a perfect gradient of sgmRNA abundance. However, in this study, the effect of two linked factors, the relative order of the body TRSs and the size differences between sgmRNAs, could not be definitely clarified.

Template switch during coronavirus transcription could be interpreted as a transition between two states. In the first one, a duplex between the nascent minus RNA strand and the genomic positive RNA used as template is formed (Fig. 5). In the second one, the nascent RNA strand is paired with the leader TRS. RNA chaperones could be involved in template switch by decreasing the energy required for the transition from the first to the second duplex.

RNA chaperones are non-specific nucleic acid binding proteins that rescue RNAs trapped in unproductive folding states (Cristofari and Darlix, 2002; Herschlag, 1995; Lorsch, 2002; Schroeder, Barta, and Semrad, 2004). RNA chaperone activity can not be predicted based on domain conservation. Nevertheless, RNA chaperones are proteins with the highest frequency of long intrinsically disordered regions (Ivanyi-Nagy et al., 2005). A series of *in vitro* assays have been used to determine the RNA chaperone activity of a protein: (i) enhancement of ribozyme cleavage, (ii) allowance of rapid and accurate RNA annealing, and (iii) facilitation of RNA strand transfer and exchange (Cristofari and Darlix, 2002). From these activities, the first one has the strongest chaperone requirements (Rajkowitsch et al., 2005).

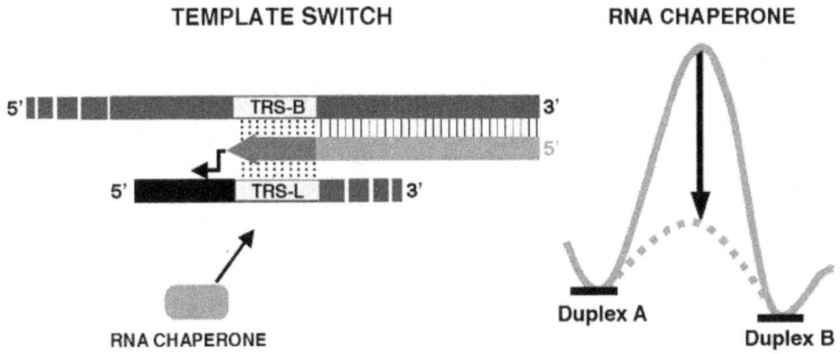

Fig. 5. Tentative RNA chaperone involvement in template switch during coronavirus transcription. Left panel, elements involved in the template switch step of coronavirus transcription. Right panel, scheme representing the action of RNA chaperones.

The importance of coronavirus N protein in RNA synthesis (Almazan, Galan, and Enjuanes, 2004), its characteristics as non-specific RNA binding protein, and its intrinsic disorder, led us to investigate the potential nucleic acid chaperone activity of TGEV N protein. This protein enhanced RNA ribozyme self-cleavage and nucleic acid annealing similarly as hnRNP A1, a cellular RNA chaperone, does (Zuñiga et al., 2007). In addition, SARS-CoV N protein also displayed RNA chaperone activity (Zuñiga et al., 2007). Therefore nucleic acid chaperone activity could be a common property of all coronavirus N proteins.

Previous to the identification of coronavirus N proteins as RNA chaperones (Zuñiga et al., 2006), three other viral RNA chaperones were already described, all of them being nucleocapsid proteins of RNA viruses: retrovirus (Bertrand and Rossi, 1994; Tsuchihashi and Brown, 1994), hepatitis delta virus (Huang et al., 2003; Huang and Wu, 1998), and hepatitis C virus (Cristofari et al., 2004). This RNA chaperone list is quickly growing with the recent inclusion of two other proteins, nucleocapsid from hantavirus (Mir and Panganiban, 2006), and 3AB from poliovirus (DeStefano and Titilope, 2006). Up to date, the activity of a virus-encoded RNA chaperone *in vivo* using a mammalian system has not been described, as the data reported have been generated using a bacterial RNA folding trap (Clodi, Semrad, and Schroeder, 1999). Next step in the characterization of Cov N protein RNA chaperone activity will be the evaluation of the effect of this activity on RNA synthesis using coronavirus-derived *in vitro* and *in vivo* systems. The identification of chaperone activity in coronaviruses, and its potential involvement in template switch during coronavirus transcription, links thermodynamic and molecular aspects of coronavirus transcription (Zuñiga et al., 2007).

EFFECT OF CORONAVIRUS INFECTION ON HOST CELL

The availability of antibodies and primers to evaluate expression of genes involved in signal transduction by quantitative RT-PCR has been essential to study the effect of virus infection into cells. In fact, many of these assays are already commercially available. Futhermore, more powerful approaches, such as the availability of microarrays (Kato-Maeda, Gao, and Small, 2001), libraries of small interfering RNAs (siRNA) (Silva et al., 2005), and proteomic techniques to evaluate protein-protein interaction (Vasilescu and Figeys, 2006), have facilitated studies on modification of cell network interactions in uninfected versus CoVs infected cells.

The effect of CoV infection on host cells has been studied mostly using the MHV system and, more recently, SARS-CoV. It was found that MHV and SARS-CoV infections induce mitogen activated kinases (MAPKs), especially p38 MAPK (Banerjee et al., 2002). In addition, activation of AP-1 and a weak induction of Akt signaling pathways was found after SARS-CoV infection (Mizutani et al., 2004a; Mizutani et al., 2004b). SARS-CoV N protein has been involved in the induction of these signaling pathways (He et al., 2003; Surjit et al., 2004a). Special attention has been dedicated to the study of the cell signaling pathways altered after CoV infection and the relation between these alterations and the effects on the host leading to the pathology.

CoV infection affects both host cell transcription and translation. Interestingly, CoVs give rise to mRNAs that are structurally similar to those of their eukaryotic hosts and this allows CoVs to parasitize the host machinery to translate the viral mRNA. The compartimentalization of CoV synthesis in virus factories could shield virus replication against the cell machinery degradation induced by the virus. Alternatively, specific factors that interfere with host translation or transcription machinery, or others that enhance viral specific synthesis must be responsible for the increase of virus-encoded molecules.

EFFECT ON CELL TRANSLATION

Infection by MHV leads to inhibition of host protein translation (Banerjee et al., 2000; Hilton et al., 1986; Tahara et al., 1994), but not to complete shut off, that is accompanied by an increase of MHV protein synthesis (Banerjee et al., 2002; Kyuwa et al., 1994; Siddell et al., 1981). The mechanism of selective CoV-specific protein synthesis, which occurs concomitantly with host protein inhibition in infected cells, is poorly characterized. It has been shown that chimeric mRNAs containing bovine coronavirus (BCoV) or MHV leader sequences are translated more efficiently in CoV infected cells (Senanayake and Brian, 1999; Tahara et al., 1994). As N protein binds to this sequence, it has been suggested that CoV mRNAs bind to N protein forming a complex that may act as a strong translation initiation signal (Tahara et al., 1998). Increased phosphorylation of eIF4E translation factor has also been described in CoV infected cells as a consequence of the activation of p38 MAPK (Banerjee et al., 2002; Mizutani et al., 2004c). Phosphorylated eIF4E has a higher affinity for cap structures and eIF4G, as a consequence, usually leads to enhanced translation rates (Gingras, Raught, and Sonenberg, 1999). Other possible mechanism of host cell protein synthesis shut off is the specific cleavage of 28S rRNA, an integral component of 60S ribosome, described in MHV-infected cells (Banerjee et al., 2000). Finally, specific host mRNAs are degraded in MHV-infected cells (Kyuwa et al., 1994) and similar results were observed after infection with SARS-CoV, being the genes involved in the host translational machinery the most down regulated ones (Leong et al., 2005).

Degradation of RNA transcripts and host endogenous mRNAs, leading to a strong host protein synthesis inhibition may be mediated by nsp-1 as the homologous protein of SARS-CoV promoted the degradation of host mRNA transcrips (Kamitani et al., 2006). In contrast, no degradation of the 28S and 18S rRNAs occurred in either nsp1-expressing cells or SARS-CoV- infected cells. Therefore, nsp1 suppresses host gene expression and may play an important role in SARS-CoV pathogenesis.

EFFECT ON CELL TRANSCRIPTION

The effect of pathogens on host-cell gene expression programs has been monitored on a broad scale using DNA microarrays, clarifying the effect of virus infection on cell transcriptosome. Alteration of the transcription pattern after CoV infection has been reported mainly for MHV and SARS-CoV infection. One of the most comprehensive studies (Tang et al., 2005) shows a higher perturbation of cellular gene transcription after SARS-CoV infection than after infection by human coronavirus (HCoV)-229E. Apart from the down-regulation of genes involved in cell host translation and maintenance of cytoskeletal network, up-regulation of genes related to stress response, pro-apoptotic, pro-inflammatory and procoagulating genes was observed (Leong et al., 2005; Ng et al., 2004; Tang et al., 2005). The balance between the genes up- and down-regulated after CoV infection would explain the pathogenesis and the differences in the severity of the illness caused by these viruses (see below).

Extracellular signal-regulated kinase (ERK) signalling pathway may affect a wide range of cellular functions including cell proliferation, differentiation and survival (Davis, 1995; Sebolt-Leopold et al., 1999). These functions may play a role in virus propagation as well, since viruses are obligate intracellular parasites. In fact, MHV infection activates the Raf-MEK-ERK signalling pathway (Cai, Liu, and Zhang, 2007), one of the three mitogen-activated protein kinase (MAPK) cascades. These kinases play important roles in regulation of cell growth and differentiation (Robinson and Cobb, 1997). The effect of MHV infection on the cells in turn affects MHV RNA synthesis (Cai, Liu, and Zhang, 2006).

EFFECT ON CELL CYCLE AND APOPTOSIS

In general, viruses use host cell cycle regulation for their own replication advantage. Each virus promotes a different pattern of interference with cell cycle. MHV, infectious bronchitis virus (IBV), transmissible gastroenteritis virus (TGEV), and SARS-CoV infection lead to the accumulation of infected cells in the Go/G1 phase (Chen and Makino, 2004; Chen et al., 2003; Yuan et al., 2005). Transcriptional profiling after SARS-CoV and HCoV-229E infection, shows alteration in genes involved in cell-cycle regulation, including up-regulation of genes that can mediate growth arrest at the G1 phase (Tang et al., 2005). An accumulation of hypophosphorylated retinoblastoma protein has been involved in the arrest of the cell cycle progression in the Go/G1 phase after MHV infection. Four viral proteins: the nsp1 from MHV (Chen et al., 2004), SARS-CoV 3b and 7a proteins (Yuan et al., 2005; Yuan et al., 2006), and the N protein (Chen et al., 2002; Wurm et al., 2001) have been proposed as responsible of cell cycle arrest.

Many viruses encode proteins that can modulate apoptosis (Barber, 2001; Benedict, Norris, and Ware, 2002; Roulston, Marcellus, and Branton, 1999). Induction of apoptosis in infected cells can contribute directly to viral pathogenesis, while inhibition of apoptosis can prevent premature death of the infected cells, allowing the virus to replicate to a high titer or allowing the establishment of a persistent infection. Infection by CoVs, such as TGEV, MHV, and SARS-CoV induces apoptosis in certain cells (An et al., 1999; Bailey et al., 1949; Cai et al., 2003; Eleouet et al., 1998; Mizutani et al., 2004b; Wang et al., 2000). Nevertheless, stimulation of both apoptotic and anti-apoptotic molecules has been described (Leong et al., 2005; Mizutani et al., 2004a; Mizutani et al., 2004b; Ng et al., 2004; Tang et al., 2005). The delicate counterbalance of pro- and anti-apoptotic molecules during CoV infection should ensure cell survival during the early phase of infection to allow rapid multiplication of progeny virus before cell lysis occurs. The apoptosis induced by CoV infection is tissue specific, and this observation may explain the pathology of the infection and the effects on the host organism. For instance, the data obtained shows that SARS-CoV infects epithelial cells of the enteric tract and develops an anti-apoptotic response. This response may be important to inhibit or delay destruction of infected enterocytes, probably leading to an extension of virus production shedding. These findings are consistent with clinical observations demonstrating a relatively normal endoscopic and microscopic appearance of the intestine in patients with SARS (Leung et al., 2003). On the other hand, CoV infection induces apoptosis in other tissues. SARS-CoV causes lymphopenia due to the depletion of T lymphocytes by apoptosis (Yang et al., 2004), or liver impairment due to hepatocyte apoptosis (Chau et al., 2004).

Over-expression of several CoV proteins, such as MHV E protein, induces apoptosis (An et al., 1999). After SARS-CoV infection T-cells undergo apoptosis probably by direct interaction with the anti-apoptotic factor Bcl-xL (Yang et al., 2005). Protein 7a from SARS-CoV also induces apoptosis by a caspase-dependent pathway in several cell lines, including those derived from lung, kidney and liver (Tan et al., 2004).

EFFECT ON HOST SYSTEMS

CoV infection affects several host systems. Functional grouping of altered genes after CoV infection showed that more genes involved in inflammation, coagulation and stress are upregulated in SARS-CoV than in HCoV-229E infection (Tang et al., 2005). Of the four CoVs known to infect human (HCoV-229E, HCoV-OC43, HCoV-NL63, and SARS-CoV), the three first are generally associated with mild upper respiratory tract infections such as the common cold in immunocompetent host. In contrast, SARS-CoV causes respiratory failure in over 60% of affected persons, with a mortality rate of 10%. The observed effect on host-gene expression (Tang et al., 2005) may help to explain the pathologies caused by these CoVs.

CoV infection leads to the induction of inflammation. Several studies showed that the proinflammatory response is increased after SARS-CoV infection (Leong et al., 2005; Ng et al., 2004; Tang et al., 2005; Wong et al., 2004). For instance, the upregulation of IL-8 expression (Cinatl et al., 2004; Tang et al., 2005), consistent with the increase of this chemokine in plasma of infected patients (Wong et al., 2004), may be of pathogenic importance as IL-8 level has been positively correlated with disease severity in pulmonary infection by respiratory syncytial virus. In fact, a positive correlation was found between IL-8 levels in blood and alveolar spaces and the number of polymorphonuclear neutrophil in bronchoalveolar lavage of patients with pneumonia and acute respiratory distress syndrome

(Wong et al., 2004). Other chemokines upregulated after SARS-CoV infection may mediate the chemotaxis of lymphocytes and neutrophils, contributing to the significant increase in neutrophil homing to the lung that may account for the localized nature of the response in SARS patients (Ng et al., 2004). Similar results were obtained with MHV, where the neurovirulence of the virus correlated with the upregulation of proinflammatory cytokines associated with the recruitment of lymphocytes and macrophages at the site of the infection, leading to the manifestation of encephalitis (Li et al., 2004).

In contrast, despite the upregulation of IL-8 in intestinal epithelial cells, biopsy specimens taken from the colon and terminal ileum of patients with SARS failed to demonstrate any inflammatory infiltrates (Leung et al., 2003). Neutrophil infiltration in the intestine of SARS patients may be limited despite neutrophilia due to changes of cytokine and chemokine levels in the intestinal environment. It has been observed that SARS-CoV infection inhibited production of IL-18 that is constitutively expressed in intestinal epithelial cells (Cinatl et al., 2004; Wong et al., 2004). Suppression of IL-18 levels reduced neutrophil accumulation in liver and lungs (Dinarello and Fantuzzi, 2003). The absence of T lymphocyte infiltration of the intestine in SARS may be a consequence of the profound decline of both $CD4^+$ and $CD8^+$ lymphocytes in the blood (Cui et al., 2003), possibly resulting from lymphocyte apoptosis. Although macrophage counts were increased in lungs, macrophage infiltration was absent from the gut of SARS patients. In agreement with this, in CaCo-2 cells, SARS-CoV down-regulates MIF, identified as a major factor produced by intestinal cells in response to microbial infection, regulating macrophage emigration, inflammation and cell metabolism (Maaser et al., 2002).

Innate immunity probably is essential to control CoV infections *in vivo*. Cytokine profiling after SARS-CoV infection suggested an early activation of the innate immunity pathway (Ng et al., 2004). The same observations were made after MHV infection (Rempel et al., 2004). SARS-CoV infection induces CCL5, CCL3, and CXCL10 cytokines in different cell lines. This induction is mediated by replicase nsp1 throughout NF-KB signalling pathway (Law et al., 2007). This indicates that nsp1 may play a role in SARS-CoV pathogenesis.

IFN-γ is critical in resolving MHV infection (Parra et al., 1999). IFNs have significant anti-SARS-CoV effects (Cinati et al., 2003; Haagmans et al., 2004; Loutfy et al., 2003), and only immunocompromised transgenic mice, with impaired IFN responses, are infected with HCoV-229E (Lassnig et al., 2005) or SARS-CoV (Hogan et al., 2004). CoVs have developed strategies to escape IFN responses. In fact, low levels of IFN have been found in SARS patients (Ng et al., 2004) probably due to the interference with interferon regulatory factor 3 (IRF-3) activation (Spiegel et al., 2005). Infection of cells with virus causes the activation of several cellular transcription factors, such as IRF-3 and NF-kB, which activate the production of IFN genes. Once IFN is synthesized and released from cells, IFN binds to IFN receptors initiating a signalling cascade of the JAK/STAT pathway that result in activated transcription factors translocating to the nucleus. These transcription factors activate genes containing an interferon-stimulated response element (ISRE) in their promoters. Activation of these genes enables the cell to combat the virus infection and can prevent viral replication (Garcia-Sastre and Biron, 2006). Many viruses such as SARS-CoV have developed mechanisms to subvert the interferon response. It has been shown that SARS-CoV proteins nsp1, 3b, 6 and N antagonize interferon. The last three proteins inhibit the expression of IFN by inhibiting IRF-3. N protein dramatically inhibited expression from an NF-KB responsive promoter. All three proteins were able to inhibit expression from an ISRE promoter after infection with Sendai virus, while only proteins 3b and 6 were able to inhibit expression from the ISRE promoter after treatment with IFN. Three IFN antagonists of SARS-CoV (proteins N, 3b, and 6) inhibit IFN response by different mechanism (Kopecky-Bromberg et al., 2007).

Blood coagulation genes are also affected by SARS-CoV infection. The upregulation of several genes has been involved in the activation of this pathway (Ng et al., 2004; Tang et al., 2005). The result leads to a pro-coagulation profile that mimics the pathological observations. In fact, at autopsy, many SARS patients have unusually disseminated small vessel thromboses in the lungs (Franks et al., 2003). Also, vascular damage in various tissues has been reported (Ding et al., 2003). Activation of procoagulation pathway has also been described in MHV infection (Ding et al., 1998), resulting in confluent hepatocellular necrosis.

EXPRESSION SYSTEMS BASED ON CORONA-VIRUSES

The recent development of reverse genetics systems for CoVs allowed the engineering of CoV-derived vectors, that have several advantages over other viral expression systems: (i) CoVs are single-stranded RNA viruses that replicate in the cytoplasm without a DNA intermediary, making integration of the virus genome into the host cell chromosome unlikely (Lai and Cavanagh, 1997); (ii) these viruses have the largest RNA virus genome and, in principle, have room for the insertion of large foreign genes (Enjuanes et al., 2001; Enjuanes et al., 2005); (iii) a pleiotropic secretory immune response is best induced by the stimulation of gut associated lymphoid tissues. Since CoVs in general infect the mucosal surfaces, both respiratory and enteric, CoVs may be used to target the antigen to the enteric and respiratory areas to induce a strong secretory immune response; (iv) the tropism of CoVs may be engineered by modifying the S gene (Ballesteros, Sanchez, and Enjuanes, 1997; Kuo et al., 2000; Sanchez et al., 1999); (v) non-pathogenic CoV strains infecting most species of interest (human, porcine, bovine, canine, feline, and avian) are available and therefore suitable to develop safe virus vectors; and (vi) infectious CoV cDNA clones are available to design expression systems.

Current CoV vectors are based on single genomes engineered by targeted recombination or by the construction of a cDNA encoding an infectious RNA. CoV-derived replicons have also been used for expression of foreign genes. Targeted recombination is being used for the introduction of specific changes into the CoV genome through recombination between a donor synthetic RNA and a recipient parental virus with some characteristics that allows its counter-selection (Masters and Rottier, 2005). This strategy led to the generation of attenuated viruses by deletion of specific genes (de Haan et al., 2002a; Haijema, Volders, and Rottier, 2004). The first non-viral protein expressed using a CoV-derived vector was green fluorescent protein (GFP) (Fischer et al., 1997). GFP gene was inserted in MHV genome by replacing the non-essential gene 4 by targeted recombination. Further studies have shown that expression of enhanced GFP under the control of an optimized

TRS, led to high and stable protein expression levels both *in vitro* and *in vivo* (Sarma et al., 2002). Multivalent expression vectors were also generated by targeted recombination of the MHV genome (de Haan et al., 2003). Thus, two unrelated luciferase genes were also expressed at several genomic positions, both individually and combined within a single MHV genome.

The first infectious CoV cDNA clone was obtained for TGEV. This cDNA was propagated as a bacterial artificial chromosome (BAC) (Almazán et al., 2000). The stability of this BAC in *E. coli* was improved by the insertion of an intron (Gonzalez et al., 2002). The same strategy has been successfully applied to obtain infectious cDNA clones for SARS-CoV and HCoV-OC43 (Almazan et al., 2006; St-Jean et al., 2006). The great advantage of this system is that CoV reverse genetics only involves standard recombinant DNA technologies performed within bacteria, with high efficiency in the rescue of recombinant viruses (Enjuanes et al., 2005).

Two alternative procedures have also been used for the generation of infectious CoV cDNAs. One of them was the assembly of a full-length CoV cDNA by *in vitro* ligation of adjoining cDNA subclones that spanned the entire genomic RNA. The cDNA is *in vitro* transcribed to generate a full-length CoV RNA that is infectious. The advantage of this system is that allows the assembly of RNA or DNA genomes which are too large, circular, or unstable in cloning vectors (Baric and Sims, 2005). This strategy was first used to obtain an infectious TGEV cDNA (Yount, Curtis, and Baric, 2000), and it was subsequently improved and used to generate infectious cDNAs for MHV (Yount et al., 2002), SARS-CoV (Yount et al., 2003), and IBV (Youn, Leibowitz, and Collisson, 2005).

Another strategy to obtain CoV infectious cDNAs was based in the *in vitro* transcription of infectious RNA, from a cDNA copy of a CoV genome that has been cloned and propagated in vaccinia virus. This strategy has been used to generate infectious clones for HCoV-229E (Thiel et al., 2001a) and IBV (Casais et al., 2001). The advantage of this system is the high stability of cloned cDNAs in the vaccinia virus and that genome modifications can be made by homologous recombination of the vaccinia virus genome (Britton et al., 2005).

Single genome CoV-derived vectors have been generated for members of CoV groups 1, 2, and 3 using the infectious cDNAs described above. Vectors based on Group 1 viruses have been derived from TGEV. Using the TGEV infectious cDNA maintained as a BAC, the GFP gene was successfully expressed by replacing the non-essential 3a and 3b genes by sequences encoding GFP. The engineered genome was very stable (>30 passages in cultured cells) and led to the production of high protein levels (50 $\mu g/10^6$cells) (Sola et al., 2003). Using this vector the dicistronic expression of two surface proteins of porcine respiratory and reproductive syndrome virus (PRRSV) virus (Gp5 and M proteins) has been shown (Fig. 6). After more than ten passages in tissue culture 80% and 100% of virus infected cells expressed Gp5 and M proteins, respectively (S. Zuñiga, C. M. Sanchez, and L. Enjuanes, unpublished results). This vector was also used for dicistronic expression of rotavirus VP2 and VP6. These viral proteins self aggregated leading to the formation of rotavirus like particles (VLPs) in the cytoplasm of infected cells (J. E. Ceriani, J. Ortego, J. Plana-Duran, and Luis Enjuanes, unpublished results). Therefore, expression levels using these vectors are similar to those described with vectors derived from other positive strand RNA viruses such as Sindbis virus (Agapov et al., 1998; Frolov et al., 1996). Using TGEV-derived vectors expressing GFP, the induction of lactogenic immunity has been demonstrated (Sola et al., 2003). Groups 2 and 3 CoV-derived vectors have been constructed for MHV and IBV (Baric and Sims, 2005; Youn, Leibowitz, and Collisson, 2005).

Propagation-deficient CoV-derived vectors have also been generated based on replicons derived from group 1 CoVs TGEV and HCoV-229E. The advantages of these systems are; (i) an increased room for the insertion of foreign genes when compared with full-length infectious cDNAs, as the structural genes were deleted, (ii) the possibility of engineering multigene expression vectors due to the transcriptional strategy of CoVs and, (iii) an increased safety as the replicons are non-infectious. TGEV-derived replicons have been generated from the infectious cDNA (Almazan, Galan, and Enjuanes, 2004). Replicons derived from HCoV-229E were also generated (Thiel et al., 2001b). A multigene vector based on HCoV-229E replicon has also been constructed that expressed GFP, chloranphenicol acetyltransferase (CAT), and firefly luciferase (LUC) (Thiel et al., 2003). Human CoV-based vector RNA can be packed into propagation-deficient pseudovirions that, in turn, can be used to transduce immature and mature human dendritic cells.

A B

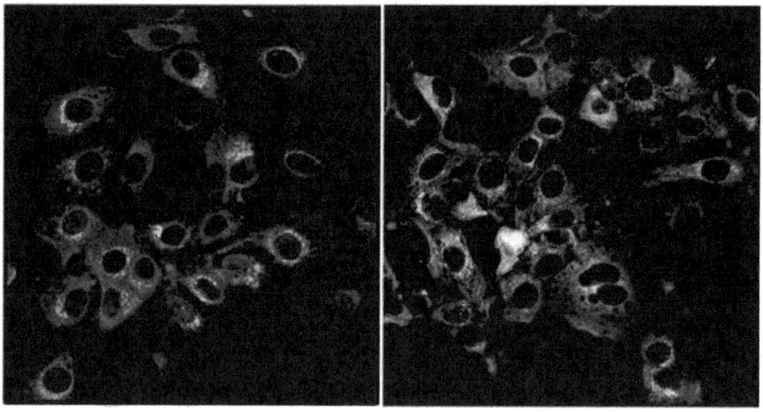

Fig. 6. Dicistronic expression of PRRSV Gp5 and M proteins using a recombinant TGEV vector. TRS from gene 3a (TRS-3a) was used to express Gp5, whereas an artificial TRS including 22 nt from 5'-TRS of N gene and 3'-TRS sequences from M gene (TRS-22N) was used to express M protein. ST cells were infected with the vector and double immunofluorescence staining was performed. TGEV N protein specific monoclonal antibodies and a secondary antibody staining red (shown as dark-grey) were used to identify virus infected cells. (A) Expression of Gp5 was detected with rabbit antiserum specific for a Gp5 peptide coupled to a secondary antibody staining green (shown as light grey). (B) Expression of M protein was detected with a rabbit antiserum specific for a M protein peptide, coupled to a secondary antibody staining green (shown as light grey).

Safe replication-competent propagation-deficient virus vectors, based on TGEV genomes deficient in the essential gene E obtained using E^+ packaging cell lines have also been developed (Ortego et al., 2002). Virus titres were directly related to E protein expression levels in the packaging cell lines (Ortego et al., 2002).

In order to achieve optimal heterologous gene expression levels from a CoV-derived vector, several factors must be considered, such as the insertion site into the genome RNA, the stability of the cloned insert, and the nature of the TRSs, as desirable expression levels could be modulated by using specific regulatory sequences preceding the gene. In general, it has been found that vector stability is highly dependent on the nature of the foreign gene, some inserts being maintained at least twenty passages, whereas others are lost at passage 2.

For TGEV-derived vectors, expression levels and virus stability were optimal when heterologous gene was inserted replacing non-essential 3a and 3b genes. In contrast, insertion of the same expression cassette between genes N and 7, located at the 3' end of the genome, led to loss of the insert with the passages in tissue culture (Fisher et al., 1999; Hsue and Masters, 1999; Sarma et al., 2002; Sola et al., 2003). A comprehensive study about the insertion site dependence of the heterologous gene expression, using a CoV-derived vector, was performed expressing two luciferase genes from an MHV genome engineered by targeted recombination (de Haan et al., 2003). In general, expression levels were a function of gene identity and insertion site, with higher expression levels the closer the gene was to the 3' end of the genome, without significantly affecting *in vitro* growth of the virus. When recombinant CoVs were unstable, location of the heterologous gene in upstream positions in the MHV genome improved the stability of the recombinant viruses.

Expression levels and stability of CoV-derived vectors are also highly dependent on the nature of the heterologous gene cloned [de Haan et al., 2003; S. Zuniga, I. Sola, and L. Enjuanes, unpublished results]. For instance, only 30% of cells infected with a TGEV-based vector expressed Gp5 protein, were observed in passage 1, and ORF5 gene was lost by passage fifth in tissue culture and *in vivo*. Interestingly, improvement of the stability and expression of PRRSV Gp5 was achieved by its co-expression with PRRSV M protein from the same TGEV genome under the control of a different TRS. Most likely, Gp5 and M proteins form heterodimers with a lower toxicity than that of Gp5 protein alone. Coexpression of Gp5 and M proteins lead to the expression of these proteins in 80% and 100%, respectively, of the transfected cells, for at least 20 passages in tissue culture and also after *in vivo* inoculation of the virus (S. Zuñiga, J. E. Ceriani, I. Sola, J. L. Moreno, and L. Enjuanes, unpublished results) (Fig. 6).

CORONAVIRUS VECTOR BIOSAFETY

Vaccines based in live attenuated viruses have the limitation of the possible reversion to wild type by mutation or by the recovery of genes deleted to attenuate the vector, by recombination with circulating CoVs. Consequently, live virus vaccine formulations should include rational approaches for minimizing the potential for reversion to wild type phenotype by recombination repair (Enjuanes et al., 2007a; Enjuanes et al., 2007b). One of those strategies is the introduction of attenuating mutations or deletions at distal domains of the genome. For instance, vaccines based on modifications of the replicase gene, as a potential target for attenuating CoVs, could in principle be generated by mutagenesis. Modifications introduced in the MHV nsp1 coding region have identified residues important for protein processing and viral RNA replication that may affect virus virulence and could be introduced in vaccine candidates (Brockway and Denison, 2005). SARS nsp1 blocks host macromolecular synthesis and abrogates interferon (IFN) signalling (Kamitani et al., 2006), providing further evidence that nsp1 coding regions represent potential virulence determinants. Similarly, deletion of the nsp2 gene in MHV and SARS-CoV yields viable attenuated mutant viruses that replicate about one log less efficiently than wildtype virus in cell culture and in animals, and may also provide the basis for the design of live vaccines (Denison et al., 2007; Graham et al., 2005). Alternatively, specific substitutions in ORF 1b-nsp14 attenuates MHV replication in mice (Sperry et al., 2005). As nsp14 and nsp2 are conserved in several CoVs, it may be possible to engineer attenuating changes common to different CoVs.

Changes in gene order within the replicase, or even relocation to the 3'-end of the genome may lead to attenuated virus phenotypes. CoVs have a strictly conserved genome organization with structural genes occurring in the order 5'-Rep-S-E-M-N-3'. MHV virus mutants with the genes encoding the structural proteins located in different order have been constructed (de Haan et al., 2002b). These recombinant viruses were tested for the ability to replicate in the natural host, the mouse. The results indicate that the canonical CoV genome organization is not essential for *in vivo* replication. Some of the mutants showed an attenuated phenotype, similarly to what has been observed for vesicular stomatitis virus (VSV) (Ball et al., 1999). Therefore, deliberate rearrangement of the viral genes may be useful in the generation of attenuated CoVs, which would make safer vaccines, due to their reduced risk of generating viable viruses by recombination with circulating field viruses.

Other options to include safeguards into the genetically engineered vaccines, particularly those that can prevent the recovery of the original virulent phenotype by recombination between the vaccine strain and viruses circulating in the field have been developed. One of them is the construction of replication-competent, propagation-defective viruses (pseudovirions) that are defective in one gene conferring an attenuated phenotype or even the ability for virus propagation (Enjuanes et al., 2005). These viruses could be grown in packaging cell lines providing *in trans* the missing protein. In the case of TGEV or SARS-CoV, vaccine candidates without the E gene have been constructed. In order to prevent the rescue of the virulent phenotype by recombination with a circulating human CoV, the deletion of an essential gene, located in a position distant from gene E, and the relocation of this gene to the position previously occupied by E gene has been proposed (Enjuanes et al., 2005). A potential recombination leading to the rescue of gene E would lead to the loss of the essential gene relocated to the position previously occupied by gene E.

An interesting approach for developing safer, recombination resistant live SARS-CoV vaccines has been described by modifying the TRS of a vaccine strain to a sequence incompatible with the TRS of any known circulating CoV. Recombinant events that could take place between wild type CoVs and TRS remodeled SARS-CoV would result in genomes containing lethal mixed regulatory sequences that block expression of subgenomic mRNAs (Yount et al., 2006).

ACKNOWLEDGEMENTS

This work has been supported by grants from the Comisión Interministerial de Ciencia y Tecnología (CICYT), the Consejería de Educación y Cultura de la Comunidad de Madrid, Fort Dodge Veterinaria, and the European Communities (Frame VI, DISSECT project SP22-CT-2004-511060 and RiBiGene SSPE-CT-2005-022639). ISG, SZ, and FA received contracts from the Department of Education and Science of Spain.

REFERENCES

Agapov, E.V., Frolov, I., Lindenbach, B.D., Pragai, B.M., Schlesinger, S., and Rice, C.M. (1998). Noncytopathic Sindbis virus RNA vectors for heterologous gene expression. Proc. Natl. Acad. Sci. USA *95*, 12989-12994.

Almazan, F., DeDiego, M.L., Galan, C., Escors, D., Alvarez, E., Ortego, J., Sola, I., Zuñiga, S., Alonso, S., Moreno, J.L., Nogales, A., Capiscol, C., and Enjuanes, L. (2006). Construction of a SARS-CoV infectious cDNA clone and a replicon to study coronavirus RNA synthesis. J. Virol. *80*, 10900-10906.

Almazan, F., Galan, C., and Enjuanes, L. (2004). The nucleoprotein is required for efficient coronavirus genome replication. J. Virol. *78*, 12683-12688.

Almazán, F., González, J.M., Pénzes, Z., Izeta, A., Calvo, E., Plana-Durán, J., and Enjuanes, L. (2000). Engineering the largest RNA virus genome as an infectious bacterial artificial chromosome. Proc. Natl. Acad. Sci. USA *97*, 5516-5521.

Alonso, S., Izeta, A., Sola, I., and Enjuanes, L. (2002). Transcription regulatory sequences and mRNA expression levels in the coronavirus transmissible gastroenteritis virus. J. Virol. *76*, 1293-1308.

An, S., Chen, C.J., Yu, X., Leibowitz, J.L., and Makino, S. (1999). Induction of apoptosis in murine coronavirus-infected cultured cells and demonstration of E protein as an apoptosis inducer. J. Virol. *73*, 7853-7859.

Bailey, O.T., Pappenheimer, A.M., Cheever, F.S., and Daniels, J.B. (1949). A murine virus (JHM) causing disseminated encephalomyelitis with extensive destruction of myelin. J. Exp. Med. *90*, 195.

Ball, L.A., Pringle, C.R., Flanagan, B., Perepelitsa, V.P., and Wertz, G.W. (1999). Phenotypic consequences of rearranging the P, M, and, G genes of vesicular stomatitis virus. J. Virol. *73*, 4705-4712.

Ballesteros, M.L., Sanchez, C.M., and Enjuanes, L. (1997). Two amino acid changes at the N-terminus of transmissible gastroenteritis coronavirus spike protein result in the loss of enteric tropism. Virology *227*, 378-388.

Banerjee, S., An, S., Zhou, A., Silverman, R.H., and Makino, S. (2000). RNase L-independent specific 28S rRNA cleavage in murine coronavirus-infected cells. J. Virol. *74*, 8793-8802.

Banerjee, S., Narayanan, K., Mizutani, T., and Makino, S. (2002). Murine coronavirus replication-induced p38 mitogen-activated protein kinase activation promotes interleukin-6 production and virus replication in cultured cells. J. Virol. *76*, 5937-5948.

Barber, G.N. (2001). Host defense, viruses and apoptosis. Cell Death Differ. *8*, 113-126.
Baric, R.S., Nelson, G.W., Fleming, J.O., Deans, R.J., Keck, J.G., Casteel, N., and Stohlman, S.A. (1988). Interactions between coronavirus nucleocapsid protein and viral RNAs: implications for viral transcription. J. Virol. *62*, 4280-4287.
Baric, R.S., and Sims, A.C. (2005). Development of mouse hepatitis virus and SARS-CoV infectious cDNA constructs. Curr. Top. Microbiol. Immunol. *287*, 229-252.
Benedict, C.A., Norris, P.S., and Ware, C.F. (2002). To kill or be killed: viral evasion of apoptosis. Nat. Immunol. *3*, 1013-1018.
Bertrand, E.L., and Rossi, J.J. (1994). Facilitation of hammerhead ribozyme catalysis by the nucleocapsid protein of HIV-1 and the heterogeneous nuclear ribonucleoprotein A1. EMBO J. *13*, 2904-2912.
Boguszewska-Chachulska, A.M., and Haenni, A.L. (2005). RNA viruses redirect host factors to better amplify their genome. Adv. Virus Res. *65*, 29-61.
Bost, A.G., Prentice, E., and Denison, M.R. (2001). Mouse hepatitis virus replicase protein complexes are translocated to sites of M protein accumulation in the ERGIC at late times of infection. Virology *285*, 21-29.
Brian, D.A., and Baric, R.S. (2005). Coronavirus genome structure and replication. Curr. Top. Microbiol. Immunol. *287*, 1-30.
Brian, D.A., Hogue, B.G., and Kienzle, T.E. (1995). The coronavirus hemagglutinin esterase glycoprotein. In The coronaviridae, S.G. Siddell, ed. (New York: Plenum Press), pp. 165-176.
Brian, D.A., and Spaan, W.J.M. (1997). Recombination and coronavirus defective interfering RNAs. Semin. Virol. *8*, 101-111.
Britton, P., Evans, S., Dove, B., Davies, M., Casais, R., and Cavanagh, D. (2005). Generation of a recombinant avian coronavirus infectious bronchitis virus using transient dominant selection. J. Virol. Methods *123*, 203-211.
Brockway, S.M., and Denison, M.R. (2005). Mutagenesis of the murine hepatitis virus nsp1-coding region identifies residues important for protein processing, viral RNA synthesis, and viral replication. Virology *340*, 209-223.
Cai, Y., Liu, Y., Yu, D., and Zhang, X. (2003). Down-regulation of transcription of the proapoptotic gene BNip3 in cultured astrocytes by murine coronavirus infection. Virology *316*, 104-115.

Cai, Y., Liu, Y., and Zhang, X. (2006). Induction of transcription factor Egr-1 gene expression in astrocytoma cells by murine coronavirus infection. Virology *355*, 152-163.

Cai, Y., Liu, Y., and Zhang, X. (2007). Suppression of Coronavirus Replication by Inhibition of the MEK Signaling Pathway. J. Virol. *81*, 446-456.

Callebaut, P., Correa, I., Pensaert, M., Jiménez, G., and Enjuanes, L. (1988). Antigenic differentiation between transmissible gastroenteritis virus of swine and a related porcine respiratory coronavirus. J. Gen. Virol. *69*, 1725-1730.

Calvo, E., Escors, D., Lopez, J.A., Gonzalez, J.M., Alvarez, A., Arza, E., and Enjuanes, L. (2005). Phosphorylation and subcellular localization of transmissible gastroenteritis virus nucleocapsid protein in infected cells. J. Gen. Virol. *86*, 2255-2267.

Carmo-Fonseca, M., Mendes-Soares, L., and Campos, I. (2000). To be or not to be in the nucleolus. Nat. Cell. Biol. *2*, E107-E112.

Casais, R., Thiel, V., Siddell, S.G., Cavanagh, D., and Britton, P. (2001). Reverse genetics system for the avian coronavirus infectious bronchitis virus. J. Virol. *75*, 12359-12369.

Cavanagh, D. (2003). Severe acute respiratory syndrome vaccine development: experiences of vaccination against avian infectious bronchitis coronavirus. Avian Pathol. *32*, 567-582.

Chang, C.K., Sue, S.C., Yu, T.H., Hsieh, C.M., Tsai, C.K., Chiang, Y.C., Lee, S.J., Hsiao, H.H., Wu, W.J., Chang, W.L., Lin, C.H., and Huang, T.H. (2006). Modular organization of SARS coronavirus nucleocapsid protein. J. Biomed. Sci. *13*, 59-72.

Chau, T.N., Lee, K.C., Yao, H., Tsang, T.Y., Chow, T.C., Yeung, Y.C., Choi, K.W., Tso, Y.K., Lau, T., Lai, S.T., and Lai, C.L. (2004). SARS-associated viral hepatitis caused by a novel coronavirus: report of three cases. Hepatology *39*, 302-310.

Chen, C.J., and Makino, S. (2004). Murine coronavirus replication induces cell cycle arrest in G0/G1 phase. J. Virol. *78*, 5658-5669.

Chen, C.J., Sugiyama, K., Kubo, H., Huang, C., and Makino, S. (2004). Murine coronavirus nonstructural protein p28 arrests cell cycle in G_0/G_1 phase. J. Virol. *78*, 10410-10419.

Chen, H., Coote, B., Attree, S., and Hiscox, J.A. (2003). Evaluation of a nucleoprotein-based enzyme-linked immunosorbent assay for the detection of antibodies against infectious bronchitis virus. Avian Pathol. *32*, 519-526.

Chen, H., Gill, A., Dove, B.K., Emmett, S.R., Kemp, C.F., Ritchie, M.A., Dee, M., and Hiscox, J.A. (2005). Mass spectroscopic characterization of the coronavirus infectious bronchitis virus nucleoprotein and elucidation of the role of phosphorylation in RNA binding by using surface plasmon resonance. J. Virol. *79*, 1164-1179.

Chen, H., Wurm, T., Britton, P., Brooks, G., and Hiscox, J.A. (2002). Interaction of the coronavirus nucleoprotein with nucleolar antigens and the host cell. J. Virol. *76*, 5233-5250.

Choi, K.S., Huang, P., and Lai, M.M. (2002). Polypyrimidine-tract-binding protein affects transcription but not translation of mouse hepatitis virus RNA. Virology *303*, 58-68.

Choi, K.S., Mizutani, A., and Lai, M.M. (2004). SYNCRIP, a member of the heterogeneous nuclear ribonucleoprotein family, is involved in mouse hepatitis virus RNA synthesis. J. Virol. *78*, 13153-13162.

Cinati, J., Morgenstern, B., Bauer, G., Chandra, P., Rabenau, H., and Doerr, H.W. (2003). Treatment of SARS with human interferons. Lancet *362*, 293-294.

Cinatl, J., Jr., Hoever, G., Morgenstern, B., Preiser, W., Vogel, J. U., Hofmann, W.K., Bauer, G., Michaelis, M., Rabenau, H.F., and Doerr, H.W. (2004). Infection of cultured intestinal epithelial cells with severe acute respiratory syndrome coronavirus. Cell. Mol. Life Sci. *61*, 2100-2112.

Clodi, E., Semrad, K., and Schroeder, R. (1999). Assaying RNA chaperone activity in vivo using a novel RNA folding trap. EMBO J. *18*, 3776-3782.

Cox, E., Hooyberghs, J., and Pensaert, M.B. (1990). Sites of replication of a porcine respiratory coronavirus related to transmissible gastroenteritis virus. Res. Vet. Sci. *48*, 165-169.

Cox, E., Pensaert, M.B., Callebaut, P., and van Deun, K. (1990). Intestinal replication of a porcine respiratory coronavirus closely related antigenically to the enteric transmissible gastroenteritis virus. Vet. Microbiol. *23*, 237-243.

Cristofari, G., and Darlix, J.L. (2002). The ubiquitous nature of RNA chaperone proteins. Prog. Nucleic Acid Res. Mol. Biol. *72*, 223-268.

Cristofari, G., Ivanyi-Nagy, R., Gabus, C., Boulant, S., Lavergne, J. P., Penin, F., and Darlix, J.L. (2004). The hepatitis C virus core protein is a potent nucleic acid chaperone that directs dimerization of the viral (+) strand RNA in vitro. Nucleic Acids Res. *32*, 2623-2631.

Cui, W., Fan, Y., Wu, W., Zhang, F., Wang, J.Y., and Ni, A.P. (2003). Expression of lymphocytes and lymphocyte subsets in patients with severe acute respiratory syndrome. Clin. Infect. Dis. *37*, 857-859.

Davis, R.J. (1995). Transcriptional regulation by MAP kinases. Mol. Reprod. Dev. *42*, 459-467.

de Haan, C.A., van Genne, L., Stoop, J.N., Volders, H., and Rottier, P.J. (2003). Coronaviruses as vectors: position dependence of foreign gene expression. J. Virol. *77*, 11312-11323.

de Haan, C.A.M., Masters, P.S., Shen, S., Weiss, S., and Rottier, P.J.M. (2002a). The group-specific murine coronavirus genes are not essential, but their deletion, by reverse genetics, is attenuating in the natural host. Virology *296*, 177-189.

de Haan, C.A.M., Volders, H., Koetzner, C.A., Masters, P.S., and Rottier, P.J.M. (2002b). Coronavirus maintain viability despite dramatic rearrangements of the strictly conserved genome organization. J. Virol. *76*, 12491-12502.

de Vries, A.A., Chirnside, E.D., Bredenbeek, P.J., Gravestein, L.A., Horzinek, M.C., and Spaan, W.J. (1990). All subgenomic mRNAs of equine arteritis virus contain a common leader sequence. Nucleic Acids Res. *18*, 3241-3247.

Denison, M.R. (1999). The common cold. Rhinoviruses and coronaviruses. In Viral infections of the respiratory tract, R. Dolin, and P. F. Wringht, eds., (New York: Marcel Dekker, Inc.) Vol. 127, pp. 253-280.

Denison, M.R., Graham, R., Sparks, J.S., and Eckerle, L.D. (2007). SARS coronavirus replicase proteins in pathogenesis. Virus Res., In press.

DeStefano, J.J., and Titilope, O. (2006). Poliovirus protein 3AB displays nucleic acid chaperone and helix-destabilizing activities. J. Virol. *80*, 1662-1671.

Dinarello, C.A., and Fantuzzi, G. (2003). Interleukin-18 and host defense against infection. J. Infect. Dis. *187*, S370-S384.

Ding, J.W., Ning, Q., Liu, M.F., Lai, A., Phillips, M.J., Peltekian, K., Fung, L.S., Hollaway, C., Leibowitz, J., and Levy, G. (1998). Expression of the FGL-2 gene and its protein product (prothrombinase) in tissues during murine hepatitis virus strain 3 (MHV-3) infection. Adv. Exp. Med. Biol. *440*, 609-618.

Ding, Y., Wang, H., Shen, H., Li, Z., Geng, J., Han, H., Cai, J., Li, X., Kang, W., Weng, D., Lu, Y., Wu, D., He, L., and Yao, K. (2003). The clinical pathology of severe acute respiratory syndrome (SARS): a report from China. J. Pathol. *200*, 282-289.

Drosten, C., Günther, S., Preiser, W., van der Werf, S., Brodt, H.-R., Becker, S., Rabenau, H., Panning, M., Kolesnikova, L.,

Fouchier, R.A.M., Berger, A., Burguiere, A.-M., Cinatl, J., Eickmann, M., Escriou, N., Grywna, K., Kramme, S., Manuguerra, J.-C., Muller, S., Rickerts, W., Sturmer, M.V., Klenk, H.-D., and Osterhaus, A.D.M.E. (2003). Identification of a novel coronavirus in patients with severe acute respiratory syndrome. N. Engl. J. Med. *348*, 1967-1976.

Eleouet, J.F., Chilmonczyk, S., Besnardeau, L., and Laude, H. (1998). Transmissible gastroenteritis coronavirus induces programmed cell death in infected cells through a caspase-dependent pathway. J. Virol. *72*, 4918-4924.

Enjuanes, L. (2005). Coronavirus replication and reverse genetics. Curr. Top. Microbiol. Immunol. (Berlin: Springer) Vol. 287.

Enjuanes, L., Almazan, F., Sola, I., and Zuniga, S. (2006). Biochemical aspects of coronavirus replication and virus-host interaction. Annu. Rev. Microbiol. *60*, 211-230.

Enjuanes, L., DeDiego, M.L., Alvarez, E., Capiscol, C., and Baric, R. (2007a). Coronavirus vaccines with special focus on severe acute respiratory syndrome virus. In Coronavirus vaccines and antivirales, S. Perlman and T.M. Gallagher, eds., in press.

Enjuanes, L., DeDiego, M.L., Alvarez, E., Deming, D., Sheahan, T., and Baric, R. (2007b). Vaccines to prevent severe and acute respiratory syndrome coronavirus-induced disease. Virus Res., In press.

Enjuanes, L., Sola, I., Almazán, F., Ortego, J., Izeta, A., González, J.M., Alonso, S., Sánchez-Morgado, J.M., Escors, D., Calvo, E., Riquelme, C., and Sánchez, C.M. (2001). Coronavirus derived expression systems. J. Biotech. *88*, 183-204.

Enjuanes, L., Sola, I., Alonso, S., Escors, D., and Zuñiga, S. (2005). Coronavirus reverse genetics and development of vectors for gene expression. In Coronavirus replication and reverse genetics, L. Enjuanes, ed. (Springer), Vol. 287, pp. 161-197.

Escors, D., Camafeita, E., Ortego, J., Laude, H., and Enjuanes, L. (2001). Organization of two transmissible gastroenteritis coronavirus membrane protein topologies within the virion and core. J. Virol. *75*, 12228-12240.

Fischer, F., Stegen, C.F., Koetzner, C.A., and Masters, P.S. (1997). Analysis of a recombinant mouse hepatitis virus expressing a foreign gene reveals a novel aspect of coronavirus transcription. J. Virol. *71*, 5148-5160.

Fisher, R., Liao, Y.C., Hoffmann, K., Schillberg, S., and Emans, N. (1999). Molecular farming of recombinant antibodies in plants. Biol. Chem. *380*, 825-839.

Franks, T.J., Chong, P.Y., Chui, P., Galvin, J.R., Lourens, R.M., Reid, A.H., Selbs, E., McEvoy, C.P., Hayden, C.D., Fukuoka, J., Taubenberger, J.K., and Travis, W.D. (2003). Lung pathology of severe acute respiratory syndrome (SARS): a study of 8 autopsy cases from Singapore. Hum. Pathol. *34*, 743-748.

Frolov, I., Hoffman, T.A., Prágai, B.M., Dryga, S.A., Huang, H.V., Schlesinger, S., and Rice, C.M. (1996). Alphavirus-based expression vectors: Strategies and applications. Proc. Natl. Acad. Sci. USA *93*, 11371-11377.

Garcia-Sastre, A., and Biron, C.A. (2006). Type 1 interferons and the virus-host relationship: a lesson in detente. Science *312*, 879-882.

Gingras, A.C., Raught, B., and Sonenberg, N. (1999). eIF4 initiation factors: effectors of mRNA recruitment to ribosomes and regulators of translation. Annu. Rev. Biochem. *68*, 913-963.

Gonzalez, J.M., Gomez-Puertas, P., Cavanagh, D., Gorbalenya, A. E., and Enjuanes, L. (2003). A comparative sequence analysis to revise the current taxonomy of the family *Coronaviridae*. Arch. Virol. *148*, 2207-2235.

Gonzalez, J.M., Penzes, Z., Almazan, F., Calvo, E., and Enjuanes, L. (2002). Stabilization of a full-length infectious cDNA clone of transmissible gastroenteritis coronavirus by the insertion of an intron. J. Virol. *76*, 4655-4661.

Gorbalenya, A.E., Enjuanes, L., Ziebuhr, J., and Snijder, E.J. (2006). Nidovirales: evolving the largest RNA virus genome. Virus Res. *117*, 17-37.

Gorbalenya, A.E., Snijder, E.J., and Spaan, W.J. (2004). Severe acute respiratory syndrome coronavirus phylogeny: toward consensus. J. Virol. *78*, 7863-7866.

Gosert, R., Kanjanahaluethai, A., Egger, D., Bienz, K., and Baker, S. C. (2002). RNA replication of mouse hepatitis virus takes place at double-membrane vesicles. J. Virol. *76*, 3697-3708.

Graham, R.L., Sims, A.C., Brockway, S.M., Baric, R.S., and Denison, M.R. (2005). The nsp2 replicase proteins of murine hepatitis virus and severe acute respiratory syndrome coronavirus are dispensable for viral replication. J. Virol. *79*, 13399-13411.

Haagmans, B.L., Kuiken, T., Martina, B.E., Fouchier, R.A., Rimmelzwaan, G.F., van Amerongen, G., van Riel, D., de Jong, T., Itamura, S., Chan, K.H., Tashiro, M., and Osterhaus, A.D. (2004). Pegylated interferon-alpha protects type 1 pneumocytes against SARS coronavirus infection in macaques. Nat. Med. *10*, 290-293.

Haijema, B.J., Volders, H., and Rottier, P.J. (2004). Live, attenuated coronavirus vaccines through the directed deletion of group-specific genes provide protection against feline infectious peritonitis. J. Virol. *78*, 3863-3871.
He, R., Dobie, F., Ballantine, M., Leeson, A., Li, Y., Bastien, N., Cutts, T., Andonov, A., Cao, J., Booth, T.F., Plummer, F.A., Tyler, S., Baker, L., and Li, X. (2004a). Analysis of multimerization of the SARS coronavirus nucleocapsid protein. Biochem. Biophys. Res. Commun. *316*, 476-483.
He, R., Leeson, A., Andonov, A., Li, Y., Bastien, N., Cao, J., Osiowy, C., Dobie, F., Cutts, T., Ballantine, M., and Li, X. (2003). Activation of AP-1 signal transduction pathway by SARS coronavirus nucleocapsid protein. Biochem. Biophys. Res. Commun. *311*, 870-876.
He, R., Leeson, A., Ballantine, M., Andonov, A., Baker, L., Dobie, F., Li, Y., Bastien, N., Feldmann, H., Strocher, U., Theriault, S., Cutts, T., Cao, J., Booth, T.F., Plummer, F.A., Tyler, S., and Li, X. (2004b). Characterization of protein-protein interactions between the nucleocapsid protein and membrane protein of the SARS coronavirus. Virus Res. *105*, 121-125.
He, Y., Li, J., Heck, S., Lustigman, S., and Jiang, S. (2006). Antigenic and immunogenic characterization of recombinant baculovirus-expressed severe acute respiratory syndrome coronavirus spike protein: implication for vaccine design. J. Virol. *80*, 5757-5767.
Herschlag, D. (1995). RNA chaperones and the RNA folding problem. J. Biol. Chem. *270*, 20871-20874.
Hilton, A., Mizzen, L., MacIntyre, G., Cheley, S., and Anderson, R. (1986). Translational control in murine hepatitis virus infection. J. Gen. Virol. *67*, 923-932.
Hiscox, J.A. (2002). The nucleolus - a gateway to viral infection? Arch. Virol. *147*, 1077-1089.
Hiscox, J. A., Wurm, T., Wilson, L., Britton, P., Cavanagh, D., and Brooks, G. (2001). The coronavirus infectious bronchitis virus nucleoprotein localizes to the nucleolus. J. Virol. *75*, 506-512.
Hogan, R. J., Gao, G., Rowe, T., Bell, P., Flieder, D., Paragas, J., Kobinger, G.P., Wivel, N.A., Crystal, R.G., Boyer, J., Feldmann, H., Voss, T.G., and Wilson, J.M. (2004). Resolution of primary severe acute respiratory syndrome-associated coronavirus infection requires Stat1. J. Virol. *78*, 11416-11421.
Holt, J., Sgro, J.Y., Zuker, M., and Palmenberg, A. (2004). Computer folding of full-length viral genomes: a new toolkit for

automated analysis of RNAs longer than 10,000 bases. Seventh International Symposium on Positive Strand RNA Viruses. San Francisco, California, USA.

Hsue, B., and Masters, P.S. (1999). Insertion of a new transcriptional unit into the genome of mouse hepatitis virus. J. Virol. *73*, 6128-6135.

Huang, C., Ito, N., Tseng, C.T., and Makino, S. (2006). Severe acute respiratory syndrome coronavirus 7a accessory protein is a viral structural protein. J. Virol. *80*, 7287-7294.

Huang, P., and Lai, M.M.C. (1999). Polypyrimidine tract-binding protein binds to the complementary strand of the mouse hepatitis virus 3' untranslated region, thereby altering RNA conformation. J. Virol. *73*, 9110-9116.

Huang, P., and Lai, M.M.C. (2001). Heterogeneous nuclear ribonucleoprotein A1 binds to the 3'-untranslated region and mediates potential 5'-3'-end cross talks of mouse hepatitis virus RNA. J. Virol. *75*, 5009-5017.

Huang, Z.S., Su, W.H., Wang, J.L., and Wu, H.N. (2003). Selective strand annealing and selective strand exchange promoted by the N-terminal domain of hepatitis delta antigen. J. Biol. Chem. *278*, 5685-5693.

Huang, Z.S., and Wu, H.N. (1998). Identification and characterization of the RNA chaperone activity of hepatitis delta antigen peptides. J. Biol. Chem. *273*, 26455-26461.

Hurst, K.R., Kuo, L., Koetzner, C.A., Ye, R., Hsue, B., and Masters, P.S. (2005). A major determinant for membrane protein interaction localizes to the carboxy-terminal domain of the mouse coronavirus nucleocapsid protein. J. Virol. *79*, 13285-13297.

Ito, N., Mossel, E.C., Narayanan, K., Popov, V. L., Huang, C., Inoue, T., Peters, C.J., and Makino, S. (2005). Severe acute respiratory syndrome coronavirus 3a protein is a viral structural protein. J. Virol. *79*, 3182-3186.

Ivanyi-Nagy, R., Davidovic, L., Khandjian, E.W., and Darlix, J.L. (2005). Disordered RNA chaperone proteins: from functions to disease. Cell. Mol. Life Sci. *62*, 1409-1417.

Joo, M., and Makino, S. (1995). The effect of two closely inserted transcription consensus sequences on coronavirus transcription. J. Virol. *69*, 272-280.

Kamitani, W., Narayanan, K., Huang, C., Lokugamage, K., Ikegami, T., Ito, N., Kubo, H., and Makino, S. (2006). Severe acute respiratory syndrome coronavirus nsp1 protein suppresses host

gene expression by promoting host mRNA degradation. Proc. Natl. Acad. Sci. USA *103*, 12885-12890.
Kato-Maeda, M., Gao, Q., and Small, P.M. (2001). Microarray analysis of pathogens and their interaction with hosts. Cell Microbiol. *3*, 713-719.
Kazi, L., Lissenberg, A., Watson, R., de Groot, R.J., and Weiss, S.R. (2005). Expression of hemagglutinin esterase protein from recombinant mouse hepatitis virus enhances neurovirulence. J. Virol. *79*, 15064-15073.
Kopecky-Bromberg, S.A., Martinez-Sobrido, L., Frieman, M., Baric, R.A., and Palese, P. (2007). Severe acute respiratory syndrome coronavirus open reading frame (ORF) 3b, ORF 6, and nucleocapsid proteins function as interferon antagonists. J. Virol. *81*, 548-557.
Krishnan, R., Chang, R.Y., and Brian, D.A. (1996). Tandem placement of a coronavirus promoter results in enhanced mRNA synthesis from the downstream-most initiation site. Virology *218*, 400-405.
Kuo, L., Godeke, G.-J., Raamsman, M.J.B., Masters, P.S., and Rottier, P.J.M. (2000). Retargeting of coronavirus by substitution of the spike glycoprotein ectodomain: crossing the host cell species barrier. J. Virol. *74*, 1393-1406.
Kyuwa, S., Cohen, M., Nelson, G., Tahara, S.M., and Stohlman, S.A. (1994). Modulation of cellular macromolecular synthesis by coronavirus: implication for pathogenesis. J. Virol. *68*, 6815-6819.
Lai, M.M.C. (1997). RNA-protein interactions in the regulation of coronavirus RNA replication and transcription. Biol. Chem. *378*, 477-481.
Lai, M.M.C. (1998). Cellular factors in the transcription and replication of viral RNA genomes: a parallel to DNA-dependent RNA transcription. Virology *244*, 1-12.
Lai, M.M.C., and Cavanagh, D. (1997). The molecular biology of coronaviruses. Adv. Virus Res. *48*, 1-100.
Lassnig, C., Sanchez, C.M., Egerbacher, M., Walter, I., Majer, S., Kolbe, T., Pallares, P., Enjuanes, L., and Muller, M. (2005). Development of a transgenic mouse model susceptible to human coronavirus 229E. Proc. Natl. Acad. Sci. USA *102*, 8275-8280.
Lau, S.K., Woo, P.C., Li, K.S., Huang, Y., Tsoi, H.W., Wong, B.H., Wong, S.S., Leung, S.Y., Chan, K.H., and Yuen, K.Y. (2005). Severe acute respiratory syndrome coronavirus-like virus in Chinese horseshoe bats. Proc. Natl. Acad. Sci. USA *102*, 14040-14045.

Laude, H., and Masters, P.S. (1995). The coronavirus nucleocapsid protein. In The coronaviridae, S.G. Siddell, ed. (New York: Plenum press), pp. 141-158.
Law, A.H., Lee, D.C., Cheung, B.K., Yim, H.C., and Lau, A.S. (2007). Role for nonstructural protein 1 of severe acute respiratory syndrome coronavirus in chemokine dysregulation. J. Virol. *81*, 416-422.
Leong, W.F., Tan, H.C., Ooi, E.E., Koh, D.R., and Chow, V.T. (2005). Microarray and real-time RT-PCR analyses of differential human gene expression patterns induced by severe acute respiratory syndrome (SARS) coronavirus infection of Vero cells. Microbes Infect. *7*, 248-259.
Leung, W.K., To, K.F., Chan, P.K., Chan, H.L., Wu, A.K., Lee, N., Yuen, K.Y., and Sung, J.J. (2003). Enteric involvement of severe acute respiratory syndrome-associated coronavirus infection. Gastroenterology *125*, 1011-1017.
Li, H.P., Huang, P., Park, S., and Lai, M.M.C. (1999). Polypyrimidine tract-binding protein binds to the leader RNA of mouse hepatitis virus and serves as a regulator of viral transcription. J. Virol. *73*, 772-777.
Li, H.P., Zhang, X., Duncan, R., Comai, L., and Lai, M.M.C. (1997). Heterogeneous nuclear ribonucleoprotein A1 binds to the transcription-regulatory region of mouse hepatitis virus RNA. Proc. Natl. Acad. Sci. USA *94*, 9544-9549.
Li, Y., Fu, L., Gonzales, D.M., and Lavi, E. (2004). Coronavirus neurovirulence correlates with the ability of the virus to induce proinflammatory cytokine signals from astrocytes and microglia. J. Virol. *78*, 3398-3406.
Lin, Y.J., Liao, C.L., and Lai, M.M.C. (1994). Identification of the cis-acting signal for minus-strand RNA synthesis of a murine coronavirus: implications for the role of minus-strand RNA in RNA replication and transcription. J. Virol. *68*, 8131-8140.
Lin, Y.J., Zhang, X., Wu, R.C., and Lai, M.M.C. (1996). The 3' untranslated region of coronavirus RNA is required for subgenomic mRNA transcription from a defective interfering RNA. J. Virol. *70*, 7236-7240.
Liu, Q., Yu, W., and Leibowitz, L. (1997). A specific host cellular protein binding element near the 3' end of mouse hepatitis virus genomic RNA. Virology *232*, 74-85.
Lorsch, J.R. (2002). RNA chaperones exist and DEAD box proteins get a life. Cell *109*, 797-800.

Loutfy, M.R., Blatt, L.M., Siminovitch, K.A., Ward, S., Wolff, B., Lho, H., Pham, D.H., Deif, H., LaMere, E.A., Chang, M., Kain, K. C., Farcas, G.A., Ferguson, P., Latchford, M., Levy, G., Dennis, J.W., Lai, E.K., and Fish, E.N. (2003). Interferon alfacon-1 plus corticosteroids in severe acute respiratory syndrome: a preliminary study. J. Am. Med. Assoc. *290*, 3222-3228.

Maaser, C., Eckmann, L., Paesold, G., Kim, H.S., and Kagnoff, M.F. (2002). Ubiquitous production of macrophage migration inhibitory factor by human gastric and intestinal epithelium. Gastroenterology *122*, 667-680.

Marra, M.A., Jones, S.J.M., Astell, C.R., Holt, R.A., Brooks-Wilson, A., Butterfield, Y.S.N., Khattra, J., Asano, J.K., Barber, S.A., Chan, S.Y., Cloutier, A., Coughlin, S.M., Freeman, D., Girn, N., Griffith, O.L., Leach, S.R., Mayo, M., McDonald, H., Montgomery, S.B., Pandoh, P.K., Petrescu, A.S., Robertson, A.G., Schein, J.E., Siddiqui, A., Smailus, D.E., Stott, J.M., Yang, G.S., Plummer, F., Andonov, A., Artsob, H., Bastien, N., Bernard, K., Booth, T.F., Bowness, D., Czub, M., Drebot, M., Fernando, L., Flick, R., Garbutt, M., Gray, M., Grolla, A., Jones, S., Feldmann, H., Meyers, A., Kabani, A., Li, Y., Normand, S., Stroher, U., Tipples, G.A., Tyler, S., Vogrig, R., Ward, D., Watson, B., Brunham, R.C., Krajden, M., Petric, M., Skowronski, D.M., Upton, C., and Roper, R.L. (2003). The genome sequence of the SARS-associated coronavirus. Science *300*, 1399-1404.

Masters, P.S. (1999). Reverse genetics of the largest RNA viruses. Adv. Virus Res. *53*, 245-264.

Masters, P.S. (2006). The molecular biology of coronaviruses. Adv. Virus Res. *66*, 193-292.

Masters, P.S., and Rottier, P.J. (2005). Coronavirus reverse genetics by targeted RNA recombination. Curr. Top. Microbiol. Immunol. *287,* 133-159.

Miller, W.A., and Koev, G. (2000). Synthesis of subgenomic RNAs by positive-strand RNA virus. Virology *273*, 1-8.

Mir, M.A., and Panganiban, A.T. (2006). The bunyavirus nucleocapsid protein is an RNA chaperone: possible roles in viral RNA panhandle formation and genome replication. RNA *12*, 272-282.

Mizutani, T., Fukushi, S., Murakami, M., Hirano, T., Saijo, M., Kurane, I., and Morikawa, S. (2004a). Tyrosine dephosphorylation of STAT3 in SARS coronavirus-infected Vero E6 cells. FEBS Lett. *577*, 187-192.

Mizutani, T., Fukushi, S., Saijo, M., Kurane, I., and Morikawa, S. (2004b). Importance of Akt signaling pathway for apoptosis in SARS-CoV-infected Vero E6 cells. Virology *327*, 169-174.

Mizutani, T., Fukushi, S., Saijo, M., Kurane, I., and Morikawa, S. (2004c). Phosphorylation of p38 MAPK and its downstream targets in SARS coronavirus-infected cells. Biochem. Biophys. Res. Commun. *319*, 1228-1234.

Mohandas, D.V., and Dales, S. (1991). Endosomal association of a protein phosphatase with high dephosphorylating activity against a coronavirus nucleocapsid protein. FEBS Lett. *282*, 419-424.

Nagy, P.D., and Simon, A.E. (1997). New insights into the mechanisms of RNA recombination. Virology *235*, 1-9.

Nanda, S.K., and Leibowitz, J.L. (2001). Mitochondrial aconitase binds to the 3' untranslated region of the mouse hepatitis virus genome. J. Virol. *75*, 3352-3362.

Nelson, G.W., Stohlman, S.A., and Tahara, S.M. (2000). High affinity interaction between nucleocapsid protein and leader/intergenic sequence of mouse hepatitis virus RNA. J. Gen. Virol. *81*, 181-188.

Ng, L.F., Hibberd, M.L., Ooi, E.E., Tang, K.F., Neo, S.Y., Tan, J., Murthy, K.R., Vega, V.B., Chia, J.M., Liu, E.T., and Ren, E.C. (2004). A human in vitro model system for investigating genome-wide host responses to SARS coronavirus infection. BMC Infect. Dis. *4*, 34.

Olsen, C.W. (1993). A review of feline infectious peritonitis virus: molecular biology, immunopathogenesis, clinical aspects, and vaccination. Vet. Microbiol. *36*, 1-37.

Olson, M.O., Dundr, M., and Szebeni, A. (2000). The nucleolus: an old factory with unexpected capabilities. Trends Cell Biol. *10*, 189-196.

Ortego, J., Escors, D., Laude, H., and Enjuanes, L. (2002). Generation of a replication-competent, propagation-deficient virus vector based on the transmissible gastroenteritis coronavirus genome. J. Virol. *76*, 11518-11529.

Parker, M.M., and Masters, P.S. (1990). Sequence comparison of the N genes of five strains of the coronavirus mouse hepatitis virus suggests a three domain structure for the nucleocapsid protein. Virology *179*, 463-468.

Parra, B., Hinton, D.R., Marten, N.W., Bergmann, C.C., Lin, M.T., Yang, C.S., and Stohlman, S.A. (1999). IFN-gamma is required for viral clearance from central nervous system oligodendroglia. J. Immunol. *162*, 1641-1647.

Pasternak, A.O., Gultyaev, A.P., Spaan, W.J., and Snijder, E.J. (2000). Genetic manipulation of arterivirus alternative mRNA leader-body junction sites reveals tight regulation of structural protein expression. J. Virol. *74*, 11642-11653.

Pasternak, A.O., Spaan, W.J., and Snijder, E.J. (2004). Regulation of relative abundance of arterivirus subgenomic mRNAs. J. Virol. *78*, 8102-8113.

Pasternak, A.O., Spaan, W.J., and Snijder, E.J. (2006). Nidovirus transcription: how to make sense? J. Gen. Virol. *87*, 1403-1421.

Pasternak, A.O., van den Born, E., Spaan, W.J.M., and Snijder, E.J. (2001). Sequence requirements for RNA strand transfer during nidovirus discontinuous subgenomic RNA synthesis. EMBO J. *20*, 7220-7228.

Pasternak, A.O., van den Born, E., Spaan, W.J.M., and Snijder, E.J. (2003). The stability of the duplex between sense and antisense transcription-regulating sequences is a crucial factor in arterivirus subgenomic mRNA synthesis. J. Virol. *77*, 1175-1183.

Pedersen, K.W., van der Meer, Y., Roos, N., and Snijder, E.J. (1999). Open reading frame 1a-encoded subunits of the arterivirus replicase induce endoplasmic reticulum-derived double-membrane vesicles which carry the viral replication complex. J. Virol. *73*, 2016-2026.

Peiris, J.S., Yuen, K.Y., Osterhaus, A.D., and Stohr, K. (2003). The severe acute respiratory syndrome. N. Engl. J. Med. *349*, 2431-2441.

Pensaert, M., Callebaut, P., and Vergote, J. (1986). Isolation of a porcine respiratory, non-enteric coronavirus related to transmissible gastroenteritis. Vet. Quart. *8*, 257-260.

Pensaert, M.B., and De Bouck, P. (1978). A new coronavirus-like particle associated with diarrhea in swine. Arch. Virol. *58*, 243-247.

Penzes, Z., González, J.M., Calvo, E., Izeta, A., Smerdou, C., Mendez, A., Sánchez, C.M., Sola, I., Almazán, F., and Enjuanes, L. (2001). Complete genome sequence of transmissible gastroenteritis coronavirus PUR46-MAD clone and evolution of the Purdue virus cluster. Virus Genes *23*, 105-118.

Perlman, S. (1998). Pathogenesis of coronavirus-induced infections: review of pathological and immunological aspects. Adv. Exp. Med. Biol. *440*, 503-514.

Poon, L.L., Chu, D.K., Chan, K.H., Wong, O.K., Ellis, T.M., Leung, Y.H., Lau, S.K., Woo, P.C., Suen, K.Y., Yuen, K.Y., Guan, Y., and Peiris, J.S. (2005). Identification of a novel coronavirus in bats. J. Virol. *79*, 2001-2009.

Prentice, E., Jerome, W.G., Yoshimori, T., Mizushima, N., and Denison, M.R. (2004a). Coronavirus replication complex formation utilizes components of cellular autophagy. J. Biol. Chem. *279*, 10136-10141.

Prentice, E., McAuliffe, J., Lu, X., Subbarao, K., and Denison, M.R. (2004b). Identification and characterization of severe acute respiratory syndrome coronavirus relicase proteins. J. Virol. *78*, 9977-9986.

Rajkowitsch, L., Semrad, K., Mayer, O., and Schroeder, R. (2005). Assays for the RNA chaperone activity of proteins. Biochem. Soc. Trans. *33*, 450-456.

Reeder, R.H., and Lang, W.H. (1997). Terminating transcription in eukaryotes: lessons learned from RNA polymerase I. Trends Biochem. Sci. *22*, 473-477.

Rempel, J.D., Murray, S.J., Meisner, J., and Buchmeier, M.J. (2004). Differential regulation of innate and adaptive immune responses in viral encephalitis. Virology *318*, 381-392.

Resta, S., Luby, J.P., Rosenfeld, C.D., and Siegel, J.D. (1985). Isolation and propagation of a human enteric coronavirus. Science *229*, 978-981.

Risco, C., Muntión, M., Enjuanes, L., and Carrascosa, J.L. (1998). Two types of virus-related particles are found during transmissible gastroenteritis virus morphogenesis. J. Virol. *72*, 4022-4031.

Robinson, M.J., and Cobb, M.H. (1997). Mitogen-activated protein kinase pathways. Curr. Opin. Cell Biol. *9*, 180-186.

Rota, P.A., Oberste, M.S., Monroe, S.S., Nix, W.A., Campganoli, R., Icenogle, J.P., Peñaranda, S., Bankamp, B., Maher, K., Chen, M.-H., Tong, S., Tamin, A., Lowe, L., Frace, M., DeRisi, J.L., Chen, Q., Wang, D., Erdman, D.d., Peret, T.C.T., Burns, C., Ksiazek, T. G., Rollin, P.E., Sanchez, A., Liffick, S., Holloway, B., Limor, J., McCaustland, K., Olsen-Rassmussen, M., Fouchier, R., Gunther, S., Osterhaus, A.D.M.E., Drosten, C., Pallansch, M.A., Anderson, L.J., and Bellini, W.J. (2003). Characterization of a novel coronavirus associated with severe acute respiratory syndrome. Science *300*, 1394-1399.

Roulston, A., Marcellus, R.C., and Branton, P.E. (1999). Viruses and apoptosis. Annu. Rev. Microbiol. *53*, 577-628.

Rowland, R.R., Kervin, R., Kuckleburg, C., Sperlich, A., and Benfield, D.A. (1999). The localization of porcine reproductive and respiratory syndrome virus nucleocapsid protein to the nucleolus of infected cells and identification of a potential nucleolar localization signal sequence. Virus Res. *64*, 1-12.

Sanchez, C.M., Izeta, A., Sánchez-Morgado, J.M., Alonso, S., Sola, I., Balasch, M., Plana-Duran, J., and Enjuanes, L. (1999). Targeted recombination demonstrates that the spike gene of transmissible gastroenteritis coronavirus is a determinant of its enteric tropism and virulence. J. Virol. *73*, 7607-7618.

Sanchez, C.M., Jiménez, G., Laviada, M.D., Correa, I., Suñe, C., Bullido, M.J., Gebauer, F., Smerdou, C., Callebaut, P., Escribano, J.M., and Enjuanes, L. (1990). Antigenic homology among coronaviruses related to transmissible gastroenteritis virus. Virology *174*, 410-417.

Sarma, J.D., Scheen, E., Seo, S.H., Koval, M., and Weiss, S.R. (2002). Enhanced green fluorescent protein expression may be used to monitor murine coronavirus spread in vitro and in the mouse central nervous system. J. Neurovirol. *8*, 381-391.

Sawicki, D.L., Wang, T., and Sawicki, S.G. (2001). The RNA structures engaged in replication and transcription of the A59 strain of mouse hepatitis virus. J. Gen. Virol. *82*, 386-396.

Sawicki, S.G., and Sawicki, D.L. (1990). Coronavirus transcription: subgenomic mouse hepatitis virus replicative intermediates function in RNA synthesis. J. Virol. *64*, 1050-1056.

Sawicki, S.G., and Sawicki, D.L. (1998). A new model for coronavirus transcription. Adv. Exp. Med. Biol. *440*, 215-219.

Sawicki, S.G., Sawicki, D.L., and Siddell, S.G. (2007). A contemporary view of coronavirus transcription. J. Virol. *81*, 20-29.

Schaecher, S.R., Mackenzie, J.M., and Pekosz, A. (2006). The ORF7b protein of SARS-CoV is expressed in virus-infected cells and incorporated into SARS-CoV particles. J. Virol. *81*, 718-731.

Schelle, B., Karl, N., Ludewig, B., Siddell, S.G., and Thiel, V. (2005). Selective replication of coronavirus genomes that express nucleocapsid protein. J. Virol. *79*, 6620-6630.

Schroeder, R., Barta, A., and Semrad, K. (2004). Strategies for RNA folding and assembly. Nat. Rev. Mol. Cell. Biol. *5*, 908-919.

Sebolt-Leopold, J.S., Dudley, D.T., Herrera, R., Van Becelaere, K., Wiland, A., Gowan, R.C., Tecle, H., Barrett, S.D., Bridges, A., Przybranowski, S., Leopold, W.R., and Saltiel, A.R. (1999). Blockade of the MAP kinase pathway suppresses growth of colon tumors in vivo. Nat. Med. *5*, 810-816.

Senanayake, S.D., and Brian, D.A. (1999). Translation from the 5' untranslated region (UTR) of mRNA 1 is repressed, but that from the 5' UTR of mRNA 7 is stimulated in coronavirus-infected cells. J. Virol. *73*, 8003-8009.

Sgro, J.Y., Holt, J., Zuker, M., and Palmenberg, A. (2004). RNA folding of the complete SARS and MHV coronavirus genomes. Seventh International Symposium on Positive Strand RNA Viruses. San Francisco, California, USA.

Shen, X., and Masters, P.S. (2001). Evaluation of the role of heterogeneous nuclear ribonucleoprotein A1 as a host factor in murine coronavirus discontinuous transcription and genome replication. Proc. Natl. Acad. Sci. USA *98*, 2717-2722.

Shi, S.T., Huang, P., Li, H.P., and Lai, M.M.C. (2000). Heterogeneous nuclear ribonucleoprotein A1 regulates RNA synthesis of a cytoplasmic virus. EMBO J. *19*, 4701-4711.

Shi, S.T., Schiller, J.J., Kanjanahaluethai, A., Baker, S.C., Oh, J.W., and Lai, M.M.C. (1999). Colocalization and membrane association of murine hepatitis virus gene 1 produts and de novo-synthesized viral RNA in infected cells. J. Virol. *73*, 5957-5969.

Shi, S.T., Yu, G.Y., and Lai, M.M. (2003). Multiple type A/B heterogeneous nuclear ribonucleoproteins (hnRNPs) can replace hnRNP A1 in mouse hepatitis virus RNA synthesis. J. Virol. *77*, 10584-10593.

Siddell, S., Wege, H., Barthel, A., and ter Meulen, V. (1981). Intracellular protein synthesis and the in vitro translation of coronavirus JHM mRNA. Adv. Exp. Med. Biol. *142*, 193-207.

Siddell, S.G., Ziebuhr, J., and Snijder, E.J. (2005). Coronaviruses, toroviruses, and arteriviruses. In Virology, 10th Ed., B.W.J.M. a. V. ter-Meulen, ed. (Washington, D.C., USA: ASM Press), Vol. 1, pp. 848-856.

Silva, J.M., Li, M.Z., Chang, K., Ge, W., Golding, M.C., Rickles, R.J., Siolas, D., Hu, G., Paddison, P.J., Schlabach, M.R., Sheth, N., Bradshaw, J., Burchard, J., Kulkarni, A., Cavet, G., Sachidanandam, R., McCombie, W.R., Cleary, M.A., Elledge, S.J., and Hannon, G.J. (2005). Second-generation shRNA libraries covering the mouse and human genomes. Nat. Genet. 37, 1281-1288.

Snijder, E.J., Brinton, M.A., Faaberg, K.S., Godeny, E.K., Gorbalenya, A.E., MacLachlan, N.J., Mengeling, W.L., and Plagemann, P.G.W. (2005). Arteriviridae. In Virus Taxonomy. Eighth Report of the International Committee on Taxonomy of Viruses, C.M. Fauquet, M.A. Mayo, J. Maniloff, U. Desselberger, and L.A. Ball, eds. (San Diego, CA, USA: Academic Press) pp. 965-974.

Snijder, E.J., van der Meer, Y., Zevenhoven-Dobbe, J., Onderwater, J.J., van der Meulen, J., Koerten, H.K., and

Mommaas, A.M. (2006). Ultrastructure and origin of membrane vesicles associated with the severe acute respiratory syndrome coronavirus replication complex. J. Virol. *80*, 5927-5940.

Sola, I., Alonso, S., Zúñiga, S., Balach, M., Plana-Durán, J., and Enjuanes, L. (2003). Engineering transmissible gastroenteritis virus genome as an expression vector inducing latogenic immunity. J. Virol. *77*, 4357-4369.

Sola, I., Moreno, J.L., Zuñiga, S., Alonso, S., and Enjuanes, L. (2005). Role of nucleotides immediately flanking the transcription-regulating sequence core in coronavirus subgenomic mRNA synthesis. J. Virol. *79*, 2506-2516.

Spaan, W.J.M., Brian, D., Cavanagh, D., de Groot, R.J., Enjuanes, L., Gorbalenya, A.E., Holmes, K.V., Masters, P., Rottier, P., Taguchi, F., and Talbot, P.J. (2005). Coronaviridae. In Virus Taxonomy. Eighth Report of the International Committee on Taxonomy of Viruses, C. M. Fauquet, M. A. Mayo, J. Maniloff, U. Desselberger, and L. A. Ball, eds. (San Diego, CA, USA: Academic Press), pp. 947-964.

Spagnolo, J.F., and Hogue, B.G. (2000). Host protein interactions with the 3' end of bovine coronavirus RNA and the requirement of the poly(A) tail for coronavirus defective genome replication. J. Virol. *74*, 5053-5065.

Sperry, S.M., Kazi, L., Graham, R.L., Baric, R.S., Weiss, S.R., and Denison, M.R. (2005). Single-amino-acid substitutions in open reading frame (ORF) 1b-nsp14 and ORF 2a proteins of the coronavirus mouse hepatitis virus are attenuating in mice. J. Virol. *79*, 3391-3400.

Spiegel, M., Pichlmair, A., Martinez-Sobrido, L., Cros, J., Garcia-Sastre, A., Haller, O., and Weber, F. (2005). Inhibition of beta interferon induction by severe acute respiratory syndrome coronavirus suggests a two-step model for activation of interferon regulatory factor 3. J. Virol. *79*, 2079-2086.

St-Jean, J.R., Desforges, M., Almazan, F., Jacomy, H., Enjuanes, L., and Talbot, P.J. (2006). Recovery of a neurovirulent human coronavirus OC43 from an infectious cDNA clone. J. Virol. *80*, 3670-3674.

Stirrups, K., Shaw, K., Evans, S., Dalton, K., Casais, R., Cavanagh, D., and Britton, P. (2000). Expression of reporter genes from the defective RNA CD-61 of the coronavirus infectious bronchitis virus. J. Gen. Virol. *81*, 1687-1698.

Stohlman, S.A., Baric, R.S., Nelson, G.N., Soe, L.H., Welter, L.M., and Deans, R.J. (1988). Specific interaction between coronavirus leader RNA and nucleocapsid protein. J. Virol. *62*, 4288-4295.
Stohlman, S.A., Fleming, J.O., Patton, C.D., and Lai, M.M.C. (1983). Synthesis and subcellular localization of the murine coronavirus nucleocapsid protein. Virology *130*, 527-532.
Surjit, M., Kumar, R., Mishra, R.N., Reddy, M.K., Chow, V.T., and Lal, S.K. (2005). The severe acute respiratory syndrome coronavirus nucleocapsid protein is phosphorylated and localizes in the cytoplasm by 14-3-3-mediated translocation. J. Virol. *79*, 11476-11486.
Surjit, M., Liu, B., Jameel, S., Chow, V.T., and Lal, S.K. (2004a). The SARS coronavirus nucleocapsid protein induces actin reorganization and apoptosis in COS-1 cells in the absence of growth factors. Biochem. J. *383*, 13-18.
Surjit, M., Liu, B., Kumar, P., Chow, V.T., and Lal, S.K. (2004b). The nucleocapsid protein of the SARS coronavirus is capable of self-association through a C-terminal 209 amino acid interaction domain. Biochem. Biophys. Res. Commun. *317*, 1030-1036.
Suñe, C., Jiménez, G., Correa, I., Bullido, M.J., Gebauer, F., Smerdou, C., and Enjuanes, L. (1990). Mechanisms of transmissible gastroenteritis coronavirus neutralization. Virology *177*, 559-569.
Tahara, S.M., Dietlin, T.A., Bergmann, C.C., Nelson, G.W., Kyuwa, S., Anthony, R.P., and Stohlman, S.A. (1994). Coronavirus translational regulation: leader affects mRNA efficiency. Virology *202*, 621-630.
Tahara, S.M., Dietlin, T.A., Nelson, G.W., Stohlman, S.A., and Manno, D.J. (1998). Mouse hepatitis virus nucleocapsid protein as a translational effector of viral mRNAs. Adv. Exp. Med. Biol. *440*, 313-318.
Tan, Y.J., Fielding, B.C., Goh, P.Y., Shen, S., Tan, T.H., Lim, S.G., and Hong, W. (2004). Overexpression of 7a, a protein specifically encoded by the severe acute respiratory syndrome coronavirus, induces apoptosis via a caspase-dependent pathway. J. Virol. *78*, 14043-14047.
Tang, B.S., Chan, K.H., Cheng, V.C., Woo, P.C., Lau, S.K., Lam, C.C., Chan, T.L., Wu, A.K., Hung, I.F., Leung, S.Y., and Yuen, K.Y. (2005). Comparative host gene transcription by microarray analysis early after infection of the Huh7 cell line by severe acute respiratory syndrome coronavirus and human coronavirus 229E. J. Virol. *79*, 6180-6193.

Thiel, V., Herold, J., Schelle, B., and Siddell, S. (2001a). Infectious RNA transcribed in vitro from a cDNA copy of the human coronavirus genome cloned in vaccinia virus. J. Gen. Virol. *82*, 1273-1281.
Thiel, V., Herold, J., Schelle, B., and Siddell, S.G. (2001b). Viral replicase gene products suffice for coronavirus discontinuous transcription. J. Virol. *75*, 6676-6681.
Thiel, V., Karl, N., Schelle, B., Disterer, P., Klagge, I., and Siddell, S.G. (2003). Multigene RNA vector based on coronavirus transcription. J. Virol. *77*, 9790-9798.
Tijms, M.A., van der Meer, Y., and Snijder, E.J. (2002). Nuclear localization of non-structural protein 1 and nucleocapsid protein of equine arteritis virus. J. Gen. Virol. *83*, 795-800.
Timani, K.A., Liao, Q., Ye, L., Zeng, Y., Liu, J., Zheng, Y., Yang, X., Lingbao, K., Gao, J., and Zhu, Y. (2005). Nuclear/nucleolar localization properties of C-terminal nucleocapsid protein of SARS coronavirus. Virus Res. *114*, 23-34.
Tsuchihashi, Z., and Brown, P.O. (1994). DNA strand exchange and selective DNA annealing promoted by the Human Immunodeficiency Virus Type I nucleocapsid protein. J. Virol. *68*, 5863-5870.
van der Hoek, L., Pyrc, K., Jebbink, M.F., Vermeulen-Oost, W., Berkhout, R.J., Wolthers, K.C., Wertheim-van Dillen, P.M., Kaandorp, J., Spaargaren, J., and Berkhout, B. (2004). Identification of a new human coronavirus. Nat. Med. *10*, 368-373.
van der Meer, Y., Snijder, E.J., Dobbe, J.C., Schleich, S., Denison, M.R., Spaan, W.J.M., and Krijnse Locker, J. (1999). Localization of mouse hepatitis virus nonstructural proteins and RNA synthesis indicates a role for late endosomes in viral replication. J. Virol. *73*, 7641-7657.
van Marle, G., Dobbe, J.C., Gultyaev, A.P., Luytjes, W., Spaan, W.J.M., and Snijder, E.J. (1999). Arterivirus discontinuous mRNA transcription is guided by base pairing between sense and antisense transcription-regulating sequences. Proc. Natl. Acad. Sci. USA *96*, 12056-12061.
van Marle, G., Luytjes, W., Van der Most, R.G., van der Straaten, T., and Spaan, W.J.M. (1995). Regulation of Coronavirus mRNA transcription. J. Virol. *69*, 7851-7856.
Vasilescu, J., and Figeys, D. (2006). Mapping protein-protein interactions by mass spectrometry. Curr. Opin. Biotechnol. *17*, 394-399.

Wang, Y., Detrick, B., Yu, Z.X., Zhang, J., Chesky, L., and Hooks, J. J. (2000). The role of apoptosis within the retina of coronavirus-infected mice. Invest. Ophthalmol. Vis. Sci. *41*, 3011-3018.

Wong, C.K., Lam, C.W., Wu, A.K., Ip, W.K., Lee, N.L., Chan, I.H., Lit, L.C., Hui, D.S., Chan, M.H., Chung, S.S., and Sung, J.J. (2004). Plasma inflammatory cytokines and chemokines in severe acute respiratory syndrome. Clin. Exp. Immunol. *136*, 95-103.

Woo, P.C., Lau, S.K., Chu, C.M., Chan, K.H., Tsoi, H.W., Huang, Y., Wong, B.H., Poon, R.W., Cai, J.J., Luk, W.K., Poon, L.L., Wong, S.S., Guan, Y., Peiris, J.S., and Yuen, K.Y. (2005). Characterization and complete genome sequence of a novel coronavirus, coronavirus HKU1, from patients with pneumonia. J. Virol. *79*, 884-895.

Woo, P.C., Lau, S.K., Li, K.S., Poon, R.W., Wong, B.H., Tsoi, H.W., Yip, B.C., Huang, Y., Chan, K.H., and Yuen, K.Y. (2006). Molecular diversity of coronaviruses in bats. Virology *351*, 180-187.

Wurm, T., Chen, H., Hodgson, T., Britton, P., Brooks, G., and Hiscox, J.A. (2001). Localization to the nucleolus is a common feature of coronavirus nucleoproteins, and the protein may disrupt host cell division. J. Virol. *75*, 9345-9356.

Yan, L., Velikanov, M., Flook, P., Zheng, W., Szalma, S., and Kahn, S. (2003). Assessment of putative protein targets derived from the SARS genome. FEBS Lett. *554*, 257-263.

Yang, M., Li, C.K., Li, K., Hon, K.L., Ng, M.H., Chan, P.K., and Fok, T.F. (2004). Hematological findings in SARS patients and possible mechanisms (review). Int. J. Mol. Med. *14*, 311-315.

Yang, Y., Xiong, Z., Zhang, S., Yan, Y., Nguyen, J., Ng, B., Lu, H., Brendese, J., Yang, F., Wang, H., and Yang, X.F. (2005). Bcl-xL inhibits T-cell apoptosis induced by expression of SARS coronavirus E protein in the absence of growth factors. Biochem. J. *392*, 135-143.

Youn, S., Leibowitz, J.L., and Collisson, E.W. (2005). In vitro assembled, recombinant infectious bronchitis viruses demonstrate that the 5a open reading frame is not essential for replication. Virology *332*, 206-215.

Yount, B., Curtis, K.M., and Baric, R.S. (2000). Strategy for systematic assembly of large RNA and DNA genomes: the transmissible gastroenteritis virus model. J. Virol. *74*, 10600-10611.

Yount, B., Curtis, K.M., Fritz, E.A., Hensley, L.E., Jahrling, P.B., Prentice, E., Denison, M.R., Geisbert, T.W., and Baric, R.S.

(2003). Reverse genetics with a full-length infectious cDNA of severe acute respiratory syndrome coronavirus. Proc. Natl. Acad. Sci. USA *100*, 12995-13000.
Yount, B., Denison, M.R., Weiss, S.R., and Baric, R.S. (2002). Systematic assembly of a full-length infectious cDNA of mouse hepatitis virus strain A59. J. Virol. *76*, 11065-11078.
Yount, B., Roberts, R.S., Lindesmith, L., and Baric, R.S. (2006). Rewiring the severe acute respiratory syndrome coronavirus (SARS-CoV) transcription circuit: engineering a recombination-resistant genome. Proc. Natl. Acad. Sci. USA *103*, 12546-12551.
Yu, W., and Leibowitz, J.L. (1995a). A conserved motif at the 3' end of mouse hepatitis virus genomic RNA required for host protein binding and viral RNA replication. Virology *214*, 128-138.
Yu, W., and Leibowitz, J.L. (1995b). Specific binding of host cellular proteins to multiple sites within the 3' end of mouse hepatitis virus genomic RNA. J. Virol. *69*, 2016-2023.
Yuan, X., Shan, Y., Zhao, Z., Chen, J., and Cong, Y. (2005). G0/G1 arrest and apoptosis induced by SARS-CoV 3b protein in transfected cells. Virol. J. *2*, 66.
Yuan, X., Wu, J., Shan, Y., Yao, Z., Dong, B., Chen, B., Zhao, Z., Wang, S., Chen, J., and Cong, Y. (2006). SARS coronavirus 7a protein blocks cell cycle progression at G0/G1 phase via the cyclin D3/pRb pathway. Virology *346*, 74-85.
Zhang, X., and Lai, M.M.C. (1995). Interactions between the cytoplasmic proteins and the intergenic (promoter) sequence of mouse hepatitis virus RNA: correlation with the amounts of subgenomic mRNA transcribed. J. Virol. *69*, 1637-1644.
Ziebuhr, J. (2005). The coronavirus replicase. In Coronavirus replication and reverse genetics. L. Enjuanes, ed. (Springer), Vol. 287, pp. 57-94.
Ziebuhr, J., Snijder, E.J., and Gorbalenya, A.E. (2000). Virus-encoded proteinases and proteolytic processing in the Nidovirales. J. Gen. Virol. *81*, 853-879.
Zuñiga, S., Sola, I., Moreno, J.L., Sabella, P., Plana-Duran, J., and Enjuanes, L. (2007). Coronavirus nucleocapsid protein is an RNA chaperone. Virology *357*, 215-227.
Zuñiga, S., Sola, I., Alonso, S., and Enjuanes, L. (2004). Sequence motifs involved in the regulation of discontinuous coronavirus subgenomic RNA synthesis. J. Virol. *78*, 980-994.
Zuñiga, S., Sola, I., Moreno, J.L., Alonso, S., and Enjuanes, L. (2006). Regulation of coronavirus transcription: viral and cellular

proteins interacting with transcription-regulating sequences. Adv. Exp. Med. Biol. *581*, 31-35.

Chapter 5

Hendra and Nipah Virus

Bevan Sawatsky, Charlene Ranadheera, Hana M. Weingartl and Markus Czub

ABSTRACT

Over the past decade, the previously unknown paramyxoviruses *Hendra virus* (HeV) and *Nipah virus* (NiV) have emerged in humans and livestock in Australia and Southeast Asia. Both viruses are contagious, highly virulent, and capable of infecting a number of mammalian species and causing potentially fatal disease. Due to the lack of a licensed vaccine or antiviral therapies, HeV and NiV are designated as biosafety level (BSL) 4 agents and are potential bioterrorist agents. The genomic structure of both viruses is that of a typical paramyxovirus. However, due to limited sequence homology and little immunological cross-reactivity with other paramyxoviruses, HeV and NiV have been classified into a new genus within the family *Paramyxoviridae* named *Henipavirus*.

BRIEF HISTORY OF HENDRA AND NIPAH VIRUS

Hendra virus (HeV) was identified as the cause of an outbreak of respiratory and neurological disease in humans and horses in Australia in the early 1990s (Murray et al., 1995a; Murray et al., 1995b). Ultrastructural analysis of the new virus isolated from the kidney of a horse trainer who died in the course of infection revealed typical paramyxovirus features such as pleomorphic shape, herringbone nucleocapsid, and surface spike projections. HeV cross-reacted with convalescent sera from horses and humans, however, antisera against measles (MeV), canine distemper (CDV), and rinderpest viruses (RPV) all failed to neutralize HeV. After the initial outbreak, which killed two humans and more than a dozen horses, HeV has re-emerged sporadically in Australia (Field et al., 2000).

Nipah virus (NiV) was discovered after several hundred cases of acute febrile encephalitis of humans occurred in Malaysia and Singapore from late 1998 until 1999 (Chua et al., 2000; Halpin et al., 1999; Parashar et al., 2000; Paton et al., 1999; Selvey et al., 1995). A close association of human cases with pigs suggested that the causative agent was Japanese encephalitis virus (JEV), which is endemic to the region. However, since vaccination against JEV and mosquito control measures failed to end the outbreak, JEV was discounted as a possible cause of the outbreak. NiV was isolated from human brain tissue obtained from a fatal case. Electron microscopy, immunofluorescence, and sequence analysis indicated that the new virus was a paramyxovirus, and was closely related, but not identical, to HeV. In the past few years, NiV emerged in several outbreaks in Bangladesh (Anonymous, 2003; Butler, 2004; Hsu et al., 2004) and India (Chadha et al., 2006).

Natural infections with HeV or NiV appear to be associated with high case fatality rates in humans (NiV: 35-75%; HeV: unknown) and lower rates in animals (NiV in pigs: ~5%, HeV in horses: unknown). However, since the exact number of exposed and infected individuals has not been determined for the natural outbreaks, these rates may not be accurate.

PATHOGENESIS

Several studies have indicated that bats, particularly fruit bats ('flying foxes') of the suborder *Megachiroptera*, serve as a reservoir for NiV and HeV in the wild (Sendow et al., 2006). Both viruses have been isolated from fruit bats (Chua et al., 2002; Halpin et al., 2000; Reynes et al., 2005). Antibodies to henipaviruses have been detected in five species of bat from Australia, Malaysia, Bangladesh, Cambodia, Indonesia, and Madagascar (Iehlé et al., 2007; Johara et al., 2001; Olson et al., 2002; Sendow et al., 2006; Young et al., 1996). Fruit bats of the suborder *Megachiroptera* have a wide geographical distribution, ranging from Australia to southern Asia to eastern African islands such as Madagascar (Field et al., 2001) and it cannot be excluded that further outbreaks of henipavirus infection in humans are possible in these areas. It is believed that transmission of HeV from bats to horses and NiV from bats to pigs and humans occurred in isolated incidents. Infected body fluids and/or tissues from bats (e.g. urine or fetuses) likely served as a source of infection. Furthermore, fruits contaminated by bats infected with NiV may have played a role in causing infection in humans and pigs (Luby et al., 2006). Bats themselves do not appear to be susceptible to clinical disease after infection with NiV or HeV (Eaton et al., 2005).

The incubation period for henipavirus infections appears to range from a few days to 3 weeks in animals and humans (Goh et al., 2000), although it is difficult to determine under natural conditions. These observations are supported by experimental data on infected pigs (Middleton et al., 2002; Weingartl et al., 2005), hamsters (Wong et al., 2003), cats (Hooper et al., 1997b; Middleton et al., 2002; Mungall et al., 2006; Westbury et al., 1996), horses (Hooper et al., 1997a), and guinea pigs (Hooper et al., 1997b; Williamson et al., 1999; Williamson et al., 2001). It is believed that the majority of humans that seroconvert to NiV exhibit clinical signs (Goh et al., 2000; Wong et al., 2002), while in pigs morbidity rates are assumed to be lower than in humans. No data are available for HeV, largely because all horses with suspected HeV infection were euthanized. During the outbreak of NiV in pigs in Malaysia, a large percentage of pigs remained asymptomatic (> 80 - 90%), an observation confirmed in animal experiments using pigs (Weingartl et al., 2005). However, NiV spread very rapidly among pigs housed on farms (Mohd Nor et al., 2000). NiV caused disease and death at high rates in hamsters, guinea pigs, and cats, even when low doses of infectivity were administered (Mungall et al., 2006; Wong et al., 2003). Mice, rats, and rabbits are not susceptible to infection with henipaviruses.

At the onset of clinical disease, henipavirus infection in humans generally presents as a severe influenza-like illness with high fever, headache and myalgia. Respiratory symptoms, and even fatal pneumonia, have also been observed. Late onset or relapse of NiV encephalitis occurred in 12 (7.5%) of the survivors, and of these 18% died during or after the Malaysian outbreak. One fatal human case of HeV infection followed a protracted disease course after an initial illness with mild meningoencephalitis. The patient continued to suffer from fatigue, developed epileptic seizures, and finally coma and death 13 months after exposure to HeV (Allworth et al., 1995; Hooper et al., 1996; O'Sullivan et al., 1997; Rogers et al., 1996). Clinical disease in animals is similar to that seen in humans, with respiratory and neurological signs. The main histopathological findings include systemic vasculitis, meningitis and encephalitis. Lymphatic and respiratory organs are also subject to direct and/or indirect virus-induced damage. Virus replication has been detected in a variety of tissues, such as blood vessels, respiratory organs including the trachea, lymph nodes, and the central nervous system (CNS) (Weingartl et al., 2005). Since neurological disease as well as damage to nervous tissue is one of the hallmarks of henipavirus infection, viral transport into the brain is essential for subsequent henipavirus replication in brain cells.

It appears that in pigs, NiV infects cells of the nasal cavity, initiating viremia and spreading further into the reticuloendothelial system, with additional replication in endothelial cells of the blood and lymphatic vessels. This cell-associated and free virus can then spill over into the smooth muscle cells of the blood vessels. Positive immunostaining of the lymphocytes and monocytes/macrophages, especially osteoclasts, indirectly indicate the presence of a viremia. The virus possibly exploits several methods of crossing the blood-brain barrier (BBB). The first possibility is that NiV is transported via nervii olfactorii into the CNS, as suggested in an experimental pig study (Weingartl et al., 2005). Furthermore, NiV could cross the choroid plexus, as indicated by positive staining of the choroid plexus stroma and the ependymal lining of the ventricules, and detection of virus in the cerebrospinal fluid (CSF). Additionally, replication of the virus in the endothelial cells of the blood vessels in the brain could facilitate the virus crossover into the parenchyma of the CNS. Virus entry into the brain by transmigration of infected cells cannot be excluded, since a number of antigen-positive lymphocytes and monocytes/macrophages were observed in the vicinity of the blood and lymphatic vessels (Weingartl et al., 2005).

A conclusive diagnosis of henipavirus infection is only possible after laboratory testing. Clinical signs, even in conjunction with epidemiological data are not sufficient for a final diagnosis. Virus isolation and/or the detection of viral gene products by means of electron microscopy, immunostaining and RT-PCR would indicate an ongoing infection. Detection of antibodies, in particular virus neutralizing antibodies or antibodies against the henipavirus glycoproteins would be indicative of an infection with NiV or HeV beyond the acute phase. Serological diagnosis does not preclude the possibility that the patient is still viremic and actively shedding virus.

GENETICS

HENIPAVIRUS GENOME STRUCTURE AND ENCODED PROTEINS

NiV and HeV have been classified into a new genus in the family *Paramyxoviridae*, named *Henipavirus*, in which HeV is the type species. Like all members of the order *Mononegavirales*, NiV and HeV are enveloped viruses with a negative-sense single-stranded RNA [(-)-ssRNA] genome. Until recently, NiV and HeV had the largest known genomes of any paramyxovirus at 18,246 nt and 18,234 nt, respectively. NiV and HeV have been superseded in terms of genome length since the full sequences of J-virus (J-V, 18,954 kb) (Jack et al., 2005) and Beilong virus (BeV, 19,212 kb) (Li et al., 2006) became available. Morbilliviruses and respiroviruses typically have genome lengths of ~15-16 kb (Harcourt et al., 2001; Harcourt et al., 2000; Wang et al., 2000). The linear gene order of henipaviruses is similar to that of the morbilliviruses and respiroviruses, and consists of the six transcriptional units 3'-N-P/V/W/C-M-F-G-L-5'. Six structural proteins are produced during infection: the nucleocapsid protein (N, 532 aa), the phosphoprotein (P, 707 aa in HeV and 709 aa in NiV), the matrix protein (M, 352 aa), the fusion glycoprotein (F, 546 aa), the attachment glycoprotein (G, 604 aa in HeV and 602 aa in NiV), and the viral RNA-dependent RNA polymerase (L, 2,244 aa). The henipavirus P gene also has the potential to produce several accessory proteins from an alternate open reading frame (ORF) (C) or via the process of co-transcriptional editing (V and W), a feature also found in morbilliviruses and respiroviruses (Harcourt et al., 2001; Harcourt et al., 2000; Vidal and Kolakofsky, 1989; Wang et al., 2000; Yu et al., 1998a; Yu et al., 1998b). Expression of C, V, W, has yet to be detected in the course of infection. The C protein is encoded by a separate smaller ORF within the P gene with two consecutive start sites, located 23 and 26 nucleotides downstream from the NiV P translational start site which produces a protein 166 aa in length (Harcourt et al., 2000). Like the P transcripts of many other paramyxoviruses (Harcourt et al., 2000), NiV and HeV P transcripts contain a conserved AG-rich region, which serves as an RNA editing site. The polymerase complex incorporates one or two additional G residues at the editing site (position 1213 of the NiV P transcript and position 1207 of the HeV transcript) to generate transcripts for V (55 aa) and W (47 aa), respectively (Harcourt et al., 2000). The RNA editing sites of NiV and HeV are identical, and are closely related to those of the morbilliviruses and respiroviruses, and more distantly related to those of the rubulaviruses (Harcourt et al., 2000).

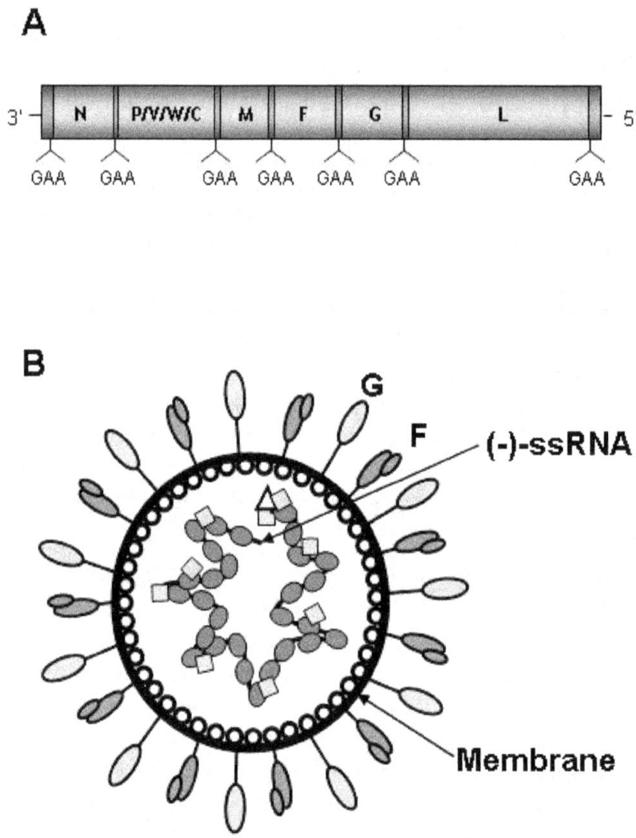

Fig. 1. Henipavirus genomic and virion structure. A) Schematic representation of the henipavirus genome along with the intergenic region trinucleotide repeats. B) Representation of the henipavirus virion structure. The surface glycoproteins (F and G) are shown on the diagram. The matrix protein (M, open circles) underlies the membrane. The (-)-ssRNA genomic RNA is encapsidated in the virion by the nucleocapsid protein (N, grey ovals) and is also associated with the phosphoprotein (P, light grey squares) and the L protein (open triangle).

The viral gene products of henipaviruses are similar in size to other viruses within the subfamily *Paramyxovirinae*, except for the HeV and NiV P proteins, which are approximately 100 aa longer than other P proteins from similar viruses (Wang et al., 1998). The size differences between henipavirus genomes and other paramyxoviruses are due primarily to the N gene 5' non-translated region [NTR] (568 nt for HeV and 586 nt for NiV vs. 43-111 nt), P gene 5' NTR (469 nt vs. 66-123 nt), the F gene 5' NTR (418 nt for HeV and 412 nt for NiV vs. 38-137 nt), G gene 3' NTR (233 nt vs. 20-78 nt), and G gene 5' NTR (516 nt for HeV and 504 nt for NiV vs. 66-111 nt) (Harcourt et al., 2001).

The intergenic region (IGR) separating each of the viral genes is identical for both the NiV and HeV genomes and consists of a consensus trimeric nucleotide sequence, GAA (Fig. 1A) (Harcourt et al., 2000; Wang et al., 2000). Morbilliviruses and respiroviruses also have the same conserved IGR. However, HeV and NiV are the only viruses which have identical IGR at all seven gene junctions. Morbilliviruses and respiroviruses show at least some deviation in the IGR consensus sequence, which usually occurs at gene junctions closer to the 5' end of the genome (Harcourt et al., 2000; Kolakofsky et al., 1998).

The transcriptional start and stop signals are located in the NTR of the viral genes. NiV and HeV have identical transcriptional start signals for the N, P, M, F and L genes (Harcourt et al., 2001; Harcourt et al., 2000; Wang et al., 2000). The transcriptional initiation signal for the G gene has a single nucleotide difference when compared to the corresponding HeV initiation sequence (Harcourt et al., 2000). The transcriptional termination signals are also identical for the NiV and HeV N, P, G and L genes (Harcourt et al., 2000). There are two nucleotide changes in both the M and F genes between NiV and HeV (Harcourt et al., 2000). Table 1 shows the close similarity between the initiation and termination sequences of henipaviruses with morbilliviruses and respiroviruses.

The genomic termini of most paramyxoviruses, including NiV and HeV, are well conserved and complementary (Halpin et al., 2004). The 3' termini of henipaviruses are 52 nucleotides long with an additional 3 nucleotide intergenic sequence, which is identical in size to that of other members of the subfamily *Paramyxovirinae*. The 3' leader sequences of NiV and HeV have 77% sequence similarity with each other. The 5' trailer sequence is 30 nucleotides long with an additional 3 nucleotide intergenic sequence located immediately after the L gene. Unlike the 3' leader sequence, there is much more variability between the sizes of 5' trailers; those of HeV and NiV are shorter than those of morbilliviruses and respiroviruses.

Table 1. Comparison of the gene start and stop signal for a variety of paramyxoviruses (Accession numbers: *Nipah virus* AF212302, *Hendra virus* AF017149, *Measles virus* NC001498, *Canine distemper virus* AF014953, *Sendai virus* NC001552, and *Human parainfluenza virus 3* D84095). Viral gene junctions that differ from the GAA consensus are indicated in bold face.

Genes	Virus	Gene Stop	Gene Junction	Gene Start
Leader-N	NiV		GAA	UCCUUGGUUCU
	HeV		GAA	UCCUUGGUUCU
	MeV		GAA	UCCUAAGUUCU
	CDV		GAA	UCCCAGUUACU
	SeV		GAA	UCCCAGUUUC
	HPIV-3		GAA	UCCUAAUUUC
N-P	NiV	AAUUCUUUUU	GAA	UCCUAGGUUCU
	HeV	AAUUCUUUUU	GAA	UCCUAGGUUCU
	MeV	AAUAUUUUUU	GAA	UCCUUGGUCCA
	CDV	AAUAUUUUUU	GAA	UCCUGGUCCA
	SeV	CAUUCUUUUU	GAA	UCCCACUUUC
	HPIV-3	CAUUCUUUUU	GAA	UCCUAAUUGC
P-M	NiV	AAUUCUUUUU	GAA	UCCUCUGUCCA
	HeV	AAUUCUUUUU	GAA	UCCUCUGUCCA
	MeV	AAUAUUUUUU	GAA	UCCUCGUUUCA
	CDV	AAUAUUUUUU	GAA	UCCUGUUCUCG
	SeV	AAUUCUUUUU	GAA	UCCCACUUUC
	HPIV-3	AAUUCUUUUU	GAA	UCCUAUUUUC
	NDV			
	RPV			
M-F	NiV	AAUGUUUUUU	GAA	UCCUCGGUUC
	HeV	AAUUCUUUUU	GAA	UCCUCGGUUC
	MeV	UUUGUUUU	GAA	UCCCGGUUCC
	CDV	AAUUAGUUUU	GAA	UCCCAGGUCC
	SeV	UAUUCUUUUU	GAA	UCCCUAUUUC
	HPIV-3	UUUAGUUUUU	GAA	UCCUAGUUUC
F-G	NiV	AAUUAUUUUUU	GAA	UCCUGGGUCCA
	HeV	AAAUGUUUUUU	GAA	UCCUGGGUUCA
	MeV	UUAAUUUU	GAA	UCCCACGUUCU
	CDV	UAAUUUCUUUU	GAA	UCCCGAGUCCA
	SeV	UAUUCUUUUU	GAA	UCCCACUUUCA
	HPIV-3	CAUGUUUUUU	GAA	UCCUUGUUUCA
G-L	NiV	UAAUUCUUUUU	GAA	UCCUGGGUCCA
	HeV	UAAUUCUUUUU	GAA	UCCUGGGUUCA
	MeV	AUUCUUUUU	**GCA**	UCCCAGGUUCA
	CDV	AAUAUUUUUUU	**GAU**	UCCUAGGUUCU
	SeV	AAUUCUUUUU	**GGG**	UCCCACUUAC
	HPIV-3	UAAUGUUUUUU	GAA	UCCUCCUUUCA
L-Trailer	NiV	UAAUUCUUUUU	GAA	
	HeV	UAAUUCUUUUU	GAA	
	MeV	UAAUUUGUUUU	GAA	
	CDV	CAAUAUGCU$_{(8)}$	**GUU**	
	SeV	UCAUUCUUUUU	GAA	
	HPIV-3	UCAUUCUUUUU	**GUA**	

NiV strain variation

Sequence analysis of NiV isolates from the recent outbreaks in Bangladesh has shed some light on the population dynamics of NiV in a natural setting. The Malaysia (18,246 nt) and Bangladesh (18,252 nt) strains of NiV differ very little, with maximal divergence at the protein level seen in the P protein (92.0% aa identity) (Harcourt et al., 2005). The N protein (98.3% aa identity), M protein (98.9% aa identity), F protein (98.4% aa identity), G protein (95.5% aa identity), and L protein (98.2% aa identity) all display a high degree of relatedness between the Malaysia and Bangladesh strains (Harcourt et al., 2005). The open reading frames exhibit greater than 90% nucleotide homology, while the 3' and 5' NTRs all show much more variable levels of homology, ranging from 75.5% for the 3' NTR of the G gene to 100% for the 3' NTR of the N gene (Harcourt et al., 2005). The NiV strain that caused an outbreak of encephalitis in India has not been isolated and therefore has not been fully sequenced, but N and M sequences obtained from patients during this outbreak were more closely related to the Bangladesh strain than the Malaysia strain of NiV (Chadha et al., 2006). The relatively small amount of sequence divergence between the Malaysia and Bangladesh strains indicates that NiV may have a high degree of genetic stability. There is also some speculation that the distinct genetic lineages of NiV that have been detected in different geographical regions (Harcourt et al., 2005; Reynes et al., 2005) are a result of co-evolution with the local animal reservoir (Chadha et al., 2006).

THE VIRAL LIFE CYCLE

The viral structural proteins all have essential roles throughout the viral life cycle. A schematic representation of the virion is shown in Fig. 1B. The viral membrane is studded with two glycoproteins. The G protein is involved in viral attachment to the cellular receptor and the F protein is involved in mediating fusion of the plasma and viral membranes and release of the viral nucleocapsid (NC) into the cytoplasm of cells. The matrix protein is membrane-associated, provides structural support for the virion, and also aids in the assembly and egress of the virion. The viral core or NC consists of the N, P, and L proteins and contains all factors necessary for viral transcription and genome replication. An overview of the henipavirus life cycle is shown in Fig. 2.

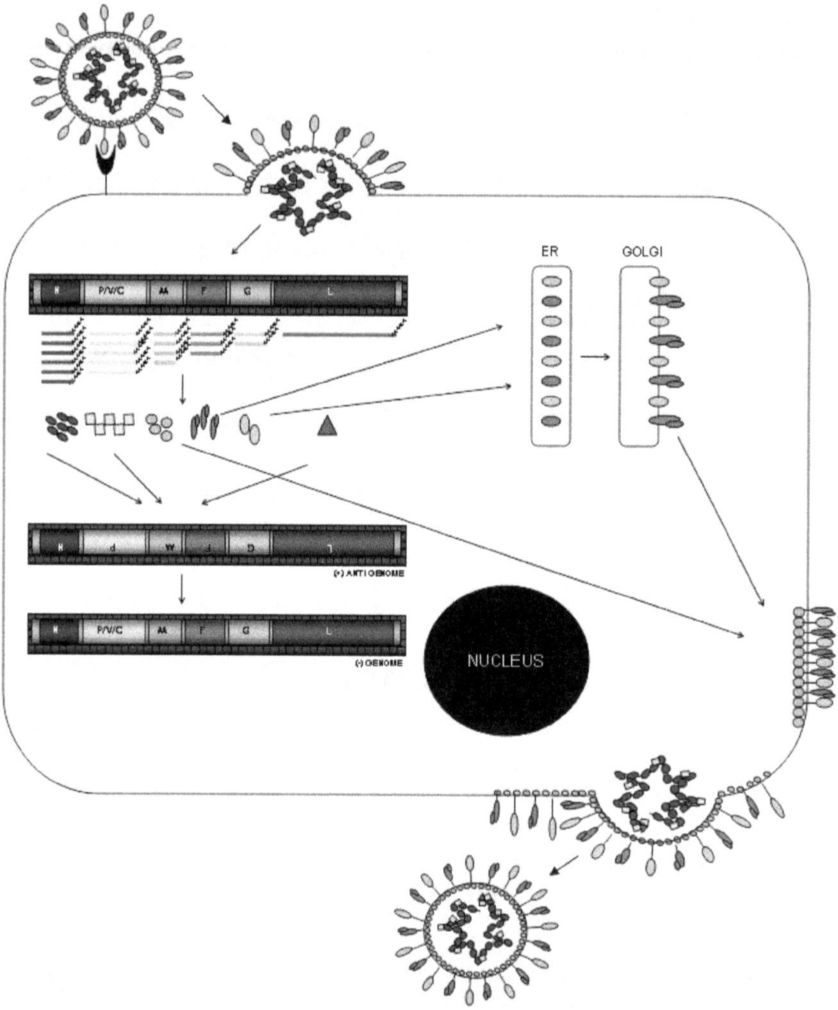

Fig. 2. The henipavirus life cycle. Virus entry occurs via interaction between the G glycoprotein and the cellular receptors (ephrin-B2 and ephrin-B3), followed by membrane fusion at neutral pH mediated by the F glycoprotein. Transcription and replication of the genome occur in the cytoplasm of infected cells. The F and G glycoprotein precursors are proteolytically processed and glycosylated in the endoplasmic reticulum (ER) and the Golgi apparatus before insertion into the cellular membrane. The newly-synthesized viral RNP (N+P+L+genome) is packaged into virions likely through interaction with the matrix protein before fully mature viral particles bud from the cellular plasma membrane.

VIRAL ENTRY

NiV and HeV entry into cells is dependent on the presence of specific cellular receptors. Recently, ephrin-B2 and ephrin-B3 have been identified as henipavirus receptors (Bonaparte et al., 2005; Negrete et al., 2005; Negrete et al., 2006). By analogy to other paramyxoviruses, it is believed that binding of the cellular receptor triggers a conformational change in G which in turn activates the F protein which then mediates fusion of the viral and plasma membranes allowing entry of the viral NC into the cytoplasm. Viral entry and surface glycoproteins will be discussed in more detail in the following sections.

PARAMYXOVIRUS REPLICATION

Studies on the replication of HeV or NiV have been limited. Based on their evolutionary relatedness to other paramyxoviruses, NiV and HeV are believed to have modes of replication similar to those of other paramyxoviruses. However, experimental support for these concepts as they pertain to henipaviruses is currently lacking.

Synthesis of viral mRNA

Upon release of the NC into the cytoplasm, the viral genes must be transcribed and the viral mRNA must be translated. The polymerase complex or more specifically the transcriptase, consisting of the P and L proteins, is believed to bind the leader region of the viral genome and initiate transcription of the first gene, N. It was previously assumed that transcription of viral mRNAs followed a 'Single Initiation Stop-Start Model' (Emerson, 1982). This model proposes that the polymerase complex positions itself at position 1 of the 3' end of the genome, and first transcribes the leader region and subsequent downstream genes (Emerson, 1982). The termination of the previous gene transcript is a prerequisite for transcription of downstream genes (Banerjee et al., 1977). Once the polymerase complex has reached the end of the gene, the polymerase complex ceases transcribing the viral RNA, scans the intergenic region, and reinitiates transcription at the next gene start consensus sequence. However, recent studies with vesicular stomatitis virus (VSV) have called into question some aspects of this well accepted model. Using a UV mapping technique, Whelen and Wertz have demonstrated that N mRNA transcripts are present in infected cells without the requirement for the transcription of the leader RNA (Whelan and Wertz, 2002). More recently, Qanungo and colleagues more directly identified the initiation point of the transcriptase by purifying viral transcriptases from infected cells and using them to synthesize viral transcripts (Qanungo et al., 2004). Viral mRNA was also generated in the absence of the leader RNA, confirming prior observations (Qanungo et al., 2004). These results led to a modified version of the previously described model. The transcriptase enters the genome at the 3' end but only initiates transcription at the transcriptional start signal of the N gene (Qanungo et al., 2004; Whelan and Wertz, 2002). Host cellular machinery is then used for translation. Production of all viral mRNAs is thought to be produced through the stuttering of the viral polymerase along the viral template in a discontinuous manner through a single site of initiation (Emerson, 1982).

This method of initiation and termination is believed to cause the formation of a gradient of viral mRNA, resulting in the production of more transcripts from the genes at the 3' end than from genes at the 5' end of the genome. This is suggested to happen because the polymerase complex has a tendency to fall off the viral template as it gets further away from the 3' end of the genome. Therefore, regulation of the expression of viral proteins is based on their position in the genome. Production of viral transcripts does not follow a linear gradient, but rather declines at points of attenuation due to differences in the gene boundaries and/or polyadenylation signals (Kato et al., 1999; Wright et al., 2005). There are only a few publications supporting the hypothesis of gradient transcript production (Cattaneo et al., 1987; Homann et al., 1990). In a more recent study of HeV transcripts, which employed quantitative real-time PCR, it appeared that two points of attenuation existed. While viral transcript levels were high for N, P, and M, a 5.5-fold decrease in transcript levels was observed for F and G, with a further 2.2-fold decrease in the level of the L transcript (Wright et al., 2005). These results indicate that a gradient of viral transcripts is present, albeit not a linear gradient, as the polymerase progresses along the viral genome, with more 3' viral gene transcripts generated than 5' viral gene transcripts.

Synthesis of viral genome

The synthesis of full-length viral (+)-sense anti-genome follows upon accumulation of sufficient amounts of viral proteins, but it is not clear at present how the transition from transcription of viral mRNA to replication of viral anti-genome and genome is initiated. Historically, the most commonly accepted model of replication has focussed around the expression of the viral N protein. It was believed that as the amount of N increased, N alone mediated a transition from viral transcription to viral replication. Supposedly, the N protein was somehow able to mask the intergenic regions of the viral template, thereby causing read-through of the intergenic regions by the viral polymerase (P and L), which should lead to the production of full-length anti-genomic RNA. At the same time, encapsidation of the full length anti-genome would occur. The encapsidated anti-genome could then serve as a template for production of negative-sense viral genomes. Recent studies with VSV have led to a modified version of this model of replication. Gupta and colleagues found that a trimeric polymerase complex, consisting of N, P, and L, was essential in the production of viral genomes (Gupta et al., 2003). This modified mechanism still fits into the historically accepted model of replication, except that rather than N acting independently, it acts in conjunction with the polymerase complex to read through the intergenic regions and produce full-length viral RNA. However, this new evidence implied that production of viral transcripts or full-length viral RNA is predetermined based upon the formation of a transcriptase (P-L) complex (Curran et al., 1992; Das et al., 1995; Gao and Lenard, 1995a; Gao and Lenard, 1995b) or a replicase (N-P-L) complex (Curran et al., 1991; Gupta et al., 2003; Horikami et al., 1992; Vidal and Kolakofsky, 1989; Vulliemoz and Roux, 2001). According to the modified version, levels of the N protein still play an important role in virus replication, since increased levels of N protein will contribute to the formation of the replicase complex, but are not solely responsible for the transition from viral transcription to viral replication (shown in Fig. 3).

Encapsidation of viral genome

Upon synthesis, full-length viral RNA is encapsidated by the N protein. Increased levels of newly expressed N protein are essential for the encapsidation of full-length viral RNA. The N protein will bind to other molecules of N, although this interaction must occur during the binding of full length viral RNA. In the absence of viral RNA, N will instead bind cellular RNA and will no longer be available for further functions in viral replication. In order to facilitate the role of N protein in replication, the P protein binds to N shortly after synthesis and maintains it in a soluble form (Fig. 4) (Howard and Wertz, 1989; Peluso and Moyer, 1988). Furthermore, the P protein acts as a chaperone to enhance the specificity of N for viral RNA rather than non-specific cellular RNA (Banerjee et al., 1989; Chenik et al., 1994; Curran et al., 1995; Das and Banerjee, 1992; Davis et al., 1986; Errington and Emmerson, 1997; Garcia et al., 1993; Nishio et al., 1996; Peluso and Moyer, 1984; Precious et al., 1995; Spehner et al., 1997; Yang et al., 1998; Zhao and Banerjee, 1995). Upon binding of N to viral RNA, P dissociates from N and either binds other N proteins or the viral polymerase (Curran et al., 1993; Curran et al., 1995; Fooks et al., 1993; Homann et al., 1991; Horikami et al., 1992; Horikami et al., 1992; Spehner et al., 1991). While specific steps of encapsidation have been elucidated for several members of the order *Mononegavirales*, no studies have been published for henipaviruses.

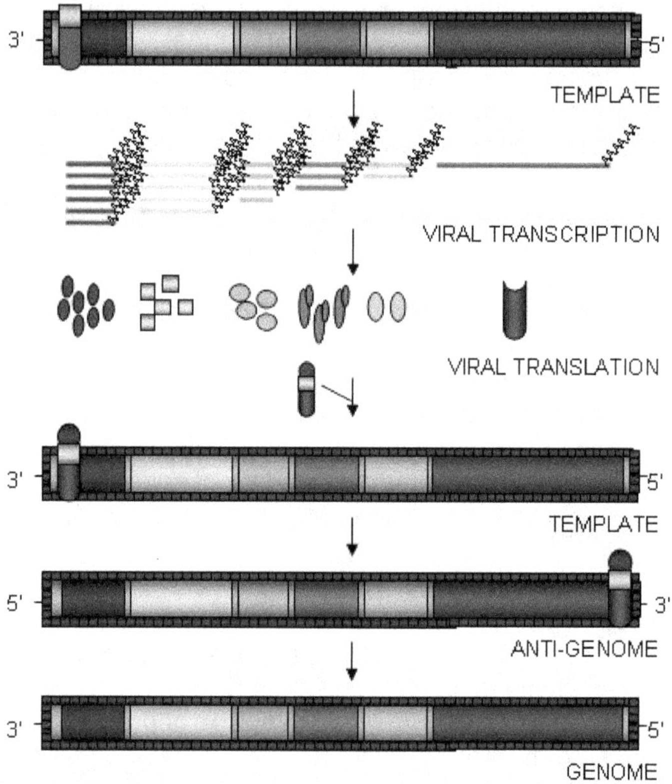

Fig. 3. Synthesis of henipavirus RNA illustrating transcription of viral genes and replication of full-length anti-genomic and genomic RNA.

The 'Rule of Six'

The encapsidation of the viral genomic template by the N protein creates a helical nucleocapsid core. It has been suggested that each molecule of N is able to interact with precisely six nucleotides of viral genomic or anti-genomic RNA (Egelman et al., 1989). Due to this particular binding of N to full length viral RNA, it is believed that the viral genome will only be replicated efficiently if the genome length is a multiple of six nucleotides (Calain and Roux, 1993). This peculiarity of paramyxovirus replication has been termed the 'Rule of Six'. This rule does not apply to all viruses in the order *Mononegavirales*, such as rhabdoviruses and filoviruses. It should also be noted that members of the subfamily *Pneumovirinae* within the family

Paramyxoviridae (e.g. respiratory syncytial virus and human metapneumovirus) also do not have hexameric genome lengths. However, the viral genomes of all members of the subfamily *Paramyxovirinae*, including NiV and HeV, have genomes with lengths in multiples of six nucleotides (Enami et al., 1989; Harcourt et al., 2001; Harcourt et al., 2000; Hausmann et al., 1996; Sidhu et al., 1994; Wang et al., 2000). Several studies employing paramyxovirus minigenome systems have demonstrated that a genome length with a multiple of six is essential for efficient viral replication and deviation from a multiple of six caused attenuation of viral replication (Calain and Roux, 1993; Engelhorn et al., 1993; Harty and Palese, 1995; Kolakofsky et al., 1998; Radecke et al., 1995; Sidhu et al., 1995). Studies using a NiV minigenome have shown that a hexameric genome length is important for NiV replication (Halpin et al., 2004). Halpin and colleagues inserted 1-6 nucleotides at the junction between the translational stop and the L gene NTR. Their results indicated that efficient reporter protein expression from the NiV minigenome occurred only when the minigenome size was a multiple of six nucleotides (Halpin et al., 2004).

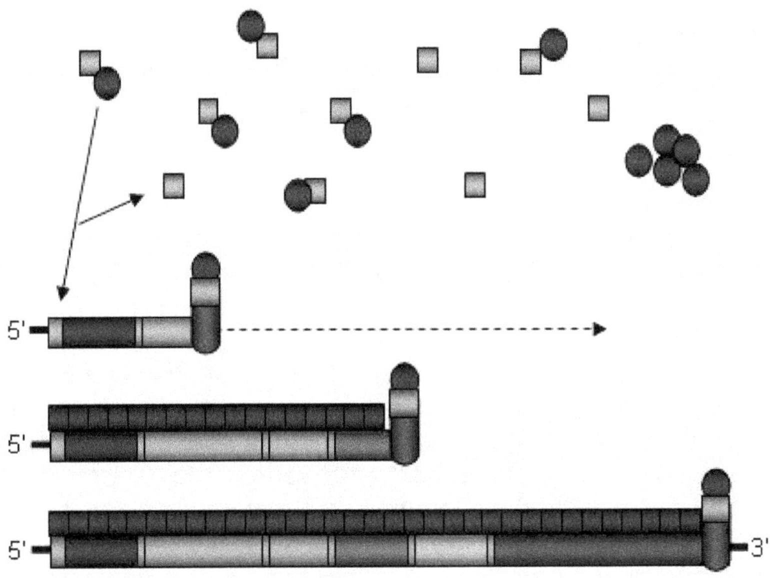

Fig. 4. Encapsidation of the henipavirus RNA by the nucleocapsid and phosphoprotein. Soluble N protein (dark circles) is complexed with either P (light squares) or P and L (light squares+dark cylinder). When it encapsidates the henipavirus genome, N (dark squares) binds nascent viral RNA as it is being synthesized by the replicase complex (N/P/L).

Proteins interacting with NiV and HeV N

The N protein has a number of roles in the viral life cycle. Two of its major functions are the ability to interact with the polymerase complex to assist in production of full-length viral genome and to encapsidate viral RNA. Additionally, the N protein aids in virion assembly and egress through its interaction with the M protein. The N protein has many interacting partners in the formation of the replication competent NC. It interacts with the P protein both in the maintenance of N in a soluble state and in the formation of the replicase complex. Interaction of NiV N with NiV P or HeV N with HeV P has been demonstrated by Chan and colleagues, who have determined that there are two independent NiV N binding regions on the NiV P protein, one located at the C terminal end (aa 636-709) and the other at the N terminal end (aa 3-220) (Chan et al., 2004). It was also found that the binding site for P is located at aa 468-496 of both HeV and NiV N (Chan et al., 2004). As mentioned earlier, the N protein has a major role in the encapsidation of viral antigenome, which involves both N-RNA and N-N interactions and subsequent assembly into the NC (Eshaghi et al., 2005b; Goldsmith et al., 2003; Ong et al., 2002) (unpublished data).

NiV L protein

The RNA-dependent RNA polymerase recognizes the viral genome only in the context of the NC, which appears to be a unique feature of (-)-ssRNA viruses. The L protein contains the enzymatic functions necessary for viral transcription, replication, mRNA capping, polyadenylation and methyltransferase activities (Abraham et al., 1975; Banerjee, 1987a; Banerjee, 1987b; Hammond and Lesnaw, 1987a; Hammond and Lesnaw, 1987b; Hercyk et al., 1988; Hunt et al., 1988; Hunt et al., 1984; Ongradi et al., 1985; Testa and Banerjee, 1977). The L proteins of members of the order *Mononegavirales* share a high degree of sequence similarity. Comparison between the sequences of many paramyxovirus L proteins revealed six conserved domains (Poch et al., 1990). Each domain is able to function on its own without the need for interaction between domains (Brown et al., 2005; Duprex et al., 2002). Domain III of the paramyxovirus L protein has been shown to contain the highly conserved GDNQ motif, which is thought to be the catalytic site for phosphodiester bond formation (Poch et al., 1989). The henipavirus L proteins have also been shown to have these six domains based on sequence analysis. However, in Domain III of henipaviruses the Gln residue of the GDNQ motif has been replaced with a Glu residue, resulting in a GDNE motif (Harcourt et al., 2001). By introducing single amino acid substitutions into the

GDNE motif, Magoffin and colleagues determined that the Glu residue was interchangeable with a Gln residue, but replacement of Glu with Lys or Pro abrogated polymerase function in the NiV minigenome system (Magoffin et al., 2007). If the Glu residue was replaced with Asp, Ala, Asn, Ile or Gly polymerase function was still observed, but at lower levels than in the wild-type L protein (Magoffin et al., 2007). These data demonstrate functional homology between henipavirus L proteins and other paramyxovirus L proteins, and show that certain changes to the Glu residue of the GDNE motif will impair NiV L function.

VIRUS ASSEMBLY AND EGRESS

Once all of the viral proteins have been produced and the viral genome has been encapsidated, the viral components are shuttled to the plasma membrane where the M protein mediates virion assembly and viral egress via budding from the plasma membrane (Lamb and Kolakofsky, 1996; Takimoto et al., 2001). The matrix protein is a membrane-associated protein which has been suggested to crosslink the viral core (the NC) and the viral envelope (the cytoplasmic tails of the surface proteins F and G) (Coronel et al., 1999; Morgan and Kingsbury, 1984; Sanderson et al., 1994; Stricker et al., 1994; Yoshida et al., 1976). In the absence of virus budding, the M protein is thought to localize to the perinuclear region, where it is believed to contact NC and subsequently travel to the plasma membrane (De et al., 1991; Ghildyal et al., 2002; Markwell and Fox, 1980; Portner and Murti, 1986; Portner and Murti, 1986; Ryan and Kingsbury, 1988; Stricker et al., 1994). The M protein is also known to concentrate the surface glycoproteins on the plasma membrane into patches (Nagai et al., 1976; Stricker et al., 1994; Yoshida et al., 1979) and, in conjunction with the NC, may provide the driving force to initiate budding from the cell (Coronel et al., 1999; Garoff et al., 1998; Mebatsion et al., 1996; Stricker et al., 1994). M protein is essential for virus budding and has the intrinsic ability to bud from the plasma membrane when expressed on its own (Ciancanelli and Basler, 2006; Pantua et al., 2006; Sugahara et al., 2004; Takimoto et al., 2001). Another recent study has indicated that NiV M may have a coordinating role in the release of virus-like particles (VLPs) from cells (Patch et al., 2007). The budding function of viral matrix proteins is derived from late domains (PPxY, PTAP, FPIV, or YxxL) which interact with cellular factors in order to mediate the release of virions from the cell membrane (Freed, 2002; Morita and Sundquist, 2004; Schmitt et al., 2005). Recently, Ciancanelli and colleagues, demonstrated that the NiV M protein has the ability to bud from transfected cells in the absence of other viral proteins (Ciancanelli and Basler, 2006). NiV M has a YxxL-type late domain (YMYL), which is similar to the late domain of equine infectious anemia virus M protein (YPDL) (Ciancanelli and Basler, 2006). Mutation of the NiV M YMYL motif abolished budding activity, but did not alter its ability to interact with other M molecules, nor did it affect its cellular localization (Ciancanelli and Basler, 2006).

REVERSE GENETICS

The advent of reverse genetics has created new opportunities to study the replication of (-)-ssRNA viruses (Conzelmann, 1996; Conzelmann, 1998). Very few tools exist for the manipulation of RNA. Therefore, the (-)-ssRNA genome is cloned as a full-length cDNA copy into a plasmid in the antigenomic orientation such that transcription from this plasmid will generate a (-)-ssRNA transcript. When expression plasmids encoding the other constituents of the viral ribonucleoprotein (RNP) complex (i.e. N, P, and L) are provided along with the genomic plasmid, functional NC can be formed and replication can begin, thereby generating *de novo* infectivity. The promoter used to generate the full-length (-)-ssRNA genome is typically recognized by the bacteriophage T7 RNA-dependent RNA polymerase (T7 pol), although promoters driving other polymerases, such as RNA polymerase II or RNA polymerase I, have also been developed. In fact, the RNA polymerase II-derived immediate early promoter from cytomegalovirus (CMV) has been used to generate fully infectious measles virus (MeV) and Borna disease virus (BDV) (Martin et al., 2006).

Minigenome systems

The minigenome system is useful for studies involving the leader, trailer, and 3' and 5' NTRs or in studying the roles of individual viral proteins in replication. Minigenome systems are a valuable alternative to working with infectious virus, in particular when a high level of bio-containment is necessary, as is the case for NiV and HeV. Minigenome systems retain the leader, and 3' NTR of N, and the trailer and 5' NTR of L of a particular virus genome. The intervening portion of the viral genome that contains viral genes is replaced with a reporter gene encoding e.g. green fluorescent protein (GFP) or chloramphenicol acetyltransferase (CAT). Several functional henipavirus minigenome systems exist, although only a minigenome driven by T7 RNA polymerase has been published for NiV (Halpin et al., 2004; Magoffin et al., 2007). However, minigenomes have a significant disadvantage since they lack the ability to maintain themselves in cell culture, because they no longer encode the viral proteins necessary for replication.

Replicon systems

A replicon system is similar to a minigenome system except that the genomic portion encodes the viral genes necessary for replication (N, P, and L) along with the reporter gene, enabling this system to perpetuate itself. This system also has the benefit that more of the viral genome is present in the system, making it more similar to the natural system than the more artificial minigenome system. There are currently no published studies that have used NiV or HeV replicon systems.

Full-length virus rescue systems

A full-length rescue system was recently published for NiV (Yoneda et al., 2006). The system used expression plasmids for NiV N, P, and L supplied in *trans*, a (-)-sense genomic plasmid, and infection of modified vaccinia virus Ankara expressing the T7 RNA polymerase (MVA-T7) to drive transcription of the genome (Yoneda et al., 2006). An additional recombinant NiV was developed that includes an eGFP gene as an additional transcriptional unit between the N and P genes. The rescued viruses grew to the same titer as the parental viruses, and rNiV was approximately as lethal as NiV in the golden hamster model. These viruses were shown to replicate to similar levels as that of wild-type NiV (Yoneda et al., 2006).

HENIPAVIRUS SURFACE PROTEINS

Paramyxoviruses have at least two surface glycoproteins, which are called F and G for NiV and HeV. Some paramyxoviruses have an additional small transmembrane protein designated SH (e.g. SeV and HRSV) that appears to be largely dispensable for virus growth. The henipavirus F and G glycoproteins are essential for virus entry and membrane fusion in the initial stages of infection. The attachment glycoprotein (G) is responsible for binding to the cellular receptor, while the fusion glycoprotein (F) is responsible for mediating virus-cell membrane fusion. Co-expression of F and G is absolutely required for NiV and HeV glycoprotein-mediated membrane fusion (Bossart et al., 2001; Bossart et al., 2002; Tamin et al., 2002). Separately, the F and G glycoproteins are not able to mediate cell-cell fusion when expressed in recombinant systems. In this respect, NiV and HeV are similar to the majority of other members of the family *Paramyxoviridae* in their fusion requirements, whereas the glycoproteins of several other members of the family (e.g. SV5 F and HPIV2 F) are able to mediate fusion independent of their homologous attachment glycoproteins. HeV F and HeV G also seem to differ from other paramyxoviruses in that their fusion activity is modulated by expression of activated RhoA, while SV5 F and SV5 HN are unaffected (Schowalter et al., 2006).

THE FUSION GLYCOPROTEIN (F)

The henipavirus fusion glycoprotein is a type I integral transmembrane glycoprotein with an extracellular N- and an intracytoplasmic C-terminus. The 546-aa F protein is initially synthesized as a ~61 kDa precursor (F_0) which is cleaved into the functional heterodimeric form F_1+F_2, and is transported to the cell surface (Bossart et al., 2001; Bossart et al., 2002; Harcourt et al., 2000). The F proteins of NiV and HeV have 88% aa homology, highlighting the high degree of genetic relatedness between the viruses (Harcourt et al., 2000; Wang et al., 2001). The F_1 subunit has a molecular weight of ~49 kDa and the F_2 subunit has a molecular weight of ~19 kDa (Bossart et al., 2001; Bossart et al., 2002). The F protein contains several distinct domains. An N-terminal signal peptide consisting of the first 28 residues directs the insertion of the nascent F_0 polypeptide into the ER membrane (Moll et al., 2004b). The F protein has an α-helical transmembrane domain located close to the C-terminus (aa 489 to 518) that anchors the protein in the viral or cellular membrane (Moll et al., 2004b). Cleavage of the F_0 subunit results in a disulfide-linked heterodimer and exposure of the fusion peptide at the N-terminus of the larger F_1 subunit. The fusion protein also has two heptad repeat regions, the N-terminal HR1 domain and the C-terminal HR2 domain, which are crucial for membrane fusion. A schematic representation of the henipavirus F protein structure is shown in Fig. 5A.

Cleavage of the fusion protein

A significant amount of work has been performed to identify the protease or proteases responsible for activation of the NiV and HeV F proteins. N-terminal Edman degradation amino acid sequencing showed that cleavage of the HeV F_0 precursor occurs at residue Lys_{109} within the sequence Asp_{107}-Val_{108}-Lys_{109}-↓-Leu_{110} (Michalski et al., 2000). The ability of many paramyxoviruses to cause systemic infections is restricted to those that have proteolytic cleavage sequences capable of being cleaved by ubiquitous cellular proteases. In many cases, the ubiquitous protease is the *trans*-Golgi network-resident protease furin, which cleaves at the consensus sequence Arg-X-Lys/Arg-Arg-↓-X. The F proteins of many members of the family

Paramyxoviridae have a dibasic cleavage site usually containing the residues Lys/Arg-Arg at the C-terminus of the F_2 subunit (Harcourt et al., 2000; Moll et al., 2004a). NiV F and HeV F do not share this motif. The F_2 subunit of HeV has the residues Val-Lys at its C-terminus, while the F_2 subunit of NiV F has the residues Val-Arg at the same positions. Alanine-scanning mutations in this region of NiV F have revealed that mutation of Lys_{109} to Ala does not affect its proteolytic cleavage (Moll et al., 2004a). Likewise, mutation of the eight residues upstream of the HeV F cleavage site do not prevent cleavage, nor does substitution of all eight upstream residues with Ala (Craft and Dutch, 2005). These data confirmed prior findings demonstrating that HeV could be propagated in the furin-deficient LoVo cell line (Michalski et al., 2000). Further studies showed that recombinant HeV F expressed in LoVo cells was efficiently processed and cleaved to the mature F_1+F_2 heterodimer (Pager et al., 2004). Chimeric SV5 F protein in which the furin cleavage site had been replaced with the Val_{105}-Gly_{106}-Asp_{107}-Val_{108}-Lys_{109} of HeV F were transported to the cell surface, but were not cleaved, while replacement with His_{102}-Asp_{103}-Leu_{104}-Val_{105}-Gly_{106}-Asp_{107}-Val_{108}-Lys_{109} resulted in partial cleavage, indicating that some of the residues upstream of the cleavage site can influence protease recognition (Craft and Dutch, 2005).

Several clues to the identity of the protease responsible for cleavage of NiV F and HeV F emerged during various biochemical studies. Cleavage of HeV F was inhibited by treatment with carbonyl cyanide 3-chlorophenylhydraze (CCCP), which blocks transport of protein from the ER (Pager et al., 2004), indicating that cleavage occurs further downstream in the secretory pathway. Likewise, treatment with monensin, an inhibitor of transport between the medial and *trans*-Golgi compartments, also inhibited HeV F cleavage, and that this cleavage event occurred after the F_0 precursor acquired resistance to EndoH treatment, an indicator of Golgi compartment residence (Pager et al., 2004). This was also confirmed by treatment with brefeldin A, which inhibited HeV F cleavage and further indicated the involvement of the *trans*-Golgi network or the cell surface in HeV F processing (Pager et al., 2004). In order to discriminate between *trans*-Golgi network and cell surface processing, treatment with NaF and $AlCl_3$ which blocks the formation of secretory vesicles in the *trans*-Golgi network also inhibited cleavage of HeV F was inhibited (Pager et al., 2004). Incubation of HeV F-expressing cells at 20°C, which prevents transport of vesicles from the *trans*-Golgi network to the cell surface, also inhibited cleavage of HeV F, further implicating the *trans*-Golgi network and/or the cell surface (Pager et al., 2004). Depletion of intracellular Ca^{2+} levels using the chelating agent EGTA or treatment with the Ca^{2+} ionophore A23187 had no effect on HeV F cleavage, indicating that the responsible protease required little to no Ca^{2+} (Pager et al., 2004). Treatment with agents that raised the intracellular pH (chloroquine, NH_4Cl, bafilomycin A1, or concanamycin A) revealed that transport of HeV F through the *trans*-Golgi network was not inhibited and that cleavage occurred in secretory vesicles budding from the *trans*-Golgi network (Pager et al., 2004). Complementary biochemical studies on NiV F cleavage were performed by another laboratory. These experiments showed that fusion and NiV F cleavage were not dependent on soluble proteases present in fetal calf serum added to culture medium (Diederich et al., 2005). This group also showed that NiV F cleavage was inhibited by treatment with brefeldin A and monensin, indicating that retention of NiV F in the ER or medial Golgi inhibited cleavage, as did inhibition of vesicle transport from the *trans*-Golgi network to the cell surface by $NaF/AlCl_3$ treatment (Diederich et al., 2005). The key finding by these studies was that endocytosis of NiV F was critical for cleavage, as determined by treatment with chlorpromazine (an inhibitor of clathrin-mediated endocytosis), or NH_4Cl, and antibody uptake experiments (Diederich et al., 2005). The NiV F protein is internalized via a specific membrane-proximal YXXΦ sequence that is important for the production of biologically-active i.e. fusion competent NiV F (Vogt et al., 2005). This group confirmed the importance of the endosome by co-expression of an endocytosis-deficient mutant NiV F protein with the NiV G attachment protein. Fusion was almost completely inhibited, even when the extracellular pH was decreased to 5, indicating that acidification alone is not sufficient to activate NiV F (Diederich et al., 2005).

Protease inhibition experiments finally revealed the identity of the protease responsible for cleaving HeV F. Protease inhibitors such as aprotinin and PMSF (serine protease inhibitors), leupeptin, antipain, and chymostatin (inhibitors of trypsin-like serine proteases and some cysteine proteases), pepstatin A (aspartic protease inhibitor), GM6001 and *o*-phenanthroline (metalloprotease inhibitors) had no effect on HeV F cleavage (Pager and Dutch, 2005). HeV F cleavage was inhibited, however, by a cocktail of serine and cysteine protease inhibitors. Lack of effect by serine protease inhibitors and the efficiency of calpeptin (a calpain and cysteine protease inhibitor) in blocking cleavage indicated that cysteine proteases are responsible (Pager and Dutch, 2005). Thus, the non-specific inhibitors of cathepsin B, H, and L (E-64, E-64d, as well as cathepsin inhibitor I

[Cath I]), the specific cathepsin L inhibitor III (Cath LIII) and the cathepsin B/L inhibitor CA-074Me blocked HeV F cleavage (Pager and Dutch, 2005). Suppression of cathepsin L expression by shRNA inhibited HeV F cleavage and fusion while cathepsin B suppression had no effect. In addition, preparations of cathepsin L were able to cleave immunopurified HeV F_0, further verifying the specificity of cathepsin L for HeV F (Pager and Dutch, 2005). Cathepsin B is able to cleave NiV F_0, but only cleavage by cathepsin L produced NiV F_1+F_2 subunits of the correct size, indicating that cathepsin L is required for the production of mature and functionally active NiV F (Pager et al., 2006).

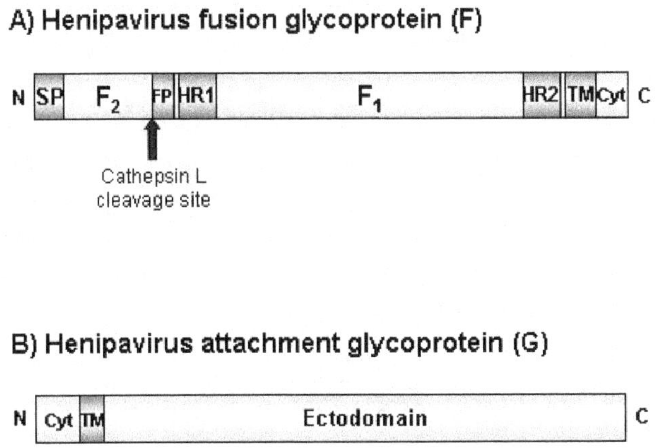

Fig. 5. The Henipavirus glycoproteins. A) Schematic representation of the fusion glycoprotein (F). The first 28 aa consist of a cleavable signal peptide (SP) that directs insertion of the F protein into the ER membrane. Cleavage of F by cathepsin L occurs after residue 109, between the F2 subunit and the fusion peptide (FP, aa 110-129) of the F1 subunit. The N-terminal (HR1, aa 137-178) and C-terminal (HR2, aa 453-485) heptad repeats pack against each other in the stable post-fusion conformation of the protein. The transmembrane domain span residues 489-518, followed by the cytoplasmic tail (aa 519-546). B) Schematic representation of the attachment glycoprotein (G). The cytoplasmic tail consists of the N-terminal 45 residues, followed by the transmembrane domain/stop anchor sequence (aa 46-70). The ectodomain (aa 71-604 for HeV G and aa 71-602 for NiV G) fold into a membrane-proximal stem followed by the C-terminal ~420 residues which fold into the paramyxovirus β-propeller globular head.

Glycosylation of the fusion protein

The henipavirus F protein contains five potential N-linked glycosylation sites in its ectodomain, two of which are located in the F_1 subunit and the remaining three in the F_2 subunit. Two of the HeV F_2 sites (Asn_{67} and Asn_{99}) and two of the HeV F_1 sites (Asn_{414} and Asn_{464}) are normally glycosylated (Carter et al., 2005), although previous studies involving lectin binding experiments had failed to show glycosylation of the HeV F_1 subunit (Michalski et al., 2000). NiV F has an identical glycosylation pattern. Removal of Asn_{67} and Asn_{99} has no effect on protein transport, cell surface expression, and fusogenicity (Moll et al., 2004b). However, mutation of Asn_{414} completely abolishes transport to the cell surface and fusion, while mutants in which the other glycosylation sites had been removed (Asn_{67}, Asn_{99}, Asn_{464}, and Asn_{64}, which is normally not glycosylated) were fusion competent, although removal of Asn_{464} does exhibit greatly decreased cell surface expression (Moll et al., 2004b). The same results have also been obtained for HeV F. Glycosylation at Asn_{414} of HeV F is critical for fusion activity, while mutant F proteins in which the glycans at Asn_{67}, Asn_{99}, and Asn_{464} had been removed retained the ability to mediate membrane fusion (Carter et al., 2005). The authors of both studies have speculated that Asn_{414} is crucial for correct folding and transport of NiV and HeV F to the cell surface (Carter et al., 2005; Moll et al., 2004b). Additionally, the glycosylation site at Asn_{464} lies within the HR2 region, which is required for function in other members of the family *Paramyxoviridae* (McGinnes et al., 2001; Zimmer et al., 2001). Mutation of this site in NiV and HeV F results in efficient cell-cell fusion despite greatly decreased cell surface expression (Carter et al., 2005; Moll et al., 2004b), indicating that there may be subtle differences between the mechanism of fusion in henipaviruses compared to other paramyxoviruses.

One interesting finding in more recent literature has involved the concept of shielding of NiV F by N-glycans. Removal of two, three, or four glycans from NiV F increased fusion in certain combinations, but pseudotyped VSVΔG particles also became more sensitive to inhibition by an HR2-derived peptide as well as anti-NiV F serum (Aguilar et al., 2006). The authors of this study speculated that increased fusion by some of the glycan-deficient mutants may be due to increased fusion kinetics via faster formation of the final stable six-helix bundle structure (Aguilar et al., 2006). Apparently, the henipavirus F protein displays much more plasticity in its function than other members of the family *Paramyxoviridae*.

The heptad repeat (HR) regions

After binding of the cellular receptor by the attachment protein, the fusion protein is triggered to undergo a series of conformational changes, ultimately leading to the formation of a stable post-fusion structure with fused membranes. In its pre-fusion form, the paramyxovirus F protein is a meta-stable intermediate with the fusion peptide buried within the protein. Triggering of the fusion protein releases the fusion peptide, which is inserted into the opposing membrane. The HR1 region lies directly adjacent to the fusion peptide and the HR2 region is directly adjacent to the transmembrane region.

Prior studies have shown that the activity of the fusion proteins from a variety of different virus families can be abolished by treatment with peptides derived from the HR1 and HR2. Peptides derived from the HR2 regions of NiV F and HeV F block fusion by NiV F and HeV F (Bossart et al., 2005b; Porotto et al., 2006). Interestingly, peptides derived from the cognate HR2 region of the HPIV3 F protein are even more effective than HeV F HR2-derived peptides in blocking HeV F-mediated fusion, though the HeV F HR2-derived peptides were ineffective against HPIV3 F (Porotto et al., 2006). The stability and activity of the HeV F HR2-derived peptides was also improved with capping by amidation and acetylation as well as with the addition of poly(ethylene)glycol (PEG) chains (Bossart et al., 2005b).

During membrane fusion, the HR regions fold back upon each other so that in the stable post-fusion conformation the fusion peptide and the transmembrane are brought together in close proximity. The HR1 and HR2 helices of NiV F and HeV F colocalize in the post-fusion conformation in a highly stable six-helix bundle, which has been crystallized (Lou et al., 2006; Xu et al., 2004b). The analysis of these crystal structures has provided a structural basis for the inhibition mediated by henipavirus F HR2-derived peptides (Xu et al., 2004a).

There is no available published data pertaining to the role of other henipavirus F protein domains such as the cytoplasmic tail or the transmembrane anchor, nor any studies determining which residues of the fusion peptide are important for function. These experiments, however, are no doubt underway in several laboratories.

THE ATTACHMENT GLYCOPROTEIN (G)

The henipavirus attachment glycoprotein is a type II integral transmembrane glycoprotein with an extracellular C-terminus and an intracytoplasmic N-terminus. The G proteins of NiV and HeV migrate at slightly different sizes in SDS gels. HeV G typically has a molecular weight of ~74-77 kDa (Bossart et al., 2001; Bossart et al., 2002), while NiV G has a molecular weight of ~72-75 kDa (Bossart et al., 2002); the reason for this difference is unclear. NiV G and HeV G exhibit 83% aa homology, which is slightly lower than what is seen for the F proteins (Harcourt et al., 2000). The attachment protein acts in concert with the fusion protein to mediate viral attachment, membrane fusion, and viral entry. The henipavirus G protein has eight potential N-linked glycosylation sites that are spread evenly throughout the ectodomain of the protein (Yu et al., 1998a). The cytoplasmic tail of NiV and HeV G consists of the N-terminal 45 residues, followed by a transmembrane domain of 25 residues (White et al., 2005; White et al., 2005; Yu et al., 1998a). The ectodomain of HeV G contains the C-terminal 534 residues, while the ectodomain of NiV G is 2 aa shorter at 532 residues (Bossart et al., 2005a; White et al., 2005). The ectodomain of the G protein encompasses the receptor binding domain which is responsible for interaction with cellular receptors ephrin-B2 and ephrin-B3 (see below). A schematic representation of the henipavirus G protein structure is shown in Fig. 5B.

Structure and antigenicity

While NiV G and HeV G exhibit >80% aa homology, they show very low homology to other members of the family *Paramyxoviridae*. Interestingly, henipavirus G proteins display higher homology of ~23-25% to members of the genus *Respirovirus* (e.g. HPIV1, HPIV3, and SeV) than the ~15% homology to the more closely-related members of the genus *Morbillivirus* (e.g. MeV and CDV) (Chan et al., 2001). Although most paramyxovirus attachment glycoproteins are largely unrelated by sequence, they share a high degree of structural homology (Langedijk et al., 1997). The paramyxovirus attachment proteins share a common fold despite their limited sequence similarity. With the crystallization of the attachment proteins of NDV, HPIV3, and SV5 (Crennell et al., 2000; Lawrence et al., 2004; Yuan et al., 2005) sequence comparisons and homology modeling are now possible between these proteins and other paramyxovirus attachment proteins. Specifically, the C-terminal ~420 residues of NiV G and HeV G have been predicted to fold in a characteristic 'β-propeller' structure consisting of six β-sheets each containing four anti-parallel β-strands (Guillaume et al., 2006b; White et al., 2005). The six β-sheets are arranged radially around the longitudinal axis of the G protein such that they form a fan-shaped structure.

The neutralizing epitopes of HeV G have been characterized in some detail. Based on mAb studies, residues important for the binding of mAbs have been found in eight different regions of HeV G (residues 183-185, 191-195, 289, 324, 385-386, 447, 517, and 570) (White et al., 2005). These residues map to the predicted globular head of HeV G and confirm certain details of the proposed 3D structure of this part of the ectodomain. Residues 183-185, 517, 570, and 447 seem to form several overlapping neutralizing epitopes which were bound by two mAbs in this study (H1 and H2.1) (White et al., 2005). Residues 183-185 are found adjacent to the predicted fourth strand of the sixth β-sheet of the globular head (β6S4), while residue 570 is located in the same β-sheet, but is located in the loop region between the second and third strands (β6S1 and β6S2) (White et al., 2005). Residue 517 is located in the loop region between the first and second strands of the fifth β-sheet (β5S1 and β5S2) and residue 447 is located in the first strand of the fourth β-sheet (β4S1) (White et al., 2005). The epitope for the mAb 17A5 is defined by residues 191-195 and is located in β6S4 and a portion of the distal loop (White et al., 2005). Residues 385-386 within an extended loop between β3S2 and β3S3 define the epitope for the mAb 3A5.D2 (White et al., 2005). Finally, the discontinuous residues 289 and 324 define the epitope for the mAb 8H4; residue 289 is located in the loop distal to β2S1 and residue 324 is located in the loop proximal to β2S4 (White et al., 2005). Mutations at these residues either greatly reduce or abolish neutralizing activity of the aforementioned mAbs, indicating that they play critical roles in the antibody binding to the HeV G ectodomain. The neutralization epitopes characterized for HeV G are located in similar regions to their morbillivirus counterparts MeV H and RPV H.

Human mAbs have also been developed against NiV G. One study described the creation and characterization of seven Fab fragments, denoted m101 through m107, which were generated and selected by a phage-panning library (Zhu et al., 2006a). All of the antibodies inhibited fusion by the HeV F and G glycoproteins, and some also displayed good inhibition of fusion mediated by NiV F and G. The Fab fragment m101 had the best HeV G-inhibiting activity, while m102 had the highest cross-inhibitory activity against NiV G (Zhu et al., 2006a). When it was converted to an IgG1 class of immunoglobulin, m101 fusion inhibitory activity was increased by 10-fold and it efficiently neutralized HeV (Zhu et al., 2006a). A mutation in Pro_{185}→Ala and double mutations in Gln_{191}→Ala and Lys_{192}→Ala reduced binding of m101 by at least 50% when compared to wild type NiV G (Zhu et al., 2006a). Likewise, the mutation Gly_{183}→Ala inhibited the binding of m102 by at least 50% compared to wild type HeV G (Zhu et al., 2006a). Binding of m102 was also significantly affected (~50-60% inhibition of binding) by the mutations at residues 185, 191, and 192, as well as at residue 184 (Zhu et al., 2006a). These residues are predicted to be located in the same regions as epitopes for the HeV G mAbs H1, H2.1, and 17A5 mentioned above. The cross-neutralization of HeV G and NiV G by the m101 and m102 antibodies further confirms that NiV G and HeV G share a significant amount of antigenicity. This parallels other studies using a soluble mutant of HeV G (sG_{HeV}) which consists only of the glycoprotein ectodomain (aa 71-604) and is able to competitively inhibit fusion mediated by both NiV and HeV glycoproteins as well as NiV and HeV infection (Bossart et al., 2005a). Using their sG_{HeV} construct, Bossart and colleagues also determined that HeV G monomers form disulfide-linked dimers and these HeV G dimers can associate non-covalently as a dimer of dimers, or a tetramer (Bossart et al., 2005a).

Receptor binding domain

Two receptors for NiV and HeV have been identified, ephrin-B2 (Bonaparte et al., 2005; Negrete et al., 2005) and ephrin-B3 (Negrete et al., 2006), which will be discussed later. The henipavirus G protein is required for attachment to the host cell receptor, which is thought to trigger membrane fusion by the F protein. For MeV the sites on G which interact with the CD46 and SLAM receptor molecules have been determined in some detail by mutation of individual residues. In particular, the MeV H residues Ser_{546} and Asn_{481} (Li and Qi, 2002), Val_{451} (Lecouturier et al., 1996), Ser_{548} and Phe_{549} (Massé et al., 2002), Ala_{428}, Leu_{464}, Ile_{487}, Ala_{527}, Tyr_{452}, and Pro_{486} (Vongpunsawad et al., 2004), and residues 473-477 (Patterson et al., 1999) have all been implicated in CD46 binding. The MeV H residues 429-438 (Hu et al., 2004), Tyr_{529}, Asp_{530}, Arg_{533}, Tyr_{553}, Thr_{531}, Phe_{552}, and Pro_{554} (Vongpunsawad et al., 2004), and Asp_{505} and Asp_{507} (Massé et al., 2004) have all been implicated in SLAM binding. There is also some evidence that the CD46 and SLAM binding sites overlap on the MeV H globular head (Massé et al., 2004; Santiago et al., 2002) and that mutations at residues 481 and 533 abrogate binding to both CD46 and SLAM (Hadac et al., 2004). The residues Glu_{401}, Arg_{416}, and Tyr_{526} of NDV HN are critical for receptor binding (Connaris et al., 2002). From this list, the majority of the residues implicated in paramyxovirus receptor binding are located in the C-terminal ~35% of the attachment glycoprotein.

The receptor binding domain of NiV G has also been characterized to some extent by mAb binding studies. Zhu and colleagues demonstrated that the Fab fragments m101 and m102, which had been shown to have significant HeV G and NiV G neutralizing activity (see previous section), also could compete with ephrin-B2 for binding to soluble HeV G (sG_{HeV}) in an ELISA-type assay (Zhu et al., 2006a). In plates coated with ephrin-B2, serial dilutions of Fab fragment m101 were mixed with sG_{HeV} and the resulting amount of bound sG_{HeV} was determined by OD measurement at 450 nm. It was found that with increasing m101 concentration, less sG_{HeV} was bound to ephrin-B2 coated on the plate (Zhu et al., 2006a). The Fab fragments m101 and m102, as well as m103, were able to compete with each other, indicating that they bound to overlapping epitopes distinct from other antibody used in the study (Zhu et al., 2006a). Two additional Fab fragments, m106 and m107, also neutralize HeV G, but do not compete with m101, m102, or m103, indicating that these antibodies neutralize HeV G by different mechanisms (Zhu et al., 2006a), although the mechanism is unclear.

Guillaume and colleagues used an approach involving systematic mutation of targeted regions of the NiV G globular head in order to abrogate the ability of NiV G to promote fusion when co-expressed with NiV F (Guillaume et al., 2006b). Mutation of Glu_{533}→Gln reduced the fusion promotion ability of NiV G by >80% compared to wild type NiV G, as did the mutation Trp_{504}→Ala (Guillaume et al., 2006b). In fact, the other mutations used in this study (Ala_{532}→Lys, Asn_{557}→Ala, Glu_{505}→Ala, Thr_{531}→Ala, and Gln_{530}→Ala) all inhibited fusion by at least 50%, indicating that this entire region is important for NiV G-mediated fusion. The authors mapped these residues onto the hypothetical structural model of NiV G and determined that there were receptor binding 'hot spots' at residues 504 and 505, which map to the extended loop connecting β4S4 and β5S1, residues 530-533, which map to the loop connecting β5S2 and β5S3, and residue 557, which is located in β6S1 (Guillaume et al., 2006b).

Our group has also shown recently that expression of NiV G is able to render cells resistant to infection by both NiV and HeV (Sawatsky et al., 2007). Additionally, we also found that this resistance occurs despite the presence of both receptors, ephrin-B2 and ephrin-B3, on the cell surface (Sawatsky et al., 2007), which contrasts with the behavior of MeV H, which downregulates cell surface expression of its receptors, CD46 and SLAM (Schneider-Schaulies et al., 1995a; Schneider-Schaulies et al., 1995b; Welstead et al., 2004).

There are currently no published studies available examining other putative functional domains of the henipavirus G protein such as the region responsible for fusion promotion, and the functional roles of the cytoplasmic tail, transmembrane domain, and the stem region.

HENIPAVIRUS RECEPTORS AND CONTRIBUTIONS TO PATHOGENESIS

Ephrin-B2 was identified by two groups in 2005 as a receptor for NiV and HeV (Bonaparte et al., 2005; Negrete et al., 2005). Negrete and colleagues used Fc-tagged NiV G as bait in immunoprecipitation experiments to identify a binding partner in permissive cell lines, which proved to be ephrin-B2 based on analysis of two tryptic peptide fragments (Negrete et al., 2005). It was also verified that ephrin-B2 was necessary for NiV F+NiV G-mediated fusion and for virus entry using NiV F/G pseudotyped VSVΔG-RFP viral particles (Negrete et al., 2005). Bonaparte and colleagues identified ephrin-B2 as a receptor through microarray analysis of fusion-permissive and fusion-deficient HeLa cell clones (Bonaparte et al., 2005). Incubation of Vero cells with soluble ephrin-B2 was able to inhibit infection by both HeV and NiV, and expression of ephrin-B2 conferred susceptibility to HeV and NiV infection to fusion-deficient HeLa cells (Bonaparte et al., 2005). More recently, ephrin-B3 has also been identified as a receptor for NiV G, although the binding affinity of NiV G for ephrin-B3 was approximately ten-fold less than for ephrin-B2 (K_d = 0.58 nM for ephrin-B3 vs. 0.06 nM for ephrin-B2) (Negrete et al., 2006). This opens the possibility that ephrin-B3 may act as a secondary or minor receptor, in particular in tissues in which ephrin-B2 levels are not as high. Ephrin-B2 is expressed to a high degree in the vascular endothelium but ephrin-B3 is not, while ephrin-B3 is expressed to a higher level in the CNS where ephrin-B2 expression is lower. This provides an interesting link between tissue expression of the henipavirus receptors and pathological consequences seen during infection.

IMMUNOLOGY

Henipavirus-infected hosts develop a significant humoral immune response against the respective virus (HeV, NiV), provided they survive the initial phase of infection (Weingartl et al., 2005; Wong et al., 2003). Experimental data (Guillaume et al., 2004; Mungall et al., 2006; Weingartl et al., 2006; Wong et al., 2003) support the assumption that neutralizing antibodies play a major role in viral clearance from the infected host and in prevention of virus-induced disease, as seen with other paramyxoviruses. Neutralizing antibodies are mainly directed towards the G protein and to a lesser extent, the F protein. It is likely that initial steps of viral entry are inhibited through binding of neutralizing antibodies to the G or F protein. After infection, antibodies directed towards other viral proteins such as the N or P proteins can also be detected in sera, but their roles in the pathogenesis of viral infection is unclear. Since some of the epitopes of the N protein appear to be strong immunogens during the course of infection, detection of antibody responses against the N protein has been proposed to be a useful test for diagnostic purposes (Chen et al., 2006; Eshaghi et al., 2005a; Kashiwazaki et al., 2004; Tan et al., 2004). However, the assays (e.g. ELISA) need to be validated prior to use in the field on samples from target species.

VACCINATION STRATEGIES

Using a hamster model of NiV infection (Wong et al., 2003), recombinant vaccinia viruses expressing either NiV F or G were used to induce immunity to NiV (Guillaume et al., 2006a). Passive immunity could also be transferred to unprotected animals via inoculation of serum from vaccinated animals (Guillaume et al., 2004). Further studies by Guillaume and colleagues have shown that anti-NiV G mAbs were more effective than anti-NiV F antibodies in protecting hamsters from NiV infection (Guillaume et al., 2006a). Administration of anti-NiV G and anti-NiV F sera before infection induced complete immunity and administration of anti-NiV G sera up to 24 hours after infection could partially protect hamsters from infection, but post-exposure administration of anti-NiV F protected a lower proportion of animals (Guillaume et al., 2006a). A similar approach has been attempted in pigs using a canarypox vector (ALVAC) expressing either NiV F or NiV G (Weingartl et al., 2006). Typically, higher NiV neutralization titers were observed in pigs vaccinated with ALVAC-NiV G than with ALVAC-NiV F, and the combination of both vaccines proved to provide better protection than individual administration (Weingartl et al., 2006). More recently, a feline model of NiV infection has been developed, and sG_{HeV} and secreted NiV G (sG_{NiV}) glycoproteins were used to successfully vaccinate animals against NiV infection (Mungall et al., 2006). These vaccination and passive antibody treatment strategies appear to promising for prevention or therapy of henipavirus infections (Bossart and Broder, 2006; Eaton et al., 2005; Zhu et al., 2006b).

INHIBITION OF INNATE IMMUNITY

The multiple proteins predicted to be encoded by the P gene of henipaviruses (C, V, and W) have been shown to inhibit the interferon (IFN) response in recombinant systems through a variety of mechanisms. Henipavirus V and W proteins have been found to inhibit double-stranded RNA (dsRNA) signaling in the cytoplasm and nucleus by different mechanisms (Shaw et al., 2005). The henipavirus P, V, and W proteins bind cellular STAT proteins through a common

N-terminal domain (Rodriguez et al., 2004; Shaw et al., 2004). This leads to sequestration of STAT proteins and abrogation of IFN signaling (Rodriguez et al., 2002). The W protein binds to STAT1 in the nucleus (Shaw et al., 2005; Shaw et al., 2004), whereas the V and P proteins interact with STAT1 in the cytoplasm (Rodriguez et al., 2002; Rodriguez et al., 2003; Shaw et al., 2004). The C protein does show detectable inhibition of IFN signaling, albeit to a lower degree than the V and W proteins (Park et al., 2003). Taken together, these studies indicate that henipaviruses have evolved multifaceted approaches to evade the innate immune system.

CONCLUSIONS

Over the past 15 years, previously unknown paramyxoviruses have emerged and have caused outbreaks of severe diseases in humans, sea mammals, and Serengeti lions. Additionally, other novel paramyxoviruses have been isolated from bats (Chua et al., 2001), tree shrews (Tidona et al., 1999), horses (Renshaw et al., 2000), rodents, and humans (van den Hoogen et al., 2001). However, infection with these viruses has not been associated with overt clinical disease. Of these, HeV, NiV, and Menangle virus (MenV) (Chua et al., 1999; Murray et al., 1995b; Philbey et al., 1998), the previously unknown causative agents of diseases in humans and livestock in Australia and Malaysia, clearly merit the term ‘emerging’. This illustrates that our knowledge of animal virology is limited and that both unknown and known paramyxoviruses such as henipaviruses pose an ongoing threat. Furthermore, the variations between the genomes of NiV isolates indicate that there is more than one strain of NiV circulating in a wide geographical area covering East Africa, Southeast Asia, and Australia.

HeV and NiV are unusual among the paramyxoviruses in that they cause potentially fatal disease in a number of host species, including humans. The natural reservoir for these viruses appears to be fruit bats, which do not develop disease upon infection. This might indicate that henipaviruses are not well-adapted to mammalian hosts (with the exception of fruit bats) and that these infections of livestock and humans are rare events. It is surprising, then, that both viruses employ various mechanisms to evade the innate immunity of mammalian hosts to which they are not well-adapted. The molecular pathogenesis of henipaviruses is still poorly understood and warrants further research.

Strategies to combat outbreaks of henipavirus infections are tightly associated with fast and reliable diagnostic capabilities. This can only be achieved through technical expertise and international efforts. As a second line of defense, vaccination of animals at risk may have to be considered. Promising vaccine candidates have been developed and further testing should be encouraged.

ACKOWLEDGEMENTS

The authors would like to acknowledge research funding from the Public Health Agency of Canada, the CBRNE Research and Technology Initiative, and Canadian Food Inspection Agency and student fellowship funding from the Manitoba Health Research Council and the University of Manitoba.

REFERENCES

Abraham, G., Rhodes, D.P., and Banerjee, A.K. (1975). The 5' terminal structure of the methylated mRNA synthesized in vitro by vesicular stomatitis virus. Cell *5*, 51-59.

Aguilar, H.C., Matreyek, K.A., Filone, C.M., Hashimi, S.T., Levroney, E.L., Negrete, O.A., Bertolotti-Ciarlet, A., Choi, D.Y., McHardy, I., Fulcher, J.A., Su, S.V., Wolf, M.C., Kohatsu, L., Baum, L.G., and Lee, B. (2006). N-glycans on Nipah virus fusion protein protect against neutralization but reduce membrane fusion and viral entry. J. Virol. *80*, 4878-4889.

Allworth, T., O'Sullivan, J., Selvey, L., and Sheridan, J. (1995). Equine morbillivirus in Queensland. Commun. Dis. Intell. *19*, 575

Anonymous (2003). Outbreaks of encephalitis due to Nipah/Hendra-like viruses, Western Bangladesh. ICDDR,B Health Sci. Bull. *1*, 1-6.

Banerjee, A.D., Abraham, G., and Colonno, R.J. (1977). Vesicular stomatitis virus: mode of transcription. J. Gen. Virol. *34*, 1-8.

Banerjee, A.K. (1987a). The transcription complex of vesicular stomatitis virus. Cell *48*, 363-364.

Banerjee, A.K. (1987b). Transcription and replication of rhabdoviruses. Microbiol. Rev. *51*, 66-87.

Banerjee, A., Masters, P., Das, T., and Chattopadhyay, D. (1989). Specific interaction of vesicular stomatitis virus nucleocapsid protein (N) with the phosphoprotein (NS) prevents its binding with non-specific RNA. In Genetics and Pathogenicity of Negative Strand Viruses. (Amsterdam: Elsevier), pp. 121-128.

Bonaparte, M.I., Dimitrov, A.S., Bossart, K.N., Crameri, G., Mungall, B.A., Bishop, K.A., Choudhry, V., Dimitrov, D.S., Wang, L.-F., Eaton, B.T., and Broder, C.C. (2005). Ephrin-B2 ligand is a functional receptor for Hendra virus and Nipah virus. Proc. Natl. Acad. Sci. USA *102*, 10652-10657.

Bossart, K.N. and Broder, C.C. (2006). Developments towards effective treatments for Nipah and Hendra virus infection. Expert Rev. Anti. Infect. Ther. *4*, 43-55.

Bossart, K.N., Crameri, G., Dimitrov, A.S., Mungall, B.A., Feng, Y.-R., Patch, J.R., Choudhary, A., Wang, L.-F., Eaton, B.T., and Broder, C.C. (2005a). Receptor binding, fusion inhibition, and induction of cross-reactive neutralizing antibodies by a soluble G glycoprotein of *Hendra virus*. J. Virol. *79* , 6690-6702.

Bossart, K.N., Mungall, B.A., Crameri, G., Wang, L.-F., Eaton, B.T., and Broder, C.C. (2005b). Inhibition of *Henipavirus* fusion and infection by heptad-derived peptides of the *Nipah virus* fusion glycoprotein. Virol. J. *2*, 57

Bossart, K.N., Wang, L.-F., Eaton, B.T., and Broder, C.C. (2001). Functional expression and membrane fusion tropism of the envelope glycoproteins of Hendra virus. Virology *290*, 121-135.

Bossart, K.N., Wang, L.-F., Flora, M.N., Chua, K.B., Lam, S.K., Eaton, B.T., and Broder, C.C. (2002). Membrane fusion tropism and heterotypic functional activities of the *Nipah virus* and *Hendra virus* envelope glycoproteins. J. Virol. *76*, 11186-11198.

Brown, D.D., Rima, B.K., Allen, I.V., Baron, M.D., Banyard, A.C., Barrett, T., and Duprex, W.P. (2005). Rational attenuation of a morbillivirus by modulating the activity of the RNA-dependent RNA polymerase. J. Virol. *79*, 14330-14338.

Butler, D. (2004). Fatal fruit bat virus sparks epidemics in southern Asia (news). Nature *429*, 7

Calain, P. and Roux, L. (1993). The Rule of Six, a basic feature for efficient replication of Sendai virus defective interfering RNA. J. Virol. *67*, 4822-4830.

Carter, J.R., Pager, C.T., Fowler, S.D., and Dutch, R.E. (2005). Role of N-linked glycosylation of the Hendra virus fusion protein. J. Virol. *79*, 7922-7925.

Cattaneo, R., Rebmann, G., Schmid, A., Baczko, K., ter Meulen, V., and Billeter, M.A. (1987). Altered transcription of a defective measles virus genome derived from a diseased human brain. EMBO J. *6*, 681-688.

Chadha, M.S., Comer, J.A., Lowe, L., Rota, P.A., Rollin, P.E., Bellini, W.J., Ksiazek, T.G., and Mishra, A.C. (2006). Nipah virus-associated encephalitis outbreak, Siliguri, India. Emerg. Infect. Dis. *12*, 235-240.

Chan, Y.P., Chua, K.B., Koh, C.L., Lim, M.E., and Lam, S.K. (2001). Complete nucleotide sequences of Nipah virus isolates from Malaysia. J. Gen. Virol. *82*, 2151-2155.

Chan, Y.P., Koh, C.L., Lam, S.K., and Wang, L.-F. (2004). Mapping of domains responsible for nucleocapsid protein-phosphoprotein interaction of henipaviruses. J. Gen. Virol. *85*, 1675-1684.

Chen, J.-M., Yu, M., Morrissy, C., Zhao, Y.-G., Meehan, G., Sun, Y.-X., Wang, Q.-H., Zhang, W., Wang, L.-F., and Wang, Z.-L. (2006). A comparative indirect ELISA for the detection of henipavirus antibodies based on a recombinant nucleocapsid protein expressed in *Escherichia coli*. J. Virol. Methods *136*, 273-276.

Chenik, M., Chebli, K., Gaudin, Y., and Blondel, D. (1994). *In vivo* interaction of rabies virus phosphoprotein (P) and nucleoprotein (N): existence of two N-binding sites on P protein. J. Gen. Virol. *75*, 2889-2896.

Chua, K.B., Goh, K.J., Wong, K.T., Kamarulzaman, A., Tan, P.S.K., Ksiazek, T.G., Zaki, S.R., Paul, G., Lam, S.K., and Tan, C.T. (1999). Fatal encephalitis due to Nipah virus among pig-farmers in Malaysia. Lancet *354*, 1257-1259.

Chua, K.B., Wang, L.-F., Lam, S.K., Crameri, G., Yu, M., Wise, T., Boyle, D., Hyatt, A.D., and Eaton, B.T. (2001). *Tioman virus*, a novel paramyxovirus isolated from fruit bats in Malaysia. Virology *283*, 215-229.

Chua, K.B., Bellini, W.J., Rota, P.A., Harcourt, B.H., Tamin, A., Lam, S.K., Ksiazek, T.G., Rollin, P.E., Zaki, S.R., Sheih, W., Goldsmith, C.S., Gubler, D.J., Roehrig, J.T., Eaton, B., Gould, A.R., Olson, J., Field, H., Daniels, P., Ling, A.E., Peters, C.J., Anderson, L.J., and Mahy, B.W. (2000). Nipah virus: a recently emergent deadly paramyxovirus. Science *288* , 1432-1435.

Chua, K.B., Koh, C.L., Hooi, P.S., Wee, K.F., Khong, J.H., Chua, B.H., Chan, Y.P., Lim, M.E., and Lam, S.K. (2002). Isolation of Nipah virus from Malaysian Island flying-foxes. Microbes Infect. *4*, 145-151.

Ciancanelli, M.J. and Basler, C.F. (2006). Mutation of YMYL in the Nipah virus matrix protein abrogates budding and alters subcellular localization. J. Virol. *80*, 12070-12078.

Conzelmann, K.-K. (1996). Genetic manipulation of non-segmented negative-strand RNA viruses. J. Gen. Virol. *77*, 381-388.

Conzelmann, K.-K. (1998). Nonsegmented negative-strand RNA viruses: Genetics and manipulation of viral genomes. Annu. Rev. Genet. *32*, 123-162.

Connaris, H., Takimoto, T., Russell, R., Crennell, S., Moustafa, I., Portner, A., and Taylor, G. (2002). Probing the sialic acid binding site of the hemagglutinin-neuraminidase of Newcastle disease virus: identification of key amino acids involved in cell binding, catalysis, and fusion. J. Virol. *76*, 1816-1824.

Coronel, E.C., Murti, K.G., Takimoto, T., and Portner, A. (1999).

Human parainfluenza virus type 1 matrix and nucleoprotein genes transiently expressed in mammalian cells induce the release of virus-like particles containing nucleocapsid-like structures. J. Virol. *73*, 7035-7038.
Craft, W.W., Jr. and Dutch, R.E. (2005). Sequence motif upstream of the Hendra virus fusion protein cleavage site is not sufficient to promote efficient proteolytic processing. Virology *341*, 130-140.
Crennell, S., Takimoto, T., Portner, A., and Taylor, G. (2000). Crystal structure of the multifunctional paramyxovirus hemagglutinin-neuraminidase. Nat. Struct. Biol. *7*, 1068-1074.
Curran, J., Boeck, R., and Kolakofsky, D. (1991). The Sendai virus P gene expresses both an essential protein and an inhibitor of RNA synthesis by shuffling modules via mRNA editing. EMBO J. *10*, 3079-3085.
Curran, J., Homann, H., Buchcholz, C., Rochat, S., Neubert, W., and Kolakofsky, D. (1993). The hypervariable C-terminal tail of the Sendai paramyxovirus nucleocapsid protein is required for template function but not for RNA encapsidation. J. Virol. *67*, 4358-4364.
Curran, J., Marq, J., and Kolakofsky, D. (1992). The Sendai virus nonstructural C protein specifically inhibit viral mRNA synthesis. Virology *189*, 647-656.
Curran, J., Marq, J.B., and Kolakofsky, D. (1995). An N-terminal domain of the Sendai paramyxovirus P protein acts as a chaperone for the NP protein during the nascent chain assembly step of genome replication. J. Virol. *69*, 849-855.
Das, T. and Banerjee, A. (1992). Role of the phosphoprotein (P) in the encapsidation of presynthesized and de novo synthesized vesicular stomatitis virus RNA by the nucleocapsid protein (N) in vitro. Cell. Mol. Biol. *38*, 17-26.
Das, T., Gupta, A.K., Sims, P.W., Gelfand, C.A., Jentoft, J.E., and Banerjee, A.K. (1995). Role of cellular casein kinase II in the function of the phosphoprotein subunit of RNA polymerase of vesicular stomatitis virus. J. Biol. Chem. *270*, 24100-24107.
Davis, N., Arnheiter, H., and Gertz, G. (1986). Vesicular stomatitis virus N and NS proteins form multiple complexes. J. Virol. *59*, 751-754.
De, B., Lesoon, A., and Banerjee, A. (1991). Human parainfluenza virus type 3 transcription in vitro: role of cellular actin in mRNA synthesis. J. Virol. *65*, 3268-3275.
Diederich, S., Moll, M., Klenk, H.-D., and Maisner, A. (2005). The Nipah virus fusion protein is cleaved within the endosomal compartment. J. Biol. Chem. *280*, 29899-29903.
Duprex, W.P., Collins, F.M., and Rima, B.K. (2002). Modulating the

function of the measles virus RNA-dependent RNA polymerase by insertion of green fluorescent protein into the open reading frame. J. Virol. *76*, 7322-7328.

Eaton, B.T., Broder, C.C., and Wang, L.-F. (2005). Hendra and Nipah viruses: pathogenesis and therapeutics. Curr. Mol. Med. *5*, 805-816.

Egelman, E.H., Wu, S.S., Amrein, M., Portner, A., and Murti, G. (1989). The Sendai virus nucleocapsid exists in at least four different helical states. J. Virol. *63*, 2233-2243.

Emerson, S.U. (1982). Reconstitution studies detect a single polymerase entry site on the vesicular stomatitis virus genome. Cell *31*, 635-642.

Enami, M., Kohama, T., and Sugiura, A. (1989). A measles virus subgenomic RNA: structure and generation mechanism. Virology *171*, 427-433.

Engelhorn, M., Stricker, R., and Roux, L. (1993). Molecular cloning and characterization of a Sendai virus internal deletion defective RNA. J. Gen. Virol. *74*, 137-141.

Errington, W. and Emmerson, P. (1997). Assembly of recombinant Newcastle disease virus nucleocapsid protein into nucleocapsid-like structures is inhibited by the phosphoprotein. J. Gen. Virol. *78*, 2335-2339.

Eshaghi, M., Tan, W.S., Ong, S.T., and Yusoff, K. (2005a). Purification and characterization of Nipah virus nucleocapsid protein produced in insect cells. J. Clin. Microbiol. *43*, 3172-3177.

Eshaghi, M., Tan, W.S., and Yusoff, K. (2005b). Identification of epitopes in the nucleocapsid protein of Nipah virus using a linear phage-displayed random peptide library. J. Med. Virol. *75*, 147-152.

Field, H.E., Barratt, P.C., Hughes, R.J., Shield, J., and Sullivan, N.D. (2000). A fatal case of Hendra virus infection in a horse in north Queensland: clinical and epidemiological features. Aust. Vet. J. *78*, 279-280.

Field, H., Young, P., Yob, J.M., Mills, J., Hall, L., and Mackenzie, J. (2001). The natural history of Hendra and Nipah viruses. Microbes Infect. *3*, 307-314.

Fooks, A., Stephenson, J., Warnes, A., Dowsett, A., Rima, B., and Wilkinson, G. (1993). Measles virus nucleocapsid protein expressed in insect cells assembles into nucleocapsid-like structures. J. Gen. Virol. *74*, 1439-1444.

Freed, E.O. (2002). Viral Late Domains. J. Virol. *76*, 4679-4687.

Gao, Y. and Lenard, J. (1995a). Cooperative binding of multimeric phosphoprotein (P) of vesicular stomatitis virus to polymerase (L) and template. J. Virol. *69*, 7718-7723.

Gao, Y. and Lenard, J. (1995b). Multimerization and transcriptional activation of the phosphoprotein (P) of vesicular stomatitis virus by casein kinase-II. EMBO J. *14*, 1240-1247.

Garcia, J., Garcia-Barreno, B., Vivo, A., and Melero, J. (1993). Cytoplasmic inclusions of respiratory syncytial virus-infected cell formation on inclusion-bodies in transfected cells that co-express the nucleoprotein, the phosphoprotein and the 22K protein. Virology *195*, 243-247.

Garoff, H., Hewson, R., and Opstelten, D.-J.E. (1998). Virus maturation by budding. Microbiol. Mol. Biol. Rev. *62*, 1171-1190.

Ghildyal, R., Murray, M., Vardaxis, N., and Meanger, J. (2002). Respiratory syncytial virus matrix protein associates with nucleocapsids in infected cells. J. Gen. Virol. *83*, 753-757.

Goh, K.J., Tan, C.T., Chew, N.K., Tan, P.S.K., Kamarulzaman, A., Sarji, S.A., Wong, K.T., Abdullah, B.J.J., Chua, K.B., and Lam, S.K. (2000). Clinical features of Nipah virus encephalitis among pig farmers in Malaysia. N. Engl. J. Med. *342*, 1229-1235.

Goldsmith, C.S., Whistler, T., Rollin, P.E., Ksiazek, T.G., Rota, P.A., Bellini, W.J., Daszak, P., Wong, W.T., Shieh, W.-J., and Zaki, S.R. (2003). Elucidation of Nipah virus morphogenesis and replication using ultrastructural and molecular approaches. Virus Res. *92*, 89-98.

Guillaume, V., Contamin, H., Loth, P., Georges-Courbot, M.-C., Lefeuvre, A., Marianneau, P., Chua, K.B., Lam, S.K., Buckland, R., Deubel, V., and Wild, T.F. (2004). Nipah virus: Vaccination and passive protection studies in a hamster model. J. Virol. *78*, 834-840.

Guillaume, V., Contamin, H., Loth, P., Grosjean, I., Georges Courbot, M.C., Deubel, V., Buckland, R., and Wild, T.F. (2006a). Antibody prophylaxis and therapy against Nipah virus infection in hamsters. J. Virol. *80*, 1972-1978.

Guillaume, V., Aslan, H., Ainouze, M., Guerbois, M., Wild, T.F., Buckland, R., and Langedijk, J.P.M. (2006b). Evidence of a potential receptor-binding site on the Nipah virus G protein (NiV-G): Identification of globular head residues with a role in fusion promotion and their localization on an NiV-G structural model. J. Virol. *80*, 7546-7554.

Gupta, AK., Shaji, D., and Banerjee, AK. (2003). Identification of a novel tripartite complex involved in replication of vesicular stomatitis virus genome RNA. J. Virol. *77*, 732-738.

Hadac, E.M., Peng, K.-W., Nakamura, T., and Russell, S.J. (2004). Reengineering paramyxovirus tropism. Virology *329*, 217-225.

Halpin, K., Young, P.L., Field, H., and Mackenzie, J.S. (1999). Newly discovered viruses of flying foxes. Vet. Microbiol. *68*, 83-87.

Halpin, K., Young, P.L., Field, H.E., and Mackenzie, J.S. (2000). Isolation of Hendra virus from pteropid bats: a natural reservoir of

Hendra virus. J. Gen. Virol. *81*, 1927-1932.
Halpin, K., Bankamp, B., Harcourt, B.H., Bellini, W.J., and Rota, P.A. (2004). Nipah virus conforms to the rule of six in a minigenome replication assay. J. Gen. Virol. *85*, 701-707.
Hammond, D.C. and Lesnaw, J.A. (1987a). The fates of undermethylated mRNA cap structures of vesicular stomatitis virus (New Jersey) during in vitro transcription. Virology *159*, 229-36.
Hammond, D.C. and Lesnaw, J.A. (1987b). Functional analysis of hypomethylation variants of the New Jersey serotype of vesicular stomatitis virus. Virology *160*, 330-5.
Harcourt, B.H., Lowe, L., Tamin, A., Liu, X., Bankamp, B., Bowden, N., Rollin, P.E., Comer, J.A., Ksiazek, T.G., Hossain, M.J., Gurley, E.S., Breiman, R.F., Bellini, W.J., and Rota, P.A. (2005). Genetic characterization of Nipah virus, Bangladesh, 2004. Emerg. Infect. Dis. *11*, 1594-1597.
Harcourt, B.H., Tamin, A., Halpin, K., Ksiazek, T.G., Rollin, P.E., Bellini, W.J., and Rota, P.A. (2001). Molecular characterization of the polymerase gene and genomic termini of Nipah virus. Virology *287*, 192-201.
Harcourt, B.H., Tamin, A., Ksiazek, T.G., Rollin, P.E., Anderson, L.J., Bellini, W.J., and Rota, P.A. (2000). Molecular characterization of Nipah virus, a newly emergent paramyxovirus. Virology *271*, 334-349.
Harty, R.N. and Palese, P. (1995). Mutations within noncoding terminal sequences of model RNAs of Sendai virus: influence on reporter gene expression. J. Virol. *69*, 5128-31.
Hausmann, S., Jacques, J.P., and Kolakofsky, D. (1996). Paramyxovirus RNA editing and the requirement for hexamer genome length. RNA *2*, 1033-45.
Hercyk, N., Horikami, S.M., and Moyer, S.A. (1988). The vesicular stomatitis virus L protein possesses the mRNA methyltransferase activities. Virology *163*, 222-5.
Homann, H.E., Hofschneider, P.H., and Neubert, W.J. (1990). Sendai virus gene expression in lytically and persistently infected cells. Virology *177*, 131-140.
Homann, H., Willenbrink, W., Buchholz, C., and Neubert, W. (1991). Sendai virus protein-protein interactions studied by a protein-blotting protein-overlay technique: mapping of domains on NP protein required for binding to P protein. J. Virol. *65*, 1304-1309.
Hooper, P.T., Gould, A.R., Russell, G.M., Kattenbelt, J.A., and Mitchell, G. (1996). The retrospective diagnosis of a second outbreak of equine morbillivirus infection. Aust. Vet. J. *74*, 244-245.
Hooper, P.T., Ketterer, P.J., Hyatt, A.D., and Russell, G.M. (1997a). Lesions of experimental equine morbillivirus pneumonia in horses.

Vet. Pathol. *34*, 312-322.
Hooper, P.T., Westbury, H.A., and Russell, G.M. (1997b). The Lesions of experimental equine morbillivirus disease in cats and guinea pigs. Vet. Pathol. *34*, 323-329.
Horikami, S.M., Curran, J., Kolakofsky, D., and Moyer, S.A. (1992). Complexes of Sendai virus NP-P and P-L proteins are required for defected interfering particle genome replication in vitro. J. Virol. *66*, 4901-4908.
Howard, M. and Wertz, G. (1989). Vesicular stomatitis virus RNA replication: a role for the NS protein. J. Gen. Virol. *70*, 2683-2694.
Hsu, V.P., Hossain, M.J., Parashar, U.D., Ali, M.M., Ksiazek, T.G., Kuzmin, I., Niezgoda, M., Rupprecht, C., Bresee, J., and Breiman, R.F. (2004). Nipah virus encephalitis reemergence, Bangladesh. Emerg. Infect. Dis. *10*, 2082-2087.
Hu, C., Zhang, P., Liu, X., Qi, Y., Zou, T., and Xu, Q. (2004). Characterization of a region involved in binding of measles virus H protein and its receptor SLAM (CD150). Biochem. Biophys. Res. Commun. *316*, 698-704.
Hunt, D.M., Mehta, R., and Hutchinson, K.L. (1988). The L protein of vesicular stomatitis virus modulates the response of the polyadenylic acid polymerase to S-adenosylhomocysteine. J. Gen. Virol. *69*, 2555-2561.
Hunt, D.M., Smith, E.F., and Buckley, D.W. (1984). Aberrant polyadenylation by a vesicular stomatitis virus mutant is due to an altered L protein. J. Virol. *52*, 515-521.
Iehlé, C., Razafitrimo, G., Razainirina, J., Andriaholinirina, N., Goodman, S.M., Faure, C., Georges-Courbot, M.-C., Rousset, D., and Reynes, J.-M. (2007). Henipavirus and Tioman virus antibodies in pteropodid bats, Madagascar. Emerg. Infect. Dis. *13*, 159-161.
Jack, P.J.M., Boyle, D.B., Eaton, B.T., and Wang, L.-F. (2005). The complete genome sequence of J virus reveals a unique genome structure in the family *Paramyxoviridae*. J. Virol. *79*, 10690-10700.
Johara, M.Y., Field, H., Rashdi, A.M., Morrissy, C., van der Heide, B., Rota, P., bin Adzhar, A., White, J., Daniels, P., Jamaluddin, A., and Ksiazek, T. (2001). Nipah virus infection in bats (order Chiroptera) in peninsular Malaysia. Emerg. Infect. Dis. *7*, 439-441.
Kashiwazaki, Y., Na, Y.N., Tanimura, N., and Imada, T. (2004). A solid-phase blocking ELISA for detection of antibodies to Nipah virus. J. Virol. Methods *121*, 259-261.
Kato, A., Kiyotani, K., Hasan, M.K., Shioda, T., Sakai, Y., Yoshida, T., and Nagai, Y. (1999). Sendai virus gene start signals are not equivalent in reinitiation capacity: moderation at the fusion protein gene. J. Virol. *73*, 9237-9246.

Kolakofsky, D., Pelet, T., Garcin, D., Hausmann, S., Curran, J., and Roux, L. (1998). Paramyxovirus RNA synthesis and the requirement for hexamer genome length: the Rule of Six revisited. J. Virol. *72*, 891-899.

Lamb, R.A. and Kolakofsky, D. (1996). *Paramyxoviridae*: The viruses and their replication. In Fields Virology. B.N. Fields, D.M. Knipe, P.M. Howley, R.M. Chanock, J.L. Melnick, T.P. Monath, B. Roizman, and S.E. Straus, eds. (Philadelphia, PA: Lippincott-Raven).

Langedijk, J.P.M., Daus, F.J., and van Oirschot, J.T. (1997). Sequence and structure alignment of *Paramyxoviridae* attachment proteins and discovery of enzymatic activity for a morbillivirus hemagglutinin. J. Virol. *71*, 6155-6167.

Lawrence, M.C., Borg, N.A., Streltsov, V.A., Pilling, P.A., Epa, C.V., Varghese, J.N., McKimm-Breschkin, J.L., and Colman, P.M. (2004). Structure of the haemagglutinin-neuraminidase from human parainfluenza virus type III. J. Mol. Biol. *335*, 1343-1357.

Lecouturier, V., Fayolle, J., Caballero, M., Carabaña, J., Celma, M.L., Fernandez-Muñoz, R., Wild, T.F., and Buckland, R. (1996). Identification of two amino acids in the hemagglutinin glycoprotein of measles virus (MV) that govern hemadsorption, HeLa cell fusion, and CD46 downregulation: Phenotypic markers that differentiate vaccine and wild-type MV strains. J. Virol. *70*, 4200-4204.

Li, L. and Qi, Y. (2002). A novel amino acid position in hemagglutinin glycoprotein of measles virus is responsible for hemadsorption and CD46 binding. Arch. Virol. *147*, 775-786.

Li, Z., Yu, M., Zhang, H., Magoffin, D.E., Jack, P.J.M., Hyatt, A., Wang, H.-Y., and Wang, L.-F. (2006). *Beilong virus*, a novel paramyxovirus with the largest genome of non-segmented negative-stranded RNA viruses. Virology *346*, 219-228.

Lou, Z., Xu, Y., Xiang, K., Su, N., Qin, L., Li, X., Gao, G.F., Bartlam, M., and Rao, Z. (2006). Crystal structures of Nipah and Hendra virus fusion core proteins. FEBS J. *273*, 4538-4547.

Luby, S.P., Rahman, M., Hossain, M.J., Blum, L.S., Husain, M.M., Gurley, E., Khan, R., Ahmed, B.-N., Rahman, S., Nahar, N., Kenah, E., Comer, J.A., and Ksiazek, T.G. (2006). Foodborne transmission of Nipah virus, Bangladesh. Emerg. Infect. Dis. *12*, 1888-1894.

Magoffin, D.E., Halpin, K., Rota, P.A., and Wang, L.-F. (2007). Effects of single amino acid substitutions at the E residue in the conserved GDNE motif of the Nipah virus polymerase (L) protein. Arch. Virol. *152*, 827-832.

Markwell, M.A.K. and Fox, C.F. (1980). Protein-protein interactions within paramyxoviruses identified by native disulfide bonding or reversible chemical cross-linking. J. Virol. *33*, 152-166.

Martin, A., Staeheli, P., and Schneider, U. (2006). RNA polymerase II-controlled expression of antigenomic RNA enhances the rescue efficacies of two different members of the *Mononegavirales* independently of the site of viral genome replication. J. Virol. *80*, 5708-5715.
Massé, N., Ainouze, M., Néel, B., Wild, T.F., Buckland, R., and Langedijk, J.P.M. (2004). Measles virus (MV) hemagglutinin: Evidence that attachment sites for MV receptors SLAM and CD46 overlap on the globular head. J. Virol. *78*, 9051-9063.
Massé, N., Barrett, T., Muller, C.P., Wild, T.F., and Buckland, R. (2002). Identification of a second major site for CD46 binding in the hemagglutinin protein from a laboratory strain of measles virus (MV): Potential consequences for wild-type MV infection. J. Virol. *76*, 13034-13038.
McGinnes, L., Sergel, T., Reittner, J., and Morrison, T. (2001). Carbohydrate modifications of the NDV fusion protein heptad repeat domains influence maturation and fusion activity. Virology *283*, 332-342.
Mebatsion, T., Koenig, M., and Conzelmann, K. (1996). Budding of rabies virus particles in the absence of the spike glycoprotein. Cell *84*, 941-951.
Michalski, W.P., Crameri, G., Wang, L.-F., Shiell, B.J., and Eaton, B. (2000). The cleavage activation and sites of glycosylation in the fusion protein of Hendra virus. Virus Res. *69*, 83-93.
Middleton, D.J., Westbury, H.A., Morrissy, C.J., van der Heide, B.M., Russell, G.M., Braun, M.A., and Hyatt, A.D. (2002). Experimental Nipah virus infection in pigs and cats. J. Comp. Pathol. *126*, 124-136.
Mohd Nor, M.N., Gan, C.H., and Ong, B.L. (2000). Nipah virus infection of pigs in peninsular Malaysia. Rev. Sci. Tech. *19*, 160-165.
Moll, M., Diederich, S., Klenk, H.-D., Czub, M., and Maisner, A. (2004a). Ubiquitous activation of Nipah virus fusion protein does not require a basic amino acid at the cleavage site. J. Virol. *78*, 9705-9712.
Moll, M., Kaufmann, A., and Maisner, A. (2004b). Influence of N-glycans on processing and biological activity of the Nipah virus fusion protein. J. Virol. *78*, 7274-7278.
Morgan, E. and Kingsbury, D. (1984). Complete sequence of the Sendai virus NP gene from a cloned insert. Virology *135*, 279-287.
Morita, E. and Sundquist, W.I. (2004). Retrovirus budding. Annu. Rev. Cell Dev. Biol. *20*, 395-425.
Mungall, B.A., Middleton, D., Crameri, G., Bingham, J., Halpin, K., Russell, G., Green, D., McEachern, J., Pritchard, L.I., Eaton, B.T., Wang, L.-F., Bossart, K.N., and Broder, C.C. (2006). A feline model of

acute Nipah virus infection and protection with a soluble glycoprotein-based subunit vaccine. J. Virol. *80*, 12293-12302.
Murray, K., Rogers, R., Selvey, L., Selleck, P., Hyatt, A., Gould, A., Gleeson, L., Hooper, P., and Westbury, H. (1995a). A novel morbillivirus pneumonia of horses and its transmission to humans. Emerg. Infect. Dis. *1*, 31-33.
Murray, K., Selleck, P., Hooper, P., Hyatt, A., Gould, A., Gleeson, L., Westbury, H., Hiley, L., Selvey, L., Rodwell, B., and Ketterer, P. (1995b). A morbillivirus that caused fatal disease in horses and humans. Science *268*, 94-97.
Nagai, Y., Ogura, H., and Klenk, H.-D. (1976). Studies on the assembly of the envelope of Newcastle disease virus. Virology *69*, 523-538.
Negrete, O.A., Levroney, E.L., Aguilar, H.C., Bertolotti-Ciarlet, A., Nazarian, R., Tajyar, S., and Lee, B. (2005). EphrinB2 is the entry receptor for Nipah virus, an emergent deadly paramyxovirus. Nature *436*, 401-405.
Negrete, O.A., Wolf, M.C., Aguilar, H.C., Enterlein, S., Wang, W., Mühlberger, E., Su, S.V., Bertolotti-Ciarlet, A., Flick, R., and Lee, B. (2006). Two key residues in ephrinB3 are critical for its use as an alternative receptor for Nipah virus. PLoS Pathog. *2*, e7
Nishio, M., Tsurudome, M., Kawano, M., Watanabe, N., Ohgimoto, S., Ito, M., Komada, H., and Ito, Y. (1996). Interaction between nucleocapsid protein (NP) and phosphoprotein (P) of human parainfluenza virus type 2: one of the two NP binding sites on P is essential for granule formation. J. Gen. Virol. *77*, 2457-2463.
O'Sullivan, J.D., Allworth, A.M., Paterson, D.L., Snow, T.M., Boots, R., Gleeson, L.J., Gould, A.R., Hyatt, A.D., and Bradfield, J. (1997). Fatal encephalitis due to novel paramyxovirus transmitted from horses. Lancet *349*, 93-95.
Olson, J.G., Rupprecht, C., Rollin, P.E., An, U.S., Niezgoda, M., Clemins, T., Walston, J., and Ksiazek, T.G. (2002). Antibodies to Nipah-like virus in bats (*Pteropus lylei*), Cambodia. Emerg. Infect. Dis. *8*, 987-988.
Ong, S.T., Tan, W.S., Hassan, S.S., Mohd Lila, M.A., and Yusoff, K. (2002). Cloning and expression of the nucleocapsid protein gene of Nipah virus in different strains of *Escherichia coli*. J. Biochem. Mol. Biol. Biophys. *6*, 347-350.
Ongradi, J., Cunningham, C., and Szilagyi, J.F. (1985). The role of polypeptides L and NS in the transcription process of vesicular stomatitis virus New Jersey using the temperature-sensitive mutant tsE1. J. Gen. Virol. *66*, 1011-1123.
Pager, C.T., Craft, W.W., Jr., Patch, J., and Dutch, R.E. (2006). A

mature and fusogenic form of the Nipah virus fusion protein requires proteolytic processing by cathepsin L. Virology *346*, 251-257.

Pager, C.T. and Dutch, R.E. (2005). Cathepsin L is involved in proteolytic processing of the Hendra virus fusion protein. J. Virol. *79*, 12714-12720.

Pager, C.T., Wurth, M.A., and Dutch, R.E. (2004). Subcellular localization and calcium and pH requirements for proteolytic processing of the Hendra virus fusion protein. J. Virol. *78*, 9154-9163.

Pantua, H.D., McGinnes, L.W., Peeples, M.E., and Morrison, T.G. (2006). Requirements for the assembly and release of Newcastle disease virus-like particles. J. Virol. *80*, 11062-11073.

Parashar, U.D., Sunn, L.M., Ong, F., Mounts, A.W., Arif, M.T., Ksiazek, T.G., Kamaluddin, M.A., Mustafa, A.N., Kaur, H., Ding, L.M., Othman, G., Radzi, H.M., Kitsutani, P.T., Stockton, P.C., Arokiasamy, J., Gary, H., Jr., and Anderson, L.J. (2000). Case-control study of risk factors for human infection with a new zoonotic paramyxovirus, Nipah virus, during a 1998-1999 outbreak of severe encephalitis in Malaysia. J. Infect. Dis. *181*, 1755-1759.

Park, M.-S., Shaw, M.L., Muñoz, J., Cros, J.F., Nakaya, T., Bouvier, N., Palese, P., García-Sastre, A., and Basler, C.F. (2003). Newcastle disease virus (NDV)-based assay demonstrates interferon-antagonist activity for the NDV V protein and the Nipah virus V, W, and C proteins. J. Virol. *77*, 1501-1511.

Patch, J.R., Crameri, G., Wang, L.-F., Eaton, B.T., and Broder, C.C. (2007). Quantitative analysis of Nipah virus proteins released as virus-like particles reveals central role for the matrix protein. Virol. J. *4*, 1

Paton, N.I., Leo, Y.S., Zaki, S.R., Auchus, A.P., Lee, K.E., Ling, A.E., Chew, S.K., Ang, B., Rollin, P.E., Umapathi, T., Sng, I., Lee, C.C., Lim, E., and Ksiazek, T.G. (1999). Outbreak of Nipah-virus infection among abbatoir workers in Singapore. Lancet *354*, 1253-1256.

Patterson, J.B., Scheiflinger, F., Manchester, M., Yilma, T., and Oldstone, M.B.A. (1999). Structural and functional studies of the measles virus hemagglutinin: Identification of a novel site required for CD46 interaction. Virology *256*, 142-151.

Peluso, R. and Moyer, S. (1984). Vesicular stomatitis virus proteins required for the in vitro replication of defective interfering particle genome RNA. In Nonsegmented negative strand viruses. (New York, N.Y.: Academic Press Inc.), pp. 153-160.

Peluso, R. and Moyer, S. (1988). Viral proteins required for the *in vitro* replication of vesicular stomatitis virus defective interfering particles genome RNA. Virology *162*, 369-376.

Philbey, A.W., Kirkland, P.D., Ross, A.D., Davis, R.J., Gleeson, A.B.,

Love, R.J., Daniels, P.W., Gould, A.R., and Hyatt, A.D. (1998). An apparently new virus (family *Paramyxoviridae*) infectious for pigs, humans, and fruit bats. Emerg. Infect. Dis. *4*, 269-271.

Poch, O., Blumberg, B.M., Bougueleret, L., and Tordo, N. (1990). Sequence comparison of five polymerases (L proteins) of unsegmented negative-strand RNA viruses: theoretical assignment of functional domains. J. Gen. Virol. *71*, 1153-1162.

Poch, O., Sauvaget, I., Delarue, M., and Tordo, N. (1989). Identification of four conserved motifs among the RNA-dependent polymerase encoding elements. EMBO J. *8*, 3867-3874.

Porotto, M., Doctor, L., Carta, P., Fornabaio, M., Greengard, O., Kellogg, G.E., and Moscona, A. (2006). Inhibition of Hendra virus fusion. J. Virol. *80*, 9837-9849.

Portner, A. and Murti, K. (1986). Localization of P, NP, and M proteins on Sendai virus nucleocapsid using immunogold labeling. Virology *150*, 469-478.

Precious, B., Young, D., Bermingham, A., Fearns, R., Ryan, M., and Randall, R. (1995). Inducible expression of the P, V, NP genes of the paramyxovirus SV5 in cell-lines and an examination of NP:P and NP:V interactions. J. Virol. *69*, 8001-8010.

Qanungo, K.R., Shaji, D., Mathur, M., and Banerjee, A.K. (2004). Two RNA polymerase complexes from vesicular stomatitis virus-infected cells that carry out transcription and replication of genome RNA. Proc. Natl. Acad. Sci. USA *101*, 5952-7.

Radecke, F., Spielhofer, P., Schneider, H., Kaelin, K., Huber, M., Dötsch, C., Christiansen, G., and Billeter, M.A. (1995). Rescue of measles virus from cloned cDNA. EMBO J. *14*, 5773-5784.

Renshaw, R.W., Glaser, A.L., Van Campen, H., Weiland, F., and Dubovi, E.J. (2000). Identification and phylogenetic comparison of Salem virus, a novel paramyxovirus of horses. Virology *270*, 417-429.

Reynes, J.-M., Counor, D., Ong, S., Faure, C., Seng, V., Molia, S., Walston, J., Georges-Courbot, M.C., Deubel, V., and Sarthou, J.-L. (2005). Nipah virus in Lyle's flying foxes, Cambodia. Emerg. Infect. Dis. *11*, 1042-1047.

Rodriguez, J.J., Cruz, C.D., and Horvath, C.M. (2004). Identification of the nuclear export signal and STAT-binding domains of the Nipah virus V protein reveals mechanisms underlying interferon evasion. J. Virol. *78*, 5358-5367.

Rodriguez, J.J., Parisien, J.-P., and Horvath, C.M. (2002). Nipah virus V protein evades alpha and gamma interferons by preventing STAT1 and STAT2 activation and nuclear accumulation. J. Virol. *76*, 11476-11483.

Rodriguez, J.J., Wang, L.-F., and Horvath, C.M. (2003). Hendra virus

V protein inhibits interferon signaling by preventing STAT1 and STAT2 nuclear accumulation. J. Virol. *77*, 11842-11845.
Rogers, R.J., Douglas, I.C., Baldock, F.C., Glanville, R.J., Seppanen, K.T., Gleeson, L.J., Selleck, P.N., and Dunn, K.J. (1996). Investigation of a second focus of equine morbillivirus infection in coastal Queensland. Aust. Vet. J. *74*, 243-244.
Ryan, K. and Kingsbury, D. (1988). Carboxyl-terminal region of Sendai virus P protein is required for binding to viral nucleocapsids. Virology *167*, 106-112.
Sanderson, C.M., Wu, H.-H., and Nayak, D.P. (1994). Sendai virus M protein binds independently to either the F or the HN glycoprotein in vivo. J. Virol. *68*, 69-76.
Santiago, C., Björling, E., Stehle, T., and Casasnovas, J.M. (2002). Distinct kinetics for binding of the CD46 and SLAM receptors to overlapping sites in the measles virus hemagglutinin protein. J. Biol. Chem. *277*, 32394-32301.
Sawatsky, B., Grolla, A., Kuzenko, N., Weingartl, H., and Czub, M. (2007). Inhibition of henipavirus infection by Nipah virus attachment glycoprotein occurs without cell surface down-regulation of ephrin-B2 or ephrin-B3. J. Gen. Virol. *88*, 582-591.
Schmitt, A.P., Leser, G.P., Morita, E., Sundquist, W.I., and Lamb, R.A. (2005). Evidence for a new viral late-domain core sequence, FPIV, necessary for budding of a paramyxovirus. J. Virol. *79*, 2988-2997.
Schneider-Schaulies, J., Dunster, L.M., Kobune, F., Rima, B., and ter Meulen, V. (1995a). Differential downregulation of CD46 by measles virus strains. J. Virol. *69*, 7257-7259.
Schneider-Schaulies, J., Schnorr, J.-J., Brinckmann, U., Dunster, L.M., Baczko, K., Liebert, U.G., Schneider-Schaulies, S., and ter Meulen, V. (1995b). Receptor usage and differential downregulation of CD46 by measles virus wild-type and vaccine strains. Proc. Natl. Acad. Sci. USA *92*, 3943-3947.
Schowalter, R.M., Wurth, M.A., Aguilar, H.C., Lee, B., Moncman, C.L., McCann, R.O., and Dutch, R.E. (2006). Rho GTPase activity modulates paramyxovirus fusion protein-mediated cell-cell fusion. Virology *350*, 323-334.
Selvey, L.A., Wells, R.M., McCormack, J.G., Ansford, A.J., Murray, K., Rogers, R.J., Lavercombe, P.S., Selleck, P., and Sheridan, J.W. (1995). Infection of humans and horses by a newly described morbillivirus. Med. J. Aust. *162*, 642-645.
Sendow, I., Field, H.E., Curran, J., Darminto, Morrissey, C., Meehan, G., Buick, T., and Daniels, P. (2006). Henipavirus in *Pteropus vampyrus* bats, Indonesia. Emerg. Infect. Dis. *12*, 711-712.

Shaw, M.L., Cardenas, W.B., Zamarin, D., Palese, P., and Basler, C.F. (2005). Nuclear localization of the Nipah virus W protein allows for inhibition of both virus- and Toll-like receptor 3-triggered signaling pathway. J. Virol. *79*, 6078-6088.

Shaw, M.L., García-Sastre, A., Palese, P., and Basler, C.F. (2004). Nipah virus V and W proteins have a common STAT1-binding domain yet inhibit STAT1 activation from the cytoplasmic and nuclear compartments, respectively. J. Virol. *78*, 5633-5641.

Sidhu, M.S., Chan, J., Kaelin, K., Spielhofer, P., Radecke, F., Schneider, H., Masurekar, M., Dowling, P.C., Billeter, M.A., and Udem, S.A. (1995). Rescue of synthetic measles virus minireplicons: measles genomic termini direct efficient expression and propagation of a reporter gene. Virology *208*, 800-807.

Sidhu, M.S., Crowley, J., Lowenthal, A., Karcher, D., Menonna, J., Cook, S., Udem, S., and Dowling, P. (1994). Defective measles virus in human subacute sclerosing panencephalitis brain. Virology *202*, 631-641.

Spehner, D., Drillien, R., and Howley, P.M. (1997). The assembly of the measles virus nucleoprotein into nucleocapsid-like particles is modulated by the phosphoprotein. Virology *232*, 260-268.

Spehner, D., Kirn, A., and Drillien, R. (1991). Assembly of nucleocapsidlike structures in animal cells infected with a vaccinia virus recombinant encoding the measles virus nucleoprotein. J. Virol. *65*, 6296-6300.

Stricker, R., Mottet, G., and Roux, L. (1994). The Sendai virus matrix protein appears to be recruited in the cytoplasm by the viral nucleocapsid to function in viral assembly and budding. J. Gen. Virol. *75*, 1031-1042.

Sugahara, F., Uchiyama, T., Watanabe, H., Shimazu, Y., Kuwayama, M., Fujii, Y., Kiyotani, K., Adachi, A., Kohno, N., Yoshida, T., and Sakaguchi, T. (2004). Paramyxovirus Sendai virus-like particle formation by expression of multiple viral proteins and acceleration of its release by C protein. Virology *325*, 1-10.

Takimoto, T., Murti, K.G., Bousse, T., Scroggs, R.A., and Portner, A. (2001). Role of matrix and fusion proteins in budding of Sendai virus. J. Virol. *75*, 11384-11391.

Tamin, A., Harcourt, B.H., Ksiazek, T.G., Rollin, P.E., Bellini, W.J., and Rota, P.A. (2002). Functional properties of the fusion and attachment glycoproteins of Nipah virus. Virology *296*, 190-200.

Tan, W.S., Ong, S.T., Eshaghi, M., Foo, S.-S., and Yusoff, K. (2004). Solubility, immunogenicity and physical properties of the nucleocapsid protein of Nipah virus produced in *Escherichia coli*. J. Med. Virol. *73*, 105-112.

Testa, D. and Banerjee, A.K. (1977). Two methyltransferase activities in the purified virions of vesicular stomatitis virus. J. Virol *24*, 786-793.
Tidona, C.A., Kurz, H.W., Gelderblom, H.R., and Darai, G. (1999). Isolation and molecular characterization of a novel cytopathogenic paramyxovirus from tree shrews. Virology *258*, 425-434.
van den Hoogen, B.G., de Jong, J.C., Groen, J., Kuiken, T., de Groot, R., Fouchier, R.A.M., and Osterhaus, A.D.M.E. (2001). A newly discovered human pneumovirus isolated from young children with respiratory tracct disease. Nat. Med. *7*, 719-724.
Vidal, S. and Kolakofsky, D. (1989). Modified model for the switch from Sendai virus transcription to replication. J. Virol. *63*, 1951-1958.
Vogt, C., Eickmann, M., Diederich, S., Moll, M., and Maisner, A. (2005). Endocytosis of the Nipah virus glycoproteins. J. Virol. *79*, 3865-3872.
Vongpunsawad, S., Oezgun, N., Braun, W., and Cattaneo, R. (2004). Selectively receptor-blind measles viruses: Identification of residues necessary for SLAM- or CD46-induced fusion and their localization on a new hemagglutinin structural model. J. Virol. *78*, 302-313.
Vulliemoz, D. and Roux, L. (2001). 'Rule of Six': how does the Sendai virus RNA polymerase keep count? J. Virol. *75*, 4506-4518.
Wang, L.-F., Harcourt, B.H., Yu, M., Tamin, A., Rota, P.A., Bellini, W.J., and Eaton, B.T. (2001). Molecular biology of Hendra and Nipah viruses. Microbes Infect. *3*, 279-287.
Wang, L.-F., Michalski, W.P., Yu, M., Pritchard, L.I., Crameri, G., Shiell, B., and Eaton, B.T. (1998). A Novel P/V/C gene in a bew member of the *Paramyxoviridae* family, which causes lethal infection in humans, horses, and other animals. J. Virol. *72*, 1482-1490.
Wang, L.-F., Yu, M., Hansson, E., Pritchard, L.I., Shiell, B., Michalski, W.P., and Eaton, B.T. (2000). The exceptionally large genome of Hendra virus: Support for creation of a new genus within the family *Paramyxoviridae*. J. Virol. *74*, 9972-9979.
Weingartl, H., Czub, S., Copps, J., Berhane, Y., Middleton, D., Marszal, P., Gren, J., Smith, G., Ganske, S., Manning, L., and Czub, M. (2005). Invasion of the central nervous system in a porcine host by Nipah virus. J. Virol. *79*, 7528-7534.
Weingartl, H.M., Berhane, Y., Caswell, J.L., Loosmore, S., Audonnet, J.-C., Roth, J.A., and Czub, M. (2006). Recombinant Nipah virus vaccines protect pigs against challenge. J. Virol. *80*, 7929-7938.
Welstead, G.G., Hsu, E.C., Iorio, C., Bolotin, S., and Richardson, C.D. (2004). Mechanism of CD150 (SLAM) down regulation from the host cell surface by measles virus hemagglutinin protein. J. Virol. *78*, 9666-9674.
Westbury, H.A., Hooper, P.T., Brouwer, S.L., and Selleck, P.W.

(1996). Susceptibility of cats to equine morbillivirus. Aust. Vet. J. *74*, 132-134.
Whelan, S.P. and Wertz, G.W. (2002). Transcription and replication initiate at separate sites on the vesicular stomatitis virus genome. Proc. Natl. Acad. Sci. USA *99*, 9178-9183.
White, J.R., Boyd, V., Crameri, G.S., Duch, C.J., van Laar, R.K., Wang, L.-F., and Eaton, B.T. (2005). Location of, immunogenicity of and relationships between neutralization epitopes on the attachment protein (G) of *Hendra virus*. J. Gen. Virol. *86*, 2839-2848.
Williamson, M.M., Hooper, P.T., Selleck, P.W., Westbury, H.A., and Slocombe, R.F. (1999). Experimental Hendra virus infection in pregnant guinea-pigs and fruit bats (*Pteropus poliocephalus*). J. Comp. Pathol. *122*, 201-207.
Williamson, M.M., Hooper, P.T., Selleck, P.W., Westbury, H.A., and Slocombe, R.F.S. (2001). A guinea-pig model of Hendra virus encephalitis. J. Comp. Pathol. *124*, 273-279.
Wong, K.T., Grosjean, I., Brisson, C., Blanquier, B., Fevre-Montagne, M., Bernard, A., Loth, P., Georges-Courbot, M.-C., Chevallier, M., Akaoka, H., Marianneau, P., Lam, S.K., Wild, T.F., and Deubel, V. (2003). A golden hamster model for human acute Nipah virus infection. Am. J. Pathol. *163*, 2127-2137.
Wong, K.T., Shieh, W.-J., Kumar, S., Norain, K., Abdullah, W., Guarner, J., Goldsmith, C.S., Chua, K.B., Lam, S.K., Tan, C.T., Goh, K.J., Chong, H.T., Jusoh, R., Rollin, P.E., Ksiazek, T.G., Zaki, S.R., and the Nipah Virus Pathology Working Group (2002). Nipah virus infection: Pathology and pathogenesis of an emerging paramyxoviral zoonosis. Am. J. Pathol. *161*, 2153-2167.
Wright, P.J., Crameri, G., and Eaton, B.T. (2005). RNA synthesis during infection by Hendra virus: an examination by quantitative real-time PCR of RNA accumulation, the effect of ribavirin and the attenuation of transcription. Arch. Virol. *150*, 521-532.
Xu, Y., Gao, S., Cole, D.K., Zhu, J., Su, N., Wang, H., Gao, G.F., and Rao, Z. (2004a). Basis for fusion inhibition by peptides: analysis of the heptad repeat regions of the fusion proteins from Nipah and Hendra viruses, newly emergent zoonotic paramyxoviruses. Biochem. Biophys. Res. Commun. *315*, 664-670.
Xu, Y., Lou, Z., Liu, Y., Cole, D.K., Su, N., Qin, L., Li, X., Bai, Z., Rao, Z., and Gao, G.F. (2004b). Crystallization and preliminary crystallographic analysis of the fusion core from two new zoonotic paramyxoviruses, Nipah virus and Hendra virus. Acta Crystallogr. D (Biol. Crystallogr.) *D60*, 1161-1164.
Yang, J., Hooper, D.C., Wunner, W.H., Koprowski, H., Dietzschold, B., and Fu, Z.F. (1998). The specificity of rabies virus RNA

encapsidation by nucleoprotein. Virology. *242*, 107-117.
Yoneda, M., Guillaume, V., Ikeda, F., Sakuma, Y., Sato, H., Wild, T.F., and Kai, C. (2006). Establishment of a Nipah virus rescue system. Proc. Natl. Acad. Sci. USA *103*, 16508-16513.
Yoshida, T., Nagai, Y., Maeno, K., Iinuma, M., Hamaguchi, M., Matsumoto, T., Nagayoshi, S., and Hoshino, M. (1979). Studies on the role of M protein in virus assembly using a ts mutant of HVJ (Sendai virus). Virology *92*, 139-154.
Yoshida, T., Nagai, Y., Yoshii, S., Maeno, K., and Matsumoto, T. (1976). Membrane (M) protein of HVJ (Sendai virus): Its role in virus assembly. Virology *71*, 143-161.
Young, P.L., Halpin, K., Selleck, P.W., Field, H., Gravel, J.L., Kelley, M.A., and Mackenzie, J.S. (1996). Serologic evidence for the presence in pteropus bats of a paramyxovirus related to equine morbillivirus. Emerg. Infect. Dis. *2*, 239-240.
Yu, M., Hansson, E., Langedijk, J.P.M., Eaton, B.T., and Wang, L.-F. (1998a). The attachment protein of Hendra virus has high structural similarity but limited primary sequence homology compared with viruses in the genus *Paramyxovirus*. Virology *251*, 227-233.
Yu, M., Hansson, E., Shiell, B., Michalski, W., Eaton, B.T., and Wang, L.-F. (1998b). Sequence analysis of the Hendra virus nucleoprotein gene: comparison with other members of the subfamily *Paramyxovirinae*. J. Gen. Virol. *79*, 1775-1780.
Yuan, P., Thompson, T.B., Wurzburg, B.A., Paterson, R.G., Lamb, R.A., and Jardetzky, T.S. (2005). Structural studies of the parainfluenza virus 5 hemagglutinin-neuraminidase tetramer in complex with its receptor, sialyllactose. Structure *13*, 803-815.
Zhao, H. and Banerjee, A. (1995). Interaction between the nucleocapsid protein and the phosphoprotein of human parainfluenza virus 3. J. Biol. Chem. *270*, 12485-12490.
Zhu, Z., Dimitrov, A.S., Bossart, K.N., Crameri, G., Bishop, K.A., Choudhry, V., Mungall, B.A., Feng, Y.-R., Choudhary, A., Zhang, M.-Y., Feng, Y., Wang, L.-F., Xiao, X., Eaton, B.T., Broder, C.C., and Dimitrov, D.S. (2006a). Potent neutralization of Hendra and Nipah viruses by human monoclonal antibodies. J. Virol. *80*, 891-899.
Zhu, Z., Dimitrov, A.S., Chakraborti, S., Dimitrova, D., Xiao, X., Broder, C.C., and Dimitrov, D.S. (2006b). Development of human monoclonal antibodies against diseases caused by emerging and biodefense-related viruses. Expert Rev. Anti. Infect. Ther. *4*, 57-66.
Zimmer, G., Trotz, I., and Herrler, G. (2001). N-glycans of F protein differentially affect fusion activity of human respiratory syncytial virus. J. Virol. *75*, 4744-4751.

Chapter 6

Avian Influenza: Molecular Mechanisms of Pathogenesis and Host Range

Hans-Dieter Klenk, Mikhail Matrosovich and Jürgen Stech

ABSTRACT

Wild aquatic birds are the natural hosts for a large variety of influenza A viruses. Occasionally viruses are transmitted from this reservoir to other species and may then cause devastating outbreaks in domestic poultry or give rise to human influenza pandemics. We show here that proteolytic activation of the hemagglutinin is an important determinant for pathogenicity and that adaptation of the receptor binding specificity of the hemagglutinin and adaptation of the polymerase to new hosts play important roles in interspecies transmission.

INTRODUCTION

Avian influenza viruses are the causative agents of periodically occurring disease outbreaks in birds with high damage to the poultry industry. In addition, these viruses provide the genetic pool from which human and other mammalian influenza viruses emerge. Avian influenza has therefore a high impact on both, animal and human health. Although it has been known for a long time that the outcome of influenza virus infection in animals and man is a multifactorial trait depending on practically all viral genes (Klenk and Rott, 1988), research has shown that some viral proteins play a particularly prominent role. These are the hemagglutinin, the polymerase, and the NS1 protein. The role of the NS1 protein as a modulator of innate immunity has been reviewed elsewhere (Noah and Krug, 2005). After a brief general overview on avian influenza viruses, we focus here on the hemagglutinin and the polymerase as determinants of interspecies transmission and pathogenesis.

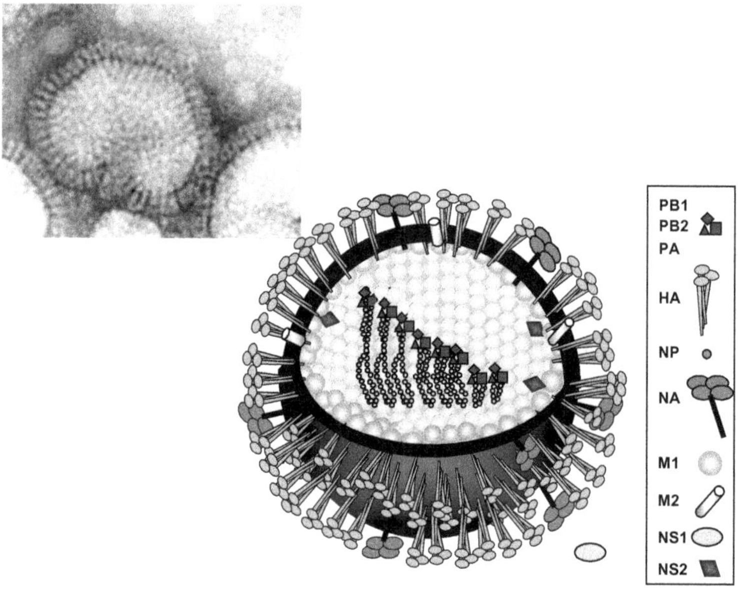

Fig. 1. Structure of influenza A virus. Upper panel: Electron micrograph of a virus particle (diameter 100 nm). The spikes at the surface (length 10nm) can be seen. Lower panel: Virus model. The envelope contains hemagglutinin (HA) and neuraminidase (NA) spikes as well the M2 protein. The interior of the particle is formed by ribonucleoproteins, consisting of the genomic RNA-segments, the nucleocapsid protein NP, and the polymerase proteins PB1, PB2, and PA. The inner side of the envelope is lined by the matrix protein M1. Virions contain also small amounts of the NS2 protein. NS1 is a non-structural protein.

CLASSIFICATION, STRUCTURE, AND LIFE CYCLE OF INFLUENZA A VIRUSES

Influenza viruses belong to the family Orthomyxoviridae. There are three genera or types: influenza A virus, influenza B virus, and influenza C virus. All avian influenza viruses belong to type A, a large group of viruses comprising 16 hemagglutinin (HA) and 9 neuraminidase (NA) subtypes. Types B and C do not occur in birds. The genome of influenza A viruses consists of eight single-stranded RNA molecules of negative polarity embedded in an enveloped virion (Fig. 1). The entire genome of influenza A viruses has a length of about 14,000 nucleotides. The individual RNA segments have a circular structure resulting from base pairing at the 3' and 5' ends. The segmentation of the genome is of particular biological significance, since it is a precondition for the high genetic variability of these viruses. Each segment encodes one or two viral proteins. The RNA genome forms together with the nucleocapsid protein NP and the polymerase proteins PB1, PB2, and PA the helical nucleocapsid core of virus particles. It is surrounded by the lipid-containing viral envelope that is lined on its inner side by the matrix protein M1 and has glycoprotein spikes on its outer surface. Influenza A viruses have hemagglutinin (HA) spikes with receptor binding and fusion functions, and neuraminidase (NA) spikes with receptor destroying activity. The fusion activity of HA depends on proteolytic cleavage by cellular proteases and a pH dependent conformational change. This activation process results in exposure of the fusion domain of HA and its penetration into the target membrane and thus in membrane fusion. The M2 protein which forms an ion channel in the viral envelope is essential for uncoating. The NS2 protein mediates the export of newly synthesized RNPs from the nucleus to the cytoplasm. The NS1 protein is not incorporated into virions. It has several functions, including inhibition of interferon, inhibition of cellular RNA splicing, and modulation of apoptosis. Some strains have another non-structural protein, PB1-F2, which also plays a role in apoptosis. The surface glycoproteins of influenza viruses are the subtype and strain specific antigenic determinants. They are the major viral components responsible for eliciting protective immunity. Matrix and nucleocapsid proteins show type specific antigenicity and allow the serological differentiation of types A, B, and C.

HA initiates infection by mediating binding to N-acetyl-neuraminic acid-containing receptors and membrane fusion following endocytosis. The M2 ion channel plays an important role in uncoating by lowering the pH within the virus particle and, thus, allowing dissociation of the internal components. The nucleocapsid complexes are then transported into the nucleus, where transcription and replication take place. The genomic RNA (vRNA) serves as template for two different RNA species: complementary RNA (cRNA) which is a complete copy of vRNA, and mRNA with a cap structure at the 5' end and with the 3' terminal nucleotides of cRNA being replaced by a poly-A-tail. The cRNA is the template for new vRNA molecules. The cap structures of the viral mRNAs are derived from cellular mRNA molecules. Viral mRNA utilizes the cellular translation machinery for the synthesis of viral proteins. The virus enhances the coding capacity of its genome by several mechanisms. These include splicing, expression of two cistrons in tandem position, and start of translation at two different initiation codons. Translation of nucleocapsid protein, polymerase proteins, matrix protein, and proteins NS1 and NS2 occurs on free polysomes. Ribonucleoproteins are assembled in the nucleus and subsequently exported into the cytoplasm. The envelope proteins are translated at the rough endoplasmic reticulum and then transported by the exocytotic apparatus to the plasma membrane where virus particles are assembled in a budding process. The neuraminidase mediates virus release by removing receptors from the infected cell, but it plays also a role in virus entry (for references see below).

HISTORY OF AVIAN INFLUENZA

Avian influenza has been known since the last decades of the 19th century when a severe and highly contagious poultry disease was observed in Northern Italy which appeared to be different from the then already well known fowl cholera (Perroncito, 1878). The new disease was called fowl plague. In 1901, Italian and Austrian scientists discovered almost simultaneously that it was caused by a filterable agent (Lode and Gruber, 1901; Centanni, 1902). Together with foot-and-mouth-disease, rabbit myxomatosis, and African horse sickness, fowl plague belongs therefore to the first animal infections recognized to have a viral etiology. The exact nature of fowl plague virus (FPV) remained obscure, however, for more than 50 years until W. Schäfer identified it in 1955 as an influenza A virus (Schaefer, 1955). Within the last 130 years, more than 30 outbreaks of fowl plague or fowl-plague-like disease have been reported world-wide (Becker, 1966; Alexander, 1986; Subbarao et al., 2006; Werner and Harder, 2006).

They were caused by viruses belonging either to subtype H5 or H7, and the birds affected with lethal illness were primarily chickens, turkeys and other species of the order Galliformes. Most of them have been confined to specific geographical areas and were eliminated in less than a year through comprehensive eradication programs. Even if the disease was enzootic for a larger period of time, geographical spread was limited. The H5N1 outbreak ongoing since 1996 has been unprecedented by its length, by its spread over large parts of the world, and by so far almost 200 million dead birds (Table 1).

Table 1: Outbreaks of highly pathogenic avian influenza

Year	Subtype	Country
1878-1935	*	Italy-enzootic
1890-1930	*	Germany-enzootic
1922	*	England
1923-1945	*	Egypt-enzootic
1924-1925	*	USA
1927	*	Indonesia
1929	*	England
1929	*	USA
1959	H5N1	Scotland
1961	H5N3	South Africa
1963	H7N3	England
1966	H5N9	Canada
1976	H7N7	Australia
1979	H7N7	England
1979	H7N7	Germany
1983-84	H5N2	USA
1983-84	H5N8	Ireland
1985	H7N7	Australia
1991	H5N1	England
1992	H7N3	Australia
1994-95	H5N2	Mexico
1994	H7N3	Australia
1994	H7N3	Pakistan
1996-ongoing	H5N1	Asia, Europe, Africa
1997	H7N4	Australia
1997	H5N2	Italy
1999	H7N1	Italy
2002	H7N3	Chile
2003	H7N7	Netherlands
2004	H7N3	Canada
2004	H5N2	USA
2004	H5N2	South Africa
2004	H7N3	Pakistan

*presumably H7

When it became clear in 1955 that FPV was an influenza A virus, other avian influenza viruses were identified. These viruses were often obtained during relatively mild disease outbreaks, such as influenza virus A/chick/Germany/N/49 (H10N7) that was originally suspected to be NDV. Shown to be "negative" when analyzed by serological assays specific for NDV it was named "virus N" (Dinter, 1949; Dinter and Bakos, 1950) and only subsequently identified as an influenza A virus (Rott and Schaefer, 1960). The number of such viruses rapidly increased in the following years. They were isolated from a wide variety of wild and domestic birds and they included all known serotypes (H1-16, N1-9). These low-pathogenic avian influenza (LPAI) viruses have therefore to be distinguished from highly virulent FPV or FPV-like viruses which are now called highly pathogenic avian influenza (HPAI) viruses.

PATHOGENESIS

HPAI viruses are defined as H5 and H7 viruses that cause 75% or higher mortality after experimental infection of chickens and that have a polybasic hemagglutinin cleavage site (see below). All viruses that do not meet these criteria are classified as LPAI viruses (Alexander, 1998; OIE, 2003). LPAI virus replication is confined to the intestinal and respiratory tract resulting in mild disease or asymptomatic infection. Virus is shed in the feces. A common route of dissemination among aquatic birds is therefore through contaminated water. HPAI viruses are also shed in feces with high concentrations. However, these viruses are more readily transmitted among birds in densely populated flocks by the nasal and oral routes through contact with virus-contaminated materials. In contrast to the local LPAI virus infection of the intestinal or respiratory tract, HPAI viruses cause systemic infection. As a result, virus can be recovered from many organs of infected animals. Large hemorrhages distributed all over the body, edema, and cutaneous ischemia are major symptoms of the disease. The final stage of the infection can be characterized by the emergence of neurological signs, such as photophobia and dullness (Gratzl and Koehler, 1968; Narayan et al., 1972). HPAI viruses have been found to specifically target lymphocytes and lymphoid tissues (Van Campen et al., 1989b; Van Campen et al., 1989a), myocytes in the heart muscle (Kobayashi et al., 1996), and endothelial cells (Feldmann et al., 2000; Suarez et al., 1998), and these cell tropisms may play an important pathogenic role in systemic virus dissemination and in the vascular leakage underlying hemorrhages and edema.

ECOLOGY AND EVOLUTION

Wild aquatic birds of the orders *Anseriformes* (waterfowl) and *Charadriiformes* (shorebirds) are the natural hosts of all LPAI viruses providing a large genetic pool from which the influenza A viruses of other host species are derived (Webster et al., 1992; Olsen et al., 2006).

Since the host barrier is not an insurmountable obstacle for LPAI viruses, they can occasionally be transmitted from their natural reservoir to terrestrial birds and mammals. Most of these transmissions are transient and do not result in a stable new lineage. On rare occasions, however, the viruses adapt to the new species and give rise to a new lineage. Adaptation requires multiple mutations and it may involve gene reassortment after co-infection with another virus. There may be intermediate hosts in the interspecies transmission including pigs, chickens and quails.

By these mechanisms, viruses with new surface glycoproteins and therefore a distinct change in antigenicity may be generated. If a new virus with such an antigenic shift occurs in man, it causes a pandemic. There were three pandemics in the last century. After the Spanish influenza (H1N1) in 1918, which presumably resulted from the introduction of a complete avian influenza virus without reassortment into the human population (Tumpey et al., 2005), an antigenic shift in HA and NA gave rise to the Asian influenza (H2N2) in 1957 and another shift in HA to the Hong Kong influenza (H3N2) in 1968. In 1977 the H1N1 subtype re-emerged and co-circulates now in man with H3N2 and influenza B virus.

HPAI viruses arise by introduction and circulation of H5 and H7 LPAI viruses in domestic poultry with subsequent mutations in HA (Roehm et al., 1995; Campitelli et al., 2004; Fouchier et al., 2004). In general, aquatic birds are not the natural host of HPAI viruses. However, there is evidence that the H5N1 HPAI viruses circulating in Asia have recently been re-introduced from poultry into wild aquatic birds (Chen et al., 2006).

RECENT TRANSMISSIONS TO MAN

For a long time, HPAI viruses have been thought to be apathogenic for man. However, within the past 10 years human infections with H5 and H7 HPAI as well as H9N2 LPAI viruses have been observed that were often associated with severe disease. As yet, none of them has adapted to man. However, since humans have not been exposed before and are therefore immunologically naïve to these viruses, they are a considerable pandemic threat.

Transmission of H5N1 HPAI virus was first observed during an epizootic in 1997 in Hong Kong, when 18 people became ill showing the symptoms of classical influenza. Six of these patients died. The virus that was transmitted without reassortment from infected chickens contained an HA gene derived from a H5N1 virus first isolated from a goose in 1996 (A/goose/Guangdong/1/96) (Claas et al., 1998; Subbarao et al., 1998; Xu et al., 1999). Related H5N1 viruses circulated in birds in Southern China from 1997 to 2001 without transmission to humans (Cauthen et al., 2000; Guan et al., 2002). Since 2003 severe outbreaks have been observed periodically in Vietnam, Thailand, Indonesia, China, and other Asian countries (WHO, 2003). Genetic comparison with the original isolates from 1997 revealed that the new viruses represented different genotypes resulting from multiple reassortments. Among these, the Z genotype was dominant. The HA of all of these viruses was similar to that of the A/goose/Guangdong/1/96 strain (Li et al., 2004). Extensive molecular-epidemiological studies revealed that domestic ducks and land-based poultry in Southern China probably had a central role in the generation and maintenance of the H5N1 virus and that wild birds might have contributed to the broad spread of the virus (Horimoto and Kawaoka, 2005). The role of migratory birds as carriers of the virus was further underlined when, following an outbreak in a bird sanctuary at Lake Qinghai in Northwest China in spring 2005 (Liu et al., 2005), the virus spread within a few months to Siberia, Europe and Africa. It has to be pointed out, however, that international poultry trade may have played an important role in virus dissemination as well. As in 1997, the current H5N1 viruses have the capacity to cross species barriers and to cause disease with high case-fatality rates in mammals, such as cats (Kuiken et al., 2004), tigers (Enserink and Kaiser, 2004), and stone martens (Weber et al., 2007). This includes man with so far (Jan. 2007) 264 reported patients and 157 deaths. Human infection is characterized by severe pneumonia with high viral load and hypercytokinemia. There is also evidence for systemic virus spread (de Jong et al., 2006).

Although sporadic human transmissions of HPAI H7N7 have been reported before (Kurtz et al., 1996; Banks et al., 1998), there has been only one poultry outbreak in 2003 in the Netherlands where a larger number of human infections have been observed. 83 cases with relatively mild influenza-like illness and conjunctivitis were observed. However, there was one human infection with acute respiratory distress syndrome and fatal outcome (Koopmans et al., 2003; van Kolfschooten, 2003; Fouchier et al., 2004). Seroepidemiological studies suggested that the number of inapparant infections was much larger, and there was also evidence for human-to-human transmission (Bosman, 2004).

Human H9N2 infections with mild respiratory disease were observed in small numbers in 1999 and 2003 (Peiris et al., 1999; Guo et al., 2000; WHO, 2003). One of these viruses appears to be related to the HPAI H5N1 virus from 1997, since both have internal proteins with the same genetic background (Lin et al., 2000). The H9N2 viruses are now enzootic in Eurasia (Choi et al., 2004).

PROTEOLYTIC ACTIVATION OF THE HEMAGGLUTININ DETERMINES PATHOGENICITY

Cleavage of the hemagglutinin is necessary for fusion activity and, therefore, for the infectivity of the virus. X-ray crystallographic studies have shown that the cleavage site is located in a circular loop projecting away from the surface of the precursor HA into the solvent (Fig. 2). The activating enzymes are provided by the host. Avian influenza viruses show large variations in cleavability, and these differences are prime determinants of pathogenicity. In the following we will give a brief outline of this concept. For more detailed information, the reader is referred to previous reviews by us (Klenk and Garten, 1994) and others (Steinhauer, 1999).

LPAI viruses, like mammalian influenza A viruses, have hemagglutinins with a single arginine at the cleavage site. They are activated by proteases secreted form epithelial cells that are present only in respiratory or intestinal tissues. Infection is therefore restricted to these organs (Fig. 2). Although a number of trypsin-like proteases, such as plasmin, tryptase Clara and factor X have been found to activate hemagglutinin with a monobasic cleavage site *in vitro* (for references see Klenk and Garten, 1994), little is known about the enzymes that activate these viruses in their natural setting. Recently two serine proteases (TMPRSS2 and HAT) from human airway epithelium have been found to activate human influenza A viruses as well as an LPAI virus (Boettcher et al., 2006). However, these enzymes have not been identified yet in avian tissues. Bacterial proteases may also activate HA's of restricted cleavability and promote the development of pneumonia in mice after combined viral-bacterial infection (Tashiro et al., 1987a). Among the viruses activated by bacterial proteases are LPAI strains (Tashiro et al., 1987b), but it is not known, whether this type of activation plays a pathogenetic role in avian influenza.

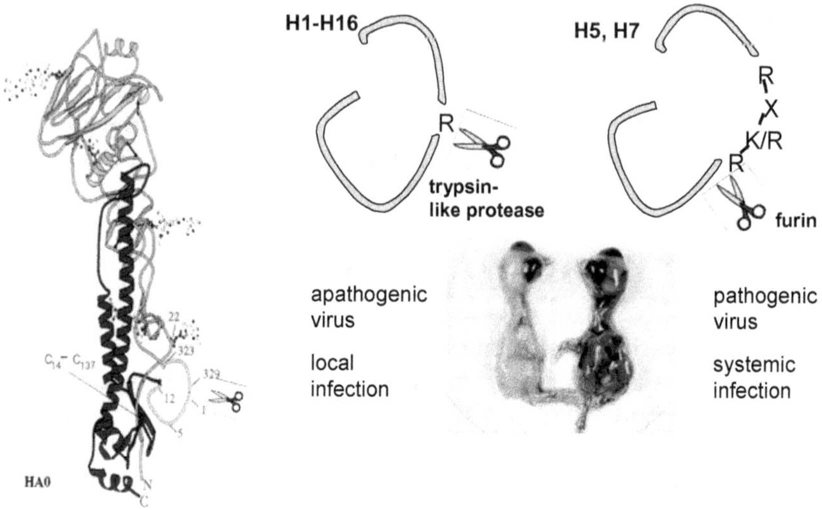

Fig. 2. The cleavage site of HA determines the pathogenicity of avian influenza viruses. HA is cleaved into subunits HA1 (grey) and HA2 (dark-grey). The cleavage site is located in a loop (light-grey) projecting from the surface of the molecule (Chen et al., 1998). LPAI viruses (subtypes H1-H16) have a single arginine at the cleavage site that is recognized by trypsin-like proteases that are present only in specific tissues, such as intestinal epithelia. These viruses cause therefore local infection. HPAI viruses (subtypes H5 and H7) are activated at a multibasic cleavage site by the ubiquitous protease furin. These viruses spread, therefore, throughout the organism.

HPAI viruses are activated by a different cleavage mechanism. Their hemagglutinins are activated at multibasic cleavage sites by furin, a member of the proprotein convertase family of eukaryotic subtilisin-like serine endoproteases (Stieneke-Groeber et al., 1992). The ubiquity of this enzyme accounts for the systemic infection typical for these viruses (Fig. 2). Furin is a constituent of the constitutive secretory pathway in practically all cells and accumulates in the trans-Golgi network, which is also the cellular compartment where this HA type is cleaved (for references see Klenk and Garten, 1994). Other proprotein convertases which resemble furin in structure and substrate specificity are PACE4, PC5/6, and LPC/PC7. Like furin, PC5/6 activates HAs with multibasic cleavage sites, whereas PACE and LPC/PC7 do not (Horimoto et al., 1994; Hallenberger et al., 1997). All four proteases have been cloned form chicken tissues, and chicken furin, which has 77% amino acid sequence homology with human furin, was shown to activate HA (Feldmann, 1998). The HAs of most HPAI viruses have the consensus sequences R-X-K/R-R (for references see Klenk and Garten, 1994) or R-X-X-R (for references see Neumann and Kawaoka, 2006) at the cleavage site, motifs that are both recognized by furin. The only exception to these rules is the HA of A/chick/Pennsylvania/83 (H5N2) which contains the unusual tetrapeptide K-K-K-R (Kawaoka and Webster, 1989). Another important determinant is a carbohydrate side chain close to the cleavage site that interferes with protease accessibility. Loss of this carbohydrate resulted in enhanced HA cleavability and pathogenicity (Kawaoka et al., 1984). However, masking of the cleavage site by this oligosaccharide was overcome when the number of basic amino acids was increased (Kawaoka and Webster, 1989; Ohuchi et al., 1989). It was also shown that HA can acquire high cleavability only if the basic sequence was introduced by insertion, but not by amino acid exchanges in the carboxy-terminus of HA1 (Ohuchi et al., 1991).

Increased pathogenicity as a consequence of insertions at the cleavage site has first been observed in laboratory studies involving sequential cell culture passages of strains A/seal/Massachusetts/1/80 (H7N7) (Li et al., 1990) and A/turkey/Oregon/71 (H7N3). In the latter case, the acquisition of the furin recognition motif resulted from recombination of the HA gene with 28S ribosomal RNA (Khatchikian et al., 1989; Morsy et al., 1994). The HA gene may not only recombine with cellular RNA but also with other viral gene segments, as has been observed recently when new HPAI viruses emerging in the field have been analyzed. Thus, comparison of A/chicken/Chile/02 (H7N3) isolates revealed that the HA genes of the highly pathogenic strains had an insertion of 30 nucleotides at the cleavage site that was presumably derived from the nucleoprotein gene of the unrelated A/gull/Maryland/704/77 (H13N6) virus (Suarez et al., 2004). Recombination between HA and matrix protein genes of the same virus generated the highly pathogenic A/chicken/BC/04 (H7N3) viruses (Pasick et al., 2005). Polymerase slippage has been suggested as an alternative strategy by which a multibasic cleavage site is generated. (Garcia et al., 1996; Perdue et al., 1997). However, there are other examples where the mechanism of insertion is not clear (Banks et al., 2001).

It is also not clear yet why the acquisition of a multibasic cleavage site and therefore the generation of HPAI viruses occurs in nature only with subtypes H5 and H7. Interestingly however, high cleavability was also observed with a subtype H3 HA after *in vitro* insertion of a multibasic cleavage site and removal of an adjacent oligosaccharide by recombinant DNA technology (Ohuchi et al., 1991). Thus, it appears that confinement of HPAI viruses to subtypes H5 and H7 cannot be attributed to structural restrictions of the HA protein, but that the responsible mechanisms are at the level of RNA replication.

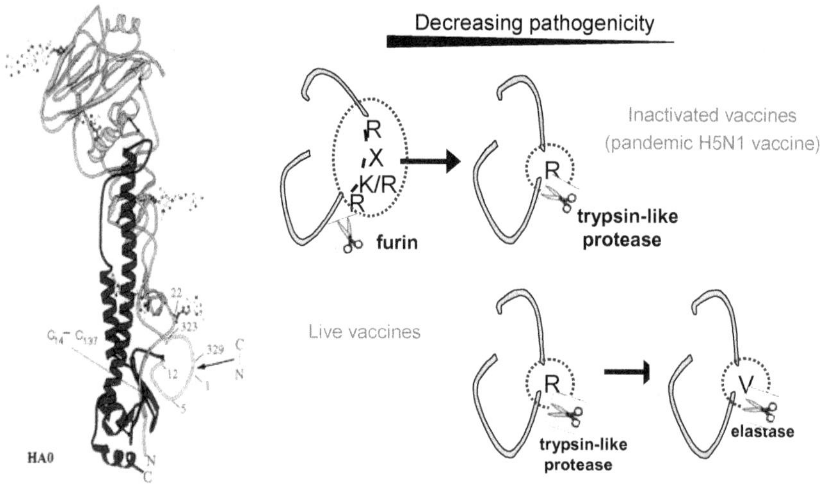

Fig. 3. Generation of vaccine strains by genetic manipulation of the HA cleavage site. HPAI viruses can be converted into LPAI viruses by exchanging the polybasic cleavage site for a single arginine. Such viruses can be used for mass production of pandemic vaccines. Viruses suitable for live vaccination can be obtained when pathogenicity is further reduced by replacing the trypsin cleavage site by an elastase cleavage site.

In contrast to natural evolution where HPAI viruses generally appear to be derived from LPAI viruses, recombinant viruses with reduced pathogenicity can be generated by *in vitro* mutation at the cleavage site (Fig. 3). Thus, replacement of the multibasic cleavage site of a highly pathogenic H5N1 virus by a single arginine resulted in an attenuated virus that can be used for mass production of an inactivated pandemic vaccine (Subbarao et al., 2003). Protease activation mutants that are cleaved by elastase, an activating enzyme not occurring in a natural setting, are even less pathogenic. Such viruses have the potential to be used as live vaccines (Stech et al., 2005).

RECEPTOR SPECIFICITY OF THE HEMAGGLUTININ: A DETERMINANT OF HOST RANGE

RECEPTORS OF INFLUENZA VIRUSES: SIALIC ACIDS

Influenza infection is initiated by virus attachment to cell-surface molecules commonly called viral receptors. Specific identity of molecules that serve as functional receptors for influenza viruses on target cells in vivo is still not defined. However, it was recognized long ago that sialic acid moieties of the receptors are essential for the binding (Gottschalk et al., 1972; Paulson, 1985; Herrler et al., 1995; Matrosovich et al., 2007).

Sialic acids (Sia) are a family of 9-carbon acid sugars which are ubiquitous at the cell-surface membranes and in sera and epithelial secretions of animals (Schauer, 1982; Kelm and Schauer, 1997; Schauer and Kamerling, 1997; Varki, 1997)). The common distinct features of sialic acids are a carboxylic group, an amido group and a polyhydroxylated "tail" attached to the pyranose ring at positions 2, 5, and 6, respectively. More than 40 known sialic acid species differ from each other by substituents at four hydroxyls (O_4, O_7, O_8, and O_9) and at the amido group. In contrast to influenza C viruses, which bind to 9-O-acetylated sialic acid species ((Herrler and Klenk, 1987)), influenza A and B viruses bind primarily to non-O-acetylated sialic acids (Matrosovich et al., 2007). Enzymatic hydroxylation of the 5-N-acetyl group of N-acetylneuraminic acid (Neu5Ac) leads to N-glycolylneuraminic acid (Neu5Gc). Neu5Ac- and Neu5Gc-containing glycans frequently occur together. Their ratio varies in different hosts and tissues. Neu5Gc is abundant in monkeys, pigs, cows, horses and mice, but it is absent in normal human tissues and apparently in birds (Schauer, 1982; Varki, 1997). Some avian, swine, equine and human viruses with H3 subtype HA were reported to bind to Neu5Gc-containing receptors (Matrosovich et al., 1997; Suzuki et al., 1997; Masuda et al., 1999; Ito et al., 2000; Suzuki et al., 2000). However, the viruses of other subtypes (H2, H4, H5, H6, H9, H11, H12 and H14) bound to Neu5Gc weakly, if at all (Gambaryan et al., 2005b). These data suggest that binding to Neu5Gc is not essential for the replication of the majority of influenza A virus subtypes.

The crystal structures of the hemagglutinins of the H1, H3, H5, and H9 subtypes complexed with natural and synthetic sialic acid compounds were determined by X-ray analysis (Skehel and Wiley,

2000; Ha et al., 2001; Ha et al., 2003; Gamblin et al., 2004). The sialic acid-binding site is a shallow pocket located on the HA globular head. The interactions between the amino acid residues in the binding site and the sialyloligosaccharide moieties of the receptors are very weak (Kdiss > 0.1 mM) (Sauter et al., 1989; Gambaryan et al., 1999; Matrosovich et al., 2000). However, the influenza virion contains about 350 to 400 HA trimers on its surface and the high binding avidity required for virus attachment to the cell is accomplished by co-operative binding of multiple HA spikes to multiple copies of receptors. Virus binding depends not only on HA affinity for the terminal sialic acid residues, but also on the structure of underlying oligosaccharide and protein or lipid moieties of the receptors, as well as on the abundance and accessibility of receptors on the cell surface. Because of this complex mode of binding, the receptor-binding properties of influenza viruses can be affected by amino acid substitutions inside the sialic acid-binding pocket, on the pocket rim, and by distant mutations resulting in altered glycosylation or altered electrostatic charge of the HA globular head (Matrosovich et al., 2007). Changes at these positions allow the virus to adapt its receptor-binding activity for a best fit in a new replication environment, for example, after introduction into a new host species.

Sialic acids are represented *in vivo* mainly as components of oligosaccharide chains of glycoproteins and gangliosides (sialic acid-containing glycosphingolipids). It is generally believed that large glycoproteins are better exposed on the cell surface and are more likely to serve for the initial influenza virus attachment, whereas subsequent binding to gangliosides could bring the viral and cell membranes into closer proximity and, thus, facilitate the membrane fusion and entry of the viral genome into the cell (Paulson, 1985; Herrler et al., 1995). Some studies suggested that gangliosides can mediate virus entry into cells even in the absence of glycoprotein receptors, as exogenous gangliosides restored susceptibility to influenza A and C virus infection of cells that have been rendered resistant by destroying endogenous receptors with bacterial sialidases (Bergelson et al., 1982; Herrler and Klenk, 1987). A common property of influenza viruses in wild aquatic birds distinguishing these viruses from most human viruses was strong avian virus binding to gangliosides with relatively short sugar cores, such as GM3, GD1a, and 3´sialylparagloboside (Matrosovich et al., 1997; Miller-Podraza et al., 2000; Gambaryan et al., 2003). A strict conservation of this property among avian viruses raised the hypothesis that binding to gangliosides could be essential for the virus replication in the intestine, the primary target tissue of influenza viruses in aquatic birds (Webster et al., 1978). At the same time, mutant laboratory cells lacking gangliosides could be infected by avian and human influenza A viruses and produced infectious virus progeny demonstrating that gangliosides were neither absolutely necessary for the early nor for the late stages of the infection (Matrosovich et al., 2007). Thus, the relative roles of glycoproteins and gangliosides as influenza virus receptors remain unclear. These roles could vary among different viral host species and target tissues.

In oligosacharide chains of glycoproteins and gangliosides, sialic acids are α2-3- or α2-6-linked to Gal and GalNAc, α2-6-linked to GlcNAc, or α2-8-linked to the second sialic acid residue. Few other types of linkage have been recognized, such as Sia(α2-4)Gal and Sia(α2-9)Sia, but they occur less often, and their binding to influenza viruses has not been studied. Several studies suggested that influenza viruses generally do not bind to α2-8-linked Neu5Ac moieties (Suzuki et al., 1986; Suzuki et al., 1992; Matrosovich et al., 1993). Therefore, variants of the sialic acid glycosidic linkage to the penultimate sugar that could play a role in influenza virus A reception in vivo are limited to Neu5Ac(α2-3/6)Gal, Neu5Ac(α2-3/6)GalNAc, and Neu5Ac(α2-6)GlcNAc. However, these disaccharide moieties can be presented in highly variable microenvironments (Schauer and Kamerling, 1997). Influenza viruses bind to sialic acids located at the terminal position of oligosaccharides and do not bind to internally located Sia residues (Suzuki, 1994; Matrosovich et al., 2007). Early binding studies demonstrated that the recognition of sialic acid-containing disaccharide moieties can be affected by the structure of more distant parts of the glycans (Paulson, 1985; Suzuki, 1994). However, these data were obtained for only a few virus strains and/or oligosaccharide sequences. The role of these core oligosaccharides in receptor binding will be discussed in more detail below. The structural diversity of sialic acid-containing natural molecules is increased by conformational flexibility of oligosaccharides, by variations in the position of oligosaccharide chains in the macromolecular receptors, and by their different steric accessibility in the context of other cell-surface molecules. The term "presentation" is commonly used to describe all structural variables that influence recognition of cell-surface carbohydrate receptors by animal and viral lectins (Weis, 1997). The presentation of influenza virus sialyloligosaccharide receptors is not well understood yet, and the current knowledge on the receptor specificity of influenza viruses is mostly limited to the recognition of the terminal sialyl-galactosyl receptor epitopes.

RECEPTOR SPECIFICITY OF AVIAN VIRUSES: RECOGNITION OF THE SIA-GAL LINKAGE

Rogers and Paulson (1983) first discovered that avian influenza viruses differentiate between the types of Sia-Gal glycosidic linkage. In these studies, duck viruses with the H3 HA bound to terminal Neu5Ac(α2-3)Gal-containing receptors significantly better than they bound to Neu5Ac(α2-6)Gal-containing receptors. By contrast, human influenza viruses preferentially bound to 6-linked sialic acids (Paulson, 1985). Later, the same receptor-binding pattern was observed in studies on larger panels of avian and human H1, H2, and H3 virus HA subtypes (Rogers and D'Souza, 1989; Connor et al., 1994). Furthermore, studies on avian virus strains of most other HA subtypes revealed that influenza viruses of wild aquatic birds, irrespective of their HA antigenic subtype, preferentially recognize Neu5Ac(α2-3)Gal-containing receptors (Nobusawa et al., 1991; Gambaryan et al., 1997). Histochemical analysis of duck intestinal epithelial cells using linkage-specific lectins revealed the presence of Neu5Ac(α2-3)Gal-containing sequences and no detectable expression of Neu5Ac(α2-6)Gal-containing receptors (Ito et al., 1998). Human viruses did not bind to plasma membranes isolated from duck intestinal cells, confirming the lack of Neu5Ac(α2-6)Gal-terminated receptors in duck intestine (Gambaryan et al., 2002). Thus, the receptor specificity of duck viruses correlates with the predominance of Neu5Ac(α2-3)Gal residues on the tissues in which the virus replicates in ducks.

To characterize the receptor specificity of the avian viruses in more detail, Matrosovich et al. (1997, 2000) analyzed viral binding to free N-acetylneuraminic acid and sialyloligosaccharides. A characteristic feature of most avian viruses was their at least tenfold higher affinity for 3'-sialyllactose (3'SL) than for free Neu5Ac. Several other oligosaccharides with terminal Neu5Ac(α2-3)Gal-moiety but different structures of the third and subsequent saccharide rings also bound stronger than free sialic acid. These effects indicated that, besides binding to the terminal Neu5Ac residue of the receptor determinant, avian HAs also bind to the penultimate 3-linked Gal residue. This property clearly differentiates avian viruses from human and swine influenza A viruses, which bind Neu5Ac and 3'SL with about the same affinity, i.e., do not appreciably interact with 3-linked Gal (Matrosovich et al., 1993; Matrosovich et al., 1997; Matrosovich et al., 2000). In contrast to 3'SL, 6'-sialyllactose (6'SL) and 6'-sialyl(N-acetyllactosamine) (6'SLN) have a weaker affinity for avian viruses than Neu5Ac, indicative of possible steric interference between the HA receptor-binding site and the 6-linked lactose/acetyllactosamine moiety.

The crystal structures of the avian HA complexes with the sialylpentasacharides LSTa and LSTc (Ha et al., 2001; Ha et al., 2003) provided insights into the molecular mechanisms of the linkage recognition by influenza viruses. For H3 and H5 avian HAs, the 3-linked sialyloligosaccharide LSTa is bound with the minimum-energy *trans* conformation of the Sia-Gal linkage. In this conformation, the glycosidic oxygen and the axial 4-OH group of the Gal form hydrogen bonds with the side chain amide and carbonyl groups of the conserved glutamine in HA position 226 (226Q). The 6-linked sialyloligosaccharide LSTc binds to avian HAs in its own lowest energy *cis* conformation. In this case, the hydrophobic C6 methylene group of the α (2-6)-linkage projects downward toward the polar atoms of the side chain of 226Q. These unfavorable polar-nonpolar interactions between 6-linked Gal and 226Q, account, at least in part, for the poor binding of avian viruses to 6-linked receptors. Thus, the receptor specificity of avian viruses is determined by the optimal fit to the HA receptor-binding site of the most abundant *trans* conformer of the Neu5Acα (2-3)Gal moiety and a poor fit of the major *cis* conformer of the Neu5Acα(2-6)Gal epitope.

Six additional amino acids were found to be highly conserved in the receptor-binding site of most avian viruses (138A, 190E, 194L, 225G, 226Q, and 228G) when compared to the binding site of human viruses (Matrosovich et al., 1997). These amino acids either directly participate in hydrogen bond formation and van der Waals interactions with Neu5Ac(α2-3)Gal moiety (190E, 194L, 225G, and 226Q), or make direct contacts with such amino acids (138A and 228G). Mutations in any of these positions were found to decrease the avian virus binding to the 3-linked galactose moiety and to lower the virus affinity for the Neu5Ac(α2-3)Gal-containing sialylglycopolymers (Matrosovich et al., 2000). Thus, the conservation of these amino acids is required for the avian virus HA recognition of "the α2-3-linkage-specific motif" (Ha et al., 2001) and efficient virus attachment to target cells in birds.

RECEPTOR SPECIFICITY AND PANDEMICS

Two human pandemics in the last century were caused by reassortant viruses that carried HA (H3N2, 1968) or both HA and NA (H2N2, 1957) from avian influenza viruses (Webster et al., 1992; Cox and Subbarao, 2000). The pandemic of 1918 likely originated from an avian virus that was transmitted to humans as a whole (Taubenberger et al., 1997). Studies on the receptor specificity of the earliest virus strains from these pandemics revealed that they acquired a typical human-virus-like receptor specificity (Connor et al., 1994; Matrosovich et al., 2000; Kobasa et al., 2004; Glaser et al., 2005). It is generally believed, therefore, that alteration of the receptor specificity is a prerequisite for the highly effective replication and human-to-human transmission which characterize pandemic virus strains (Cox and Subbarao, 2000; Horimoto and Kawaoka, 2001; Baigent and McCauley, 2003; Matrosovich et al., 2007).

Mutation E190D, either alone or in combination with substitution G225D, was critical for the alteration of the receptor specificity of the avian precursor and the emergence of the H1N1 pandemic virus (Reid et al., 1999; Matrosovich et al., 2000; Kobasa et al., 2004; Glaser et al., 2005). In the case of H2 and H3 pandemic viruses, substitution Q226L was essential for the acquisition of the human virus-like specificity by the avian virus HA, whereas an additional mutation, G228S, had only a marginal effect (Rogers et al., 1983; Connor et al., 1994; Vines et al., 1998; Matrosovich et al., 2000). In particular, some H2 human viruses isolated in 1957 had 226L but still maintained the avian virus-like 228G; these viruses might have represented the earliest step of adaptation of the avian H2 HA to humans.

This alteration of receptor specificity of the avian virus HAs in humans correlates with the abundant expression of 6-linked sialic acid epitopes and with an apparent deficiency of 3-linked epitopes in human airway epithelium (Baum and Paulson, 1990; Couceiro et al., 1993; Shinya et al., 2006). The differences in receptor specificity of avian and human influenza viruses raised the hypothesis that avian viruses cannot directly transmit to humans and that an intermediate animal host is required for such transmission. Because pigs support replication of both avian and human viruses, they were considered to be a plausible intermediate host for the generation of human pandemic strains by gene reassortment (Scholtissek et al., 1998). The potential role of pigs as "mixing vessels" in zoonotic transmission of influenza viruses was supported by the findings that both 3-linked and 6-linked sialic acids are present in the pig tracheal epithelium ((Ito et al., 1998)) and that swine influenza viruses preferentially bind to 6-linked receptors (Rogers and D'Souza, 1989; Ito et al., 1998; Matrosovich et al., 2000; Gambaryan et al., 2005b).

The human infections caused by H5N1 chicken viruses in Hong Kong in 1997 (Claas et al., 1998; Subbarao et al., 1998) demonstrated for the first time that avian viruses can directly infect humans without preceding evolution in pigs or any other intermediate host. The H5N1/97 and other H5N1 poultry viruses that cause sporadic cases of severe disease in humans after 1997 typically have an avian-virus-like preference for Neu5Ac(α2-3)Gal-terminated receptors (Matrosovich et al., 1999; Gambaryan et al., 2006). This observation did not support the concept that such receptors are absent in human airways. Indeed, experiments in differentiated cultures of human airway epithelium demonstrated that ciliated cells express sufficient amounts of 3-linked sialic acids to allow infection of these cells with avian viruses (Matrosovich et al., 2004; Thompson et al., 2006). However, despite their ability to infect human airway epithelium, avian viruses appeared to replicate less efficiently than human viruses (Matrosovich et al., 2004; Shinya et al., 2006; Thompson et al., 2006). To elucidate the role of the receptor specificity in the restricted replication of avian viruses in humans, Matrosovich et al. (Matrosovich et al., 2006) employed recombinant viruses that differed solely by two amino acid substitutions in the HA receptor-binding site. They found that the virus with the avian-virus-like HA grew to 2-10-fold lower titers than its human counterpart. These results for the first time estimated the level of receptor-dependent restriction of avian influenza viruses in human airway epithelium. They support the concept that alteration of the receptor specificity of an avian virus is essential for its human-to-human transmission and pandemic spread.

DISTINCTIVE RECEPTOR SPECIFICITY OF VIRUSES FROM DIFFERENT AVIAN SPECIES

Direct transmissions of avian H5N1, H9N2, and H7N7 viruses to humans attracted attention to avian influenza viruses in different bird species and prompted studies on their receptor specificity. Matrosovich et al. (1999) compared H5N1 viruses isolated from poultry and wild aquatic birds. They found that despite the fact that all these viruses bound to 3-linked receptors and did not bind to 6-linked receptors, the fine receptor specificity of poultry viruses and enzymatic activity of their neuraminidases differed from those of closely related viruses isolated from wild aquatic birds. These authors also compared HA and NA amino acid sequences of H5 and H7 viruses from distinct avian species and revealed that poultry viruses differ from duck viruses by a presence of additional N-linked glycans at the top of their HA and by large deletions in the stalk of the NA. These characteristic changes in HA and/or NA were found to occur in many independent lineages of poultry viruses, (Matrosovich et al., 1999; Banks et al., 2001; Campitelli et al., 2004) suggesting their functional role in the adaptation of viruses in wild aquatic birds to terrestrial domestic birds. These initial findings led to the hypothesis that sialic acid-containing receptors in different birds are not identical and that distinctions in the receptors are reflected in the receptor specificity and NA activity of influenza viruses circulating in these birds. Studies performed in the following years confirmed this hypothesis.

In 1999, H9N2 poultry viruses caused several cases of human infection in the Guangdong province of China and in Hong Kong, SAR (Peiris et al., 1999). Both the human and closely related poultry viruses harboured mutations in the conserved positions of the receptor binding site, among them, in positions 190 and 226 (Lin et al., 2000). Furthermore, H9N2 poultry viruses displayed preferential binding to 6-linked sialic acids and had mutations in the hemadsorbing site of their neuraminidases, features that are typical for swine and human influenza viruses (Matrosovich et al., 2001). These findings provided the first example of a stable lineage of avian influenza viruses with human virus-like receptor specificity and an ability to infect different species of poultry (quail, chicken, pheasant, pigeon, etc.). They showed that a strict receptor specificity, which is typical of influenza viruses maintained in wild aquatic birds, may not be required for influenza virus replication in other avian species.

Gambaryan et al. (2002) compared the binding of duck, chicken and human influenza viruses to cell membranes and gangliosides from epithelial tissues of duck, chicken and African green monkey (as a model of humans). Human viruses bound to cell membranes of monkey and chicken but not to those of duck, suggesting that chicken cells, similar to monkey cells and in contrast to duck cells, express Sia(α2-6)Gal-containing receptors. Duck virus bound to gangliosides with short sugar chains that were abundant in duck intestine. Human and chicken viruses did not bind to these gangliosides and bound more strongly than duck virus to gangliosides with long sugar chains that were found in chicken intestinal and monkey lung tissues. These data indicated that sialylglycoconjugates of chickens are more similar to those in the respiratory epithelium of monkeys than to those in the epithelial tissues of duck. This report provided a possible explanation for the existence of avian H9N2 viruses with human virus-like receptor specificity (Matrosovich et al., 2001) and emphasized the role of the chicken as a potential intermediate host for the transmission of viruses from aquatic birds to humans.

Wan and Perez (2005) performed a similar study on distribution of 3-linked and 6-linked sialic acids in quail. Using lectins, *Sambucus nigra* agglutinin (SNA) and *Maackia amurensis* agglutinin (MAA), which bind, respectively, to 3-linked and 6-linked Sias, these authors found both types of sialic acid epitopes in quail trachea and intestine. Furthermore, both avian and human influenza viruses bound to plasma membrane preparations from quail tracheal and intestinal epithelial cells. These data suggested that quails, similar to chickens and pigs, could provide the environment for the replication and gene mixing between avian and human influenza viruses.

Using SNA and MAA lectins together with a panel of human and avian influenza viruses with well-characterized receptor specificity, Gambaryan et al. (2006b) compared expression of influenza virus receptors in the intestinal and respiratory tissues of about 20 different species from the 6 orders of birds, including *Anseriformes*, *Charadriiformes*, and *Galliformes*. All avian viruses required sialic acid for binding to cellular receptors, as treatment of cell membranes with bacterial neuraminidase abolished binding. Ducks (*Anatidae*) were found to differ from other birds by a reduced expression of receptors recognized by both SNA and MAA in duck intestine. However, duck viruses efficiently bound to the duck intestinal cells. The cells from intestinal epithelium of gulls efficiently bound MAA and avian influenza viruses, but did not bind SNA and human viruses. These results suggested that although both ducks and gulls predominantly express 3-linked receptors and lack 6-linked receptors, the fine structure of the 3-linked sialyloligosaccharides differs between ducks and gulls. In accord with the previous reports on chicken and quail (Gambaryan et al., 2002; Wan and Perez, 2006), the intestinal cells of several *Galliformes* were found to bind SNA, MAA and both avian and human influenza viruses. A few members of *Cicorniiformes, Podicipediformes,* and *Gruiformes* tested bound MAA but showed only weak binding of avian influenza viruses. This study revealed that sialic acid receptors can vary substantially among different avian species, and that receptor specificity of influenza viruses circulating in these species appears to reflect this variation. In particular, viruses of ducks, gulls and chickens displayed distinct binding patterns.

A significant recent advance in the knowledge on the receptor specificity of the avian influenza viruses was associated with the use of synthetic sialylglycopolymers (SGPs) as mono-specific structurally-defined molecular probes which comprise multiple copies of sialyloligosaccharide moieties attached to soluble hydrophilic polymeric carrier (Bovin et al., 1993; Bovin, 1998). By comparing the viral binding to a panel of SGPs that harboured the same Neu5Ac(α2-3)Gal moiety in a context of different penultimate oligosaccharide sequences, Gambaryan, Bovin and colleagues were able to evaluate the role of the more distant parts of sialyloligosaccharides in the receptor recognition by avian influenza viruses (Gambaryan et al., 2003; Yamnikova et al., 2003; Gambaryan et al., 2004; Gambaryan et al., 2005a; Gambaryan et al., 2006). These studies revealed notable distinctions between duck, gull and chicken viruses.

Thus, duck viruses irrespective of their HA subtype typically bound to Neu5Ac(α2-3)Gal(α1-3)GalNAc- and Neu5Ac(α2-3)Gal(α1-3) GlcNAc-containing polymers with significantly higher affinity than to Neu5Ac(α2-3)Gal(α1-4)GlcNAc-containing ones. Fucose and sulfo substituents at the third saccharide residue had either low or negative effect on duck virus binding. These findings indicated, that duck viruses prefer the beta-1–3 bond between the terminal Neu5Ac(α2-3) Gal moiety and penultimate saccharide residue, and that the receptor epitopes in ducks are likely neither fucosylated nor sulphated. This receptor specificity of the duck viruses agreed with their strong binding to ganglioside GD1a and to gangliosides from duck intestine raising a hypothesis that binding to gangliosides can be essential for infection of duck intestinal cells *in vivo* (Matrosovich et al., 1997; Gambaryan et al., 2003).

Unlike duck viruses, highly pathogenic H5N1 human and chicken viruses isolated in Hong Kong in 1997 preferred SGPs with a beta-1-4 bond between Neu5Ac(α2-3)Gal and the next sugar residue (for example, Neu5Ac(α2-3)Gal(α1-4)GlcNAc). In addition, these H5N1 viruses showed particularly high affinity for a sulphated analogue Neu5Ac(α2-3)Gal(α1-4) (6-O-Su)GlcNAc (Su-3'SLN) (Gambaryan et al., 2004). Subsequent detailed analysis of H5 subtype viruses from different species (Gambaryan et al., 2006) confirmed these observations and revealed evolutionary changes in the receptor specificity of H5 viruses that accompanied their circulation in poultry. In particular, Asian 2003–04 H5 isolates from chickens and humans, similar to HK/97 viruses, displayed very strong binding to Su-3'SLN and additionally acquired high affinity for the fucosylated Su-3'SLN receptor. Importantly, many H5 virus isolates from North American poultry tested in that study also showed increased affinity for Su-3'SLN, suggesting that acquisition of this property occurred independently in two lineages of H5 poultry viruses. As judged by their staining with MAA, human airway epithelial cells harbour 3'SLN-moieties (Matrosovich et al., 2004), and some of them can be sulphated (Lamblin et al., 2001). Thus, high-affinity receptor determinants recognized by H5N1 chicken influenza viruses should be present in the human airway epithelium. It remains to be determined whether this feature could facilitate occasional transmissions of highly pathogenic poultry viruses to humans.

Only some gull viruses can infect ducks (Kawaoka et al., 1988), suggesting a degree of host-range restriction between these species. In particular, viruses with H13 HA are widely distributed among gulls and shorebirds but are very rarely isolated from ducks. The H13 viruses as well as recently discovered gull viruses of the H16 subtype (Fouchier et al., 2005) harbour distinctive substitutions in the receptor-binding site, which separate them from other avian viruses. Accordingly, the H13 viruses were shown to differ from duck viruses by their relatively weak interactions with the 3-linked Gal moiety (Matrosovich et al., 1997). A typical feature of the viruses isolated from gulls was a high-affinity binding of these viruses to SGPs bearing fucosylated sialyloligosaccharide moieties Neu5Ac(α2-3)Gal(α1-4)[Fuc(α1-3)]GlcNAc (SLex), whereas most duck viruses preferentially bound to non-fucosylated analogues (Yamnikova et al., 2003; Gambaryan et al., 2005a). Thus, adaptation for the recognition of fucosylated sialyloligosaccharide receptors may be required for the efficient replication of duck viruses in gulls. It is interesting in this respect that Makarova et al. (1999) described transmission of viruses with H2 HA from Eurasian ducks to North American shorebirds and gulls, and that virus replication in these species was accompanied by fast accumulation of mutations in the HA receptor-binding site.

THE VIRAL RNA POLYMERASE: A DETERMINANT OF HOST RANGE AND PATHOGENICITY

THE POLYMERASE MEDIATES TRANSCRIPTION AND REPLICATION

Early studies revealed that influenza virus infected cells contain an RNA dependent RNA polymerase activity (Page et al., 1968; Bishop et al., 1971b; Bishop et al., 1971a; Skehel, 1971; Bishop et al., 1972; Compans and Caliguiri, 1973), that its products are complementary to genomic RNA (Scholtissek, 1969), that the polymerase is present within the virions (Chow and Simpson, 1971; Penhoet et al., 1971), and that this enzyme is encoded by three large polypeptides (Inglis et al., 1976; Inglis et al., 1977; Palese et al., 1977). Later it was shown that polymerase is a heterotrimeric complex consisting of three subunits, the PB2, PB1, and PA proteins (Detjen et al., 1987; Digard et al., 1989; Area et al., 2004). The polymerase is active in the nucleus and nuclear localization sequences have been identified on each subunit: PB1 (Nath and Nayak, 1990), PB2 (Mukaigawa and Nayak, 1991), and PA (Nieto et al., 1992). Recent results suggest that cytoplasmically expressed PB1 and PA may be imported as a sub-complex, which then assembles with separately imported PB2 (Fodor and Smith, 2004). Several nuclear and cytoplasmic host proteins have been shown to be cofactors required by the polymerase (Engelhardt et al., 2005; Deng et al., 2006a; Ye et al., 2006).

The viral polymerase catalyzes both the transcription of genomic RNA, i.e. viral RNA (vRNA), to mRNA and the replication of vRNA to complementary RNA (cRNA), the intermediate template which is necessary for copying new vRNA molecules (Hay et al., 1977).

Transcription is initiated by binding m^7GpppGm cap structures at the 5' ends of cellular mRNAs by the PB2 and PB1 subunits (Ulmanen et al., 1981; Blaas et al., 1982b; Blaas et al., 1982a; Li et al., 1998; Fechter et al., 2003). Capped 5' mRNA ends are then cut off by an endonuclease activity of the PB1 and PA subunits (Shi et al., 1995; Li et al., 2001; Hara et al., 2006) and used as primers for further elongation which is eventually terminated due to stuttering (Poon et al., 1999; Zheng et al., 1999) at the uridine stretch adjacent to the 5' end of the vRNA (Luo et al., 1991; Li and Palese, 1994) and due to the bound polymerase complex at the 5' end (Poon et al., 1998) leading to polyadenylation. In contrast, the replication of vRNA is primer-independent and involves the de novo initiation from nascent RNA chains requiring first the synthesis of cRNA, which then serves as template for new vRNA (Hay et al., 1977; Hay et al., 1982). Two different initiation mechanisms have been proposed for replication of vRNA and cRNA. For the vRNA promoter, the primer-independent initiation occurs at the 3' terminus of the cRNA followed by elongation (Deng et al., 2006b) whereas for the cRNA promoter an internal initiation (nucleotides 4 and 5) is followed by realignment at the 3' terminal nucleotides 1 and 2 and then subsequent elongation. The promoters are composed of both the 5' and 3' ends of vRNA (genomic promoter) or cRNA (antigenomic promoter) which are highly conserved in all segments of all influenza A strains (Skehel and Hay, 1978; Desselberger et al., 1980). Three promoter structures have been proposed: the panhandle model (Hsu et al., 1987), the fork model (Fodor et al., 1993; Kim et al., 1997) and the corkscrew model (Flick and Hobom, 1999). According to the panhandle model, the entire promoter region is partially double-stranded whereas the fork model suggests the double-stranded configuration only at the distal promoter element. The corkscrew model predicts base-pairing at each 5' or 3' end rather than of the 5' with the 3' end and is considered the current model of the secondary structures of both the vRNA and cRNA promoters (Flick et al., 1996; Azzeh et al., 2001; Crow et al., 2004).

It has been proposed that the polymerase is first engaged in transcription, and that there is a switch to replication when enough nucleoprotein is made (Hay et al., 1977; Taylor et al., 1977; Barrett et al., 1979; Hatada et al., 1989). However, the molecular mechanism of such a switch has remained uncertain. Remarkably, cRNA synthesis early in infection could be demonstrated (Vreede et al., 2004). Therefore, a stabilization model was proposed which does not assume different synthesis modes of the viral polymerase. The vRNPs of the infecting virions were demonstrated to synthesize both mRNA and cRNA. However, the mRNA is protected from degradation due to its 5' cap and 3' poly A tail whereas the naked cRNA is presumably subjected to degradation by cellular nucleases. Later, it becomes protected by the newly synthesized polymerase proteins and is then encapsidated by the nucleoprotein. Thus, although the newly synthesized polymerase proteins and the NP are necessary for replication, they do not induce a switch to another synthesis mode but facilitate the replication indirectly by stabilizing replicative intermediates (Vreede et al., 2004; Vreede and Brownlee, 2006).

THE POLYMERASE OPERATES WITH HIGH MUTATION RATES

As an RNA dependent RNA polymerase, the influenza virus polymerase lacks any proof-reading function. Thus, point mutations are introduced at a high rate, a prerequisite for high genomic plasticity upon selective pressure. This mutation rate (defined as point mutations per site per replication) of the viral quasispecies is limited by an error threshold; the viral polymerase operates close to that limit allowing high variability but avoiding extinction (Eigen, 1993; Eigen, 2002; Biebricher and Eigen, 2005). The few available experimental studies on mutation rate of influenza A viruses are consistent with the existence of such an upper limit (Parvin et al., 1986; Suarez-Lopez and Ortin, 1994; Stech et al., 1999; Nobusawa and Sato, 2006). Beside point mutations, major genetic changes such as deletions or insertions were reported with circulating viruses in the field. Deletions within the stalk region of the neuraminidase (Steuler et al., 1984; Deshpande et al., 1985; Els et al., 1985; Castrucci and Kawaoka, 1993; Luo et al., 1993) are associated with altered virulence. The insertion of an 28S ribosomal RNA sequence into the haemagglutinin yielded a polybasic cleavage site (Khatchikian et al., 1989; Orlich et al., 1990) which is well-known to increase the virulence of avian influenza A virus strains considerably (Garten and Klenk, 1999). Moreover, there are also some examples of intragenic recombination with other influenza gene segments (Bergmann et al., 1992; Orlich et al., 1994; Mitnaul et al., 1996; Suarez et al., 2004; Pasick et al., 2005). However, insertion, deletion, and recombination in influenza A viruses are rare as compared to genetic changes due to point mutations and reassortment.

REASSORTMENT OF THE POLYMERASE COMPLEXES MAY ALTER VIRULENCE

With the knowledge of a segmented genome (Scholtissek and Rott, 1964; McGeoch et al., 1976; Palese and Schulman, 1976; Scholtissek et al., 1976), animal studies based on comparison of reassortants could be undertaken to elucidate the functions of viral proteins. Reassortant studies demonstrated that the virulence is determined by the constellation of polymerase and nucleoprotein genes. Two HPAI strains reassorted to a virus low pathogenic in chickens (Rott et al., 1979). Remarkably, double infection with two LPAI strains may yield a virus highly neuro- and pneumovirulent in mice (Scholtissek et al., 1979; Vallbracht et al., 1980). Moreover, studies in squirrel monkeys underlined that reassortant viruses with a mixed constellation of polymerase genes from an avian and a human strain (or carrying the nucleoprotein from the avian virus) were attenuated in contrast to their parental viruses. Such heterologous polymerase proteins can interact with each other and with cellular factors efficiently in avian tissue, however, less efficiently in mammalian cells. These observations indicate that an optimized interaction of the polymerase complex with host factors is involved in efficient replication *in vivo* (Snyder et al., 1987). Taken together, these findings demonstrate that circulating strains with a low pathogenic phenotype may carry viral genes with highly pathogenic potential and, thus, are able to give rise to highly virulent strains upon reassortment.

SINGLE POLYMERASE MUTATIONS INCREASE BOTH VIRULENCE AND TRANSCRIPTION/REPLICATION ACTIVITY

Pathogenicity studies in mice and in ferrets found several polymerase mutations responsible for increased virulence and enhanced replication. A single amino acid substitution in the PB2 protein, Glu-Lys at position 627, enables a single gene reassortant with a PB2 gene from an avian virus to replicate in mammalian cells. Strikingly, whereas in every avian influenza A virus PB2 protein, the amino acid at position 627 is glutamic acid, in every human influenza A strain, lysine is prevalent (Subbarao et al., 1993). This mutation increases the polymerase activity in mammalian cells. Moreover, polymerase complexes derived from avian viruses exhibited cold sensitivity in mammalian cells mostly due to the residue 627 in PB2 (Massin et al., 2001). The amino acid substitution PB2 627Lys was found to contribute to enhanced virulence of several highly pathogenic avian influenza viruses of the

H5N1 subtype from Southeast Asia both in mice (Gao et al., 1999; Hatta et al., 2001; Shinya et al., 2004; Salomon et al., 2006) and in ferrets (Govorkova et al., 2005). However, several but not all H5N1 viruses isolated from humans carry this substitution (Puthavathana et al., 2005; de Jong et al., 2006). The relevance of this mutation in the field was underlined by its occurrence in another unrelated group of HPAIV of the H7N7 subtype. This mutation was found in an isolate from a deceased veterinarian during the fowl plague outbreak in the Netherlands in 2003 (Fouchier et al., 2004). The characterization of the reconstructed PB2 sequence of the 1918 virus revealed the presence of 627Lys as well (Taubenberger et al., 2005). These findings indicate convergent evolution of different influenza A viruses in mammals. Moreover, several studies demonstrated the substitution PB2 701 Glu-Lys to increase the virulence of H7N7 and H5N1 HPAIV in mice (Hiromoto et al., 2000; Brown et al., 2001; Gabriel et al., 2005; Li et al., 2005). Other residues correlating with high pathogenicity are PB2 714Arg, PA 615Arg, and NP 319Lys (Gabriel et al., 2005). All these virulence enhancing mutations found in H7N7 and H5N1 viruses were then shown by minireplicon assays to increase polymerase activity in mammalian cells (Naffakh et al., 2000; Gabriel et al., 2005; Salomon et al., 2006). Moreover, this enhanced activity correlated with increased virulence in mice and ferrets (Gabriel et al., 2005; Salomon et al., 2006). Remarkably, it appears that an optimum of polymerase activity exists because virus variants with extreme polymerase activity were not of highest virulence; highest levels of replication/transcription might be detrimental in the field (Gabriel et al., 2005; Salomon et al., 2006).

POLYMERASE MUTATIONS MEDIATE ADAPTATION OF AVIAN INFLUENZA VIRUSES TO MAMMALIAN HOSTS

The mutations PB2 627Lys, PB2 701Asn, PB2 714Arg, PA 615Asn, and NP 319Lys shown to enhance the polymerase activity and to increase virulence in mice occurred independently in unrelated groups of mammalian isolates (Naffakh et al., 2000; Gabriel et al., 2005; Li et al., 2005; de Jong et al., 2006; Salomon et al., 2006). All other viruses including avian strains carry PB2 627Glu, PB2 701Asp, PB2 714Ser, PA 615Lys, and NP 319 Asn (Subbarao et al., 1993; Gabriel et al., 2005). The substitution PB2 627Lys is well-known in human strains (Subbarao et al., 1993) including the 1918 virus (Taubenberger et al., 2005) and in H5N1 as well as in an H7N7 HPAIV isolated from humans (Hatta et al., 2000; Fouchier et al., 2004; Shinya et al., 2004). Similarly, PB2

701Asn was found in several phylogenetic clades of H1N1, H1N2, and H3N2 swine isolates, in H3N8 and H7N7 equine isolates, and in several H5N1 HPAIV from Southeast Asia including human isolates such as A/Hk/488/97, A/HK/97/98, and A/Vietnam/3046/2004 (Gabriel et al., 2005). Remarkably, among these strains is the avian-like swine H1N1 virus A/Swine/Germany/2/81 (Ludwig et al., 1995) which belongs to a lineage originating from an avian precursor that was transmitted as a whole to swine (Schultz et al., 1991).

PB2 714Arg was found in four mammalian strains; whereas 16 avian H5N1 isolates (2001) highly-pathogenic for chicken and quails carry isoleucine at this position. These strains replicate well in mouse lung in contrast to their precursors, the Goose/Guangdong-like strains (Guan et al., 2002). One of these strains, A/Goose/Guangdong/1/96 (Xu et al., 1999), carries Ser instead of Ile at position 714 in PB2. In contrast to the Goose/Guangdong-like viruses, the H5N1 isolates appear to have an increased pathogenicity for mammals (Guan et al., 2002) suggesting that PB2 714Ile is a host range marker (Gabriel et al., 2005).

Instead of Lys at position 615 within the PA protein, Arg was found in several phylogenetic clades of human H1N1, H5N1, and H9N2 isolates including H5N1 HPAIV isolates of known pathogenicity like A/HK/483/97, A/HK/485/97, and A/HK/491/97 (Gao et al., 1999). The avian strains A/Teal/HK/W312/97 (H6N1) and A/Quail/HK/G1/97 (H9N2), proposed donors of internal genes of H5N1 viruses (Guan et al., 1999; Hoffmann et al., 2000; Choi et al., 2004), were found as well, emphasizing the relevance of PA 615Arg for host change (Gabriel et al., 2005).

NP 319Lys was found in human and equine strains as well as in several H5N1 HPAIV. Remarkably, some human H5N1 HPAIV (A/HK/156/97, A/HK/482/97, A/HK/486/97, and A/HK/538/97) isolates harbour both NP 319Lys and PA 615Arg (Gabriel et al., 2005).

A recent study in 18 individuals infected with H5N1 HPAIV indicated that high viral load and the resulting intense inflammatory response are central to pathogenesis. In five of eight isolates from fatal cases and in three of four isolates from surviving patients, the PB2 627Lys substitution was found. There was no association between the presence of Lys627 and lethal outcome. Notably, three of four viruses containing PB2 627Glu common to avian strains, but none carrying PB2 627Lys contained PB2 701Asn. This finding suggests that this substitution may compensate for the absence of PB2 627Lys in conferring enhanced polymerase activity. In addition, all viruses contained PA 615Arg (de Jong et al., 2006).

Several polymerase mutations were identified and shown to increase the activity and virulence in mice. These substitutions occurred in several independent phylogenetic clades of influenza virus strains transmitted recently from birds and still undergoing adaptation. These findings indicate convergent evolution of the polymerase complex in nature (Gabriel et al., 2005).

CONCLUSIONS

Pathogenicity and interspecies transmission of avian influenza viruses are controlled to a large extent by the hemagglutinin and the polymerase. Important molecular mechanisms governing the interplay between these viral compounds and host factors have been elucidated, but there are still many open questions:

The pivotal role of cleavage activation of the hemagglutinin for the pathogenicity of avian influenza viruses is well established. However, our knowledge on the proteases involved, particularly on those activating LPAI viruses, is still incomplete. Further information on these enzymes would be helpful for the design of protease inhibitors to be used as antiviral agents and for the development of protease activation mutants that already proved to be promising vaccine candidates. It also remains to be elucidated why the insertion of a furin cleavage site and, thus, the acquisition of high pathogenicity is restricted to subtypes H5 and H7.

Increasing evidence suggests that the receptor specificity varies among avian influenza viruses depending on the sialic acid patterns of their hosts. Further studies are warranted to determine whether these distinctions restrict virus transmissions among different birds and whether viruses of some avian species, such as chickens or quails, may have a higher affinity for receptors in epithelial respiratory tissues of mammals including humans and thus, have an increased pandemic potential.

Several studies demonstrated that mutations in the polymerase proteins increase virulence and mediate adaptation to a new host. These substitutions were shown to enhance the polymerase activity in mammalian cells and to occur in unrelated virus strains recently transmitted from birds to mammalian hosts. Thus, it appears that, in addition to an altered receptor specificity, adaptation of the polymerase complex to host factors is also necessary for effective interspecies transmission.

ACKNOWLEDGEMENTS

We acknowledge support of our own work from the Deutsche Forschungsgemeinschaft (SFB 286 and SFB 593); Fonds der Chemischen Industrie; VIRGIL European Network of Excellence on Antiviral Drug Resistance, EU, and European Surveillance Network for Influenza in Pigs 2, EU.

REFERENCES

Alexander, D.J. (1986). Avian Influenza - Historical Aspects. Paper presented at: Second International Symposium on Avian Influenza (Georgia Center for Continuing Education, The University of Georgia, Athens, Georgia, USA).

Alexander, D.J. (1998). Control strategies of the International Office of Epizooties, the European Union and the harmonization of international standards for the diagnosis of avian influenza. Paper presented at: Fourth International Symposium on Avian Influenza. (Richmond, Virginia, United States Animal Health Association, D.E. Swayne, R.D. Slemons).

Area, E., Martin-Benito, J., Gastaminza, P., Torreira, E., Valpuesta, J.M., Carrascosa, J.L., and Ortin, J. (2004). 3D structure of the influenza virus polymerase complex: localization of subunit domains. Proc. Natl. Acad. Sci. USA *101*, 308-313.

Azzeh, M., Flick, R., and Hobom, G. (2001). Functional analysis of the influenza A virus cRNA promoter and construction of an ambisense transcription system. Virology *289*, 400-410.

Baigent, S.J., and McCauley, J.W. (2003). Influenza type A in humans, mammals and birds: determinants of virus virulence, host-range and interspecies transmission. Bioessays *25*, 657-671.

Banks, J., Speidel, E., and Alexander, D.J. (1998). Characterisation of an avian influenza A virus isolated from a human - is an intermediate host necessary for the emergence of pandemic influenza viruses? Arch. Virol. *143*, 781-787.

Banks, J., Speidel, E.S., Moore, E., Plowright, L., Piccirillo, A., Capua, I., Cordioli, P., Fioretti, A., and Alexander, D.J. (2001). Changes in the haemagglutinin and the neuraminidase genes prior to the emergence of highly pathogenic H7N1 avian influenza viruses in Italy. Arch. Virol. *146*, 963-973.

Barrett, T., Wolstenholme, A.J., and Mahy, B.W. (1979). Transcription and replication of influenza virus RNA. Virology *98*, 211-225.

Baum, L.G., and Paulson, J.C. (1990). Sialyloligosaccharides of the respiratory epithelium in the selection of human influenza virus receptor specificity. Acta Histochem. Suppl. *40*, 35-38.

Becker, W.B. (1966). The isolation and classification of Tern virus: influenza A-Tern South Africa--1961. J. Hyg. (Lond). *64*, 309-20.

Bergelson, L.D., Bukrinskaya, A.G., Prokazova, N.V., Shaposhnikova, G.I., Kocharov, S.L., Shevchenko, V.P., Kornilaeva, G.V., and Fomina-Ageeva, E.V. (1982). Role of gangliosides in reception of influenza virus. Eur. J. Biochem. *128*, 467-474.

Bergmann, M., Garcia-Sastre, A., and Palese, P. (1992). Transfection-mediated recombination of influenza A virus. J. Virol. *66*, 7576-7580.

Biebricher, C.K., and Eigen, M. (2005). The error threshold. Virus Res. *107*, 117-127.

Bishop, D.H., Obijeski, J.F., and Simpson, R.W. (1971a). Transcription of the influenza ribonucleic acid genome by a virion polymerase. I. Optimal conditions for in vitro activity of the ribonucleic acid-dependent ribonucleic acid polymerase. J. Virol. *8*, 66-73.

Bishop, D.H., Obijeski, J.F., and Simpson, R.W. (1971b). Transcription of the influenza ribonucleic acid genome by a virion polymerase. II. Nature of the in vitro polymerase product. J. Virol. *8*, 74-80.

Bishop, D.H., Roy, P., Bean, W.J., Jr., and Simpson, R.W. (1972). Transcription of the influenza ribonucleic acid genome by a virion polymerase. 3. Completeness of the transcription process. J. Virol. *10*, 689-697.

Blaas, D., Patzelt, E., and Kuechler, E. (1982a). Cap-recognizing protein of influenza virus. Virology *116*, 339-348.

Blaas, D., Patzelt, E., and Kuechler, E. (1982b). Identification of the cap binding protein of influenza virus. Nucleic Acids Res. *10*, 4803-4812.

Boettcher, E., Matrosovich, T., Beyerle, M., Klenk, H.D., Garten, W., and Matrosovich, M. (2006). Proteolytic activation of influenza

viruses by serine proteases TMPRSS2 and HAT from human airway epithelium. J. Virol. *80*, 9896-9898.
Bosman, A. (2004). Avian flu epidemic 2003. RIVM Report.
Bovin, N.V. (1998). Polyacrylamide-based glycoconjugates as tools in glycobiology. Glycoconj. J. *15*, 431-446.
Bovin, N.V., Korchagina, E.Y., Zemlyanukhina, T.V., Byramova, N.E., Galanina, O.E., Zemlyakov, A.E., Ivanov, A.E., Zubov, V.P., and Mochalova, L.V. (1993). Synthesis of polymeric neoglycoconjugates based on N-substituted polyacrylamides. Glycoconj. J. *10*, 142-151.
Brown, E.G., Liu, H., Kit, L.C., Baird, S., and Nesrallah, M. (2001). Pattern of mutation in the genome of influenza A virus on adaptation to increased virulence in the mouse lung: identification of functional themes. Proc. Natl. Acad. Sci. USA *98*, 6883-6888.
Campitelli, L., Mogavero, E., De Marco, M.A., Delogu, M., Puzelli, S., Frezza, F., Facchini, M., Chiapponi, C., Foni, E., Cordioli, P., et al. (2004). Interspecies transmission of an H7N3 influenza virus from wild birds to intensively reared domestic poultry in Italy. Virology *323*, 24-36.
Castrucci, M.R., and Kawaoka, Y. (1993). Biologic importance of neuraminidase stalk length in influenza A virus. J. Virol. *67*, 759-764.
Cauthen, A.N., Swayne, D.E., Schultz-Cherry, S., Perdue, M.L., and Suarez, D.L. (2000). Continued circulation in China of highly pathogenic avian influenza viruses encoding the hemagglutinin gene associated with the 1997 H5N1 outbreak in poultry and humans. J. Virol. *74*, 6592-6599.
Centanni, E. (1902). Zentbl. Bakt. Parasitkde. Abt. I Orig. *31*, 182-201.
Chen, H., Smith, G.J., Li, K.S., Wang, J., Fan, X.H., Rayner, J.M., Vijaykrishna, D., Zhang, J.X., Zhang, L.J., Guo, C.T., et al. (2006). Establishment of multiple sublineages of H5N1 influenza virus in Asia: implications for pandemic control. Proc. Natl. Acad. Sci. USA *103*, 2845-2850.
Chen, J., Lee, K.H., Steinhauer, D.A., Stevens, D.J., Skehel, J.J., and Wiley, D.C. (1998). Structure of the hemagglutinin precursor cleavage site, a determinant of influenza pathogenicity and the origin of the labile conformation. Cell *95*, 409-417.
Choi, Y.K., Ozaki, H., Webby, R.J., Webster, R.G., Peiris, J.S., Poon, L., Butt, C., Leung, Y.H., and Guan, Y. (2004). Continuing evolution of H9N2 influenza viruses in Southeastern China. J. Virol. *78*, 8609-8614.

Chow, N.L., and Simpson, R.W. (1971). RNA-dependent RNA polymerase activity associated with virions and subviral particles of myxoviruses. Proc. Natl. Acad. Sci. USA *68*, 752-756.
Claas, E.C., Osterhaus, A.D., van Beek, R., De Jong, J.C., Rimmelzwaan, G.F., Senne, D.A., Krauss, S., Shortridge, K.F., and Webster, R.G. (1998). Human influenza A H5N1 virus related to a highly pathogenic avian influenza virus. Lancet *351*, 472-477.
Compans, R.W., and Caliguiri, L.A. (1973). Isolation and properties of an RNA polymerase from influenza virus-infected cells. J. Virol. *11*, 441-448.
Connor, R.J., Kawaoka, Y., Webster, R.G., and Paulson, J.C. (1994). Receptor specificity in human, avian, and equine H2 and H3 influenza virus isolates. Virology *205*, 17-23.
Couceiro, J.N., Paulson, J.C., and Baum, L.G. (1993). Influenza virus strains selectively recognize sialyloligosaccharides on human respiratory epithelium; the role of the host cell in selection of hemagglutinin receptor specificity. Virus Res. *29*, 155-165.
Cox, N.J., and Subbarao, K. (2000). Global epidemiology of influenza: past and present. Annu. Rev. Med. *51*, 407-421.
Crow, M., Deng, T., Addley, M., and Brownlee, G.G. (2004). Mutational analysis of the influenza virus cRNA promoter and identification of nucleotides critical for replication. J. Virol. *78*, 6263-6270.
de Jong, M.D., Simmons, C.P., Thanh, T.T., Hien, V.M., Smith, G.J., Chau, T.N., Hoang, D.M., Chau, N.V., Khanh, T.H., Dong, V.C., et al. (2006). Fatal outcome of human influenza A (H5N1) is associated with high viral load and hypercytokinemia. Nat. Med. *12*, 1203-1207.
Deng, T., Engelhardt, O.G., Thomas, B., Akoulitchev, A.V., Brownlee, G.G., and Fodor, E. (2006a). Role of ran binding protein 5 in nuclear import and assembly of the influenza virus RNA polymerase complex. J. Virol. *80*, 11911-11919.
Deng, T., Vreede, F.T., and Brownlee, G.G. (2006b). Different de novo initiation strategies are used by influenza virus RNA polymerase on its cRNA and viral RNA promoters during viral RNA replication. J. Virol. *80*, 2337-2348.
Deshpande, K.L., Naeve, C.W., and Webster, R.G. (1985). The neuraminidases of the virulent and avirulent A/Chicken/Pennsylvania/83 (H5N2) influenza A viruses: sequence and antigenic analyses. Virology *147*, 49-60.
Desselberger, U., Racaniello, V.R., Zazra, J.J., and Palese, P. (1980). The 3' and 5'-terminal sequences of influenza A, B and C

virus RNA segments are highly conserved and show partial inverted complementarity. Gene *8*, 315-328.

Detjen, B.M., St-Angelo, C., Katze, M.G., and Krug, R.M. (1987). The three influenza virus polymerase (P) proteins not associated with viral nucleocapsids in the infected cell are in the form of a complex. J. Virol. *61*, 16-22.

Digard, P., Blok, V.C., and Inglis, S.C. (1989). Complex formation between influenza virus polymerase proteins expressed in Xenopus oocytes. Virology *171*, 162-169.

Dinter, Z. (1949). Eine Variante des Virus der Geflügelpest in Bayern. Tieraerztl. Umschau *4*, 185-186.

Dinter, Z., and Bakos, K. (1950). Über die Beziehungen des Virus N zu dem Virus der klassischen Geflügelpest. Berl. Muench. Tieraerztl. Wochenschr. *63*, 101-105.

Eigen, M. (1993). Viral quasispecies. Sci. Am. *269*, 42-49.

Eigen, M. (2002). Error catastrophe and antiviral strategy. Proc. Natl. Acad. Sci. USA *99*, 13374-13376.

Els, M.C., Air, G.M., Murti, K.G., Webster, R.G., and Laver, W.G. (1985). An 18-amino acid deletion in an influenza neuraminidase. Virology *142*, 241-247.

Engelhardt, O.G., Smith, M., and Fodor, E. (2005). Association of the influenza A virus RNA-dependent RNA polymerase with cellular RNA polymerase II. J. Virol. *79*, 5812-5818.

Enserink, M., and Kaiser, J. (2004). Virology. Avian flu finds new mammal hosts. Science *305*, 1385.

Fechter, P., Mingay, L., Sharps, J., Chambers, A., Fodor, E., and Brownlee, G.G. (2003). Two aromatic residues in the PB2 subunit of influenza A RNA polymerase are crucial for cap binding. J. Biol. Chem. *278*, 20381-20388.

Feldmann, A. (1998). Gezielte Infektion von Endothelzellen durch das Virus der klassischen Geflügelpest A/FPV/Rostock/34 (H7N1) im Hühnerembryo. PhD thesis. Philipps-Universität Marburg.

Flick, R., and Hobom, G. (1999). Interaction of influenza virus polymerase with viral RNA in the 'corkscrew' conformation. J. Gen. Virol. *80*, 2565-2572.

Flick, R., Neumann, G., Hoffmann, E., Neumeier, E., and Hobom, G. (1996). Promoter elements in the influenza vRNA terminal structure. RNA *2*, 1046-1057.

Fodor, E., Seong, B.L., and Brownlee, G.G. (1993). Photochemical cross-linking of influenza A polymerase to its virion RNA promoter defines a polymerase binding site at residues 9 to 12 of the promoter. J. Gen. Virol. *74*, 1327-1333.

Fodor, E., and Smith, M. (2004). The PA subunit is required for efficient nuclear accumulation of the PB1 subunit of the influenza A virus RNA polymerase complex. J. Virol. *78*, 9144-9153.

Fouchier, R.A., Munster, V., Wallensten, A., Bestebroer, T.M., Herfst, S., Smith, D., Rimmelzwaan, G.F., Olsen, B., and Osterhaus, A.D. (2005). Characterization of a novel influenza A virus hemagglutinin subtype (H16) obtained from black-headed gulls. J. Virol. *79*, 2814-2822.

Fouchier, R.A., Schneeberger, P.M., Rozendaal, F.W., Broekman, J.M., Kemink, S.A., Munster, V., Kuiken, T., Rimmelzwaan, G.F., Schutten, M., Van Doornum, G.J., et al. (2004). Avian influenza A virus (H7N7) associated with human conjunctivitis and a fatal case of acute respiratory distress syndrome. Proc. Natl. Acad. Sci. USA *101*, 1356-1361.

Gabriel, G., Dauber, B., Wolff, T., Planz, O., Klenk, H.D., and Stech, J. (2005). The viral polymerase mediates adaptation of an avian influenza virus to a mammalian host. Proc. Natl. Acad. Sci. USA *102*, 18590-18595.

Gambaryan, A., Tuzikov, A., Pazynina, G., Bovin, N., Balish, A., and Klimov, A. (2006). Evolution of the receptor binding phenotype of influenza A (H5) viruses. Virology *344*, 432-438.

Gambaryan, A., Webster, R., and Matrosovich, M. (2002). Differences between influenza virus receptors on target cells of duck and chicken. Arch. Virol. *147*, 1197-1208.

Gambaryan, A., Yamnikova, S., Lvov, D., Tuzikov, A., Chinarev, A., Pazynina, G., Webster, R., Matrosovich, M., and Bovin, N. (2005a). Receptor specificity of influenza viruses from birds and mammals: new data on involvement of the inner fragments of the carbohydrate chain. Virology *334*, 276-283.

Gambaryan, A.S., Karasin, A.I., Tuzikov, A.B., Chinarev, A.A., Pazynina, G.V., Bovin, N.V., Matrosovich, M.N., Olsen, C.W., and Klimov, A.I. (2005b). Receptor-binding properties of swine influenza viruses isolated and propagated in MDCK cells. Virus Res. *114*, 15-22.

Gambaryan, A.S., Robertson, J.S., and Matrosovich, M.N. (1999). Effects of egg-adaptation on the receptor-binding properties of human influenza viruses. Virology *258*, 232-239.

Gambaryan, A.S., Tuzikov, A.B., Bovin, N.V., Yamnikova, S.S., Lvov, D.K., Webster, R.G., and Matrosovich, M.N. (2003). Differences between influenza virus receptors on target cells of duck and chicken and receptor specificity of the 1997 H5N1 chicken and human influenza viruses from Hong Kong. Avian Dis. *47* (3 Suppl), 1154-1160.

Gambaryan, A.S., Tuzikov, A.B., Pazynina, G.V., Webster, R.G., Matrosovich, M.N., and Bovin, N.V. (2004). H5N1 chicken influenza viruses display a high binding affinity for Neu5Acalpha2-3Galbeta1-4(6-HSO3)GlcNAc-containing receptors. Virology *326*, 310-316.

Gambaryan, A.S., Tuzikov, A.B., Piskarev, V.E., Yamnikova, S.S., Lvov, D.K., Robertson, J.S., Bovin, N.V., and Matrosovich, M.N. (1997). Specification of receptor-binding phenotypes of influenza virus isolates from different hosts using synthetic sialylglycopolymers: non-egg-adapted human H1 and H3 influenza A and influenza B viruses share a common high binding affinity for 6'-sialyl(N-acetyllactosamine). Virology *232*, 345-350.

Gamblin, S.J., Haire, L.F., Russell, R.J., Stevens, D.J., Xiao, B., Ha, Y., Vasisht, N., Steinhauer, D.A., Daniels, R.S., Elliot, A., et al. (2004). The Structure and Receptor-Binding Properties of the 1918 Influenza Hemagglutinin. Science *303*, 1838-1842.

Gao, P., Watanabe, S., Ito, T., Goto, H., Wells, K., McGregor, M., Cooley, A.J., and Kawaoka, Y. (1999). Biological heterogeneity, including systemic replication in mice, of H5N1 influenza A virus isolates from humans in Hong Kong. J. Virol. *73*, 3184-3189.

Garcia, M., Crawford, J.M., Latimer, J.W., Rivera-Cruz, E., and Perdue, M.L. (1996). Heterogeneity in the haemagglutinin gene and emergence of the highly pathogenic phenotype among recent H5N2 avian influenza viruses from Mexico. J. Gen. Virol. *77*, 1493-1504.

Garten, W., and Klenk, H.D. (1999). Understanding influenza virus pathogenicity. Trends Microbiol. *7*, 99-100.

Glaser, L., Stevens, J., Zamarin, D., Wilson, I.A., Garcia-Sastre, A., Tumpey, T.M., Basler, C.F., Taubenberger, J.K., and Palese, P. (2005). A single amino acid substitution in 1918 influenza virus hemagglutinin changes receptor binding specificity. J. Virol. *79*, 11533-11536.

Gottschalk, A., Belyavin, G., and Biddle, F. (1972). Glycoproteins as influenza virus hemagglutinin inhibitors and as cellular virus receptors. In Glycoproteins. Their composition, structure and function. Second edition. Part A, A. Gottschalk, ed. (Amsterdam-London-New York: Elsevier Publishing Company), pp. 1082-1096.

Govorkova, E.A., Rehg, J.E., Krauss, S., Yen, H.L., Guan, Y., Peiris, M., Nguyen, T.D., Hanh, T.H., Puthavathana, P., Long, H.T., et al. (2005). Lethality to ferrets of H5N1 influenza viruses isolated from humans and poultry in 2004. J. Virol. 79, 2191-2198.

Gratzl, E., and Koehler, H. (1968). Gefluegelpest. In Spezielle Pathologie und Therapie der Gefluegelkrankheiten. (Stuttgart: Ferdinand Enke Verlag).

Guan, Y., Peiris, J.S., Lipatov, A.S., Ellis, T.M., Dyrting, K.C., Krauss, S., Zhang, L.J., Webster, R.G., and Shortridge, K.F. (2002). Emergence of multiple genotypes of H5N1 avian influenza viruses in Hong Kong SAR. Proc. Natl. Acad. Sci. USA *99*, 8950-8955.

Guan, Y., Shortridge, K.F., Krauss, S., and Webster, R.G. (1999). Molecular characterization of H9N2 influenza viruses: were they the donors of the "internal" genes of H5N1 viruses in Hong Kong? Proc. Natl. Acad. Sci. USA *96*, 9363-9367.

Guo, Y.J., Krauss, S., Senne, D.A., Mo, I.P., Lo, K.S., Xiong, X.P., Norwood, M., Shortridge, K.F., Webster, R.G., and Guan, Y. (2000). Characterization of the pathogenicity of members of the newly established H9N2 influenza virus lineages in Asia. Virology *267,* 279-288.

Ha, Y., Stevens, D.J., Skehel, J.J., and Wiley, D.C. (2001). X-ray structures of H5 avian and H9 swine influenza virus hemagglutinins bound to avian and human receptor analogs. Proc. Natl. Acad. Sci. USA *98*, 11181-11186.

Ha, Y., Stevens, D.J., Skehel, J.J., and Wiley, D.C. (2003). X-ray structure of the hemagglutinin of a potential H3 avian progenitor of the 1968 Hong Kong pandemic influenza virus. Virology *309*, 209-218.

Hallenberger, S., Moulard, M., Sordel, M., Klenk, H.D., and Garten, W. (1997). The role of eukaryotic subtilisin-like endoproteases for the activation of human immunodeficiency virus glycoproteins in natural host cells. J. Virol. *71*, 1036-1045.

Hara, K., Schmidt, F.I., Crow, M., and Brownlee, G.G. (2006). Amino acid residues in the N-terminal region of the PA subunit of influenza A virus RNA polymerase play a critical role in protein stability, endonuclease activity, cap binding, and virion RNA promoter binding. J. Virol. *80*, 7789-7798.

Hatada, E., Hasegawa, M., Mukaigawa, J., Shimizu, K., and Fukuda, R. (1989). Control of influenza virus gene expression: quantitative analysis of each viral RNA species in infected cells. J. Biochem. (Tokyo) *105*, 537-546.

Hatta, M., Asano, Y., Masunaga, K., Ito, T., Okazaki, K., Toyoda, T., Kawaoka, Y., Ishihama, A., and Kida, H. (2000). Mapping of functional domains on the influenza A virus RNA polymerase PB2 molecule using monoclonal antibodies. Arch. Virol. *145*, 1947-1961.

Hatta, M., Gao, P., Halfmann, P., and Kawaoka, Y. (2001). Molecular basis for high virulence of Hong Kong H5N1 influenza A viruses. Science *293*, 1840-1842.

Hay, A.J., Lomniczi, B., Bellamy, A.R., and Skehel, J.J. (1977). Transcription of the influenza virus genome. Virology *83*, 337-355.
Hay, A.J., Skehel, J.J., and McCauley, J. (1982). Characterization of influenza virus RNA complete transcripts. Virology *116*, 517-522.
Herrler, G., Hausmann, J., and Klenk, H.-D. (1995). Sialic acid as receptor determinant of ortho- and paramyxoviruses. In Biology of the sialic acids, A. Rosenberg, ed. (New York, Plenum), pp. 315-336.
Herrler, G., and Klenk, H.D. (1987). The surface receptor is a major determinant of the cell tropism of influenza C virus. Virology *159*, 102-108.
Hiromoto, Y., Saito, T., Lindstrom, S., and Nerome, K. (2000). Characterization of low virulent strains of highly pathogenic A/Hong Kong/156/97 (H5N1) virus in mice after passage in embryonated hens' eggs. Virology *272*, 429-437.
Hoffmann, E., Stech, J., Leneva, I., Krauss, S., Scholtissek, C., Chin, P.S., Peiris, M., Shortridge, K.F., and Webster, R.G. (2000). Characterization of the influenza A virus gene pool in avian species in southern China: was H6N1 a derivative or a precursor of H5N1? J. Virol. *74*, 6309-6315.
Horimoto, T., and Kawaoka, Y. (2001). Pandemic threat posed by avian influenza A viruses. Clin. Microbiol. Rev. *14*, 129-149.
Horimoto, T., and Kawaoka, Y. (2005). Influenza: lessons from past pandemics, warnings from current incidents. Nat. Rev. Microbiol. *3*, 591-600.
Horimoto, T., Nakayama, K., Smeekens, S.P., and Kawaoka, Y. (1994). Proprotein-processing endoproteases PC6 and furin both activate hemagglutinin of virulent avian influenza viruses. J. Virol. *68*, 6074-6078.
Hsu, M.T., Parvin, J.D., Gupta, S., Krystal, M., and Palese, P. (1987). Genomic RNAs of influenza viruses are held in a circular conformation in virions and in infected cells by a terminal panhandle. Proc. Natl. Acad. Sci. USA *84*, 8140-8144.
Inglis, S.C., Carroll, A.R., Lamb, R.A., and Mahy, B.W. (1976). Polypeptides specified by the influenza virus genome I. Evidence for eight distinct gene products specified by fowl plague virus. Virology *74*, 489-503.
Inglis, S.C., McGeoch, D.J., and Mahy, B.W. (1977). Polypeptides specified by the influenza virus genoma. 2. Assignment of protein coding functions to individual genome segments by in vitro translation. Virology *78*, 522-536.
Ito, T., Couceiro, J.N.S.S., Kelm, S., Baum, L.G., Krauss, S., Castrucci, M.R., Donatelli, I., Kida, H., Paulson, J.C., Webster,

R.G., and Kawaoka, Y. (1998). Molecular basis for the generation in pigs of influenza A viruses with pandemic potential. J. Virol. *72*, 7367-7373.

Ito, T., Suzuki, Y., Suzuki, T., Takada, A., Horimoto, T., Wells, K., Kida, H., Otsuki, K., Kiso, M., Ishida, H., and Kawaoka, Y. (2000). Recognition of N-glycolylneuraminic acid linked to galactose by the alpha2,3 linkage is associated with intestinal replication of influenza A virus in ducks. J. Virol. *74*, 9300-9305.

Kawaoka, Y., Chambers, T.M., Sladen, W.L., and Webster, R.G. (1988). Is the gene pool of influenza viruses in shorebirds and gulls different from that in wild ducks? Virology *163*, 247-250.

Kawaoka, Y., Naeve, C.W., and Webster, R.G. (1984). Is virulence of H5N2 influenza viruses in chickens associated with loss of carbohydrate from the hemagglutinin? Virology *139*, 303-316.

Kawaoka, Y., and Webster, R.G. (1989). Interplay between carbohydrate in the stalk and the length of the connecting peptide determines the cleavability of influenza virus hemagglutinin. J. Virol. *63*, 3296-3300.

Kelm, S., and Schauer, R. (1997). Sialic acids in molecular and cellular interactions. Intern. Rev. Cytol. *175*, 137-240.

Khatchikian, D., Orlich, M., and Rott, R. (1989). Increased viral pathogenicity after insertion of a 28S ribosomal RNA sequence into the haemagglutinin gene of an influenza virus. Nature *340*, 156-157.

Kim, H.J., Fodor, E., Brownlee, G.G., and Seong, B.L. (1997). Mutational analysis of the RNA-fork model of the influenza A virus vRNA promoter in vivo. J. Gen. Virol. *78*, 353-357.

Klenk, H.D., and Garten, W. (1994). Host cell proteases controlling virus pathogenicity. Trends Microbiol. *2*, 39-43.

Klenk, H.D., and Rott, R. (1988). The molecular biology of influenza virus pathogenicity. Adv. Virus Res. *34*, 247-281.

Kobasa, D., Takada, A., Shinya, K., Hatta, M., Halfmann, P., Theriault, S., Suzuki, H., Nishimura, H., Mitamura, K., Sugaya, N., et al. (2004). Enhanced virulence of influenza A viruses with the haemagglutinin of the 1918 pandemic virus. Nature *431*, 703-707.

Kobayashi, Y., Horimoto, T., Kawaoka, Y., Alexander, D.J., and Itakura, C. (1996). Pathological studies of chickens experimentally infected with two highly pathogenic avian influenza viruses. Avian Pathol. *25*, 285-304.

Koopmans, M., Vennema, H., Heersma, H., van Strien, E., van Duynhoven, Y., Brown, D., Reacher, M., and Lopman, B. (2003). Early identification of common-source foodborne virus outbreaks in Europe. Emerg. Infect. Dis. *9*, 1136-1142.

Kuiken, T., Rimmelzwaan, G., van Riel, D., van Amerongen, G., Baars, M., Fouchier, R., and Osterhaus, A. (2004). Avian H5N1 influenza in cats. Science *306*, 241.
Kurtz, J., Manvell, R.J., and Banks, J. (1996). Avian influenza virus isolated from a woman with conjunctivitis. Lancet *348*, 901-902.
Lamblin, G., Degroote, S., Perini, J.M., Delmotte, P., Scharfman, A., Davril, M., Lo-Guidice, J.M., Houdret, N., Dumur, V., Klein, A., and Rousse, P. (2001). Human airway mucin glycosylation: a combinatory of carbohydrate determinants which vary in cystic fibrosis. Glycoconj J. *18*, 661-684.
Li, K.S., Guan, Y., Wang, J., Smith, G.J., Xu, K.M., Duan, L., Rahardjo, A.P., Puthavathana, P., Buranathai, C., Nguyen, T.D., et al. (2004). Genesis of a highly pathogenic and potentially pandemic H5N1 influenza virus in eastern Asia. Nature *430*, 209-213.
Li, M.L., Ramirez, B.C., and Krug, R.M. (1998). RNA-dependent activation of primer RNA production by influenza virus polymerase: different regions of the same protein subunit constitute the two required RNA-binding sites. EMBO J. *17*, 5844-5852.
Li, M.L., Rao, P., and Krug, R.M. (2001). The active sites of the influenza cap-dependent endonuclease are on different polymerase subunits. EMBO J. *20*, 2078-2086.
Li, S.Q., Orlich, M., and Rott, R. (1990). Generation of seal influenza virus variants pathogenic for chickens, because of hemagglutinin cleavage site changes. J. Virol. *64*, 3297-3303.
Li, X., and Palese, P. (1994). Characterization of the polyadenylation signal of influenza virus RNA. J. Virol. *68*, 1245-1249.
Li, Z., Chen, H., Jiao, P., Deng, G., Tian, G., Li, Y., Hoffmann, E., Webster, R.G., Matsuoka, Y., and Yu, K. (2005). Molecular basis of replication of duck H5N1 influenza viruses in a mammalian mouse model. J. Virol. *79*, 12058-12064.
Lin, Y.P., Shaw, M., Gregory, V., Cameron, K., Lim, W., Klimov, A., Subbarao, K., Guan, Y., Krauss, S., Shortridge, K., et al. (2000). Avian-to-human transmission of H9N2 subtype influenza A viruses: relationship between H9N2 and H5N1 human isolates. Proc. Natl. Acad. Sci. USA *97*, 9654-9658.
Liu, J., Xiao, H., Lei, F., Zhu, Q., Qin, K., Zhang, X.W., Zhang, X.L., Zhao, D., Wang, G., Feng, Y., et al. (2005). Highly pathogenic H5N1 influenza virus infection in migratory birds. Science *309*, 1206.
Lode, G., and Gruber, F. (1901). Zentbl. Bakt. Parasitkde. Abt. I *30*, 593-604.

Ludwig, S., Stitz, L., Planz, O., Van, H., Fitch, W.M., and Scholtissek, C. (1995). European swine virus as a possible source for the next influenza pandemic? Virology *212*, 555-561.
Luo, G., Chung, J., and Palese, P. (1993). Alterations of the stalk of the influenza virus neuraminidase: deletions and insertions. Virus Res. *29*, 141-153.
Luo, G.X., Luytjes, W., Enami, M., and Palese, P. (1991). The polyadenylation signal of influenza virus RNA involves a stretch of uridines followed by the RNA duplex of the panhandle structure. J. Virol. *65*, 2861-2867.
Massin, P., van der Werf, S., and Naffakh, N. (2001). Residue 627 of PB2 is a determinant of cold sensitivity in RNA replication of avian influenza viruses. J. Virol. *75*, 5398-5404.
Masuda, H., Suzuki, T., Sugiyama, Y., Horiike, G., Murakami, K., Miyamoto, D., Jwa Hidari, K.I., Ito, T., Kida, H., Kiso, M., et al. (1999). Substitution of amino acid residue in influenza A virus hemagglutinin affects recognition of sialyl-oligosaccharides containing N-glycolylneuraminic acid. FEBS Lett. *464*, 71-74.
Matrosovich, M., Matrosovich, T., Uhlendorff, J., Garten, W., and Klenk, H.D. (2007). Avian-virus-like receptor specificity of the hemagglutinin impedes influenza virus replication in cultures of human airway epithelium. Virology *361*, 384-390.
Matrosovich, M., Tuzikov, A., Bovin, N., Gambarian, A., Klimov, A., Cox, N., Castrucci, M., Donatelli, I., and Kawaoka, Y. (2000). Alterations of receptor-binding properties of H1, H2, and H3 avian influenza virus hemagglutinins upon introduction into mammals. J. Virol. *74*, 8502-8512.
Matrosovich, M., Zhou, N., Kawaoka, Y., and Webster, R. (1999). The surface glycoproteins of H5 influenza viruses isolated from humans, chickens, and wild aquatic birds have distinguishable properties. J. Virol. *73*, 1146-1155.
Matrosovich, M.N., Gambaryan, A.S., Teneberg, S., Piskarev, V.E., Yamnikova, S.S., Lvov, D.K., Robertson, J.S., and Karlsson, K.A. (1997). Avian influenza A viruses differ from human viruses by recognition of sialyloligosaccharides and gangliosides and by a higher conservation of the HA receptor-binding site. Virology *233*, 224-234.
Matrosovich, M.N., Gambaryan, A.S., Tuzikov, A.B., Byramova, N.E., Mochalova, L.V., Golbraikh, A.A., Shenderovich, M.D., Finne, J., and Bovin, N.V. (1993). Probing of the receptor-binding sites of the H1 and H3 influenza A and influenza B virus hemagglutinins by synthetic and natural sialosides. Virology *196*, 111-121.

Matrosovich, M.N., Klenk, H.-D., and Kawaoka, Y. (2006). Receptor specificity, host range and pathogenicity of influenza viruses. In Influenza Virology: Current Topics, Y. Kawaoka, ed. (Wymondham, England: Caister Academic Press), pp. 95-137.

Matrosovich, M.N., Krauss, S., and Webster, R.G. (2001). H9N2 Influenza A Viruses from Poultry in Asia Have Human Virus-like Receptor Specificity. Virology *281*, 156-162.

Matrosovich, M.N., Matrosovich, T.Y., Gray, T., Roberts, N.A., and Klenk, H.D. (2004). Human and avian influenza viruses target different cell types in cultures of human airway epithelium. Proc. Natl. Acad. Sci. USA *101*, 4620-4624.

McGeoch, D., Fellner, P., and Newton, C. (1976). Influenza virus genome consists of eight distinct RNA species. Proc. Natl. Acad. Sci. USA *73*, 3045-3049.

Miller-Podraza, H., Johansson, L., Johansson, P., Matrosovich, M., and Karlsson, K.-A. (2000). A strain of human influenza A virus binds to extended but not short gangliosides as assayed by thin-layer chromatography overlay. Glycobiology *10*, 975-982.

Mitnaul, L.J., Castrucci, M.R., Murti, K.G., and Kawaoka, Y. (1996). The cytoplasmic tail of influenza A virus neuraminidase (NA) affects NA incorporation into virions, virion morphology, and virulence in mice but is not essential for virus replication. J. Virol. *70*, 873-879.

Morsy, J., Garten, W., and Rott, R. (1994). Activation of an influenza virus A/turkey/Oregon/71 HA insertion variant by the subtilisin-like endoprotease furin. Virology *202*, 988-991.

Mukaigawa, J., and Nayak, D.P. (1991). Two signals mediate nuclear localization of influenza virus (A/WSN/33) polymerase basic protein 2. J. Virol. *65*, 245-253.

Naffakh, N., Massin, P., Escriou, N., Crescenzo-Chaigne, B., and van der Werf, S. (2000). Genetic analysis of the compatibility between polymerase proteins from human and avian strains of influenza A viruses. J. Gen. Virol. *81*, 1283-1291.

Narayan, O., Thorsen, J., Hulland, T.J., Ankeli, G., and Joseph, P.G. (1972). Pathogenesis of lethal influenza virus infection in turkeys. I. Extraneural phase of infection. J. Comp. Pathol. *82*, 129-137.

Nath, S.T., and Nayak, D.P. (1990). Function of two discrete regions is required for nuclear localization of polymerase basic protein 1 of A/WSN/33 influenza virus (H1 N1). Mol. Cell. Biol. *10*, 4139-4145.

Nieto, A., de la Luna, S., Barcena, J., Portela, A., Valcarcel, J., Melero, J.A., and Ortin, J. (1992). Nuclear transport of influenza virus polymerase PA protein. Virus Res. *24*, 65-75.
Noah, D.L., and Krug, R.M. (2005). Influenza virus virulence and its molecular determinants. Adv. Virus Res. *65*, 121-145.
Nobusawa, E., Aoyama, T., Kato, H., Suzuki, Y., Tateno, Y., and Nakajima, K. (1991). Comparison of complete amino acid sequences and receptor-binding properties among 13 serotypes of hemagglutinins of influenza A viruses. Virology *182*, 475-485.
Nobusawa, E., and Sato, K. (2006). Comparison of the mutation rates of human influenza A and B viruses. J. Virol. *80*, 3675-3678.
Ohuchi, M., Orlich, M., Ohuchi, R., Simpson, B.E., Garten, W., Klenk, H.D., and Rott, R. (1989). Mutations at the cleavage site of the hemagglutinin after the pathogenicity of influenza virus A/chick/Penn/83 (H5N2). Virology *168*, 274-280.
Ohuchi, R., Ohuchi, M., Garten, W., and Klenk, H.D. (1991). Human influenza virus hemagglutinin with high sensitivity to proteolytic activation. J. Virol. *65*, 3530-3537.
OIE (2003). Manual standards for diagnostic tests and vaccines, 2000. (Paris, Office International de Epizooties.).
Olsen, B., Munster, V.J., Wallensten, A., Waldenstrom, J., Osterhaus, A.D., and Fouchier, R.A. (2006). Global patterns of influenza a virus in wild birds. Science *312*, 384-388.
Orlich, M., Gottwald, H., and Rott, R. (1994). Nonhomologous recombination between the hemagglutinin gene and the nucleoprotein gene of an influenza virus. Virology *204*, 462-465.
Orlich, M., Khatchikian, D., Teigler, A., and Rott, R. (1990). Structural variation occurring in the hemagglutinin of influenza virus A/turkey/Oregon/71 during adaptation to different cell types. Virology *176*, 531-538.
Page, M.G., Ruck, B.J., and Brammer, K.W. (1968). Characterization of the ribonucleic acid product synthesized by on influenza A2 virus-induced ribonucleic acid polymerase in vitro. Biochem. J. *109*, 43P.
Palese, P., Ritchey, M.B., and Schulman, J.L. (1977). Mapping of the influenza virus genome. II. Identification of the P1, P2, and P3 genes. Virology *76*, 114-121.
Palese, P., and Schulman, J.L. (1976). Differences in RNA patterns of influenza A viruses. J. Virol. *17*, 876-884.
Parvin, J.D., Moscona, A., Pan, W.T., Leider, J.M., and Palese, P. (1986). Measurement of the mutation rates of animal viruses: influenza A virus and poliovirus type 1. J. Virol. *59*, 377-383.

Pasick, J., Handel, K., Robinson, J., Copps, J., Ridd, D., Hills, K., Kehler, H., Cottam-Birt, C., Neufeld, J., Berhane, Y., and Czub, S. (2005). Intersegmental recombination between the haemagglutinin and matrix genes was responsible for the emergence of a highly pathogenic H7N3 avian influenza virus in British Columbia. J. Gen. Virol. *86*, 727-731.

Paulson, J.C. (1985). Interactions of animal viruses with cell surface receptors. In The receptors. Vol. 2, M. Conn, ed. (Orlando, FL: Academic Press), pp. 131-219.

Peiris, M., Yuen, K.Y., Leung, C.W., Chan, K.H., Ip, P.L., Lai, R.W., Orr, W.K., and Shortridge, K.F. (1999). Human infection with influenza H9N2. Lancet *354*, 916-917.

Penhoet, E., Miller, H., Doyle, M., and Blatti, S. (1971). RNA-dependent RNA polymerase activity in influenza virions. Proc. Natl. Acad. Sci. USA *68*, 1369-1371.

Perdue, M.L., Garcia, M., Senne, D., and Fraire, M. (1997). Virulence-associated sequence duplication at the hemagglutinin cleavage site of avian influenza viruses. Virus Res. *49*, 173-186.

Perroncito, E. (1878). Epizoozia tifoide nei gallinacei. Ann. Accad. Agric. Torino *21*, 87-126.

Poon, L.L., Pritlove, D.C., Fodor, E., and Brownlee, G.G. (1999). Direct evidence that the poly(A) tail of influenza A virus mRNA is synthesized by reiterative copying of a U track in the virion RNA template. J. Virol. *73*, 3473-3476.

Poon, L.L., Pritlove, D.C., Sharps, J., and Brownlee, G.G. (1998). The RNA polymerase of influenza virus, bound to the 5' end of virion RNA, acts in cis to polyadenylate mRNA. J. Virol. *72*, 8214-8219.

Puthavathana, P., Auewarakul, P., Charoenying, P.C., Sangsiriwut, K., Pooruk, P., Boonnak, K., Khanyok, R., Thawachsupa, P., Kijphati, R., and Sawanpanyalert, P. (2005). Molecular characterization of the complete genome of human influenza H5N1 virus isolates from Thailand. J. Gen. Virol. *86*, 423-433.

Reid, A.H., Fanning, T.G., Hultin, J.V., and Taubenberger, J.K. (1999). Origin and evolution of the 1918 "Spanish" influenza virus hemagglutinin gene. Proc. Natl. Acad. Sci. USA *96*, 1651-1656.

Roehm, C., Horimoto, T., Kawaoka, Y., Suss, J., and Webster, R.G. (1995). Do hemagglutinin genes of highly pathogenic avian influenza viruses constitute unique phylogenetic lineages? Virology *209*, 664-670.

Rogers, G.N., and D'Souza, B.L. (1989). Receptor binding properties of human and animal H1 influenza virus isolates. Virology *173*, 317-322.

Rogers, G.N., Paulson, J.C., Daniels, R.S., Skehel, J.J., Wilson, I.A., and Wiley, D.C. (1983). Single amino acid substitutions in influenza haemagglutinin change receptor binding specificity. Nature *304*, 76-78.
Rott, R., Orlich, M., and Scholtissek, C. (1979). Correlation of pathogenicity and gene constellation of influenza A viruses. III. Non-pathogenic recombinants derived from highly pathogenic parent strains. J. Gen. Virol. *44*, 471-477.
Rott, R., and Schaefer, W. (1960). Physikalisch-chemische und biologische Eigenschaften des Virus N und seine Beziehung zur Influenza-A-Untergruppe der Myxoviren. Zbl. Vet. Med. *7*, 237-248.
Salomon, R., Franks, J., Govorkova, E.A., Ilyushina, N.A., Yen, H.L., Hulse-Post, D.J., Humberd, J., Trichet, M., Rehg, J.E., Webby, R.J., et al. (2006). The polymerase complex genes contribute to the high virulence of the human H5N1 influenza virus isolate A/Vietnam/1203/04. J. Exp. Med. *203*, 689-697.
Sauter, N.K., Bednarski, M.D., Wurzburg, B.A., Hanson, J.E., Whitesides, G.M., Skehel, J.J., and Wiley, D.C. (1989). Hemagglutinins from two influenza virus variants bind to sialic acid derivatives with millimolar dissociation constants: a 500-MHz proton nuclear magnetic resonance study. Biochemistry *28*, 8388-8396.
Schaefer, W. (1955). Vergleichende seroimmunologische Untersuchungen über die Viren der Influenza und der klassischen Geflügelpest. Z. Naturforsch. *10b*, 81-91.
Schauer, R. (1982). Chemistry, metabolism, and biological functions of sialic acids. Adv. Carbohydr. Chem. Biochem. *40*, 131-234.
Schauer, R. and Kamerling, J.P. (1997). Chemistry, biochemistry and biology of sialic acids. In: Glycoproteins II (New Comprehensive Biochemistry, vol. 29b), J. Montreuil, J.F.G. Vliegenthart, and H. Schachter, eds. (Elsevier Science B.V.), pp. 243-402.
Scholtissek, C. (1969). Synthesis in vitro of RNA complementary to parental viral RNA by RNA polymerase induced by influenza virus. Biochim. Biophys. Acta *179*, 389-397.
Scholtissek, C., Harms, E., Rohde, W., Orlich, M., and Rott, R. (1976). Correlation between RNA fragments of fowl plague virus and their corresponding gene functions. Virology *74*, 332-344.
Scholtissek, C., Hinshaw, V.S., and Olsen, C.W. (1998). Influenza in pigs and their role as the intermediate host. In Textbook of influenza, K.G. Nicholson, R.G. Webster, and A.J. Hay, eds. (London: Blackwell Science), pp. 137-145.

Scholtissek, C., and Rott, R. (1964). Behavior of virus-specific activities in tissue cultures infected with myxoviruses after chemical changes of the viral ribonucleic acid. Virology *22*, 169-176.
Scholtissek, C., Vallbracht, A., Flehmig, B., and Rott, R. (1979). Correlation of pathogenicity and gene constellation of influenza A viruses. II. Highly neurovirulent recombinants derived from non-neurovirulent or weakly neurovirulent parent virus strains. Virology *95*, 492-500.
Schultz, U., Fitch, W.M., Ludwig, S., Mandler, J., and Scholtissek, C. (1991). Evolution of pig influenza viruses. Virology *183*, 61-73.
Shi, L., Summers, D.F., Peng, Q., and Galarz, J.M. (1995). Influenza A virus RNA polymerase subunit PB2 is the endonuclease which cleaves host cell mRNA and functions only as the trimeric enzyme. Virology *208*, 38-47.
Shinya, K., Ebina, M., Yamada, S., Ono, M., Kasai, N., and Kawaoka, Y. (2006). Avian flu: influenza virus receptors in the human airway. Nature *440*, 435-436.
Shinya, K., Hamm, S., Hatta, M., Ito, H., Ito, T., and Kawaoka, Y. (2004). PB2 amino acid at position 627 affects replicative efficiency, but not cell tropism, of Hong Kong H5N1 influenza A viruses in mice. Virology *320*, 258-266.
Skehel, J.J. (1971). RNA-dependent RNA polymerase activity of the influenza virus. Virology *45*, 793-796.
Skehel, J.J., and Hay, A.J. (1978). Nucleotide sequences at the 5' termini of influenza virus RNAs and their transcripts. Nucleic Acids Res. *5*, 1207-1219.
Skehel, J.J., and Wiley, D.C. (2000). Receptor binding and membrane fusion in virus entry: the influenza hemagglutinin. Annu. Rev. Biochem. *69*, 531-569.
Snyder, M.H., Buckler-White, A.J., London, W.T., Tierney, E.L., and Murphy, B.R. (1987). The avian influenza virus nucleoprotein gene and a specific constellation of avian and human virus polymerase genes each specify attenuation of avian-human influenza A/Pintail/79 reassortant viruses for monkeys. J. Virol. *61*, 2857-2863.
Stech, J., Garn, H., Wegmann, M., Wagner, R., and Klenk, H.D. (2005). A new approach to an influenza live vaccine: modification of the cleavage site of hemagglutinin. Nat. Med. *11*, 683-689.
Stech, J., Xiong, X., Scholtissek, C., and Webster, R.G. (1999). Independence of evolutionary and mutational rates after transmission of avian influenza viruses to swine. J. Virol. *73*, 1878-1884.

Steinhauer, D.A. (1999). Role of hemagglutinin cleavage for the pathogenicity of influenza virus. Virology *258*, 1-20.

Steuler, H., Rohde, W., and Scholtissek, C. (1984). Sequence of the neuraminidase gene of an avian influenza A virus (A/parrot/Ulster/73, H7N1). Virology *135*, 118-124.

Stieneke-Groeber, A., Vey, M., Angliker, H., Shaw, E., Thomas, G., Roberts, C., Klenk, H.D., and Garten, W. (1992). Influenza virus hemagglutinin with multibasic cleavage site is activated by furin, a subtilisin-like endoprotease. EMBO J. *11*, 2407-2414.

Suarez, D.L., Senne, D.A., Banks, J., Brown, I.H., Essen, S.C., Lee, C.W., Manvell, R.J., Mathieu-Benson, C., Moreno, V., Pedersen, J.C., et al. (2004). Recombination resulting in virulence shift in avian influenza outbreak, Chile. Emerg. Infect. Dis. *10*, 693-699.

Suarez-Lopez, P., and Ortin, J. (1994). An estimation of the nucleotide substitution rate at defined positions in the influenza virus haemagglutinin gene. J. Gen. Virol. *75*, 389-393.

Subbarao, E.K., London, W., and Murphy, B.R. (1993). A single amino acid in the PB2 gene of influenza A virus is a determinant of host range. J. Virol. *67*, 1761-1764.

Subbarao, K., Chen, H., Swayne, D., Mingay, L., Fodor, E., Brownlee, G., Xu, X., Lu, X., Katz, J., Cox, N., and Matsuoka, Y. (2003). Evaluation of a genetically modified reassortant H5N1 influenza A virus vaccine candidate generated by plasmid-based reverse genetics. Virology *305*, 192-200.

Subbarao, K., Klimov, A., Katz, J., Regnery, H., Lim, W., Hall, H., Perdue, M., Swayne, D., Bender, C., Huang, J., et al. (1998). Characterization of an avian influenza A (H5N1) virus isolated from a child with a fatal respiratory illness. Science *279*, 393-396.

Subbarao, K., Swayne, D.E., and Olsen, C.W. (2006). Epidemiology and control of human and animal viruses. In: Influenza Virology, Current Topics, Y. Kawaoka, ed. (Norfolk, UK: Caister Academic Press), pp. 229-279.

Suzuki, T., Horiike, G., Yamazaki, Y., Kawabe, K., Masuda, H., Miyamoto, D., Matsuda, M., Nishimura, S.I., Yamagata, T., Ito, T., et al. (1997). Swine influenza virus strains recognize sialylsugar chains containing the molecular species of sialic acid predominantly present in the swine tracheal epithelium. FEBS Lett. *404*, 192-196.

Suzuki, Y. (1994). Gangliosides as influenza virus receptors. Variation of influenza viruses and their recognition of the receptor sialo-sugar chains. Progr. Lipid Res. *33*, 429-457.

Suzuki, Y., Ito, T., Suzuki, T., Holland, R.E., Jr., Chambers, T.M., Kiso, M., Ishida, H., and Kawaoka, Y. (2000). Sialic acid species as a determinant of the host range of influenza A viruses. J. Virol. *74*, 11825-11831.

Suzuki, Y., Nagao, Y., Kato, H., Matsumoto, M., Nerome, K., Nakajima, K., and Nobusawa, E. (1986). Human influenza A virus hemagglutinin distinguishes sialyloligosaccharides in membrane-associated gangliosides as its receptor which mediates the adsorption and fusion processes of virus infection. Specificity for oligosaccharides and sialic acids and the sequence to which sialic acid is attached. J. Biol. Chem. *261*, 17057-17061.

Suzuki, Y., Nakao, T., Ito, T., Watanabe, N., Toda, Y., Xu, G., Suzuki, T., Kobayashi, T., Kimura, Y., and Yamada, A. (1992). Structural determination of gangliosides that bind to influenza A, B, and C viruses by an improved binding assay: strain-specific receptor epitopes in sialo-sugar chains. Virology *189*, 121-131.

Tashiro, M., Ciborowski, P., Klenk, H.D., Pulverer, G., and Rott, R. (1987a). Role of Staphylococcus protease in the development of influenza pneumonia. Nature *325*, 536-537.

Tashiro, M., Ciborowski, P., Reinacher, M., Pulverer, G., Klenk, H.D., and Rott, R. (1987b). Synergistic role of staphylococcal proteases in the induction of influenza virus pathogenicity. Virology *157*, 421-430.

Taubenberger, J.K., Reid, A.H., Krafft, A.E., Bijwaard, K.E., and Fanning, T.G. (1997). Initial genetic characterization of the 1918 "Spanish" influenza virus. Science *275*, 1793-1796.

Taubenberger, J.K., Reid, A.H., Lourens, R.M., Wang, R., Jin, G., and Fanning, T.G. (2005). Characterization of the 1918 influenza virus polymerase genes. Nature *437*, 889-893.

Taylor, J.M., Illmensee, R., Litwin, S., Herring, L., Broni, B., and Krug, R.M. (1977). Use of specific radioactive probes to study transcription and replication of the influenza virus genome. J. Virol. *21*, 530-540.

Thompson, C.I., Barclay, W.S., Zambon, M.C., and Pickles, R.J. (2006). Infection of human airway epithelium by human and avian strains of influenza a virus. J. Virol. *80*, 8060-8068.

Tumpey, T.M., Basler, C.F., Aguilar, P.V., Zeng, H., Solorzano, A., Swayne, D.E., Cox, N.J., Katz, J.M., Taubenberger, J.K., Palese, P., and Garcia-Sastre, A. (2005). Characterization of the reconstructed 1918 Spanish influenza pandemic virus. Science *310*, 77-80.

Ulmanen, I., Broni, B.A., and Krug, R.M. (1981). Role of two of the influenza virus core P proteins in recognizing cap 1 structures (m7GpppNm) on RNAs and in initiating viral RNA transcription. Proc. Natl. Acad. Sci. USA *78*, 7355-7359.

Vallbracht, A., Scholtissek, C., Flehmig, B., and Gerth, H.J. (1980). Recombination of influenza A strains with fowl plague virus can change pneumotropism for mice to a generalized infection with involvement of the central nervous system. Virology *107*, 452-460.

Van Campen, H., Easterday, B.C., and Hinshaw, V.S. (1989a). Destruction of lymphocytes by a virulent avian influenza A virus. J. Gen. Virol. *70*, 467-472.

Van Campen, H., Easterday, B.C., and Hinshaw, V.S. (1989b). Virulent avian influenza A viruses: their effect on avian lymphocytes and macrophages in vivo and in vitro. J. Gen. Virol. 70, 2887-2895.

van Kolfschooten, F. (2003). Dutch veterinarian becomes first victim of avian influenza. Lancet *361*, 1444.

Varki, A. (1997). Sialic acids as ligands in recognition phenomena. FASEB J. *11*, 248-255.

Vines, A., Wells, K., Matrosovich, M., Castrucci, M.R., Ito, T., and Kawaoka, Y. (1998). The role of influenza A virus hemagglutinin residues 226 and 228 in receptor specificity and host range restriction. J. Virol. *72*, 7626-7631.

Vreede, F.T., and Brownlee, G.G. (2007). Influenza virion-derived viral ribonucleoproteins synthesize both mRNA and cRNA in vitro. J. Virol. *81*, 2196-204.

Vreede, F.T., Jung, T.E., and Brownlee, G.G. (2004). Model suggesting that replication of influenza virus is regulated by stabilization of replicative intermediates. J. Virol. *78*, 9568-9572.

Wan, H., and Perez, D.R. (2006). Quail carry sialic acid receptors compatible with binding of avian and human influenza viruses. Virology *346*, 278-286.

Weber, S., Harder, T., Starick, E., Beer, M., Werner, O., Hoffmann, B., Mettenleiter, T.C., and Mundt, E. (2007). Molecular analysis of highly pathogenic avian influenza virus of subtype H5N1 isolated from wild birds and mammals in northern Germany. J. Gen. Virol. *88*, 554-558.

Webster, R.G., Bean, W.J., Gorman, O.T., Chambers, T.M., and Kawaoka, Y. (1992). Evolution and ecology of influenza A viruses. Microbiol. Rev. *56*, 152-179.

Webster, R.G., Yakhno, M., Hinshaw, V.S., Bean, W.J., and Murti, K.G. (1978). Intestinal influenza: replication and characterization of influenza viruses in ducks. Virology *84*, 268-278.

Weis, W.I. (1997). Cell-surface carbohydrate recognition by animal and viral lectins. Curr. Opin. Struct. Biol. *7*, 624-630.

Werner, O. and Harder, T.C. (2006). Avian Influenza. In Influenza Report 2006. www.Influenzareport.com. B.S. Kamps, C. Hoffmann, and W. Preiser, eds. (Flying Publisher), pp. 48-86.

WHO (2003). Disease alert: 10 December 2003. Influenza A (H9N2) in Hong Kong Special Administrative Region of China. (World Health Organization).

Xu, X., Subbarao, K., Cox, N.J., and Guo, Y. (1999). Genetic characterization of the pathogenic influenza A/Goose/Guangdong/1/96 (H5N1) virus: similarity of its hemagglutinin gene to those of H5N1 viruses from the 1997 outbreaks in Hong Kong. Virology *261*, 15-19.

Yamnikova, S.S., Gambaryan, A.S., Tuzikov, A.B., Bovin, N.V., Matrosovich, M.N., Fedyakina, I.T., Grinev, A.A., Blinov, V.M., Lvov, D.K., Suarez, D.L., and Swayne, D.E. (2003). Differences between HA receptor-binding sites of avian influenza viruses isolated from Laridae and Anatidae. Avian Dis. *47*, 1164-1168.

Ye, Q., Krug, R.M., and Tao, Y.J. (2006). The mechanism by which influenza A virus nucleoprotein forms oligomers and binds RNA. Nature *444*, 1078-1082.

Zheng, H., Lee, H.A., Palese, P., and Garcia-Sastre, A. (1999). Influenza A virus RNA polymerase has the ability to stutter at the polyadenylation site of a viral RNA template during RNA replication. J. Virol. *73*, 5240-5243.

Chapter 7

Molecular Dissection of Bluetongue Virus

Polly Roy

ABSTRACT

Bluetongue virus (BTV), a member of Orbivirus genus within the *Reoviridae* family causes serious disease in livestock (sheep, goat, cattle). Partly due to this BTV has been in the forefront of molecular studies for last three decades and now represents one of the best understood viruses at the molecular and structural levels. BTV, like the other members of the family is a complex non-enveloped virus with seven structural proteins and a RNA genome consisting of 10 double-stranded (ds) RNA segments of different sizes. This article will focus on the molecular dissection of BTV with a view to understanding the role of each protein in the virus replication cycle. Data obtained from studies over a number of years have defined the key players in BTV entry, replication, assembly and exit and have increasingly found roles for host proteins at each stage. Specifically, it has been possible to determine the complex nature of the virion through 3D structure reconstructions (diameter ~ 800 Å); the atomic structure of proteins and the internal capsid (~ 700 Å, the first large highly complex structure ever solved); the definition of the virus encoded enzymes required for RNA replication; the ordered assembly of the capsid shell and the protein sequestration required for it; and the role of host proteins in virus entry and virus release. These areas are important for BTV replication but they also indicate the pathways that may be used by related viruses, which include viruses that are pathogenic to man and animals, thus providing the basis for developing strategies for intervention or prevention.

INTRODUCTION

Bluetongue virus (BTV) is the type species of the genus Orbivirus within the family *Reoviridae*. The *Reoviridae* family is one of the largest families of viruses and includes major human pathogens (e.g., rotavirus) as well as other vertebrate, plant and insect pathogens. Like the other members of the family, Orbiviruses which encompass, besides BTV, the agents causing African horse sickness (AHSV) and epizootic hemorrhagic disease of deer (EHDV), have the characteristic double-stranded and segmented features of their RNA genomes (Verwoerd, 1969; Verwoerd *et al*., 1970, 1972). However, unlike the mammalian reoviruses, Orbiviruses comprising 14 serogroups, are vectored to a variety of vertebrates by arthropod species (e.g., gnats, mosquitoes and ticks) and replicate in both hosts (Mertens *et al*., 2005). BTV, the etiological agent of Bluetongue disease of animals, is transmitted by *Culicoides* species. In sheep BTV causes an acute disease with high morbidity and mortality. BTV also infects goats, cattle and other domestic animals as well as wild ruminants (e.g., blesbuck, white-tailed deer, elk, pronghorn antelope, etc.). The disease was first described in the late 18th century and was believed for many decades to be confined to Africa. However, to date BTV has been isolated in many tropical, subtropical and temperate zones and 24 serotypes have been identified from different parts of the world. Due to its economic significance BTV has been the subject of extensive molecular, genetic and structural studies. As a consequence it now represents one of the best characterised viruses.

Unlike the reovirus and rotavirus particles, the mature BTV particle is relatively fragile and the infectivity of BTV is lost easily in mildly acidic conditions. BTV virions (550S) are architecturally complex structures composed of 7 discrete proteins that are organised into two concentric shells, the outer and inner capsids, and a genome of 10 dsRNA segments (Verwoerd, 1969; Verwoerd *et al*., 1972; Mertens *et al*., 1987; Hewat *et al*., 1992b). The outer capsid, which is composed of two major structural proteins (VP2 and VP5), is involved in cell attachment and virus penetration during the initial stages of infection (Eaton and Crameri, 1989; Hassan and Roy, 1999; Hassan *et al.*, 2001). Shortly after infection, BTV is uncoated, i.e. VP2 and VP5 are removed, to yield a transcriptionally active 470S core particle which is composed of two major proteins VP7 and VP3, and the three minor proteins VP1, VP4 and VP6 in addition to the dsRNA genome (Huismans *et al.*, 1987b). There is no evidence that any trace of the outer capsid remains associated with these cores, as has been described for reovirus (Silverstein *et al.*, 1972). The cores may be further uncoated to form 390S subcore particles that lack VP7, also in contrast to reovirus. Subviral particles are probably akin to cores derived *in vitro* from virions by physical or proteolytic treatments that remove the outer capsid and causes activation of the BTV transcriptase (Van Dijk and Huismans, 1980). In addition to the seven structural proteins, three non-structural (NS) proteins, NS1, NS2, NS3 (and a related NS3A) are synthesised in BTV-infected cells (Huismans and Els, 1979; Roy, 1996, 2000). Of these, NS3/NS3A is involved in the egress of the progeny virus. The two remaining non-structural proteins, NS1 and NS2, are produced at high levels in the cytoplasm and are believed to be involved in virus replication, assembly and morphogenesis. This chapter summarises the current understanding of the three-dimensional structure of BTV particles, sub viral particles and proteins and the role of each structural and NS proteins in the various stages of the virus life cycle.

THE BTV OUTER CAPSID STRUCTURE AND THEIR ROLE IN VIRUS ENTRY

BTV and other members of the *Reoviridae* family are non-enveloped particles and thus lack the virally encoded integral membrane proteins of enveloped viruses that are responsible for membrane fusion. Therefore, the proteins that constitute the outer surface of the members of the *Reoviridae* must adopt a structural organisation such that they can perform essentially the same function as the lipid embedded proteins of enveloped viruses (e.g. the transport of virus through the lipid bilayer from the extracellular medium and delivery of the virus capsid or genomes into the host cell).

A combination of three-dimensional structural determination and molecular/biochemical analysis of the BTV outer capsid has given some insight into how the two outer capsid proteins could function in cell entry and membrane penetration processes in an analogous manner to proteins of enveloped viruses. Cryo-electron microscopy (Cryo-EM) and image reconstruction of BTV particles have been used to deduce the general relationships of the two outer capsid proteins within the virus particle. Initially, a low-resolution (45 Å) 3D reconstruction followed by a higher resolution (22 Å) reconstruction of BTV particles was obtained. These studies revealed a well-ordered morphology of the complete virion with icosahedral symmetry and provided details of the essential organization of the two outer capsid proteins (Hewat *et al.*, 1992b, 1994; Nason *et al.*, 2004). An accurate position of the two proteins, VP2 and VP5, in relation to each other and to the underlying protein layer of the core was clearly defined (Fig. 1). In addition, their juxtaposition and structural features gave an insight into their roles in virus attachment to the host cell receptor and membrane penetration that are supported further by biochemical data. The outer capsid of BTV has an icosahedral configuration, and exhibits a unique organization made up of two distinct shapes, triskelion and globular. The outer layer does not conform to a T=13 icosahedral symmetry like that of rotaviruses and reoviruses and consists of a total of sixty triskelion densities and 120 globular densities (Fig. 1). This is in contrast to the morphology deduced by earlier negative staining EM methods which gave a characteristically fuzzy appearance to the outer capsid suggesting it is highly sensitive to sample preparation conditions (Els and Verwoerd, 1969).

The most external part of the outer capsid is the propeller-shaped triskelion motif. Each blade of the propeller is ~75 Å in length and 28 Å wide. At the tip of each blade, the molecule broadens to a width of 60 Å and, at this point, bends upwards perpendicular to the plane of the virus. These bent tips give the entire virion a diameter of ~880 Å and extend from the main body of the particle by 30 Å. Interspersed between the triskelions and lying more internally between 380 Å and 420 Å radii are the globular densities of ~60 Å in diameter (Fig. 1). These are also entirely exposed in the virion. Both proteins make extensive contacts with the underneath core surface protein VP7 (discussed later) indicating that the two proteins assemble independently on the preformed core surface. The currently accepted model on how the core particles of BTV gain access to the cytoplasm, where they become transcriptionally active, involves receptor mediated endocytosis and pH-triggered conformational change inside endosomal vesicles which activates BTV membrane-permeabilizing protein.

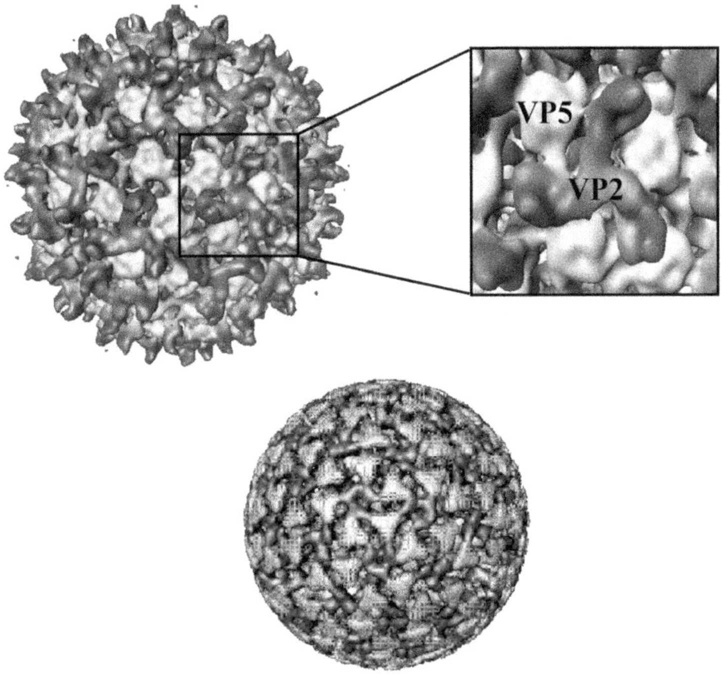

Fig. 1. Surface representations of the 3D cryo-EM structures (22 Å resolution) of BTV. **(A)** Whole particle showing sail-shaped triskelion propellers (VP2 trimers) and globular domains (VP5 trimers). **(B)** An amplified cut-off section of whole particle. **(C)** A particle with cut-off VP2 density showing arrangement of globular domains close with the surface VP7 layer of core.

Early suggestion of BTV entry via clathrin-mediated endocytosis has recently been substantiated by a comprehensive investigation using biochemical and confocal microscopy studies as well as various specific inhibitors and RNA interference (Eaton *et al.*, 1990). The data obtained from these studies conclusively demonstrated that BTV enters cells by clathrin-mediated endocytosis and requires pH-dependent penetration (Forzan *et al.*, 2007).

The two outer capsid proteins, particularly the larger protein VP2 (110 kDa), are easily removed and degraded (Verwoerd *et al.*, 1972; Huismans *et al.*, 1987a). Further, VP2 binds and mediates attachment of the virion to the cellular receptor and its subsequent internalization via early endosome where it is most likely proteolytically degraded (Eaton and Crameri, 1989; 1990; Hassan and Roy, 1999). The receptor binding VP2 is the most variable of the BTV proteins, eliciting virus neutralising antibody, and is responsible both for serotype-specificity and virus hemagglutination activity (Huismans and Erasmus, 1981; Kahlon *et al.*, 1983; Fukusho *et al.*, 1987; Ghiasi *et al.*, 1987; Inumaru and Roy, 1987; French and Roy, 1990; Gould and Eaton, 1990; Roy *et al.*, 1990; Roy, 1992). The functional features of VP2 correlate well with the structural features of the triskelion motifs on the outer capsid as the sail-shaped VP2 is the most accessible protein on the virion surface. VP2 oligomerises into a trimer when expressed as a recombinant protein (Hassan and Roy, 1999) and the volume calculation of the structure closely matches that of each observed triskelion; thus, each is composed of three VP2 molecules.

Interspersed between the VP2 triskelions but situated more internally are the VP5 globular densities that are also entirely exposed in the virion (Nason *et al.*, 2004). Three molecules of VP5 (~60 kDa) conform to the globular density observed. Since recombinant VP5 in solution also forms trimers, the 120 globular densities per virus particle comprise a total of 360 molecules (Hassan *et al.*, 2001). The structural features of VP5 and its location support its involvement in membrane permeabilizing activity. Indeed, recent studies on the biological activity of VP5 confirm this role unequivocally (Hassan *et al.*, 2001; Forzan *et al.*, 2004). At the sequence level VP5 is essentially divided into an amino terminal coiled-coil domain and a carboxyl terminal globular domain with a flexible hinge region in between (Forzan *et al.*, 2004) (Fig. 2A). VP5 does not have an N-terminal myristoyl group nor is it cleaved autocatalytically like the equivalent proteins of reoviruses and rotaviruses. However, the amino terminus of VP5 could also potentially form two amphipathic helices (helix 1 and helix 2), each with approximately equal separation of hydrophobic and hydrophilic faces followed by a putative fusion peptide (Fig. 2A). Thus, VP5 shares the characteristic features of a fusion protein with an N-terminal membrane-inserting hydrophobic region followed by a long heptad repeat region, thereby emphasising the concept that VP5 could induce membrane destabilization and permeabilization (Purdy *et al.*, 1986). Direct evidence has also demonstrated that VP5 has intrinsic membrane 'pore-formation' ability in keeping with its supposed biological role in penetration (Hassan *et al.*, 2001). The cytotoxicity of VP5 was mapped to the N-terminal sequences that potentially form the amphipathic helices.

Direct evidence for the involvement of VP5 in membrane penetration has been confirmed by a recent study (Forzan *et al.*, 2004). Using an experimental system for cell-surface expression of VP2 and VP5 it was shown that VP5 has the ability to interact with host cell membranes and to induce cell-cell fusion in a manner similar to that seen for the fusion proteins of enveloped viruses (Fig. 2B). VP5 exhibits its membrane-interacting properties only after it has undergone a low-pH triggered activation step, which presumably alters the conformation to fusion-competence. Deletion of the first 40 residues, which encompass the two amphipathic helices, abolished fusion activity and, although the precise mechanism is unknown, it is clear that VP5 can functionally substitute for a typical viral fusion protein. Although VP5 shows the same arrangement of membrane-inserting peptide and heptad repeat region as class I fusion proteins (Colman and Lawrence, 2003), it does not require a proteolytic activation step to render it fully functional. The fusion-peptide analogue is already located at the N-terminus of the native protein. The maturation of the functional homologue of VP5 in the related reovirus however, includes the proteolytic activation of a precursor protein which results in exposure of an internal fusion peptide at the N-terminus of the C-terminal cleavage product, very similar to the situation observed in class I fusion proteins.

In the BTV particle the VP5 protein makes strong contact with the underlying VP7 layer and also with the spike protein VP2 which protrudes above VP5 (Nason *et al.*, 2004). Thus combining the structural and biochemical data, a plausible hypothesis would be that the attachment of VP2 to the receptor and subsequent low pH inside endosomes induce rearrangement and conformational changes in VP2 and VP5 which result in loosening the interactions of VP5 and VP7, and simultaneously allowing VP5 to form a protein layer with intrinsic outside-in curvature. Proteolytic removal or pH-induced conformational change of VP2 then allows the VP5 amphipathic helices to freely interact with membranes and initiate the permeabilization process. This allows core particles that lack both VP2 and VP5 proteins to be released into the cytoplasm from the endosomal vesicles.

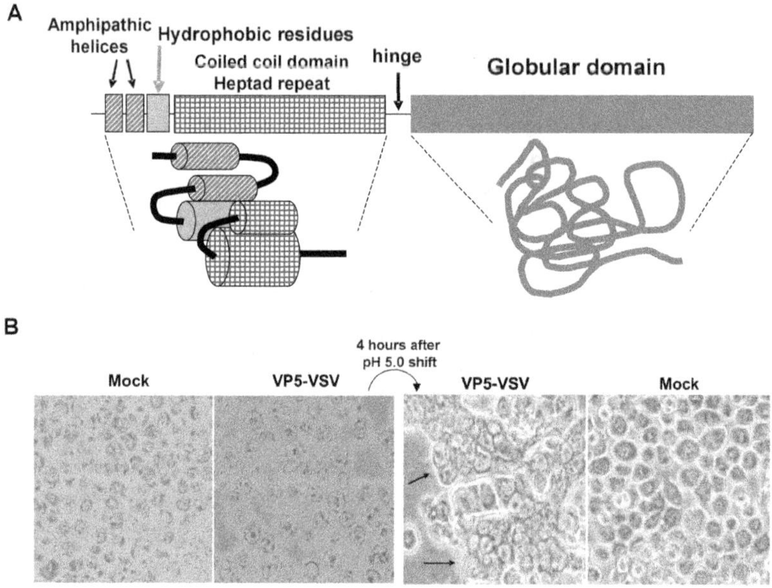

Fig. 2. Arrangement of VP5 domains and fusion activity. (A) Schematic organization of VP5 protein showing the predicted positions of two amphipathic helices, coiled-coil domain and the globular domain that are linked by a hinged region. (B) Confocal microscopy showing the fusion activity of VP5 in *Sf*9 cells after low pH shift (lower panel) and no fusion activity before the pH shift (upper panel). Note the formation of syncytium in the lower panel.

STRUCTURE OF CORE AND ITS COMPONENTS

THREE-DIMENSIONAL STRUCTURE OF CORE

Unlike the whole virion particle, the BTV core appears as an icosahedral capsomeric structure by negative staining. Considerable information is available on core structures and individual proteins due to both Cryo-EM analysis and atomic resolution X-ray crystallography studies (Prasad *et al.*, 1992, Grimes *et al*, 1997, 1998). The 470S core with a diameter of 70 nm consists of two concentric protein layers surrounding a central density. The surface layer which is made up of clusters of 260 VP7 (38 kDa) trimers, conforms to an icosahedral symmetry with a triangulation number of 13 (T=13) in a left-handed configuration and occurs between 260 Å and 345 Å from the centre of the core. The VP7 trimers sit on a thin layer formed by VP3 (100 kDa) and are easily visible as protrusions by negative staining of the particles under EM, giving a bristly appearance to cores described by Els and Verwoerd as capsomers arranged in ring-like structures (Els and Verwoerd, 1969). These ring structures were alluded to in the genus name 'Orbis' (ring or circle). These trimers are arranged around 132 distinctive channels as six-member rings, with five-member rings at the vertices of the icosahedrons (Fig. 3). The organization of the five quasi-equivalent trimers (T13) that form a protomeric unit (P, Q, R, S, T) could be seen clearly. The aqueous channels that are surrounded by these trimers can be grouped into three types; type I channels run along the icosahedral fivefold axes, type II channels surround the fivefold axes, and the type III channels are located around the icosahedral threefold axes. Some of these channels extend inward, penetrating to the inner layer. The threefold axis of the trimer (85 Å long) is perpendicular to the core surface, and the broader base of the trimer (65 Å) contacts the inner shell. The channels or 'pores' are small (7 Å diameter) at the icosahedral threefold axes (type III) and slightly larger (9 Å diameter) at the fivefold axes (type I). The latter act as 'portals' to release the newly synthesized transcripts. While the 'T' trimers of VP7 sit tightly on the 'pores' at the three-fold axes, the 'P' trimers are located loosely above the pores at fivefold axes.

3D STRUCTURE OF VP7

The VP7 trimer reveals a unique molecular architecture. Each monomer has 2 domains, an outer (distal from core centre) and an inner broader domain. The outer domain of one monomer is twisted such that it rests upon the inner domain of a threefold related next monomer in a clockwise direction (Basak *et al.*, 1992, 1996, 1997; Grimes *et al.*, 1995, 1998). The interactions between subunits are extensive, both between the 'outer' and 'inner' domains involving hydrophobic as well as specific hydrogen bond interactions. The smaller upper domain of VP7 forms the 'head region' of the trimer which appears as a knobbly projection on the outer surface of the core by Cryo-EM and contains the central one-third of the polypeptide chain of the molecule (aa 121-249) that is folded into an anti-parallel β-sandwich. The three monomers of this domain form a very stable trimer. Two other X-ray structures of VP7 outer domains, one from BTV-10 (aa 123-253) and one from AHSV-4 that were solved subsequently, also revealed a similar structural arrangement indicating that it is likely that the VP7 of all orbiviruses have similar structures (Basak *et al.*, 1996). The inner, broader domain is composed of nine α-helices and long extended loops and is formed by both the N (aa 1-120) and the C (aa 250-349) terminii of the molecule. The inter-junction region between the two domains is exposed, and is easily accessible to proteolysis. A single lysine residue of BTV-10 VP7 is situated at this junction between the two domains. Mutation at this site has a deleterious effect on core assembly (Le Blois and Roy, 1993) and indicates that the interactions between VP7-VP7 monomers are rigid and can easily be destabilized by a single point mutation at key positions (Le Blois and Roy, 1993; Limn *et al.*, 2000). Despite this, the low resolution VP7 structure (5.4 Å) indicated a very different orientation of the helical inner domain from that of the high-resolution structure suggesting an 'unwound' or 'open' form compared to the tightly packed helices found in the high resolution form (Basak *et al.*, 1997). Two radically different structures indicate that VP7 has the potential for some internal flexibility in the lower domain that attaches to the VP3 scaffold and suggests that substantial conformational change in VP7 occurs at some stage in the viral life cycle.

The interaction between various VP7 trimers and the underneath VP3 layer is extensive and relatively tight. The inner domains of VP7 pack down against six corners of each VP3 molecule such that twelve monomers of VP7 (4 trimers) are directly in association with the VP3 molecules. However, the thirteenth monomer of VP7 (organized in the icosahedron T=13) is not directly in contact with the VP3 molecule, but is raised slightly and in contact with monomers of adjacent trimers.

3D STRUCTURE OF VP3

The VP3 shell that surrounds the inner genomic RNA and the three minor proteins serves as a scaffold for the deposition of VP7 layer (Fig. 3). The 120 molecules of the VP3 form this scaffold that is organized with icosahedral symmetry, on a pseudo T=2 lattice (Grimes *et al.*, 1998). In each core 120 VP3 molecules are arranged as two sets ('A' and 'B') of 60 subunits (Grimes *et al.*, 1998). Five of the 'A' group of molecules are arranged as a five-pointed star around each fivefold axis. Five additional 'B' group molecules are positioned one between each of the points of the star, but at a further distance from the fivefold axis. Thus, VP3 molecules form a convex disc shape in which twelve decamers, positioned one at each of the fivefold axes, are arranged end to end, so as to form the complete VP3 shell. The VP3 shell has a maximum diameter of 59 nm, forming a central space with an internal diameter of 38 nm that contains both the genome as well as the minor core proteins.

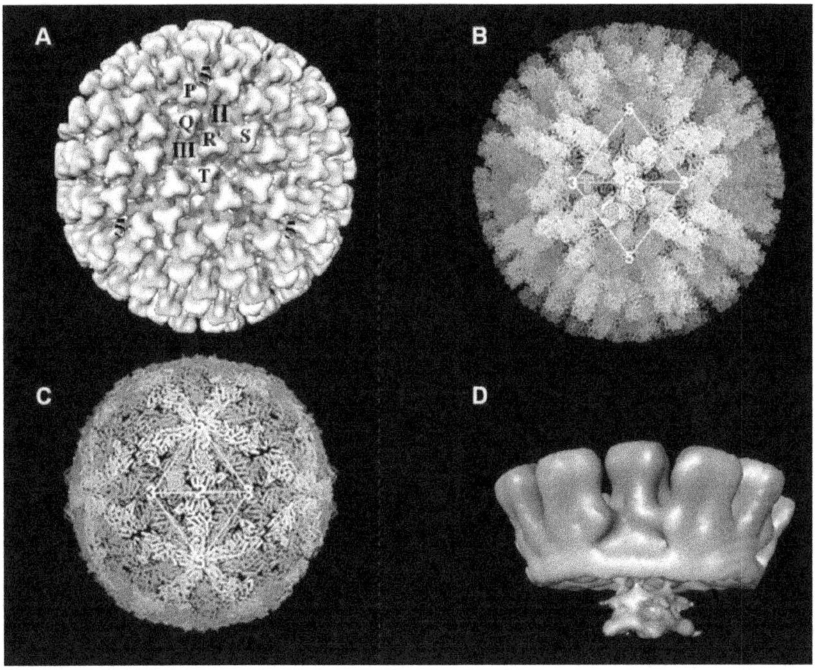

Fig. 3. 3D structure of BTV core. **(A)** Surface representations of the 3D cryo-EM structures of the whole core (700 Å in diameter) viewed along the icosahedral threefold axis showing the trimers of VP7. The five quasi-equivalent trimers (P, Q, R, S, and T) and the locations of channels II and III are marked. **(B)** Surface of the core x-ray crystallography structure showing the arrangement of 260 trimers. **(C)** The structure of the VP3 layer showing the arrangement of 120 copies of two conformationally distinct types of VP3 molecules, 'A' light gray and 'B' dark gray. 60 copies of each of the two molecules together form 12 decamers that are linked together forming a thin protein shell. The icosahedral fivefold, threefold and symmetry axes are indicated. **(D)** Cryo-EM structure of a recombinant core-like particle showing the inside view of the CLP reconstruction with VP3, VP7, VP1 and VP4. A flower-shaped density features attached to the inside surface of VP3 at all the 5-fold axes is clearly seen. (Prasad *et al.*, 1992; Grimes *et al.*, 1998; Nason *et al.*, 2004).

Each 'A' and 'B' VP3 molecule forms a flat, elongated triangular wedge shape (13-35 Å thick x 75 Å wide x 130 Å long) structure with three distinct domains and contains many α helices as well as β strands. The three domains have quite different folds from each other. The 'carapace' domain (aa 7-297, aa 588-698 and aa 855-901) forms a rigid plate that represents the majority of the surface of the subcore shell while the 'apical' domain (aa 298-587) is situated closest to the fivefold axes in the intact particle and the 'dimerization' domain (aa 699-854) is situated furthest away from the fivefold axes. This domain forms quasi twofold interactions between individual 'A' group molecules, and appears to be also responsible for the interactions of 'A' and 'B' subunits between different decamers. The inner surface of the VP3 shell has relatively few charged residues and has a series of shallow grooves.

STRUCTURE OF THE RNA GENOME

The X-ray structure of the core exhibited electron density distributed throughout the internal space of the core within the VP3 shell as had been visualized by Cryo-EM analyses of BTV and other dsRNA viruses (Prasad *et al.*, 1992, 1996; Shaw *et al.*, 1996; Dryden *et al.*, 1998; Grimes *et al.*, 1998; Gouet *et al.*, 1999). The density map showed features consistent with the layers of RNA organized in a highly ordered fashion (Gouet *et al.*, 1999). Approximately 80% of the entire genome of 19,219 base pairs (Fukusho *et al.*, 1989; Roy, 1989) can be modelled as four distinct concentric layers that have center to center spacing between RNA strands of 26-30 Å. As a consequence of the spacing of the 'A' and 'B' forms of VP3, the grooves in the inner surface of the subcore shell form a spiral around the fivefold axis. The outer dsRNA layer appears to interact with these grooves. It is uncertain how the RNA molecules pass inwards to the next and subsequent concentric shells. The density detected in the inner layers of RNA becomes progressively weaker, suggesting that its organization and icosahedral symmetry is progressively less rigid, although each layer maintains an overall spiral organization (Gouet *et al.*, 1999).

THE ARRANGEMENT OF INTERNAL MINOR PROTEINS

The structure of the enzymatic minor proteins, that are closely associated with the viral genome, could not be deduced from the crystal structure of the BTV core (Grimes *et al.*, 1995; Gouet *et al.*, 1999). However, recently by examining the structures of recombinant core-like particles (CLPs) of BTV containing the internal proteins but lacking the viral genome, it has been possible to obtain the overall arrangement of two of the three minor internal proteins, the viral polymerase protein VP1 and the mRNA capping enzyme, VP4. Cryo-EM reconstructions of CLPs with and without the internal VP1 and VP4 proteins revealed a flower-shaped density formed by these two enzymes within the VP3 shell of particles generated with VP1, VP3, VP4 and VP7 (Nason *et al.*, 2004). The VP1 and VP4 flower-shaped density, similar to that of the rotavirus particle, was clearly observed directly beneath the icosahedral 5-fold axes and attached to the underside of the VP3 layer (Prasad *et al.*, 1996). Although the individual contribution of VP1 and VP4 to the flower-shaped density was difficult to delineate from the reconstruction of CLPs, based on recent results from reovirus (Zhang *et al.*, 2003) it is likely that the central mass is predominantly due to the VP1 protein and that some of the sideward projecting density is due to VP4. By co-immunoprecipitation studies it was further confirmed that VP1 and VP4 not only interact with one another strongly but also interact with VP3. Such close interactions between the polymerase and the capping enzyme may be required for capping the nascent mRNAs as they emerge from the polymerase.

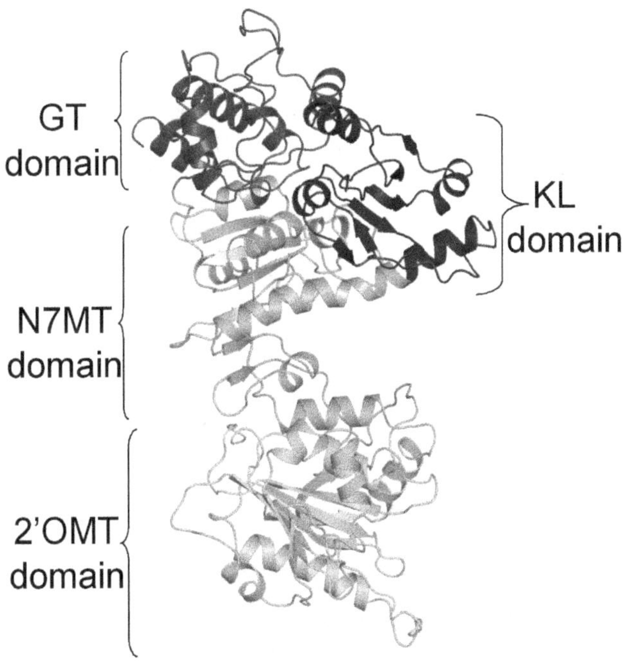

Fig. 4. Cartoon representation of the structure of the VP4 molecule. Ribbon diagram of the VP4 structure with enzymatic domains: guanylyl transferase (GT), kinase-like guk domain (guk), guanine-N7-methyltransferase (N7MT) and nucleoside-2'-O-methyltransferase (2'OMT) domain (Sutton et al., 2007).

The location of the polymerase complex is consistent with the model proposed for genome organization in BTV suggested from X-ray crystallographic studies (Gouet *et al.*, 1999). It is noteworthy that in the presence of magnesium ions and NTP substrates the BTV core-associated enzymes are activated and conformational changes can be seen around the fivefold axis of the BTV core. This results in outward movement of VP3 and VP7 allowing opening of a pore in the VP3 layer at the fivefold axes (Gouet *et al.*, 1999). The mRNAs synthesized by the core are extruded through this pore and subsequently pass out through the central space in the pentameric rings of VP7 trimers ('P' trimers). There are also pores in the VP3 layer at the threefold axes but in the intact core particle these are completely blocked by the 'T' trimer of VP7 as discussed above.

Although the location of the polymerase complex has been unequivocally demonstrated only in BTV and rotavirus using recombinant particles, similar density is seen in the reconstructions of several other dsRNA viruses including viruses from outside the *Reoviridae* family (Reinisch *et al.*, 2000; Pesavento *et al.*, 2001; Zhou *et al.*, 2001; Nakagawa *et al.*, 2003). Thus, in all these viruses the overall structural mechanism of endogenous transcription is likely to be very similar. The unique icosahedral organization of the innermost layer, found in the majority of the dsRNA viruses including those of bacterial (Butcher *et al.*, 1997) and fungal origin (Naitow *et al.*, 2002), may be a critical requirement for the appropriate positioning of the transcription enzyme complex and genome organization. In viruses with multiple dsRNA segments a consensus model consistent with the available structural and biochemical data is that each segment is associated with a transcription complex at the 5-fold axis to allow for its independent and simultaneous transcription.

THE ENZYMATIC ACTIVITIES OF MINOR PROTEINS THAT FORM THE TRANSCRIPTION COMPLEX

The final product of disassembly for all members of the *Reoviridae* is the transcriptionally active core structure that initiates the transcription of genomic RNAs. It is presumed that the overall strategy of BTV and related orbivirus mRNA transcription is similar to that of reoviruses and rotaviruses. The newly synthesized single-stranded RNA species are released into the cytosol and in turn serve both as templates for viral dsRNA genome synthesis and also act as mRNAs for the synthesis of viral proteins within the cytoplasm. In infected cells BTV transcripts are not produced at equimolar amounts from the 10 segments: the smaller genome segments are generally the most frequently transcribed, although certain segments (e.g., segment 6 RNA) are synthesized more abundantly while other in lesser amount (the smallest segment, segment 10 RNA). The molar ratios of the 10 different BTV mRNAs remain the same throughout the infection cycle. The ratios of mRNAs synthesised *in vivo* and *in vitro* are similar.

The newly synthesised mRNA species that are extruded from the core particles through the channels, are capped and methylated, and program the synthesis of virus proteins in the cytoplasm. *In vitro*, BTV mRNA synthesis can be activated by proteolytic digestion of the outer capsid (Van Dijk and Huismans, 1980, 1988). The structural integrity of the core particle appears to be essential for maintaining an efficient transcriptional activity. The delineation of the specific roles of the proteins of the transcription complex within the core, as summarized below, has been derived from the individual purified recombinant proteins.

HELICASE ACTIVITY

Transcription of the dsRNA genome by members of the *Reoviridae* occurs by a fully conservative process, i.e. dsRNA is separated, nascent RNA synthesis is initiated and the newly synthesised RNA strand is separated from the template allowing the parent strands to re-anneal (Bannerjee and Shatkin, 1970). It is logical that this process would involve a helicase protein either to unwind the dsRNA ahead of the transcriptase protein, or to separate the parental and newly synthesized RNAs following transcription. BTV is the only member of the *Reoviridae* in which helicase activity has been located to a minor structural component (VP6, 35.7 kDa) of the core (Stauber *et al.*, 1997). For reovirus this activity is associated with the major structural protein λ1 (Bisaillon *et al.*, 1997), but for rotavirus no similar activity has been described yet. It was found that incubation of purified BTV VP6 with dsRNA in the presence of ATP leads to the unwinding of the dsRNA template. Unusually, this activity was detected not only for templates which were complete duplexes but also for those which had short 5' or 3' overhangs (Stauber *et al.*, 1997).

Most models explaining helicase activity require two features: ATP driven conformational rearrangements of the protein that, in turn, generate kinetic energy; and possession of at least two nucleic acid binding sites in order to maintain contact with nucleic acid (Lohman *et al.*, 1998). This requirement seems indispensable for the unwinding reaction, since it allows the protein to bind both single strand and duplex regions of a polynucleotide simultaneously, or two complementary single strands in an unwinding fork. The majority of known helicase proteins function as a hexamer. VP6 also exists as a hexamer in solution (Kar *et al.*, 2004). Furthermore, incubation of purified VP6 with RNA resulted in the formation of protein-RNA complexes that are visible as a ring-like structure by electron microscopy (Kar *et al.*, 2004). By sequence analysis it was possible to identify regions of VP6 that have homology to the nucleic acid binding and ATPase domains of known helicase proteins (Kar *et al.*, 2004). These comparisons provided a basis for mutagenesis and the data obtained demonstrated that mutation of the putative RNA binding motif at the C terminus of VP6 (R205Q VP6) was sufficient to severely reduce RNA binding. Similarly, the mutation of the conserved lysine in the putative ATPase domain (AxxGxGK110V) was found to reduce the ability of VP6 to bind ATP. Unsurprisingly, the same mutant was unable to convert ATP to ADP. Helicase activity was altered in both mutants.

RNA DEPENDENT RNA POLYMERASE ACTIVITY

The replication of viral dsRNA occurs in two distinct steps. First, plus strand RNA (mRNA) is transcribed, using the dsRNA genome segments as templates, and extruded from the core particle (Bannerjee and Shatkin, 1970; Cohen, 1977; Van Dijk and Huismans, 1980). Second, the plus strand RNAs serve as templates for the synthesis of new minus strands at an undefined stage in the assembly of a new virus core particle. Cis-acting sequences required for the viral RNA dependent RNA polymerase (RdRP) to replicate the plus strand RNAs are believed to be located at their 3' end (Chen *et al.*, 1994; Patton *et al.*, 1996; Chen and Patton, 1998).

Studies of replicase activity in other members of the *Reoviridae* have utilized a particulate replicase, in which the catalytic subunit is proposed to possess replicase activity only in the context of a subviral particle (Patton *et al.*, 1997; Chen and Patton, 1998; Chen *et al.*, 1994, 2001; Tortorici *et al.*, 2003). Purified rotavirus VP1 has been shown to bind to the 3' end of *in vitro* synthesized rotavirus plus strand RNA, but replicase activity was only detected when the remaining core proteins were present or in the context of subviral particles (Patton, 1996; Patton *et al.*, 1997). Purified recombinant reovirus λ3 has been shown to be a poly(C)-dependent poly(G) polymerase, but does not possess replicase activity (Starnes and Joklik, 1993). The processive replicase activity of reovirus λ3 has only been reported in λ3 crystals containing short oligonucleotide templates (Tao *et al.*, 2002). It has been suggested that the generation of a functional viral replicase only occurs within the shell of the assembling virus core in order to avoid the triggering of an antiviral response, induced via dsRNA surveillance pathways.

The largest viral protein, VP1, with a Mr of 149,588 Da, is a highly basic protein and is present in a low molar ratio in the virion (Roy *et al.*, 1988). Based on its size, location, molar ratio in the core and predicted amino acid sequence, this protein is the prime candidate for the virion RdRP. The primary sequence of VP1 contains a GDD motif (position 287 to 289), characteristic of other RNA polymerases (Urakawa *et al.*, 1989). In contrast to rotavirus, in the absence of other viral proteins, soluble recombinant BTV VP1 exhibited a processive replicase activity synthesising complete complementary RNA strands of *in vitro* synthesised BTV ssRNA templates (Boyce *et al.*, 2004). Moreover, in addition to elongating the nascent minus strand, the synthesis of a unit length labelled strand was achieved and thus demonstrating that purified BTV VP1 initiates minus strand synthesis *de novo*, rather than extending a fortuitous secondary structure present at the 3' end of the plus strand template. It is unlikely that there is a fundamental difference in the way that core assembly and genome replication occur in different genera of the *Reoviridae* and this apparent discrepancy may be attributable to the varied experimental systems used to assay replicase activity. Since replicase activity associated with recombinant BTV VP1 is low, it implies that the activity of VP1 may be modulated by other viral proteins present in the assembling core particle.

In the absence of other core components BTV VP1 exhibited little sequence specificity for BTV 'plus' strand template suggesting that specificity for viral template comes from location within the viral core. The specificity of the replicase activity for virion 'plus' strands may be determined by other viral core proteins or non-structural proteins, or the specificity may occur during packaging of the 'plus' strands, without further discrimination occurring during replication.

The latter possibility seems unlikely. It seems more likely that the replicase activity of VP1 is modulated *in vivo*, as has been shown *in vitro* with hepatitis C polymerase and the viral proteins NS3, NS4A, and NS5A (Piccininni *et al.*, 2002; Shirota *et al.*, 2002). Such modulation may take the form of inhibition of the replicase activity of soluble VP1 in the cytosol, in order to prevent the formation of free dsRNA. Template specificity may be added to the replicase activity of VP1 by viral proteins present in the assembling virus core, as shown in a reconstituted influenza virus replicative transcription complex (Lee *et al.*, 2002). After dsRNA synthesis is complete viral core proteins may trigger the switch from replicase activity to further transcriptase activity.

However, much less is known about the *in vivo* RNA replication mechanisms of BTV. It is believed that, like reoviruses and rotaviruses, the packaged 'plus' strand RNA serves as a template for synthesis of a 'minus' strand, and once the 'minus' strand is synthesized, the dsRNA remains within the nascent progeny particle.

To support further that VP1 alone is the virus polymerase, a model structure of VP1 has been recently generated (Wehrfritz *et al.*, 2007). The model revealed a typical RdRP organisation that consists of three distinct domains, an N-terminal domain (NTD), a C-terminal (CTD) domain and a central polymerase domain (PD). Further by mutagenesis it was demonstrated that the signature catalytic GDD motif (aa 763-765) within the PD is indeed the active domain of the protein. It is therefore highly likely that this fragment will be folded into a typical polymerase structure as the modelling data predicts. When each of the three domains were synthesised individually, only the purified PD fragment but not the NTD or CTD fragment was able to bind NTP similar to that of the full-length protein. None of the domains had catalytic activity when tested individually or in tandem but when all three were mixed together *in vitro*, the RdRP activity was reconstituted. It is possible that although PD possesses the catalytic activity, the other two domains are needed to stabilise the protein. It is noteworthy that in reovirus λ3, the large N and C-terminal domains form a cage around the centrally placed polymerase domain holding it rigid and preventing movement during the catalytic cycle (Tao *et al.*, 2002). It is likely therefore that the VP1 and λ3 share similar structural organisation as seen in the VP1 model.

BTV mRNA CAPPING ACTIVITY

The transcripts produced by BTV cores have 'cap' structures at their 5' termini. All the catalytic activities necessary for a 'cap' 1 structure formation at the 5' end of BTV transcripts are performed by a single minor protein, VP4 protein (76.4 kDa) (Martinez-Costas *et al.*, 1998; Ramadevi *et al.*, 1998b; Ramadevi and Roy, 1998). The soluble protein VP4 possesses RNA 5'-triphosphatase activity which catalyses the first step in the RNA-capping sequence and covalently binds G via a phosphoamide linkage, a characteristic feature of guanylyltransferase (GT) enzyme. VP4 also catalyses a GTP-PPi exchange reaction indicating that protein is the GT of BTV. Additionally it has the ability to catalyse the conversion of unmethylated GpppG or *in vitro*-produced uncapped BTV RNA transcripts to m7GpppGm in the presence of S-adenosyl-L-methionine demonstrating that VP4 is also a methyltransferase enzyme. Analysis of the methylated products identified both methyltransferase type 1 (N7MTase) and type 2 (2'OMTase) activities associated with VP4, thus the complete BTV cap reaction is associated with this single core protein. These data are particularly significant as the capping activities associated with other viruses often require more than one protein. VP4 therefore represents a model enzyme, which has both methyltransferase and GT in addition to RNA triphosphatase activities in a single protein that is distinct from the viral RdRP. Purified VP4 is dimeric in solution, and the dimeric form is essential for integration into the CLP (Ramadevi *et al.*, 1998a). How the VP4 dimer is formed and how VP4 can perform multiple enzymatic activities consecutively became more clearer from a recent atomic structure of the protein which has been solved at a 2.5 Å resolution (Sutton *et al.*, 2007). The protein has an elongated structure with an array of various active sites organised in a manner to allow the sequential catalytic activities required to produce the 'cap' structure on a nascent RNA molecule. The polypeptide chain folds into four discrete domains, delineating each enzymatic reaction. The N-terminal 108 residues form a kinase-like (KL) domain, and a 30 residue α-helix then leads into the N-terminal part of the N7MTase domain. The polypeptide chain then forms the entire 2'OMTase domain before completing the N7MT domain and finally forming the putative GT domain. The KL, N7MT and 2'OMT domains are mixed α/β structures with a central β sheet and associated helices. The C-terminal GT domain is almost entirely α-helical and includes a number of relatively poorly ordered loops. There are extensive interaction surfaces between some of the domains, as shown in Fig. 4. The GT and

N7MT domains are closely associated, whilst the KL domain is reasonably strongly associated with both of these domains (indeed it is linked to the N7MT domain by a continuous α helix). In contrast, the 2'OMT domain is rather flexibly attached to the remainder of the molecule. These interactions are reflected in the flexibility of the molecule as judged from the flexing of domains observed in different crystals of VP4.

In summary, each of the three minor proteins of the BTV core has the ability to function on its own. Together they constitute a molecular motor that can unwind dsRNA templates, synthesise ssRNAs of both polarities and modify the 5' termini of the newly synthesised mRNA molecules.

PROTEIN SYNTHESIS AND REPLICATION

In tissue culture the first BTV-specific proteins are detectable 2h post-infection, and the rate of protein synthesis increases rapidly until 12 to 13h post-infection after which it slows down but continues until cell death (Huismans, 1979). Most viral proteins are synthesized throughout the infection cycle and accumulate during infection until cell lysis. Infection of mammalian cells with BTV, in contrast to insect cells, leads to a rapid inhibition of cellular macromolecular synthesis and the induction of a robust apoptotic response triggered via multiple apoptotic pathways (Mortola *et al.*, 2004).

As mentioned earlier, three NS proteins are also synthesised in infected cells in addition to the seven structural proteins. Two of these NS proteins, NS1 (64kDa) and NS2 (41kDa) are synthesized abundantly early in BTV infection and coincide with two virus-specific intracellular structures, tubules and viral inclusion bodies (VIBs), respectively. By contrast, synthesis of NS3 and related NS3A (26 and 25 kDa) varies from barely detectable to highly expressed, depending on the host cells, and may correlate with virus release. Current data suggest that NS proteins are necessary either for virus replication or during morphogenesis.

NS1 PROTEIN AND ITS ASSEMBLY AS TUBULES IN INFECTED CELLS

The 64 kDa protein NS1 of BTV is synthesized in abundance, up to 25% of the virus-specified proteins, in virus infected cells and readily assembles as tubular structures within the cytoplasm that are biochemically and morphologically distinct from the microtubules and neurofilaments of normal cells (Huismans and Els, 1979). Tubules are characteristic features not only in BTV, but in all orbivirus infection. These high-molecular-weight tubular structures are formed by helically coiled ribbons of NS1 dimers with a diameter of 52 nm and lengths of up to 1000 nm. The early and abundant synthesis of NS1 and tubules suggests their involvement in virus replication and/or virus translocation.

In a recent study (Owens *et al.*, 2004) intracellular expression of a single chain antibody to the viral NS1 protein (scFv-αNS1) demonstrated that high levels of NS1 in BTV infected cells are critical to the morphology and trafficking of virus particles. Four major changes were apparent: first, there was a tremendous reduction in virus-induced cytopathic effect; second, there was a more than 10-fold increase in the amount of virus released into the culture medium; third, there was a shift from lytic release of virus to budding from the plasma membrane; and forth, NS1 tubule formation was completely inhibited by scFv-αNS1 expression. Each of these changes, except for the lack of tubule formation, is reminiscent of what occurs during BTV infection of insect cells in culture. Based on these findings it can be hypothesised that the NS1 protein is probably a determinant of pathogenesis in the vertebrate host, and that its mechanism of action is the augmentation of virus-cell association (but not transport of virus to the cell surface), which ultimately leads to lysis of the infected cell. It is possible that NS1 is somehow involved in virus trafficking. Differential virus release in different cells suggests the involvement of host proteins. Indeed a yeast two-hybrid analysis has identified NS1 interaction with the cellular protein SUMO-1 (Noad and Roy, unpublished data). This protein is conjugated to cellular targets as a post-translational modification involved in the intracellular trafficking. Thus, it may be NS1 is involved the trafficking of immature or mature virus particles in infected cells through the interaction with cellular proteins such as SUMO-1.

VIRUS ASSEMBLY SITE, INCLUSION BODIES AND ROLE OF NS2

For many viruses assembly intermediates accumulate in specific locations within the host cell in structures described as 'virus assembly factories' or 'virus inclusion bodies'. Also in BTV infected cells, shortly after release from the endosome, core particles become very rapidly associated with a matrix that gradually surrounds the particles to form virus inclusion bodies (VIBs). Early in infection the VIBs appear as granular material scattered throughout the cell, which then coalesce to form a prominent inclusion with a perinuclear location (Hyatt and Eaton, 1988; Eaton and Hyatt, 1989; Brookes *et al.*, 1993; Kar *et al.*, 2007). As the infection progresses, these VIBs increase both in size and number (Eaton *et al.*, 1988; Hyatt and Eaton, 1988; Brookes *et al.*, 1993).

The 41 kDa NS2 is synthesized at high levels early in BTV infected cells and the predominant component of VIBs (Browne, 1967; Hyatt and Eaton, 1988; Eaton and Hyatt, 1989; Thomas *et al.*, 1990; Brookes *et al.*, 1994). In addition, the newly synthesized viral transcripts, all four subcore proteins (VP3 and 3 enzymatic proteins) as well as assembled cores and subcores can also be identified within these VIBs; thus, they appear to be the site of BTV replication (Hyatt and Eaton, 1988). Expression of phosphoprotein NS2, without other viral proteins in both insect and mammalian cells results in the formation of inclusion bodies that are indistinguishable from the VIBs found in virus-infected cells indicating that VIBs are formed mainly by NS2 (Modrof *et al.*, 2005; Thomas *et al.*, 1990). Their make-up is dependent only on a full-length NS2 with the C terminus of the protein being the major functional domain involved (Zhao *et al.*, 1994; Kar *et al.*, 2007).

Co-expression of core structural proteins has shown that while VP1, VP3, VP4, and VP6 all co-localise with NS2 inclusions without other viral proteins, the surface core protein VP7 requires the co-expression of VP3 to be recruited to these structures. This may be an early indication that partially assembled core structures in which VP3 and VP7 are already interacting, are the building blocks of core formation. Further confocal microscopic studies using recombinant NS2 and virion proteins in different combination confirmed the association of NS2 with the structural proteins and assembling particles (Modrof *et al.*, 2005; Kar *et al.*, 2007). These data together with the observation of the presence of electron-dense virus-derived immature particles inside VIBs has led to the belief that core particles may be synthesized within VIBs. It is likely that the further stages of mature virion development occur either in the cytosol or after binding of the particles to the cytoskeleton. The data emphasise the role of NS2 VIBs in capsid assembly.

NS2 is the only virus-specific protein that is phosphorylated in BTV-infected cells although the significance of the modification is not known (Modrof *et al.*, 2005). Generally, phosphorylation has been shown to play a role in the multimerization and protein-protein interactions of a number of viral proteins (Montenarh and Muller, 1987; Gao and Lenard, 1995; Spadafora *et al.*, 1996). However, recent data suggests that while the phosphorylation of NS2 is important for VIB formation, it is not responsible for interaction with other viral proteins (Modrof *et al.*, 2005).

Viruses with a segmented RNA genome face a challenging task in the recruitment and assortment of specific viral-coded RNA species in the infected cells. How are the 10-12 RNA segments specifically recruited and transported to the virus replication and assembly sites within the cytoplasm and are any specific sequences involved? It is commonly believed that each of the viral segments must possess specific sequences or structures, which are recognized by one or more virally encoded proteins to facilitate these processes. NS2 has a strong affinity for ssRNA but not for dsRNA suggesting it may have a role in the recruitment of RNA for replication (Thomas *et al.*, 1990). The activities of NS2 (formation of VIBs, phosphorylation, RNA-binding) suggest that it might be involved in recruiting and packaging of BTV ssRNA prior to encapsidation. Moreover, BTV RNAs are preferentially bound over non-specific RNAs via the amino terminal residues of the protein supporting that NS2 may have a role in the recruitment of mRNA for replication (Theron and Nel, 1997; Lymperopoulos *et al.*, 2003; Markotter *et al.*, 2004). These observations are significant since the mechanism by which the 10 dsRNA segments, that make up the viral genome, are selectively recruited and packaged into newly assembling virus particles is unresolved and represents one of the most enduring questions in the field. Recent data suggested that NS2 has a high affinity for specific BTV RNA structures that are unique in each RNA segment (Lymperopoulos *et al.*, 2003, 2006). The sequences that represent the putative binding partners for NS2 are neither the conserved 5' terminal octanucleotides nor 3' terminal hexanucleotides of all BTV RNA segments or the potential panhandle structures formed by the partial complimentary sequences of the 5' and 3' termini. Instead, these are distributed throughout the coding and non-coding regions of the different genome segments (Lymperopoulos *et al.*, 2003). Chemical and enzymatic structure probing of regions bound preferentially by NS2 has shown that the NS2 binds BTV RNA transcripts at unique hairpin-loops. The NS2-hairpin interaction has been further confirmed by using hairpin mutants which alter the sequence or the structure of the RNA independently. At this stage, the significance of these interactions for genome packaging is unclear and it is worth pointing out that, other than BTV, there are several systems where viral proteins recognise packaging signals that are contained in hairpin structures with no apparent sequence homology (Bae *et al*., 2001; Beasley and Hu, 2002). Although the RNA binding activity of NS2 can explain how BTV mRNAs are selected from the pool of cellular messages for incorporation into assembling virus particles, it remains unclear how only a single copy of each genome segment is included in newly formed core particles. NS2 could bring the viral RNAs into close proximity, and inter-segment RNA interactions may allow the formation of an RNA complex that is the basis of core assembly. Alternatively, different RNA subsets that have already interacted may be bound by NS2. The latter hypothesis is consistent with the observations that NS2 may form decameric complexes (Butan *et al*., 2004), and that each NS2 protein subunit may have several RNA binding domains (Fillmore *et al*., 2002). Interestingly, NS2 has phosphohydrolase activity, and could bind and hydrolyse ATP and GTP to their corresponding NMPs with a higher efficiency for the hydrolysis of ATP in the presence of divalent cations (Horscroft and Roy, 1997). The NTPase activity described for NS2 may play a role in providing energy for the assortment, movement, packaging or condensation of bound ssRNA similar to that found in bacteriophage ϕ6 P4 protein (Pirttimaa *et al*., 2002; Poranen and Tuma, 2004).

Although inclusion bodies have obvious advantages in providing a site for core assembly and genome packaging as well as preventing premature shut-off of core transcription by preventing the addition of viral outer capsid proteins, they may also represent a ‘trap’ for newly assembled core particles. In order for newly assembled core to be released from inclusions it is necessary that these structures must be dynamic and easily disassembled as well as assembled. Recent data suggest that protein kinase casein kinase II (CKII), interacts with NS2 and is able to phosphorylate it via two serine residues, aa 249 and 259 (Modrof *et al*., 2005). Mutagenesis of these residues to alanine in NS2 abrogates the formation of distinct VIBs whilst mutation of the same residues to aspartic acid, mimicking a constitutively phosphorylated state, results in the formation of normal inclusion body structures (Modrof *et al*., 2005). Furthermore the expression of the non-phosphorylated mutant of NS2 was sufficient to disrupt the formation of VIBs in virus-infected cells, demonstrating the potential of the mutant to act in a dominant negative fashion (Modrof *et al.*, 2005). The sequence context of the phosphorylated serine 249 (248LSDDDDQ254) in NS2 conforms to a conserved sequence which generally stabilises helix unfolding (Zetina, 2001) suggesting that phosphorylation of NS2 in this position could be involved in stabilising protein folding within its C-terminal domain. Thus, phosphorylation may act to control the propensity of NS2 to form large viral inclusion bodies and thus provide a mechanism by which newly assembled core particles can be released from VIBs. The newly assembled cores that are released from VIBs would then be free to acquire the outer layers (VP5 and VP2) outside of the VIB structure. It is noteworthy that none of the two outer capsid proteins are associated with NS2 or NS2 IBs.

CAPSID ASSEMBLY PATHWAYS

Viral capsid assembly is a precise process that is fundamental to the success of the viral infection. In recent years the assembly of the virion particles, particularly the inner capsid or core of various members of the *Reoviridae* family has been a major focus at both the structural and molecular levels. The assembly of multilayered viral cores is especially intriguing as both protein-protein and protein-RNA interactions are required to complete the core particle.

Considerable progress has been made in recent years towards understanding the role of specific residues necessary to preserve the stable BTV capsid structure and of the intermediates of the assembly pathway that achieve the final spherical icosahedral structure. This has been mainly possible due to the generation of virus-like particles (VLPs) and CLPs that are produced by co-expression of BTV proteins using the baculovirus expression system (French *et al.*, 1990; French and Roy, 1990). Assembly of BTV subviral particles by recombinant proteins in different combinations have given us the first indication of how the four major capsid proteins are assembled (Fig. 5). These assay systems, in combination with extensive site-specific and deletion mutagenesis of selected capsid proteins, have mapped the essential assembly pathway of the capsid proteins, particularly the assembly of VP3 and VP7 and their compatibility between related orbiviruses.

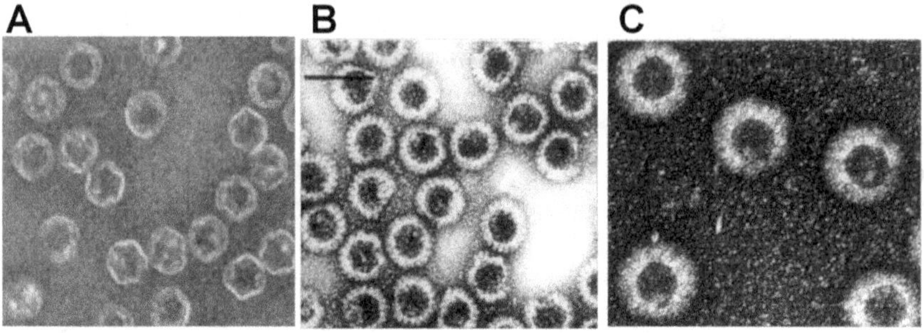

Fig. 5. Baculovirus expressed BTV like particles. Negatively stained electron micrographs of VP3 subcore-like particles, VP3 + VP7 core-like particles, and virus-like particles of 3 or 4 proteins. Note each type of particles is essentially empty.

VP3 SHELL PLAYS THE MAJOR ROLE IN CORE ASSEMBLY

VP3 interacts with all the core components and serves as a scaffold for surface protein VP7 trimers indicating that it plays the major role in capsid assembly. The highly conservative nature of the VP3 (100 kDa) sequence across 24 BTV serotypes as well as considerable similarities with the sequence of other related Orbiviruses highlights the structural constraints that may be essential for virus assembly (Iwata *et al.*, 1992; Roy, 1996). The functional importance of the VP3 layer is emphasised further by the conservative nature of the overall architecture of this shell, which is shared by all members of the *Reoviridae* as well as other viruses with segmented dsRNA genomes (Hewat *et al.*, 1992b; Prasad *et al.*, 1992, 1996; Grimes *et al.*, 1998; Reinisch *et al.*, 2000). In the absence of any other core protein, VP3 alone can assemble into an icosahedral subcore-like structure (see Fig. 5) indicating that assembly of the VP3 layer initiates the formation of the core. The equivalent protein of rotavirus (VP2) also shares this feature (Labbe *et al.*, 1991).

VP3 has a particularly high content of hydrophobic amino acids (Purdy *et al.*, 1984), but several internal hydrophobic stretches can be deleted without much effect on CLP formation (i.e. when co-expressed in presence of VP7; Tanaka and Roy, 1994). Moreover, these regions can accommodate insertion of short foreign sequences implying flexibility in VP3 folding. Similarly, deletion of 10 residues at the carboxy terminus did not affect the VP3-VP3 and

VP3-VP7 interaction or formation of CLPs (Tanaka *et al.*, 1995). In contrast, specific single substitution mutations (e.g. methionine at amino acid 500, arginine at amino acid 502) appear to be critical for core assembly emphasizing their critical position in the molecule. Interestingly, these residues are within the 'apical' domain of the molecule, situated nearer to the 5-fold axes in the particle.

The arrangements of the VP3 molecules within the core, as seen in the final structure, showed that the amino terminus of the VP3 A molecule is disordered. This has been hypothesised to accommodate the minor rotation of the apical domain abut the main body of the protein that in turn allows assembly via a proposed decameric intermediate. The crucial role of the amino terminus in BTV core assembly was verified by biochemical analysis of N-terminal deletion mutants (Kar *et al.*, 2004). The deletion of the first 10 residues significantly affected the efficiency of VP3 assembly and its interaction with VP7 trimers indicating that the N terminus is responsible for protein-protein interactions. Similarly, extension of N terminus affected the stability of the core assembly. This is in contrast to other members of the *Reoviridae* where N-terminal residues in a number of homologous proteins proved to be dispensable (Zeng *et al.*, 1998; Kim *et al.*, 2002).

An important question in VP3 assembly is whether the decamer present in the assembled particle is an identifiable intermediate of the assembly process or only arises upon assembly. In a recent study it was shown that deletion of the internal dimerisation domain abolished formation of the subcore but did not perturb the stable decamer formation and its ability to interact with the internal minor proteins (Kar *et al.*, 2004). By contrast, BTV RNAs interacted with the intact VP3 *in vitro* very efficiently, but failed to associate with VP3 decamers under the same conditions. These data suggest that assembly of the BTV core may start with the complex formed by internal proteins and the VP3 decamers, and these assembly intermediates subsequently recruit the viral RNA and possibly VP6 complexes prior to completion of the assembly of the VP3 subcore.

INTRA- AND INTERMOLECULAR INTERACTIONS THAT REQUIRED FOR ASSEMBLY OF VP7 MOLECULES ONTO THE VP3 SHELL

In the core particle 780 VP7 molecules interact with 120 VP3 molecules. The mismatch between the number of molecules in the VP3 and VP7 layers poses an interesting problem as to how these layers reconcile to form an intact icosahedral structure. In the absence of VP3, VP7 forms trimers but these trimers do not assemble as icosahedral particles. In solution, VP7 forms hexamers and pentamers of trimers. Although both VP7 domains contribute substantially to the formation of the trimer, it is the inner helical domain that controls exclusively both the formations of hexamers and pentamers and their interaction with the scaffolding layer of VP3. The precise juxtaposition of the five quasi-equivalent trimers varies slightly in accordance with the requirements of quasi-equivalence. Structural and sequence information have guided an investigation of how such a complex structure is achieved during virus assembly and what residues are required to form a stable capsid (Limn *et al.*, 2000; Limn and Roy, 2003). Extensive site-directed mutagenesis in combination with various biological assay systems have given insight into the order of the assembly pathway of VP7 and stable core formation that subsequently serve as a foundation for the deposition of the outer capsid.

The first set of mutations were targeted at the inner, helical domain that appeared to be involved in intra- (within the VP7 subunit) and inter-molecular (between the VP7 subunits) interactions (Limn *et al.*, 2000). The effects of these mutations on solubility of VP7 as well as its ability to trimerize and form CLPs in the presence of the VP3 scaffold were investigated. Some of these mutations severely affected the stability of CLP, while other mutations had either a lesser or no apparent effect on the formation of the stable capsid. Several mutations caused complete insolubility of the protein implying an acute alteration in the folding of the molecule despite the prediction that such a change would be accommodated. However, all soluble mutants, including those that did not form stable CLPs, assembled into stable trimers, indicating that CLP assembly is not an automatic consequence of trimerization. It is clear that a single substitution at a key position could destabilize BTV core assembly without drastically disrupting the trimer formation. The effects of these VP7 mutations on CLP formation also demonstrated that VP7 trimer formation could not accommodate most of the designed point mutations without any functional effect on capsid assembly (Limn *et al.*, 2000). The instability of the CLPs in these experiments were likely due to the unstable interactions between the various mutant trimers indicating that subtle alterations (such as bonding angle or subunit conformation) within a trimer could have a profound effect on trimer-trimer interaction although formation of stable trimers was still achieved. Thus, the formation of the VP7 lattice on the core surface requires exact fitting of 260 VP7 trimers. In addition, the mutations might also have influenced indirectly the interactions between the VP7 trimers and the VP3 subcore by slightly changing the flat surface of the bottom of the trimer.

The studies also highlight some of the key residues that are crucial for BTV core assembly and illustrate how the structure of VP7 in isolation under-represents the dynamic nature of the assembly process at the biological level.

Since VP7 molecules oligomerise into trimers when expressed alone in the absence of VP3 it is possible that attachment of pre-formed trimers onto the VP3 subcore and the subsequent formation of the VP7 layer would be directed by side-to-side interaction of adjacent trimers. Interactions between neighbouring trimers at the 2-fold axis appear to be mediated through a thin band around the lower domains.

A series of single, dual and multiple substitution mutants of VP7 targeting these regions showed that while single-site mutants could still form reasonably stable CLPs, the double mutations resulted in poor CLP morphology with only ~60 VP7 molecules retained by the VP3 shell in presence of high salt. But in both cases trimer formation per se was not altered suggesting that trimer assembly alone is not sufficient for completion of VP7 attachment to the subcore.

An additional set of substitution mutations was created that consisted of either substitution of sequences by those of a related orbivirus or by non-related sequences, predominantly substituting hydrophobic to charged residues. These changes were designed to maintain the folding pattern of the molecule but were expected to interfere with the protein-protein contacts (Limn and Roy, 2003).

When CLP assembly was examined with each of these VP7 mutants, all those that mimicked AHSV sequence behaved much like BTV and had no detectable effect on overall folding or the molar ratio of VP7 to VP3 in CLPs, emphasising the conservative structural nature of the two molecules. In contrast, the mutations that changed the character of the residue (e.g., hydrophobic to charged or vice versa) resulted in sometimes severe instability of CLPs. CLPs were more stable when isolated in high salt suggesting that solvation of the charged groups can prevent a substantial mismatch between the VP7 and VP3 proteins, while the assembly is driven by the hydrophobic interactions.

In summary, from a detailed analysis of an extensive range of targeted VP7 mutations it was possible to precisely identify the residues that are critical for core assembly and to reveal that assembly of a stable core depends not only on trimer formation, but more importantly on the precise 'shape' of the trimers which is necessary to drive the formation of the tight lattice of 260 VP7 trimers on the core surface. The data also suggest that initially multiple sheets of VP7 form around different nucleation sites instead of a cascade of trimer associations originating from a single nucleation site. Thus, it is likely that a number of strong VP7 trimer-VP3 contacts act as multiple equivalent initiation sites and that a second set of weaker interactions then "fill the gaps" to complete the outer layer of the core. There is a clear sequential order of trimer attachment on the VP3 scaffold. The 'T' trimers, which are at the 3-fold axis of the icosahedron locate first, while 'P' trimers that are further from the 3-fold axis and near to the 5-fold axis, are the last to attach. The distinction between those VP7 trimers that first occupy the subcore and those that follow is intriguingly subtle as it must be sufficient to allow variation in packing yet not prevent the overall biological purpose of virus assembly.

INTERACTIONS BETWEEN THE OUTER CAPSID LAYER AND THE SURFACE VP7 LAYER OF CORE

Once the stable core is assembled, the two outer capsid proteins VP2 and VP5 are acquired by interaction with the core surface, both proteins being attached to the VP7 layer of the core. The assembly of VP2 and/or VP5 to the core is easily mimicked either by baculovirus expression systems (see Fig. 5) or by *in vitro* transcription-translation system (French *et al.*, 1990; Loudon and Roy, 1991; Liu *et al.*, 1992). Both proteins can be incorporated into the nascent particle independent of each other, implying that the two proteins are directly attached to the core and do not intimately interact with each other. Once VP2 and VP5 are attached, the particles are no longer transcriptionally active. Instead, they are available for release from the infected cell.

The structural organization of the outer layer represents a drastic mismatch with the underlying VP7 layer. The VP7 trimers with a triangular shaped top portion serve as the platform for the deposition of the VP2 and VP5. Each triskelion motif of VP2 trimers essentially interacts with four VP7 trimers by their underside in four places (Nason *et al.*, 2004). The base of the triskelion interacts with the Q type VP7, whereas the propeller-like arms make three other connections with the P, R, and S VP7 trimers. By docking the X-ray structure of the VP7 trimer into the Cryo-EM virion reconstruction it was possible to identify the VP7 residues that are in contact with VP2. The base of the VP2 trimer interacts exclusively with the upper flattish surface of the VP7 trimer, which includes amino acid residues 141-143, 164-166, 195-205, and 238-241. All the VP7 molecules of the core are covered by the connections that are made with the tips of the propeller except the VP7 at the icosahedral 3-fold axis (T type). The top of the VP7 trimer at this position is thus clearly exposed to the exterior.

The globular densities (VP5 trimers) on the other hand sit right above the type II and type III channels, between the VP7 trimers, making contacts with the sides of the VP7 trimers which face these channels (Nason *et al.*, 2004). Thus, the channels are filled with triangle-shaped densities that are connected to the VP5 trimers. The inner triangle-shaped density interacts mainly with the lower portion of the β-barrel domain and includes residues 168-173, 210-215, and 226-234 of the distal β-barrel domain of VP7. Consequently, VP5 has much stronger interaction with VP7 which is consistent with the biochemical data that demonstrated that VP2 can be dissociated from the virion easily without removal of VP5 (Verwoerd and Huismans, 1972; Huismans and Cloete, 1987).

BTV EGRESS AND THE INVOLVEMENT OF HOST PROTEINS IN VIRUS EGRESS

Unlike enveloped viruses, BTV and other orbiviruses lack lipid envelopes and are released predominantly by cell lysis of the infected vertebrate hosts (Fig. 6A). By contrast, BTV release from insect vector cells is non-lytic, and the virus particles leave the infected cell by passage, either individually or in groups, through a locally disrupted plasma membrane (Fig. 6A) (Hyatt *et al.*, 1989). Biologically, a non-lytic infection of the insect maintains its ability to act as a vector but how the virus controls this differential egress is one of the most interesting aspects of BTV replication.

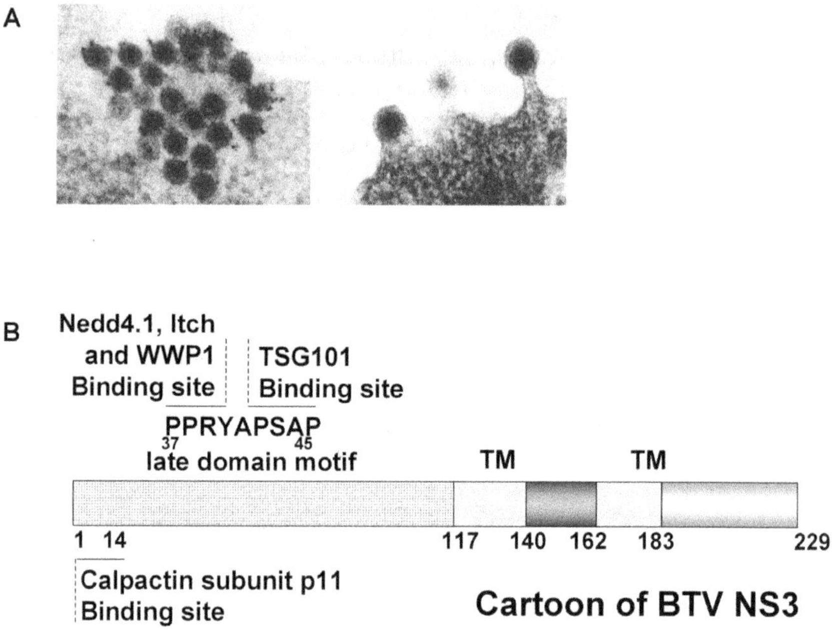

Fig. 6. BTV egress from host cells and how NS3 may facilitate BTV release. **(A)** Electron micrograph of two types of cells infected with bluetongue virus and expression level of NS3 and NS3A in these cells. **(B)** A schematic diagram showing the organisation of NS3 protein domains. BTV NS3 is an integral membrane protein with two transmembrane domain (TM) and contains the two most commonly observed late domain motifs PPXY and PS/TAP that recruit the components of cellular ESCRT (endosomal complex required for transport) complexes. In addition to facilitate BTV budding the N-terminal (cytoplasmic) 14 residues interact with cellular release factors calpactin p11. Thus NS3 utilises both p11 and Tsg101 (via late domain motifs) in order to aid in release of newly assembled virus particles from infected cells. (Beaton *et al.*, 2002; Wirblich *et al.*, 2006).

NS3 (229 aa) and its shorter form NS3A (216 aa) are the only membrane proteins encoded by BTV (Lee and Roy, 1986; French *et al.*, 1989). Both proteins can be found associated with smooth-surfaced, intracellular vesicles (Hyatt *et al.*, 1993) but do not form part of the stable structure of the mature virus. NS3/NS3A proteins have a long N-terminal and a shorter C-terminal cytoplasmic domain which are connected by two transmembrane domains and a short extracellular domain (Fig. 6B) (Bansal *et al.*, 1998; Wu *et al.*, 1999; Beaton *et al.*, 2002). A single glycosylation site is present in the extracellular domain of BTV NS3 but is missing in NS3 of AHSV,and therefore does not seem to be essential for the function of the protein (van Staden and Huismans, 1991; Bansal *et al.*,

1998). The NS3/NS3A proteins accumulate to only very low levels in BTV-infected mammalian cells but in invertebrate vector cells the expression levels of these proteins are high (French *et al.*, 1989; Guirakhoo *et al.*, 1995). The correlation between high NS3/NS3A expression and non-lytic virus release has suggested a significant functional role for NS3 in virus egress from invertebrate cells. In addition NS3 and NS3A facilitate the release of baculovirus-expressed VLPs in heterologous insect cells. The NS3 protein is localised at the site of the membrane where VLPs are released (Hyatt *et al.*, 1993), confirming that the integration of NS3/NS3A into the plasma membrane may trigger the release of mature virions. The NS3 protein of BTV exhibit cytoxicity when expressed individually in mammalian or insect cells (French *et al.*, 1989; van Staden *et al.*, 1995). The cytoxicity of NS3 requires membrane association of the protein and depends on the presence of the N terminal transmembrane domain, suggesting that NS3 might function as a viroporin and facilitate virus release by inducing membrane permeabilization (Han and Harty, 2004). It is possible that this permeabilization activity causes local disruption of the plasma membrane allowing virus particles to be extruded through a membrane pore without acquiring a lipid envelope.

The first 13 amino acids of NS3 have the potential to form an amphipathic helix (Fig. 6B). These amino terminal residues of NS3 which are absent in NS3A, interact with the calpactin light chain (p11) of the cellular annexin II complex (Beaton *et al.*, 2002), a complex that has been implicated in membrane related events along the endocytic and regulated secretory pathways including the trafficking of vesicles (for reviews see Creutz, 1992; Raynal and Pollard, 1994). The interaction of NS3 with p11 is highly specific and an NS3 peptide representing a mimic of the sequences of the p11 binding domain inhibits the release of progeny virions from BTV-infected insect cells. Although the exact physiological role of this interaction is still unknown, it is likely that interaction with p11 may direct NS3 to sites of active cellular exocytosis, or that NS3 could become part of an active extrusion process. There is some indication that cytoskeletal material is released at sites of BTV egress (Hyatt *et al.*, 1991) which may be annexin II being drawn through the membrane during the extrusion process while it is still associated with NS3. The significance of this interaction to BTV egress becomes more apparent in the light of the observation that the other cytoplasmic domain of the protein, situated at the C-terminal end interacts specifically with the BTV outer capsid protein

VP2 (Beaton *et al.*, 2002). It is likely that interaction between NS3 and assembling viruses is part of an active transportation process.

In addition to its interaction with the p11 component of the annexin II complex, NS3 is also capable of interaction with Tsg101, a cellular protein implicated in the intracellular trafficking and release of a number of enveloped viruses (Freed, 2004). It is intriguing that NS3 and NS3A of both BTV and AHSV bind *in vitro* with similar affinity to human Tsg101 and also to its ortholog from *Drosophila melanogaster* (Wirblich *et al.*, 2006). This interaction is mediated by the conserved PSAP motif in NS3 of BTV (Fig. 6B) and seems to play a role in virus release, as knockdown of Tsg101 with siRNA inhibits release of BTV from mammalian cells (Wirblich *et al.*, 2006). Like most other viral proteins which recruit Tsg101, NS3 also binds host NEDD4-like ubiquitin ligases *in vitro* via a resident PPXY late domain motif. However, the late domain motifs in NS3 do not function as effectively as the late domains of other enveloped viruses (Wirblich *et al.*, 2006). This appears to be mainly due to the presence of a unique arginine at position 3 of the PPXY motif in NS3. The low activity of the NS3 late domain motifs can be reversed by converting the PPRY motif of NS3 into a more universal PPPY, rendering NS3 as effective as Gag in facilitating retroviral VLP release as the late domains of enveloped viruses. The ability to usurp the vacuolar protein sorting pathway is likely to be more important in insect hosts as orbiviruses establish persistent infections in insect cells without causing significant cytopathic effect. Identification of insect proteins that interact with NS3 should shed more light on the question of whether NS3 is better adapted to engage insect proteins so facilitating improved virus release.

In addition to binding to p11 and Tsg101, NS3 also seems to interact with the dynein light chain 2A (Beaton & Roy, unpublished data). It is intriguing that dynein also plays an important role in endosomal/lysosomal membrane dynamics (Harrison *et al.*, 2003). Thus, it seems that NS3 acts as a platform that recruits the cellular machinery. To play a role in virus release NS3 must also associate with the maturing virions (via VP2) during morphogenesis. Current data suggest that NS3 makes use of host proteins and acts as intermediate to facilitate the release of newly synthesis progeny virus across the cell membrane.

STRUCTURALLY DESIGNED VACCINES FOR BTV

Bluetongue disease affects mainly sheep and cattle and is economically an important disease in many parts of the world, both in developed and developing countries. Since it is a vector-borne disease and transmitted to healthy livestock from infected animals by blood feeding *Culicoides* spp., attenuated, live virus vaccines are not desirable. The flexibility of baculovirus expression vectors and the capacity of the baculovirus genome to accommodate large amounts of foreign DNA allowed us to exploit the system for the simultaneous expression of multiple BTV genes by a single recombinant virus. As discussed above, based on the 3D structural data, recombinant viruses were prepared to express either VP3 and VP7, the two major core proteins, or VP3 and VP7 together with VP2 and VP5, i.e. all four major structural proteins. Expression of VP3 and VP7 resulted in CLPs that were similar in size and appearance to cores prepared from BT virions(see Fig. 5; French and Roy, 1990). Similarly, the simultaneous expression of four proteins resulted in the assembly of virtually homogeneous double-capsid particles (see Fig. 5, French *et al.*, 1990). When the 3D structure of CLPs and VLPs were analysed by Cryo-EM both types of particles were clearly comparable to authentic cores and virions, and exhibited essentially the same basic features and full complement of the two or four proteins (Hewat *et al.*, 1992a, 1992b, 1992c; Prasad *et al.*, 1992). VLPs synthesized by recombinant baculoviruses were also characterized further at the biological and immunological levels and compared to those of the native virion. VLPs exhibited high levels of haemagglutination activity similar to those of authentic BTV. Further antibodies raised to the expressed particles contained high titres of neutralizing activity against the homologous BTV serotype emphasizing their authenticity at functional level (French *et al.*, 1990).

Since VLPs elicited strong neutralizing antibodies in guinea pigs, it can be anticipated that VLPs should also elicit protective responses in sheep against BTV infection. Consequently, a number of experiments were performed to examine the protective efficacy of VLPs (10-200 μg per sheep) in 1-year old BTV free sheep divided in groups (4-8 sheep/group). All sheep were challenged by subcutaneous inoculation of 1 ml of infective sheep blood containing virulent virus at day 117 or at 14 months. The challenged sheep developed neither clinical signs nor viraemia, indicating a suppression of BTV replication. The post-challenge blood samples of the sheep that did not receive vaccine and only received saline, contained infectious BTV and these sheep developed high neutralizing antibody titres indicative of a primary infection. Protective immunity to BTV disease was obtained by vaccinating sheep with doses of 10 µg or more of BTV VLPs which resulted in long-lasting protection against homologous BTV challenge. Some preliminary evidence was obtained for cross-protection, depending on the challenge virus and the amounts of antigen used for vaccination (Roy, 1995).

CONCLUDING REMARKS AND FUTURE PERSPECTIVES

Bluetongue disease emerged in Europe only recently but is becoming one of the major threats to livestock industry world-wide. Although our knowledge of host-virus interactions and viral pathogenesis is limited, BTV is well characterised both genetically (the sequence was completed in 1989) and structurally. Understanding of molecular biology of the virus and mapping the role of each protein in virus life cycle has been benefitted significantly from the availability of recombinant BTV proteins and sub-viral particles. In addition the structure of BTV proteins, core and virion particles have contributed greatly to understanding the mechanism of protein-protein interaction in the virus assembly pathway of BTV and other orbiviruses. Most importantly, information gained from these studies has laid sound foundation for the generation of safe BTV vaccines with the possibility of use in animals in the near future. Latterly, studies have concentrated on the fundamental mechanisms that are used by the virus to invade, replicate in and escape from susceptible host cells. Progress has been made in understanding the structure and entry of intact virus particles, the role of each enzymatic protein in the transcription complex, the critical interactions that occur between the viral non-structural proteins and viral RNA and the role of cellular proteins in non-enveloped virus egress (see cartoon, Fig. 7).

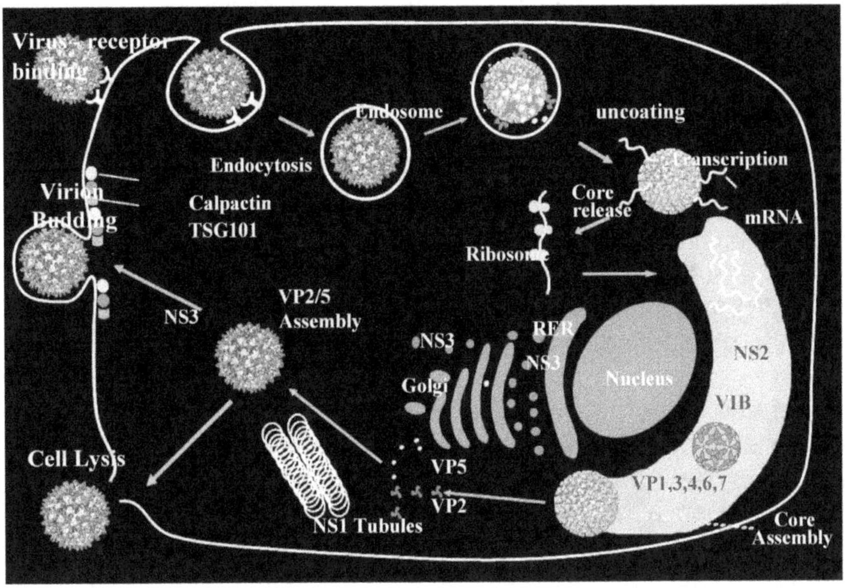

Fig. 7. Replication cycle of BTV. Schematic diagram of an BTV replication cycle based on current molecular data.

Despite these advances, some critical questions remain unanswered for the BTV life cycle and a more complete understanding of the interactions between the virus and the host cell are required for these to be addressed. For example, although progress has been made in the identification of signals for the recruitment of BTV RNA segments into the virion assembly site in the host cell cytoplasm, it has not been possible yet to determine exactly how each genome segment is packaged into the progeny virus. It is also not apparent when and how these genome segments wrap around the polymerase complex once the RNA has been recruited. Very little is known about the intracellular trafficking of newly generated virions although there are some indications of involvement of the cytoskeleton, intermediate filaments and vimentin during BTV morphogenesis. Host-virus interactions during virus trafficking will be one of the future areas needing intense attention. Recent work has revealed unexpected and striking parallels between the entry and release pathways of BTV and pathways involved in entry and release of enveloped viruses. These parallels may be the result of an enveloped ancestor virus;

or because there are a limited number of cellular pathways that can be useful for the egress of large protein complexes from cells. It is notable that the NS3 glycoprotein of BTV is an integral membrane protein that is functionally involved in virus egress by bridging between the outer capsid protein VP2 and the cellular export machinery. Although no cell-free enveloped form of BTV has been isolated, budding of BTV particles from infected cells at the plasma membrane are quite apparent. The exact role of NS3 in this process and the role of host proteins (Annexin II and p11, Tsg101 and MVB) in the release of non-enveloped viruses, such as BTV, remain to be clarified.

One of the major drawbacks of research with BTV and other members of the *Reoviridae* has been the lack of a suitable reverse genetics system which allows genetic manipulation of the virus. This has been a major obstacle in understanding the replication processes of these viruses. However, recently it has been possible to recover infectious BTV wholly from ssRNA, which suggests a means for establishing helper virus-independent reverse genetics systems for members of the *Reoviridae* (Boyce and Roy, 2007). The discovery that BTV ssRNA is infectious brings the realisation of a reverse genetics system for orbiviruses a step closer. There is no doubt that this will be one of the future dominant research areas for the members of the *Reoviridae*. Once in place, molecular and structural studies of individual BTV proteins can be placed in the context of the whole virus.

REFERENCES

Bae, S.H., Cheong, H.K., Lee, J.H., Cheong, C., Kainosho, M., and Choi, B.S. (2001). Structural features of an influenza virus promoter and their implications for viral RNA synthesis. Proc. Natl. Acad. Sci. USA *98*, 10602-10607.

Bannerjee, A.K., and Shatkin, A.J. (1970). Transcription *in vitro* by reovirus associated ribonucleic acid-dependent polymerase. J. Virol. *6*, 1-11.

Bansal, O.B., Stokes, A., Bansal, A., Bishop, D.H.L., and Roy, P. (1998). Membrane organization of bluetongue virus non-structural glycoprotein NS3. J. Virol. *72*, 3362-3369.

Basak, A., Grimes, J., Burroughs, N., Mertens, P.P.C., Roy, P., and Stuart, D. (1992). Crystallographic studies on the structure of the Bluetongue virus inner capsid, In Bluetongue, African Horse

Sickness and Related Orbiviruses., T.E. Walton, and B.I. Osburn, eds. (Boca Raton, FL: CRC Press), pp. 483-490.
Basak, A.K., Gouet, P., Grimes, J., Roy, P., and Stuart, D. (1996). Crystal structure of the top domain of African horse sickness virus VP7: comparisons with bluetongue virus VP7. J. Virol. *70*, 3797-3806.
Basak, A.K., Grimes, J.M., Gouet, P., Roy, P., and Stuart, D.I. (1997). Structures of orbivirus VP7: implications for the role of this protein in the viral life cycle. Structure *5*, 871-883.
Beasley, B.E., and Hu, W.S. (2002). cis-Acting Elements Important for Retroviral RNA Packaging Specificity. J. Virol. *76*, 4950-4960.
Beaton, A.R., Rodriguez, J., Reddy, Y.K., and Roy, P. (2002). The membrane trafficking protein calpactin forms a complex with bluetongue virus protein NS3 and mediates virus release. Proc. Natl. Acad. Sci. USA *99*, 13154-13159.
Bisaillon, M., Bergeron, J., and Lemay, G. (1997). Characterization of the nucleoside triphosphate phosphohydrolase and helicase activities of the reovirus lambda1 protein. J. Biol. Chem. *272*, 18298-18303.
Browne, J.G. and Jochim, M.M. (1967). Cytopathologic changes and development of inclusion bodies in cultured cells infected with bluetongue virus. Am. J. Vet. Res. *28*, 1091-1105.
Boyce, M., and Roy, P. (2007). Recovery of Infectious Bluetongue Virus from RNA. J. Virol. *81*, 2179-2186.
Boyce, M., Wehrfritz, J., Noad, R., and Roy, P. (2004). Purified recombinant bluetongue virus VP1 exhibits RNA replicase activity. J. Virol. *78*, 3994-4002.
Brookes, S.M., Hyatt, A.D., and Eaton, B.T. (1993). Characterization of virus inclusion bodies in bluetongue virus infected cells. J. Gen. Virol. *74*, 525-530.
Brookes, S.M., Hyatt, A.D., and Eaton, B.T. (1994). The use of immuno-gold silver staining in bluetongue virus adsorption and neutralisation studies. J. Virol. Methods *46*, 117-132.
Butan, C., Van Der Zandt, H., and Tucker, P.A. (2004). Structure and assembly of the RNA binding domain of bluetongue virus non-structural protein 2. J. Biol. Chem. *279*, 37613-37621.
Butcher, S.J., Dokland, T., Ojala, P.M., Bamford, D.H., and Fuller, S.D. (1997). Intermediates in the assembly pathway of the double-stranded RNA virus phi6. EMBO J. *16*, 4477-4487.
Chen, D., Barros, M., Spencer, E., and Patton, J.T. (2001). Features of the 3' Consensus Sequence of Rotavirus mRNAs Critical to Minus Strand Synthesis. Virology *282*, 221-229.

Chen, D., and Patton, J.T. (1998). Rotavirus RNA replication requires a single-stranded 3' end for efficient minus-strand synthesis. J. Virol. *72*, 7387-7396.
Chen, D., Zeng, C.Q., Wentz, M.J., Gorziglia, M., Estes, M.K., and Ramig, R.F. (1994). Template-dependent, *in vitro* replication of rotavirus RNA. J. Virol. *68*, 7030-7039.
Cohen, J. (1977). Ribonucleic acid polymerase activity associated with purified calf rotavirus. J. Gen. Virol. *36*, 395-402.
Colman, P.M., and Lawrence, M.C. (2003). The structural biology of type I viral membrane fusion. Nat. Rev. Mol. Cell Biol. *4*, 309-319.
Creutz, C.E. (1992). The annexins and exocytosis. Science *258*, 924-931.
Dryden, K.A., Farsetta, D.L., Wang, G., Keegan, J.M., Fields, B.N., Baker, T.S., and Nibert, M.L. (1998). Internal/structures containing transcriptase-related proteins in top component particles of mammalian orthoreovirus. Virology *245*, 33-46.
Eaton, B.T., and Crameri, G.S. (1989). The site of bluetongue virus attachment to glycophorins from a number of animal erythrocytes. J. Gen. Virol. *70*, 3347-3353.
Eaton, B.T., and Hyatt, A.D. (1989). Association of bluetongue virus with the cytoskeleton. Subcell Biochem. *15*, 233-273.
Eaton, B.T., Hyatt, A.D., and Brookes, S.M. (1990). The replication of bluetongue virus. Curr. Top. Microbiol. Immunol. *162*, 89-118.
Eaton, B.T., Hyatt, A.D., and White, J.R. (1988). Localization of the nonstructural protein NS1 in bluetongue virus-infected cells and its presence in virus particles. Virology *163*, 527-537.
Els, H.J., and Verwoerd, D.W. (1969). Morphology of bluetongue virus. Virology *38*, 213-219.
Fillmore, G.C., Lin, H., and Li, J.K.-K. (2002). Localization of the single-stranded RNA-binding domains of bluetongue virus nonstructural protein NS2. J. Virol. *76*, 499-506.
Forzan, M., Marsh, M., and Roy, P. (2007). Bluetongue virus entry into the cells. J. Virol. *81*, 4819-4827.
Forzan, M., Wirblich, C., and Roy, P. (2004). A capsid protein of nonenveloped Bluetongue virus exhibits membrane fusion activity. Proc. Natl. Acad. Sci. USA *101*, 2100-2105.
Freed, E.O. (2004). Mechanisms of enveloped virus release. Virus Res. *106*, 85-86.
French, T.J., Inumaru, S., and Roy, P. (1989). Expression of two related nonstructural proteins of bluetongue virus (BTV) type 10 in insect cells by a recombinant baculovirus: production of polyclonal ascitic fluid and characterization of the gene product in BTV-infected BHK cells. J. Virol. *63*, 3270-3278.

French, T.J., Marshall, J.J., and Roy, P. (1990). Assembly of double-shelled, virus-like particles of bluetongue virus by the simultaneous expression of four structural proteins. J. Virol. *64*, 5695-5700.
French, T.J., and Roy, P. (1990). Synthesis of bluetongue virus (BTV) corelike particles by a recombinant baculovirus expressing the two major structural core proteins of BTV. J. Virol. *64*, 1530-1536.
Fukusho, A., Ritter, G.D., and Roy, P. (1987). Variation in the bluetongue virus neutralization protein VP2. J. Gen. Virol. *68*, 2967-2973.
Fukusho, A., Yu, Y., Yamaguchi, S., and Roy, P. (1989). Completion of the sequence of bluetongue virus serotype 10 by the characterization of a structural protein, VP6, and a non-structural protein, NS2. J. Gen. Virol. *70*, 1677-1689.
Gao, Y., and Lenard, J. (1995). Cooperative binding of multimeric phosphoprotein (P) of vesicular stomatitis virus to polymerase (L) and template: pathways of assembly. J. Virol. *69*, 7718-7723.
Ghiasi, H., Fukusho, A., Eshita, Y., and Roy, P. (1987). Identification and characterization of conserved and variable regions in the neutralization VP2 gene of bluetongue virus. Virology *160*, 100-109.
Gouet, P., Diprose, J.M., Grimes, J.M., Malby, R., Burroughs, J.N., Zientara, S., Stuart, D.I., and Mertens, P.P. (1999). The highly ordered double-stranded RNA genome of bluetongue virus revealed by crystallography. Cell *97*, 481-490.
Gould, A.R., and Eaton, B.T. (1990). The amino acid sequence of the outer coat protein VP2 of neutralizing monoclonal antibody-resistant, virulent and attenuated bluetongue viruses. Virus Res. *17*, 161-172.
Grimes, J., Basak, A.K., Roy, P., and Stuart, D. (1995). The crystal structure of bluetongue virus VP7. Nature *373*, 167-170.
Grimes, J.M., Burroughs, J.N., Gouet, P., Diprose, J.M., Malby, R., Zientara, S., Mertens, P.P.C., and Stuart, D.I. (1998). The atomic structure of the bluetongue virus core. Nature *395*, 470-478.
Guirakhoo, F., Catalan, J.A., and Monath, T.P. (1995). Adaptation of bluetongue virus in mosquito cells results in overexpression of NS3 proteins and release of virus particles. Arch. Virol. *140*, 967-974.
Han, Z., and Harty, R.N. (2004). The NS3 protein of bluetongue virus exhibits viroporin-like properties. J. Biol. Chem. *279*, 43092-43097.

Harrison, R.E., Bucci, C., Vieira, O.V., Schroer, T.A., and Grinstein, S. (2003). Phagosomes fuse with late endosomes and/or lysosomes by extension of membrane protrusions along microtubules: role of Rab7 and RILP. Mol. Cell. Biol. *23*, 6494-6506.

Hassan, S.H., and Roy, P. (1999). Expression and functional characterization of bluetongue virus VP2 protein: Role in cell entry. J. Virol. *73*, 9832-9842.

Hassan, S.H., Wirblich, C., Forzan, M., and Roy, P. (2001). Expression and functional characterization of bluetongue virus VP5 protein: role in cellular permeabilization. J. Virol. *75*, 8356-8367.

Hewat, E.A., Booth, T.F., Loudon, P.T., and Roy, P. (1992a). Three-dimensional reconstruction of baculovirus expressed bluetongue virus core-like particles by cryo-electron microscopy. Virology *189*, 10-20.

Hewat, E.A., Booth, T.F., and Roy, P. (1992b). Structure of bluetongue virus particles by cryoelectron microscopy. J. Struct. Biol. *109*, 61-69.

Hewat, E.A., Booth, T.F., Wade, R.H., and Roy, P. (1992c). 3-D reconstruction of bluetongue virus tubules using cryoelectron microscopy. J. Struct. Biol. *108*, 35-48.

Hewat, E.A., Booth, T.F., and Roy, P. (1994). Structure of correctly self-assembled bluetongue virus-like particles. J. Struct. Biol. *112*, 183-191.

Horscroft, N.J., and Roy, P. (1997). Thermal denaturation of proteins for SDS-PAGE analysis by microwave irradiation. Biotechniques *22*, 224-226.

Huismans, H. (1979). Protein synthesis in bluetongue virus-infected cells. Virology *92*, 385-396.

Huismans, H., and Cloete, M. (1987). A comparison of different cloned bluetongue virus genome segments as probes for the detection of virus-specified RNA. Virology *158*, 373-380.

Huismans, H., and Els, H.J. (1979). Characterization of the tubules associated with the replication of three different orbiviruses. Virology *92*, 397-406.

Huismans, H., and Erasmus, B.J. (1981). Identification of the serotype-specific and group-specific antigens of bluetongue virus. Onderstepoort J. Vet. Res. *48*, 51-58.

Huismans, H., Van der Walt, N.T., Cloete, M., and Erasmus, B.J. (1987a). Isolation of a capsid protein of bluetongue virus that induces a protective immune response in sheep. Virology *157*, 172-179.

Huismans, H., Van Dijk, A.A., and Els, H.J. (1987b). Uncoating of parental bluetongue virus to core and subcore particles in infected L cells. Virology *157*, 180-188.
Hyatt, A.D., and Eaton, B.T. (1988). Ultrastructural distribution of the major capsid proteins within bluetongue virus and infected cells. J. Gen. Virol. *69*, 805-815.
Hyatt, A.D., Eaton, B.T., and Brookes, S.M. (1989). The release of bluetongue virus from infected cells and their superinfection by progeny virus. Virology *173*, 21-34.
Hyatt, A.D., Gould, A.R., Coupar, B., and Eaton, B.T. (1991). Localization of the non-structural protein NS3 in bluetongue virus-infected cells. J. Gen. Virol. *72*, 2263-2267.
Hyatt, A.D., Zhao, Y., and Roy, P. (1993). Release of bluetongue virus-like particles from insect cells is mediated by BTV nonstructural protein NS3/NS3A. Virology *193*, 592-603.
Inumaru, S., and Roy, P. (1987). Production and characterization of the neutralization antigen VP2 of bluetongue virus serotype 10 using a baculovirus expression vector. Virology *157*, 472-479.
Iwata, H., Yamagawa, M., and Roy, P. (1992). Evolutionary relationships among the gnat-transmitted orbiviruses that cause African horse sickness, bluetongue, and epizootic hemorrhagic disease as evidenced by their capsid protein sequences. Virology *191*, 251-261.
Kahlon, J., Sugiyama, K., and Roy, P. (1983). Molecular basis of bluetongue virus neutralization. J. Virol. *48*, 627-632.
Kar, A.K., Bhattacharya, B., and Roy, P. (2007). Bluetongue virus RNA binding protein NS2 is a modulator of viral replication and assembly. BMC Mol. Biol. *8,* 4.
Kar, A.K., Ghosh, M., and Roy, P. (2004). Mapping the assembly of Bluetongue virus scaffolding protein VP3. Virology *324*, 387-399.
Kim, J., Zhang, X., Centonze, V.E., Bowman, V.D., Noble, S., Baker, T.S., and Nibert, M.L. (2002). The hydrophilic amino-terminal arm of reovirus core shell protein lambda1 is dispensable for particle assembly. J. Virol. *76*, 12211-12222.
Labbe, M., Charpilienne, A., Crawford, S.E., Estes, M.K., and Cohen, J. (1991). Expression of rotavirus VP2 produces empty corelike particles. J. Virol. *65*, 2946-2952.
Le Blois, H., and Roy, P. (1993). A single point mutation in the VP7 major core protein of Bluetongue virus prevents the formation of core-like particles. J. Virol. *67*, 353-359.
Lee, J.W., and Roy, P. (1986). Nucleotide sequence of a cDNA clone of RNA segment 10 of bluetongue virus (serotype 10). J. Gen. Virol. *67*, 2833-2837.

Lee, M.T., Bishop, K., Medcalf, L., Elton, D., Digard, P., and Tiley, L. (2002). Definition of the minimal viral components required for the initiation of unprimed RNA synthesis by influenza virus RNA polymerase. Nucleic Acids Res. *30*, 429-438.

Limn, C.-H., Staeuber, N., Monastyrskaya, K., Gouet, P., and Roy, P. (2000). Functional dissection of the major structural protein of bluetongue virus: identification of key residues within VP7 essential for capsid assembly. J. Virol. *74*, 8658-8669.

Limn, C.K., and Roy, P. (2003). Intermolecular interactions in a two-layered viral capsid that requires a complex symmetry mismatch. J. Virol. *77*, 11114-11124.

Liu, H.M., Booth, T.F., and Roy, P. (1992). Interactions between bluetongue virus core and capsid proteins translated *in vitro*. J. Gen. Virol. *73*, 2577-2584.

Lohman, T.M., Thom, K., and Vale, R.D. (1998). Staying on track: common features of DNA helicases and microtubule motors. Cell *93*, 9-12.

Loudon, P.T., and Roy, P. (1991). Assembly of five bluetongue virus proteins expressed by recombinant baculoviruses: inclusion of the largest protein VP1 in the core and virus-like proteins. Virology *180*, 798-802.

Lymperopoulos, K., Noad, R., Tosi, S., Nethisinghe, S., Brierley, I., and Roy, P. (2006). Specific binding of Bluetongue virus NS2 to different viral plus-strand RNAs. Virology *25*, 25.

Lymperopoulos, K., Wirblich, C., Brierley, I., and Roy, P. (2003). Sequence specificity in the interaction of Bluetongue virus non-structural protein 2 (NS2) with viral RNA. J. Biol. Chem. *278*, 31722-31730.

Markotter, W., Theron, J., and Nel, L.H. (2004). Segment specific inverted repeat sequences in bluetongue virus mRNA are required for interaction with the virus non structural protein NS2. Virus Res. *105*, 1-9.

Martinez-Costas, J., Sutton, G., and Roy, P. (1998). Guanylyltranferase and RNA5-triphosphatase activities of the purified expressed VP4 protein of bluetongue virus. J. Mol. Biol. *280*, 859-866.

Mertens, P.P., Burroughs, J.N., and Anderson, J. (1987). Purification and properties of virus particles, infectious subviral particles, and cores of bluetongue virus serotypes 1 and 4. Virology *157*, 375-386.

Mertens, P.P.C., Maan, S., Samuel, A., and Attoui, H. (2005). Orbivirus, *Reoviridae*., In Virus taxonomy, VIIIth Report of the ICTV,

C.M. Fauquet, M.A. Mayo, J. Maniloff, U. Desselberger, and L.A. Ball, eds. (London: Elsevier/Academic Press), pp. 466-483.
Modrof, J., Lymperopoulos, K., and Roy, P. (2005). Phosphorylation of Bluetongue Virus Nonstructural Protein 2 Is Essential for Formation of Viral Inclusion Bodies. J. Virol. *79*, 10023-10031.
Montenarh, M., and Muller, D. (1987). The phosphorylation of Thr 124 of simian virus 40 large T antigen is crucial in cell entry. J. Virol. *221*, 199-204.
Mortola, E., Noad, R., and Roy, P. (2004). Bluetongue virus outer capsid proteins are sufficient to trigger apoptosis in mammalian cells. J. Virol. *78*, 2875-2883.
Naitow, H., Tang, J., Canady, M., Wickner, R.B., and Johnson, J.E. (2002). L-A virus at 3.4 A resolution reveals particle architecture and mRNA decapping mechanism. Nat. Struct. Biol. *9*, 725-728.
Nakagawa, A., Miyazaki, N., Taka, J., Naitow, H., Ogawa, A., Fujimoto, Z., Mizuno, H., Higashi, T., Watanabe, Y., Omura, T., *et al.* (2003). The atomic structure of *Rice dwarf* virus reveals the self-assemby mechanism of component proteins. Structure *11*, 1227-1238.
Nason, E., Rothnagel, R., Muknerge, S.K., Kar, A.K., Forzan, M., Prasad, B.V.V., and Roy, P. (2004). Interactions between the Inner and Outer Capsids of Bluetongue Virus. J. Virol. *78*, 8059-8067.
Owens, R.J., Limn, C., and Roy, P. (2004). Role of an arbovirus nonstructural protein in cellular pathogenesis and virus release. J. Virol. *78*, 6649-6656.
Patton, J.T. (1996). Rotavirus VP1 alone specifically binds to the 3' end of viral mRNA, but the interaction is not sufficient to initiate minus-strand synthesis. J. Virol. *70*, 7940-7947.
Patton, J.T., Jones, M.T., Kalbach, A.N., He, Y.W., and Xiaobo, J. (1997). Rotavirus RNA polymerase requires the core shell protein to synthesize the double-stranded RNA genome. J. Virol. *71*, 9618-9626.
Patton, J.T., Wentz, M., Xiaobo, J., and Ramig, R.F. (1996). cis-Acting signals that promote genome replication in rotavirus mRNA. J. Virol. *70*, 3961-3971.
Pesavento, J.B., Lawton, J.A., Estes, M.E., and Venkataram Prasad, B.V. (2001). The reversible condensation and expansion of the rotavirus genome. Proc. Natl. Acad. Sci. USA *98*, 1381-1386.
Piccininni, S., Varaklioti, A., Nardelli, M., Dave, B., Raney, K.D., and McCarthy, J.E. (2002). Modulation of the hepatitis C virus RNA-dependent RNA polymerase activity by the non-structural

(NS) 3 helicase and the NS4B membrane protein. J. Biol. Chem. *277*, 45670-45679.

Pirttimaa, M., Paatero, A., Frilander, M., and Bamford, D. (2002). Nonspecific nucleoside triphosphatase P4 of double-stranded RNA bacteriophage phi6 is required for single-stranded RNA packaging and transcription. J. Virol. *76*, 10122-10127.

Poranen, M., and Tuma, R. (2004). Self-assembly of double-stranded RNA bacteriophages. Virus Res. *101*, 93-100.

Prasad, B.V.V., Rothnagel, R., Zeng, C.Q., Jakana, J., Lawton, J. A., Chiu, W., and Estes, M.K. (1996). Visualization of ordered genomic RNA and localization of transcriptional complexes in rotavirus. Nature *382*, 471-473.

Prasad, B.V.V., Yamaguchi, S., and Roy, P. (1992). Three-dimensional structure of single-shelled bluetongue virus. J. Virol. *66*, 2135-2142.

Purdy, M., Petre, J., and Roy, P. (1984). Cloning of the bluetongue virus L3 gene. J. Virol. *51*, 754-759.

Purdy, M.A., Ritter, G.D., and Roy, P. (1986). Nucleotide sequence of cDNA clones encoding the outer capsid protein, VP5, of bluetongue virus serotype 10. J. Gen. Virol. *67*, 957-962.

Ramadevi, N., Burroughs, J.N., Mertens, P.P.C., Jones, I.M., and Roy, P. (1998a). Capping and methylation of mRNA by purified recombinant VP4 protein of Bluetongue virus. Proc. Natl. Acad. Sci. USA *95*, 13537-13542.

Ramadevi, N., Rodriguez, J., and Roy, P. (1998b). A leucine zipper-like domain is essential for dimerization and encapsidation of bluetongue virus nucleocapsid protein VP4. J. Virol. *72*, 2983-2990.

Ramadevi, N., and Roy, P. (1998). Bluetongue virus core protein VP4 has nucleoside triphosphate phosphohydrolase activity. J. Gen. Virol. *79*, 2475-2480.

Raynal, P., and Pollard, H.B. (1994). Annexins: the problem of assessing the biological role for a gene family of multifunctional calcium- and phospholipid-binding proteins. Biochem. Biophys. Res. Acta *1197*, 63-93.

Reinisch, K.M., Nibert, M.L., and Harrison, S.C. (2000). Structure of the reovirus core at 3.6 A resolution. Nature *404*, 960-967.

Roy, P. (1989). Bluetongue virus genetics and genome structure. Virus Res. *13*, 179-206.

Roy, P. (1992). Bluetongue virus proteins. J. Gen. Virol. *73*, 3051-3064.

Roy, P. (1995). Synthesis of particulate structures as Bluetongue virus vaccines and their use as multiple immunogen delivery systems. Bulletin de l'Institut Pasteur *93*, 3-20.

Roy, P. (1996). Orbiviruses and their replication, In Field's Virology, B.N. Fields, ed. (Philadelphia: Lippincott-Raven), pp. 1709-1734.

Roy, P. (2000). Orbivirus (Reoviridae). In The Springer Index of Viruses. (Heidelberg: Springer Verlag), pp. 957-963.

Roy, P., Adachi, A., Urakawa, T., Booth, T.F., and Thomas, C.P. (1990). Identification of bluetongue virus VP6 protein as a nucleic acid-binding protein and the localization of VP6 in virus-infected vertebrate cells. J. Virol. *64*, 1-8.

Roy, P., Fukusho, A., Ritter, G.D., and Lyon, D. (1988). Evidence for genetic relationship between RNA and DNA viruses from the sequence homology of a putative polymerase gene of bluetongue virus with that of vaccinia virus: conservation of RNA polymerase genes from diverse species. Nucleic Acids Res. *16*, 11759-11767.

Shaw, A.L., Samal, S.K., Subramanian, K., and Prasad, B.V.V. (1996). The structure of aquareovirus shows how the different geometries of the two layers of the capsid are reconciled to provide symmetrical interactions and stabilization. Structure *4*, 957-967.

Shirota, Y., Luo, H., Qin, W., Kaneko, S., Yamashita, T., Kobayashi, K., and Murakami, S. (2002). Hepatitis C virus (HCV) NS5A binds RNA-dependent RNA polymerase (RdRP) NS5B and modulates RNA-dependent RNA polymerase activity. J. Biol. Chem. *277*, 11149-11155.

Silverstein, S.C., Astell, C., Levin, D.H., Schonberg, M., and Acs, G. (1972). The mechanisms of reovirus uncoating and gene activation *in vivo*. Virology *47*, 797-806.

Spadafora, D., Canter, D.M., Jackson, R.L., and Perrault, J. (1996). Constitutive phosphorylation of the vesicular stomatitis virus P protein modulates polymerase complex formation but is not essential for transcription or replication. J. Virol. *70*, 4538-4548.

Starnes, M.C., and Joklik, W.K. (1993). Reovirus protein lambda 3 is a poly(C)-dependent poly(G) polymerase. Virology *193*, 356-366.

Stauber, N., Martinez-Costas, J., Sutton, G., Monastyrskaya, K., and Roy, P. (1997). Bluetongue virus VP6 protein binds ATP and exhibits an RNA-dependent ATPase function and a helicase activity that catalyze the unwinding of double-stranded RNA substrates. J. Virol. *71*, 7220-7226.

Sutton, G., Grimes, J.M., Stuart, D., and Roy, P. (2007). The structure of bluetongue virus VP4; a multifunctional RNA-capping production-line. Nat. Struct. Mol. Biol. *14*, 449-451.

Tanaka, S., Mikhailov, M., and Roy, P. (1995). Synthesis of bluetongue virus chimeric VP3 molecules and their interactions with VP7 protein to assemble into virus core-like particles. Virology *214*, 593-601.

Tanaka, S., and Roy, P. (1994). Identification of domains in bluetongue virus VP3 molecules essential for the assembly of virus cores. J. Virol. *68*, 2795-2802.

Tao, Y., Farsetta, D.L., Nibert, M.L., and Harrison, S.C. (2002). RNA Synthesis in a Cage-Structural Studies of Reovirus Polymerase lambda3. Cell *111*, 733-745.

Theron, J., and Nel, L.H. (1997). Stable protein-RNA interaction involves the terminal domains of bluetongue virus mRNA, but not the terminally conserved sequences. Virology *229*, 134-142.

Thomas, C.P., Booth, T.F., and Roy, P. (1990). Synthesis of bluetongue virus-encoded phosphoprotein and formation of inclusion bodies by recombinant baculovirus in insect cells: it binds the single-stranded RNA species. J. Gen. Virol. *71*, 2073-2083.

Tortorici, M.A., Broering, T.J., Nibert, M.L., and Patton, J.T. (2003). Template recognition and formation of initiation complexes by the replicase of a segmented double-stranded RNA virus. J. Biol. Chem. *3*, 3.

Urakawa, T., Ritter, D.G., and Roy, P. (1989). Expression of largest RNA segment and synthesis of VP1 protein of bluetongue virus in insect cells by recombinant baculovirus: association of VP1 protein with RNA polymerase activity. Nucleic Acids Res. *17*, 7395-7401.

Van Dijk, A.A., and Huismans, H. (1980). The *in vitro* activation and further characterization of the bluetongue virus-associated transcriptase. Virology *104*, 347-356.

Van Dijk, A.A., and Huismans, H. (1988). *In vitro* transcription and translation of bluetongue virus mRNA. J. Gen. Virol. *69*, 573-581.

van Staden, V., and Huismans, H. (1991). A comparison of the genes which encode non-structural protein NS3 of different orbiviruses. J. Gen. Virol. *72*, 1073-1079.

van Staden, V., Stoltz, M.A., and Huismans, H. (1995). Expression of nonstructural protein NS3 of African horsesickness virus (AHSV): evidence for a cytotoxic effect of NS3 in insect cells, and characterization of the gene products in AHSV infected Vero cells. Arch. Virol. *140*, 289-306.

Verwoerd, D.W. (1969). Purification and characterization of bluetongue virus. Virology *38*, 203-212.

Verwoerd, D.W., Els, H.J., De Villiers, E.M., and Huismans, H. (1972). Structure of the bluetongue virus capsid. J. Virol. *10*, 783-794.

Verwoerd, D.W., and Huismans, H. (1972). Studies on the *in vitro* and the *in vivo* transcription of the bluetongue virus genome. Onderstepoort J. Vet. Res. *39*, 185-191.

Verwoerd, D.W., Louw, H., and Oellermann, R.A. (1970). Characterization of bluetongue virus ribonucleic acid. J. Virol. *5*, 1-7.

Wehrfritz, J., Boyce, M., Mirza, S., and Roy, P. (2007). Reconstitution of bluetongue virus polymerase activity from isolated domains based on a three dimensional structural model. Biopolymers *86*, 83-94.

Wirblich, C., Bhattacharya, B., and Roy, P. (2006). Nonstructural protein 3 of bluetongue virus assists virus release by recruiting ESCRT-I protein Tsg101. J. Virol. *80*, 460-473.

Wu, C.L., Leu, T.S., Chang, T.T., and Shiau, A.L. (1999). Hepatitis C virus core protein fused to hepatitis B virus core antigen for serological diagnosis of both hepatitis C and hepatitis B infections by ELISA. J. Med. Virol. *57*, 104-110.

Zeng, C.Q., Estes, M.K., Charpilienne, A., and Cohen, J. (1998). The N terminus of rotavirus VP2 is necessary for encapsidation of VP1 and VP3. J. Virol. *72*, 201-208.

Zetina, C.R. (2001). A conserved helix-unfolding motif in the naturally unfolded proteins. Proteins *44*, 479-483.

Zhang, X., Walker, S.B., Chipman, P.R., Nibert, M.L., and Baker, T.S. (2003). Reovirus polymerase lambda3 localized by cryo-electron microscopy of virions at a resolution of 7.6 A. Nat. Struct. Biol. *10*, 1011-1018.

Zhao, Y., Thomas, C., Bremer, C., and Roy, P. (1994). Deletion and mutational analyses of bluetongue virus NS2 protein indicate that the amino but not the carboxy terminus of the protein is critical for RNA-protein interactions. J. Virol. *68*, 2179-2185.

Zhou, Z.H., Baker, M.L., Jiang, W., Dougherty, M., Jakana, J., Dong, G., Lu, G., and Chiu, W. (2001). Electron cryomicroscopy and bioinformatics suggest protein fold models for rice dwarf virus. Nat. Struct. Biol. *8*, 868-873.

Chapter 8

Molecular Biology of Porcine Circoviruses

Annette Mankertz

ABSTRACT

Porcine Circoviruses (PCV) are the smallest viruses replicating autonomously in eukaryotic cells. The virions are non-enveloped and spherical with a diameter of 16-18 nm and the covalently closed and single-stranded DNA genomes comprise less than 1800 nucleotides. The genomes encode only two major open reading frames. The gene products Rep, Rep' and Cap are involved in viral replication, regulation of transcription and capsid formation. Due to their highly limited coding capacity, circoviruses are supposed to rely principally on the host's machinery for synthesis of macro-molecules.

Two types of PCV are known, which differ with respect to their pathogenicity. Porcine circovirus type 1 (PCV1) is not linked with a disease, while porcine circovirus type 2 (PCV2) is the etiological agent of Postweaning Multisystemic Wasting Syndrome (PMWS), a new emerging and multifactorial disease in swine. PCV1 and PCV2 show a high degree of sequence homology and a similar genomic organisation; nevertheless, the basis of the distinct pathogenicity has not yet been unravelled. Here the molecular biology of PCV1 and PCV2 is described.

PORCINE CIRCOVIRUS – AN OVERVIEW

CIRCOVIRUS TAXONOMY

Circoviruses contain **cir**cular **co**valently closed single-stranded DNA (ssDNA) genomes with sizes between 1759 and 2319 nucleotides (nt). ssDNA molecules are rarely seen as virus genomes, since the majority of viruses is characterised by either dsDNA or ss/dsRNA genomes. The distinct nature of their genomes has led to the family being termed *Circoviridae*. Differences in organisation of the viral genomes and capsid morphology are reflected by the classification of the *Circoviridae* into two different genera (Fig. 1). The genus *Gyrovirus* consists of only one single virus, Chicken anemia virus (CAV), while the genus *Circovirus* currently comprises Porcine circovirus type 1 (PCV1) and type 2 (PCV2), Psittacine beak and feather disease virus (BFDV), Pigeon circovirus (PiCV), Canary circovirus (CaCV) and Goose circovirus (GoCV). Duck circovirus (DuCV), Finch circovirus (FiCV) and Gull circovirus (GuCV) are considered as tentative species in the genus, while the circoviruses of raven (RaCV) and starling (StCV) have been discovered only recently and are therefore not yet included in the current taxonomic classification (Todd et al., 2005).

Phylogenetic analysis of the family *Circoviridae* revealed that the two genera display only limited relationship. The reported size for the CAV virion ranges from 19 to 26 nm, the three open reading frames (ORF) are arranged partially overlapping within a genome of (-) strand polarity and 2298 or 2319 nt in size. Only one mRNA molecule is transcribed from a promoter/enhancer region located in the untranslated region (UTR) of the genome. Splicing of the transcript has not been detected. The UTR shows a high GC content and inverted repeats, which may be able to form hairpin structures. CAV shows homology to the newly identified ssDNA viruses in humans, TT virus and the related TTM virus, all members of the floating genus *Anellovirus*.

Members of the genus *Circovirus* differ in several aspects from CAV, because i. the particles are smaller (12 – 20 nm) and have a distinct morphology, ii. circoviruses are characterised by an ambisense genome, which is divergently transcribed and iii. splicing has been reported for at least two transcripts. The genome sizes vary from 1759 to 2319 nt. Circoviruses show similarities in genome organisation, protein sequence and protein function to the

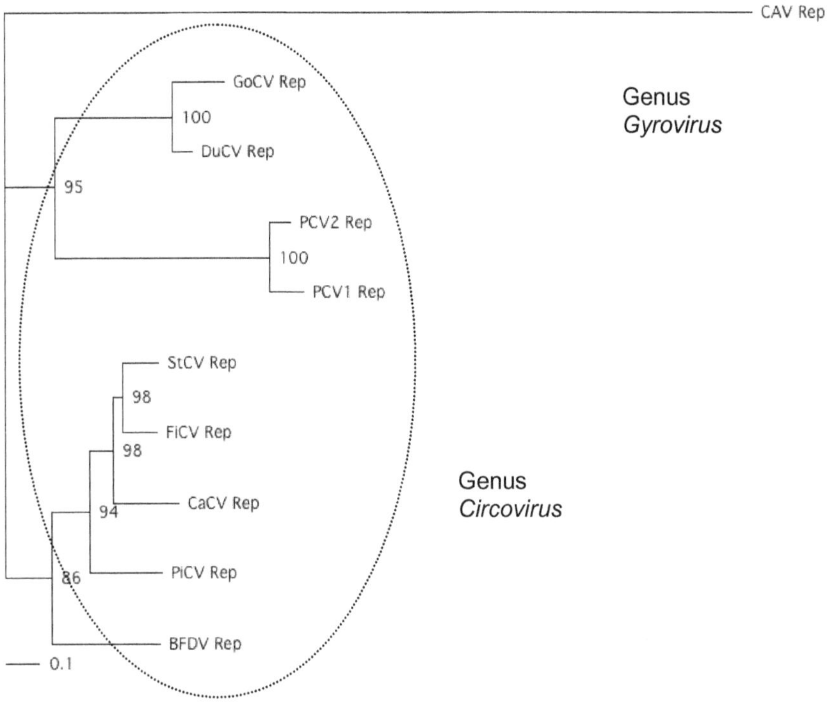

Fig. 1. Phylogenetic tree of the family *Circoviridae*. The amino acid sequences of the Rep proteins of the viruses of the genus *Circovirus* (GoCV, DuCV, BFDV, PiCV, StCV, FiCV, CaCV, PCV1, PCV2) and of the genus *Gyrovirus* (CAV) were analysed using the MacVector program package (100 data sets).

plant pathogenic *Nanoviridae* and *Geminiviridae*. It has been suggested, that the rep genes of circoviruses have evolved by a recombination event between the replication initiator domain of a nanovirus Rep protein and a helicase domain which apparently originated from an RNA-binding protein acquired from a ssRNA virus, probably a calicivirus (Gibbs and Weiller, 1999).

VIRION STRUCTURE

Circoviruses do not possess a membrane, that surrounds the capsid. The capsid of PCV1 and PCV2 is built of only one structural protein (Cap), for which a size of 30 kDa has been estimated. A basic N-terminal region contains several positively charged arginine and lysine residues, which are supposed to interact with the negatively charged DNA. This composition is seen also in capsid proteins of other circoviruses. Investigation of the structure of

PCV1, PCV2 and CAV revealed that they display an icosahedral T=1 structure containing 60 capsid protein molecules arranged in 12 pentamer clustered units (Crowther et al., 2003). PCV1 and PCV2 show similar capsid structures with flat pentameric morphological units, whereas CAV displayed protruding pentagonal trumpet-shaped units, indicating that the viruses of the same genus are structurally related, while those of viruses in different genera are not.

PATHOGENESIS OF PCV1 AND PCV2

Circoviruses are supposed to be host-specific or to have narrow host ranges. The fecal-oral route of transmission is likely, but vertical transmission has been reported in some cases. With the exception of PCV1, all known circoviruses are pathogens, which cause immune suppression and damage of lymphoreticular tissues. PCV1 and PCV2 are ubiquitously spread, but they seem to be restricted to pigs. Consensus primer-PCR-based search for a human circovirus was not successful (Hattermann et al., 2004). PCV1 does not induce a cytopathogenic effect in cultured cells and experimentally infected animals did not display any disease symptoms. In contrast, PCV2 is the etiological agent of a new disease in swine, the so-called Postweaning Multisystemic Wasting Syndrome (PMWS). PMWS was first recognised in North America in 1991. Since then it has been described as a major economic concern in most of the pig-producing areas of the world. PMWS primarily occurs in young pigs. Typical symptoms of PMWS are wasting, respiratory distress, enlargement of superficial inguinal lymph nodes, diarrhoea, paleness of the skin or icterus, but clinical signs are highly variable (Segales et al., 2006). The most consistent feature of PMWS is a generalised depletion of lymphocytes. Secondary infections with opportunistic organisms are common, indicating that the immune system is involved in PMWS pathogenesis.

Since PCV2 is also prevalent in healthy pigs, diagnosis of PMWS must concurrently meet three criteria: (i) the presence of compatible clinical signs, (ii) the presence of moderate to severe characteristic microscopic lymphoid lesions, and (iii) the presence of moderate to high amounts of PCV2 within these lesions. On affected farms, mortality may reach up to 40%, but the situation can be improved by implementation of special management plans. In first attempts, experimental reproduction of the disease according to Koch's postulates has led to a puzzling variety of results, since either no symptoms at all, only minor lesions or a full-blown PMWS were observed. Symptoms of the disease were aggravated in piglets, in which the immune system had been stimulated either by a prior vaccination or by a co-infection with porcine parvovirus (PPV) or porcine reproductive and respiratory syndrome virus (PRRSV). Later, it was demonstrated that activation of the immune system is a pivotal step in induction of PMWS (Krakowka et al., 2001). PMWS is a multifactorial disease and intrinsic factors like the status of the immune system, genetic predisposition but also practical aspects like vaccination policy and nutrition may influence the onset of the disease as well as the severity of the symptoms.

The distinct pathogenicity of PCV1 and PCV2 is of special interest for molecular virologists. To identify molecular determinants for pathogenesis, the genome nucleotide sequences of PCV1 and PCV2 isolates have been compared. Their genomes share inter-group homology of 68-76%, whereas PCV1 and PCV2 intra-group isolates share 97-99% and 95-99% sequence identity. The origin of replication (80% sequence homology) and the rep gene (82%) of the two viruses are highly conserved and exchange of the replication factors of PCV1 and PCV2 did not reveal differences: the Rep protein of PCV1 replicated the PCV1 origin with the same activity as the origin of PCV2 and vice versa, leading to the assumption that the replication factors do not contribute to the distinct pathogenesis (Mankertz et al., 2003). Only 62% homology is found between the cap genes of PCV1 and PCV2. Chimeras of PCV1 and PCV2 have been tested for their potential to stimulate the immune answer and to induce PMWS (Fenaux et al., 2004). Chimera PCV2/1 (containing the PCV1 capsid gene cloned into the backbone of the pathogenic PCV2 genome) was compared to chimera PCV1/2 (PCV2 capsid gene in PCV1). Both chimeric viruses displayed similar growth characteristics *in vitro*, but PMWS symptoms were not observed *in vivo* with either of the two chimeras. When pigs were subsequently challenged with wild-type PCV2, PCV1-2 induced protective immunity, recommending it as a vaccine candidate. ORF3 may also contribute to the pathogenesis of PCV2, because its sequence and length differs significantly in PCV1 and PCV2. ORF3(PCV2) has been described to be involved in induction of apoptosis.

MOLECULAR BIOLOGY OF PCV

GENOME ORGANISATION OF PCV

The genomes of PCV1 and PCV2 are composed of a circular ssDNA molecule with a size of 1759 and 1768 nt. They are the smallest genomes found in any mammalian virus. Two major ORFs encode proteins which carry out the two elementary functions of a virus, genome replication (rep) and packaging (cap). Circoviruses can be regarded as a paradigm for the reduction of the molecular equipment to the absolute essentials. The rep and cap genes are arranged divergently. This results in an ambisense genomic organisation and creates two intergenic regions, a larger one between the 5'-ends of rep and cap, and a shorter one between their 3'-ends. The non-coding region between the ATGs of the rep and the cap genes of PCV1 and PCV2 comprises the origin of viral genome replication (ori). Similar genomic arrangements are present in the *Geminiviridae* and *Nanoviridae*. The origin of replication is characterised by a putative stem-loop structure with a nonamer in its apex (Fig. 2). This sequence and structure element is conserved in viruses, plasmids and bacteriophages, and most of them are known to replicate via rolling circle replication (RCR; Mankertz et al., 1997). Adjacent to the stem-loop, hexamer and pentamer repeats are found, but their arrangement differs slightly between PCV1 and PCV2.

VIRAL ORFs AND PROTEINS

Two major ORFs are encoded by the genomes of PCV1 and PCV2 (Fig. 2): ORF1 (also termed V1) is located on the viral plus-strand and encodes the two proteins necessary for viral replication, Rep and Rep', which are produced from differentially spliced transcripts. A full-length transcript directs synthesis of the Rep protein (PCV1: 312 aa, 35.6 kD), in which three motifs typical for enzymes mediating replication in the RCR mode and a dNTP-binding domain have been identified. Splicing results in removal of an intron (nt 1176 to 1558) and leads to expression of Rep´. Rep´ is truncated to 168 aa (19.2 kD), it does not comprise a dNTP-binding domain and the last 48 aa are expressed in a different reading frame. The frameshift from Rep to Rep' coincides with the suggested point of recombination between the Rep protein of the nanoviruses and the RNA-directed RNA polymerase (2C protein) of the caliciviruses (Gibbs and Weiller, 1999). Synthesis of more than one Rep protein

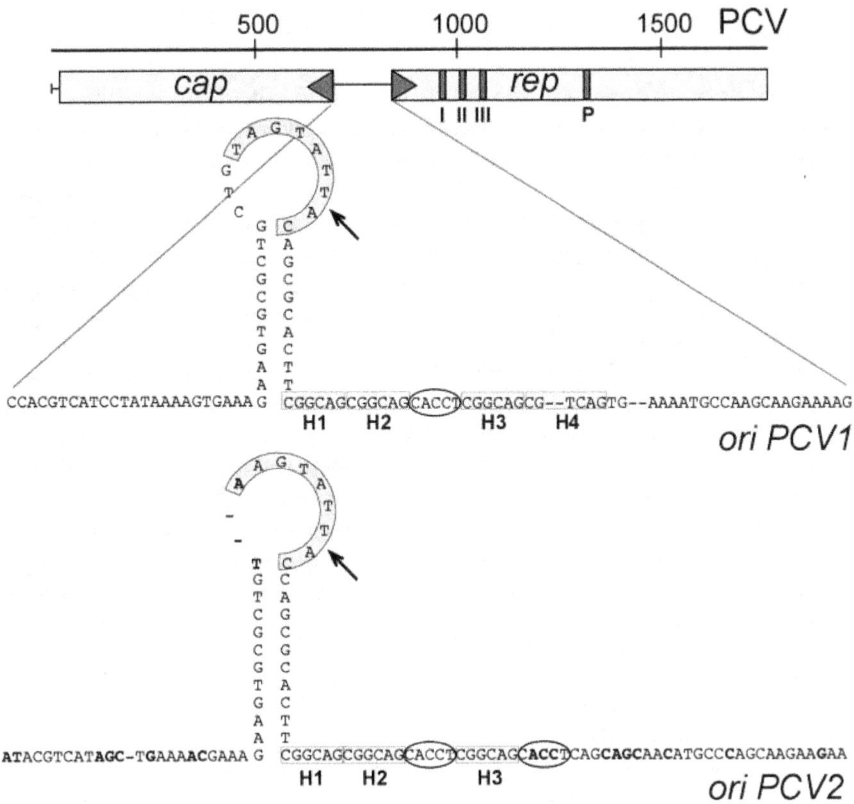

Fig. 2. A map of PCV. A linear map of the PCV genome indicates the two major ORF, rep and cap. Three motifs conserved in RCR enzymes (I-III) and a dNTP binding domain (P) are indicated within the rep gene. The lower part shows a comparison of the two origin regions of PCV1 and PCV2. Sequences have been aligned, hexamer repeats 1-4 are marked by open boxes, pentamer repeats by open ovals. The conserved nonamer within the ss loop of the hairpin is indicated by a grey box, the nicking site by an arrow.

variant by splicing is a well-established feature in other ssDNA viruses, e.g. in Maize streak virus or Adeno-associated virus type 2, but normally only one of the splice variants is the essential master Rep for initiation of replication. In contrast, Rep protein of PCV alone is unable to promote replication. It was demonstrated, that Rep and Rep' together comprise the functional replication initiator factor of PCV (Mankertz and Hillenbrand, 2001). Interestingly, replication via the PCV-origin did occur in presence of only Rep, when replication of PCV-based constructs was analysed in E.coli (Cheung, 2006). ORF2, the second largest ORF of the circoviruses is located on the counter-clockwise strand (C1) and encodes the major structural protein Cap protein (234 aa). Cap displays an N-terminus rich in basic amino acids such as arginine and lysine, and this region of the protein is assumed to bind to the viral ssDNA. After expression in bacteria or insect cells, Cap protein monomers of PCV assembled into virus-like particles when viewed by electron microscopy.

Smaller ORFs have been identified for all circoviruses, but, with the exception of ORF3, their expression has not been demonstrated. ORF3 is transcribed counterclockwise from ORF1 of PCV1 and PCV2, but the promoter has not yet been mapped. This ORF differs in sequence and size between the two PCV variants: a longer coding region of 621 codons is present in PCV1, while it is truncated in PCV2 to 315 codons. Up to now, only the PCV2-encoded protein has been studied: Its expression is not essential for PCV2 replication, but it induces apoptosis via caspase 8 and caspase 3 (Liu et al., 2005). ORF3 is assumed to be involved in viral pathogenesis via its apoptotic activity. Seroconversion was seen with both variants, when wild-type PCV2 and a mutant virus deprived of the ORF3 protein were studied in vivo (Liu et al., 2006). The copy number in serum was much higher with the wildtype PCV2, while the mutant failed to induce microscopic lesions similar to PMWS. This suggests, that ORF3 may play a role in pathogenesis of PCV2 induced diseases as PMWS. Nevertheless, this hypothesis is in conflict with the results obtained with chimeric viruses (Fenaux et al., 2004). When the cap gene of PCV1 has been inserted into PCV2, no pathogenic effects were observed. Since ORF3(PCV2) was present in this construct, it cannot be the only factor contributing to PCV2-linked pathogenesis.

TRANSCRIPTION OF PCV

The promoter of the cap gene of PCV1 has been mapped to a fragment within the rep gene (pos. 1168-1428, Fig. 3). Since Rep, Rep' and Cap did not influence transcription initiated at Pcap, it does not seem to be regulated by virus-encoded proteins (Mankertz and Hillenbrand, 2002). The cap transcript starts at pos. 1238, and an untranslated leader sequence of 119 nt (1238 to 1120) is joined to exon 2 of the ORF1 transcript at nt 737 immediately adjacent to the ATG (Mankertz et al., 1998). Processing of this RNA may have evolved to counteract synthesis of an alternative protein by use of an internal start codon in the intron. The promoter of the rep gene, Prep, is comprised within a fragment between nt 640-796, which overlaps with the origin of replication. Mapping of the rep transcript in PCV1 revealed that the starting point is located at position 766. The ATG of the rep gene is situated at position 819 and a TATA box is found at position 739. S1 analysis, RACE and PCR revealed that two differentially spliced transcripts are produced in PK15 cells persistently infected with PCV. A real-time PCR discriminating between the two transcripts indicated a variation of the ratio of the two transcripts in correlation to time (Mankertz and Hillenbrand, 2001). When the influence of virus-encoded proteins upon Prep was investigated, it became apparent that Prep is repressed by Rep but not by Rep' or Cap. Rep protein exerts negative control towards its own transcription by binding to hexamers H1/H2, the same sequence elements which have been mapped as essential for initiation of replication. Rep´ also binds to H1/H2, but this step apparently does not result in repression of Prep activity. Since mutagenesis of H1/H2 decreases but does not abolish Prep transcription, other features must contribute to repression, e.g. Rep protein may interact with transcription factors.

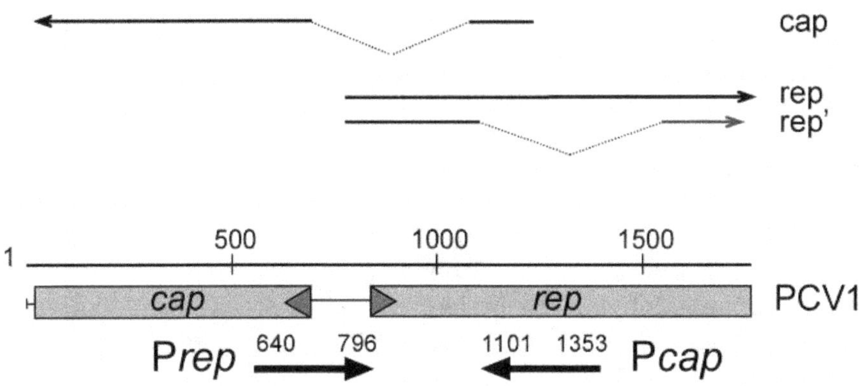

+ Rep	repression	no influence
+ Rep'	no influence	no influence
+ Cap	no influence	no influence

Fig. 3. Transcription of PCV. The upper part shows the three major transcripts cap, rep and rep', which are expressed from PCV1 and PCV2. Rep' results from differential splicing. Promoters of the rep and cap gene of PCV1 have been mapped. Prep is negatively regulated by Rep, but not by Rep' or Cap, while Pcap is not subject to autoregulation by PCV-encoded proteins.

ENTRY INTO THE CELL

Several studies have been carried out to identify the main target cell for viral replication in the infected animal, yet it is still unknown. Since PCV2 was seen *in vivo* in a broad range of cell types including hepatocytes, enterocytes, epithelial and endothelial cells, lymphocytes and fibroblasts, PCV2 may not enter the cell via a rarely expressed receptor. When binding of PCV2 to monocytic cells was investigated, it became evident that cell surface glycosaminoglycans heparan sulfate and chondroitin sulfate B are attachment receptors for PCV2. This result is supported by the finding that a heparan sulfate-binding motif (XBBXBX, B = basic amino acid, X = neutral/hydrophobic amino acid) is present on the PCV2 capsid protein (Misinzo et al., 2006). When the entry of PCV2 into monocytic cells was studied, infection was decreased significantly by chemical inhibitors that specifically blocked actin-dependent processes and clathrin-mediated endocytosis or with lysosomotropic bases (Misinzo et al., 2005). A similar effect was not observed when macropinocytosis and caveolae-dependent endocytosis were inhibited, suggesting that PCV2 enters the cells predominantly via clathrin-mediated endocytosis and that the infectious entry requires an acidic environment.

REPLICATION FACTORS

PCV are assumed to replicate their genomes via a circular dsDNA molecule, the replicative intermediate (RI), which is synthesised by host cell enzymes during the S phase of the cell cycle. For initiation of the subsequent virus-controlled replication step, a *cis*-acting factor (the origin of replication, ori) and a *trans*-acting factor (the replicase, Rep) are required. Replication activity was observed, when the putative ori and the potential rep gene were cloned and cotransfected into porcine kidney cells (PK15). This set-up allowed fine-mapping of the *cis*- and *trans*-acting factors by truncation and site directed mutagenesis. Accordingly, the origin of replication (ori) of PCV has been mapped to the intergenic region between the start points of the two major ORFs (pos. 691-816, Fig. 2). The origin of replication comprises a potential stem-loop structure with a nonamer (5´-TAGTATTAC) in its apex, whose mutagenesis resulted in inactivation of PCV replication (Mankertz et al., 1997; Cheung, 2004). Adjacent to the nonamer, hexamer and pentamer motifs are located. The hexamers serve as the binding site for the rep gene products, Rep and Rep'. Rep and Rep' exhibit three amino acid motifs (Fig. 2) which bear resemblance to the signatures of RCR proteins (Koonin and Ilyina, 1993). Whereas the function of motif I (FTLNN) has not yet been unravelled, motif II (HLQGF) is involved in metal ion binding and motif III (YCSK) comprises the catalytic site for DNA cleavage. Nonamer, stem-loop and hexamer repeats on the DNA sequence as well as the motifs of Rep are conserved not only within the circoviruses and the family *Nanoviridae* and *Geminiviridae*, but, although to a lesser extent, in several bacterial plasmids and bacteriophages replicating via RCR. These similarities have strongly suggested, that circoviruses replicate via a RCR-like replication mechanism. Besides the three RCR motifs, which are situated in the N-terminal part of Rep, a dNTP-binding motif has been identified in the centre of the protein. Site-directed mutagenesis of motif II and motif III which are responsible for DNA cleavage, abolished replication activity (Steinfeldt et al., 2006). Alteration of the first motif (FTLNN to LTLKN) reduced replication activity, while mutagenesis of the dNTP-binding motif GKS to GRI decreased the replication activity to the limits of detection. Although deletion of the C-terminus retained the four conserved motifs, the shortened Rep proteins were not able to promote replication, indicating that not only the four conserved motifs but also the C-terminus of Rep are essential for replication of PCV (Mankertz and Hillenbrand, 2001).

LOCALISATION OF VIRAL PROTEINS

Subcellular distribution of the viral proteins was investigated either by immune fluorescence using antisera against Rep, Rep', and Cap or by expression of viral proteins which were fused to a fluorescent protein (Finsterbusch et al., 2005). When these fusion proteins were studied in PK15 cells, Rep and Rep' colocalised within the nucleoplasm, although distinct subnuclear compartments were free of label. Subsequently Cap was found to be located in these areas which by staining with a specific antibody were identified as nucleoli. Nucleoli are globular, electron-dense particles of 1-2 µm. They consist of nascent rRNA transcripts transcribed from hundreds of rRNA genes in tandem arrays, which are located at nucleolar organising regions on chromosomes 13, 14, 21 and 22 (Carmo-Fonseca et al., 2002). Protein components of the nucleolus such as nucleolin, nucleophosmin and fibrillarin have been identified. They are involved in ribosome biogenesis, gene silencing, mRNA metabolism, translation and cell cycle regulation. Localisation of viral proteins in the nucleoli has been described for several RNA and DNA viruses (Hiscox, 2002) and it has been proposed that this supports either viral transcription or influences the cell cycle. When protein localisation was analysed in the context of an infection, localisation of Rep and Rep' was unaltered, while a distinct localisation pattern was observed for the Cap protein. Since Cap was seen either in the nucleoli or throughout the entire nucleus, a redistribution of Cap during different phases of viral infection was assumed. A kinetic analysis revealed that newly synthesised or internalised Cap protein gave weak fluorescence signals in the cytoplasm at 4 h pi. At 12 h and 16 h pi, the Cap protein was located predominantly in the nucleoli of infected cells, while redistribution to the nucleoplasm was seen at 24 h pi. At 48 h pi, Cap was localised in the nucleoplasm in the majority of the infected cells but also occasionally in the cytoplasm, whereas almost no Cap-specific signal was observed in the nucleoli. This change of subnuclear localisation may coincide with the viral DNA replication or virion assembly. When localisation of Cap was investigated in human and simian cells, Cap was seen in the nucleus but not in the nucleoli. Since PCV infection does not lead to synthesis of virus particles in these cells, although protein expression and genome replication was demonstrated (Hattermann et al., 2004), protein localisation may be an important prerequisite for virus replication activity.

Cap did not influence nuclear targeting of Rep and Rep' in PK15 cells. Their common N-terminus contains three clusters of basic amino acids with homology to nuclear localisation signals (NLS), termed NLS1 ($_{12}$**KR**WVFTLNNPSEEE**K**N**K**I**R**ELP$_{33}$; bold letters indicate basic aa), NLS2 ($_{63}$**KK**QTFN**K**V**K**WYFGA**R**CHIE**K**A**K**-GTD$_{87}$) and NLS3 ($_{107}$P**R**NQG**KR**SDL$_{116}$) (Finsterbusch et al. 2005). Truncation analysis revealed that aa 14 to 29 of NLS1 are essential for nuclear localisation of Rep and that NLS2 plus NLS3 are unable to traffic the truncated protein to the nucleus. Mutagenesis of basic aa in NLS1 or in NLS2 abolished nuclear localisation of the fusion proteins, while exchange of basic aa in NLS3 did not interfere with nuclear accumulation of the protein. Therefore, it is assumed that NLS1 and NLS2 mediate nuclear import, while NLS3 serves as an enhancer of nuclear import of Rep and Rep'.

REP AND REP' PROTEIN MEDIATE DNA REPLICATION

Binding to DNA

Binding of the Rep and Rep‘ proteins of PCV1 to dsDNA fragments of the origin of replication containing the stem-loop structure plus the conserved nonamer and the four hexamer repeats has been demonstrated in vitro (Steinfeldt et al., 2001).Truncation of the origin fragments identified a minimal binding site (MBS) for each protein (Fig. 4C): Rep protein binds to the right part of the stem-loop plus the two inner hexamer repeats H1/H2, while Rep' protein bound also to sequences in which the stem-loop was mutagenised, indicating that Rep and Rep' may have different requirements with respect to the sequence and the length of their MBS. It is assumed that recruitment of the replication proteins for initiation of replication at the double-stranded origin is mediated by binding to the MBS. Rep/Rep' bind alternatively to H1/H2 or to H3/H4 and binding of Rep and Rep' proteins to the two inner (H1/H2) or outer hexamers (H3/H4) resulted in complexes that were retarded differently: A larger complex was observed if Rep/Rep' bound to H1/H2, while a smaller complex was seen when Rep/Rep' bound to H3/H4. The difference in electrophoretic mobility was attributed to a variable protein stoichiometry of the two complexes: Provided that H1/H2 and H3/H4 are not equivalent, it is assumed that a higher number of Rep/Rep‘ monomers can be loaded onto H1/H2 than onto H3/H4 (Fig. 4A and B). Thus, the following model for Rep/Rep' binding has been proposed: Rep/Rep' binds preferentially to the MBS, which is composed of the hexamers H1/H2 and in case of Rep the right leg of the stem-loop. If H1/2 is not available, the proteins bind alternatively to H3/4. Since more Rep/Rep' molecules can bind to H1/2 than to H3/4, the protein:DNA ratio is different for H1/2 and H3/4.

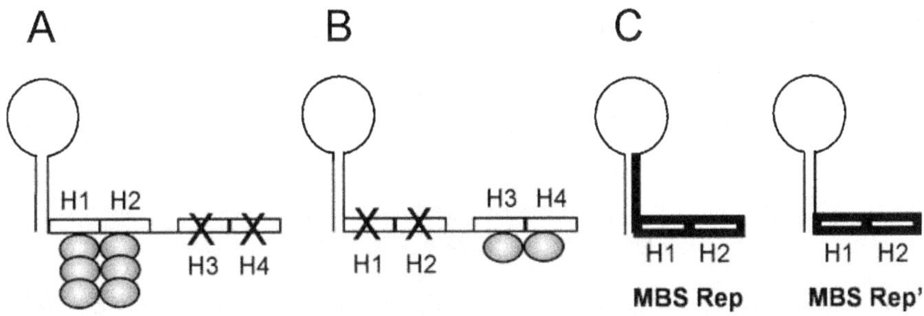

Fig. 4. A model for Rep and Rep' binding to the origin of PCV1. (A) Rep/Rep' bind preferentially to the MBS. (B) If H1/2 is not available, the proteins bind alternatively to H3/4. More Rep/Rep' molecules can be loaded upon H1/2 than to H3/4. The different MBS of Rep and Rep' are given in 4C.

Restriction of DNA

The introduction of a single-strand break generates a free 3′-hydroxyl group, which serves as a primer for subsequent DNA synthesis. A DNA-nicking capability is therefore a prerequisite for replication of PCV by RCR. The ability of PCV1 Rep and Rep′ to restrict DNA was demonstrated in vitro with synthetic ss oligonucleotides corresponding to the origin of replication of PCV1. Rep/Rep' is able to cleave the viral strand between nucleotides 7 and 8 (Fig. 2) within the conserved nonanucleotide motif (5' TAGTATT/AC 3') located at the apex of the putative stem-loop structure (Steinfeldt et al., 2006). $Mg2^{+}$ or other bivalent cations are necessary for in vitro cleavage activity of PCV1 Rep, while ATP-binding or hydrolysis is not. When the sequence and the secondary structure required for the restriction by Rep/Rep' was determined, mutation of the entire nonamer sequence or the first four nucleotides resulted in total loss of cleavage activity, while alteration of the first three nucleotides within the nonamer did not impede cleavage activity of Rep and Rep'. Thus, the canonical nonamer sequence $T^{1}AGTATTAC^{9}$ can be condensed to the hexamer $T^{4}ATTAC^{9}$ sufficient for cleavage and ligation of origin fragments in vitro, while six nucleotides within the octamer $A^{2}xTAxTAC^{9}$ are essential for PCV DNA replication in vivo (Cheung, 2004). Since oligonucleotides disabled in formation of a hairpin structure by sequence alteration of only one part of the inverted repeat were cut efficiently, formation of this secondary structure element is not a prerequisite for cleavage of origin fragments in vitro. The MBS of PCV1 comprising the right part of the putative stem-loop and the two adjacent hexamers is essential for recruiting Rep and Rep' to ds origin fragments, but its specific sequence is not required for recognition and cleavage of ss origin fragments in vitro. These findings have been corroborated by results for PCV1 replication in cultured cells, suggesting specific interaction of the replication proteins with the nonamer. Nevertheless, in vitro cleavage activity relies also upon sequences flanking the nonamer upstream and downstream, because sequence alterations enclosing both parts led to complete inhibition of cleavage despite the presence of the hexamer sequences.

Although Rep and Rep' bind to ds templates, cleavage activity was observed only for ss templates. It was assumed that cruciform extrusion of dsDNA templates provides the ssDNA conformation necessary for cleavage in vivo, like in the case of plasmids of the pT181 family (Khan, 2003). Alternatively, local melting at the viral origin of replication by a helicase activity of the Rep protein may facilitate cruciform extrusion. Since a putative helicase domain has been identified in the C-terminus of Rep, which is related to the helicases of other ssDNA and dsDNA viruses, one may speculate that this task could be performed by the Rep protein. Although the impact of the C-terminally located GKS box of PCV1 Rep on replication in cultured cells has been demonstrated, ATP-binding or hydrolysis is not necessary for in vitro cleavage activity of the origin of replication of PCV1 by Rep.

Termination of DNA replication

Rep does not only initiate the replication of PCV, but is also involved in its termination. The dual "nicking/joining" activities associated with Rep and Rep´ are pivotal events underlying the RCR-based replication of PCV in mammalian cells (Steinfeldt et al., 2006). Termination of viral replication is accomplished by ligation of the 5'-phosphate to the newly created 3'-hydroxyl group and release of a circular ssDNA molecule. Rep and Rep' are nucleotidyltransferases, which catalyze the joining of ssDNA fragments of the viral origin in vitro. The nicking/closing activity is strictly dependent on preceding substrate cleavage, while ATP is not needed for this reaction. Joining of origin fragments depends on the close proximity of the origin fragments, which is accomplished by base pairing of the stem-loop structure, since mutations affecting the right or the left arm of the hairpin reduced joining activity. This effect was abolished when the specific sequence of the hairpin was replaced by an unrelated sequence which, however, was still capable of base pairing. These results signify that base pairing is essential for joining of the origin fragments in vitro.

Cleavage is a prerequisite for joining, since oligonucleotides mimicking the cleavage products were not ligated by Rep and Rep'. The activity was restored by addition of the uncleaved origin fragment. This led to the assumption that Rep and Rep' are linked to the 5'-phosphate of the 3'-part of the cleaved origin fragment. Such a complex would result in conservation of energy and would moreover provide an explanation for the independence of the cleavage / joining reaction from ATP.

Mapping of the catalytical centre for DNA restriction

After one round of replication (or more), a second Rep/Rep' cleavage terminates DNA synthesis and the nucleotidyltransfer results in the release of a newly synthesised circular ssDNA molecule. This second cleavage can be performed by a second reactive amino acid of the same molecule, as in case of the RepA protein of bacteriophage ØX174, which contains two closely spaced tyrosyl groups at its active site, that participate alternating during initiation and termination of replication. Likewise, the second catalytic center can be provided by a second Rep protein in a dimer or multimer. For instance, Tyr-103 of the Rep protein of Tomato Yellow Leaf Curl Virus has been identified to mediate the cleavage and joining reaction in a multimeric complex. Mutation of Tyr-93 in motif III of the Rep protein of PCV1 led to the complete loss of nicking and closing activity. Therefore, Tyr-93 is assumed to be the catalytical amino acid responsible for cleaving the phosphodiester bond (T. Steinfeldt, unpublished). The hypothesis, that Rep/Rep' form a dimer or multimer which is responsible for initiation and termination of RCR has been supported by the detection of homo- and heterooligomeric complexes between Rep and Rep' in vitro and in vivo (T. Finsterbusch, unpublished).

INTERACTION OF PCV PROTEINS WITH HOST FACTORS

Studies investigating the interaction of PCV encoded proteins led to puzzling results, because the outcome was depending on the system used for expression. While two-hybrid analysis in bacteria revealed interaction of Rep and Cap (Timmusk et al., 2006), this was not observed in yeast cells, suggesting that post-translational modifications of Rep and Cap may significantly modulate their function. Rep and Rep' have been observed to interact in yeast cells which was verified by coimmunoprecipitation from mammalian cells, the natural target of PCV (T. Finsterbusch, unpublished). When the interaction of viral proteins with that of the host was investigated, the PCV2 replicase protein was found to interact with the intermediate filament protein syncoilin and c-myc, a key regulator of transcription which is involved in many cellular processes. The capsid protein interacted with the complement factor C1qB, which is a subunit of the C1 enzyme complex activating the complement cascade. Another interaction partner of Cap protein identified was P-selectin, a protein that is rapidly expressed at the surface of activated or injured cells. The ORF3 product bound to a sequence similar to the RGS16 (regulator of G-protein signalling 16) protein, suggesting that the ORF3 protein may interfere with the G signalling of infected cells.

A MODEL FOR RCR REPLICTION OF PCV

PCV enters the cells via attachment to GAG and clathrin-mediated endocytosis. In permissive cells, the viral ssDNA is converted by host enzymes into the ds replicative intermediate (Fig. 5). Rep and Rep' are recruited to the dsDNA origin by binding to the MBS. Rep inherent helicase activity induces a cruciform extrusion, providing the ss conformation of the nonamer, which is necessary for cleavage. The site-specific cleavage of the viral strand between nucleotides 7 and 8 within the conserved nonamer is the initial step for the subsequent RCR of the viral DNA. Restriction results in generation of a free 3'-hydroxyl group, that serves as a primer of plus strand DNA synthesis. Replication is terminated by another Rep/Rep' molecule in a dimer or multimer.

Transcripts are expressed from the dsDNA, mRNA is exported from the nucleus, virus-encoded proteins are synthesised in the cytoplasm and re-imported into the nucleus via their NLS. Rep represses the transcription of Prep. Interaction of Rep, Rep' and Cap with cellular factors provides an interface to the host. The steps leading to assembly and release of the virus have not yet been studied in detail.

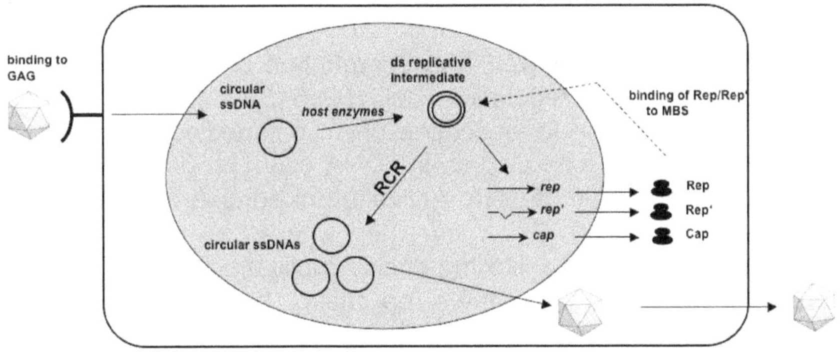

Fig. 5. Lifecycle of PCV. PCV uses glycosaminoglycans as attachment molecule. The ssDNA genome is transported into the nucleus and converted presumably by host enzymes into a dsDNS intermediate. rep and cap mRNAs are transcribed, the proteins are synthesised and reimported from the cytoplasm. Rep/Rep' bind to the dsDNA and initiate rolling circle replication by introduction of a nick that serves as primer. Events leading to assembly and release of virions are not yet understood.

CONCLUSION

During the last few years, we have learned a lot about the molecular life-cycle of PCV. PCV1 and PCV2 are doubtlessly interesting objects of molecular virology: due to their limited coding capacity, they must rely highly on the information content of the host cell. Moreover the two strains display a distinct pathogenicity: PCV1 is not linked to a disease, while PCV2 induces PMWS, a severe multifactorial disease in swine, which compromises and unbalances the immune system. Although several studies have been aimed at identification of a molecular trigger for the onset of PMWS, the results provided more questions than answers. The attraction of PCV can perhaps be seen in the apparent simplicity of the circoviral genome that contrasts highly with the poorly understood pathogenesis.

ACKNOWLEDGEMENTS

I thank all members of the Mankertz Lab at the Robert Koch-Institut for their commitment to our work. Research in the PCV research group was supported by grants from the Robert Koch-Institut, the Freie Universitaet Berlin (Nafoeg Programm), the Deutsche Forschungsgemeinschaft (Ma2126/2-1) and the European Union (5th FP, QLK2-CT-1999-00307, and 6th FP, 513928, PCVD).

REFERENCES

Carmo-Fonseca, M. (2002). The contribution of nuclear compartmentalization to gene regulation. Cell *108*, 513-521.

Cheung, A.K. (2004). Identification of an octanucleotide motif sequence essential for viral protein, DNA, and progeny virus biosynthesis at the origin of DNA replication of porcine circovirus type 2. Virology *324*, 28-36

Cheung, A.K. (2006). Rolling-circle replication of an animal circovirus genome in a theta-replicating bacterial plasmid in Escherichia coli. J. Virol. *80,* 8686-8694.

Crowther, R.A., Berriman, J.A., Curran, W.L., Allan, G.M., and Todd, D. (2003). Comparison of the structures of three circoviruses: chicken anemia virus, porcine circovirus type 2, and beak and feather disease virus. J. Virol. *77*, 13036-13041.

Fenaux, M., Opriessnig, T., Halbur, P.G., Elvinger, F., and Meng, X.J. (2004). A chimeric porcine circovirus (PCV) with the immunogenic capsid gene of the pathogenic PCV type 2 (PCV2) cloned into the genomic backbone of the nonpathogenic PCV1 induces protective immunity against PCV2 infection in pigs. J. Virol. *78*, 6297-6303.

Finsterbusch, T., Steinfeldt, T., Caliskan, R., and Mankertz, A. (2005). Analysis of the subcellular localization of the proteins Rep, Rep' and Cap of porcine circovirus type 1. Virology *343*, 36-46. Erratum in: Virology *349*, 492 (2006).

Gibbs M.J., and Weiller G.F. (1999). Evidence that a plant virus switched hosts to infect a vertebrate and then recombined with a vertebrate-infecting virus. Proc. Natl. Acad. Sci. USA *96*, 8022-8027.

Hattermann, K., Maerz, A., Slanina, H., Schmitt, C., and Mankertz, A. (2004). Assessing the risk potential of porcine circoviruses for xenotransplantation: consensus primer-PCR-based search for a human circovirus. Xenotransplantation *6*, 547-50.

Hattermann, K., Roedner, C., Schmitt, C., Finsterbusch, T., Steinfeldt, T., and Mankertz, A. (2004). Infection studies on human cell lines with porcine circovirus type 1 and porcine circovirus type 2. Xenotransplantation *11*, 284-294.

Hiscox, J.A. (2002). The nucleolus--a gateway to viral infection? Arch. Virol. *147*, 1077-1089.

Khan, S.A. (2003). DNA-protein interactions during the initiation and termination of plasmid pT181 rolling-circle replication. Prog. Nucleic Acid Res. Mol. Biol. *75*, 113-37.

Krakowka, S., Ellis, J.A., McNeilly, F., Ringler, S., Rings, D.M., and Allan, G. (2001). Activation of the immune system is the pivotal event in the production of wasting disease in pigs infected with porcine circovirus-2 (PCV-2). Vet. Pathol. *38*, 31-42.

Koonin, E.V., and Ilyina, T.V. (1993). Computer-assisted dissection of rolling circle DNA replication. Biosystems *30*, 241-268.

Liu, J., Chen, I., and Kwang, J. (2005). Characterization of a previously unidentified viral protein in porcine circovirus type 2-infected cells and its role in virus-induced apoptosis. J. Virol. *79*, 8262-8274.

Liu, J., Chen, I., Du, Q., Chua, H., and Kwang, J. (2006). The ORF3 protein of porcine circovirus type 2 is involved in viral pathogenesis in vivo. J. Virol. *80*, 5065-5073.

Mankertz, A., Persson, F., Mankertz, J., Blaess, G., and Buhk, H.J. (1997). Mapping and characterization of the origin of DNA replication of porcine circovirus. J. Virol. *71*, 2562-2566.

Mankertz, A., Mankertz, J., Wolf, K., and Buhk, H.J. (1998). Identification of a protein essential for replication of porcine circovirus. J. Gen. Virol. *79*, 381-384.

Mankertz, A., and Hillenbrand, B. (2001). Replication of porcine circovirus type 1 requires two proteins encoded by the viral rep gene. Virology *279*, 429-438.

Mankertz, A., and Hillenbrand, B. (2002) Analysis of transcription of Porcine circovirus type 1. J. Gen. Virol. *83*, 2743-2751.

Mankertz, A., Mueller, B., Steinfeldt, T., Schmitt, C., and Finsterbusch, T. (2003). New reporter gene-based replication assay reveals exchangeability of replication factors of porcine circovirus types 1 and 2. J. Virol. *77*, 9885-9893.

Misinzo, G., Meerts, P., Bublot, M., Mast, J., Weingartl, H.M., and Nauwynck, H.J. (2005). Binding and entry characteristics of porcine circovirus 2 in cells of the porcine monocytic line 3D4/31. J. Gen. Virol. *86*, 2057-2068.

Misinzo, G., Delputte, P.L., Meerts, P., Lefebvre, D.J., and Nauwynck, H.J. (2006). Porcine circovirus 2 uses heparan sulfate and chondroitin sulfate B glycosaminoglycans as receptors for its attachment to host cells. J. Virol. *80*, 3487-3494.

Segales J., Allan G.M., and Domingo M. (2005). Porcine circovirus diseases. Anim. Health Res. Rev. *2*, 119-142.

Steinfeldt, T., Finsterbusch, T., and Mankertz, A. (2001). Rep and Rep' protein of porcine circovirus type 1 bind to the origin of replication in vitro. Virology *291*, 152-160.

Steinfeldt, T., Finsterbusch, T., and Mankertz, A. (2006). Demonstration of nicking/joining activity at the origin of DNA replication associated with the rep and rep' proteins of porcine circovirus type 1. J. Virol. *80*, 6225-6234.

Timmusk, S., Fossum, C., and Berg, M. (2006). Porcine circovirus type 2 replicase binds the capsid protein and an intermediate filament-like protein. J. Gen. Virol. *87*, 3215-3223.

Todd, D., Bendinelli, M., Biagini, P., Hino, S., Mankertz, A., Mishiro, S., Niel, C., Okamoto, H., Raidal, S., Ritchie, B. and Teo, G. (2005). Circoviridae. In Virus Taxonomy, Eighth report of the International Committee on Taxonomy of Viruses. Fauquet C., Mayo M.A., Maniloff J., Desselberger U. and Ball L.A. eds. (Amsterdam, The Netherlands: Elsevier), pp. 327-334.

Chapter 9

Molecular Biology of Animal Herpesviruses

Thomas C. Mettenleiter, Günther M. Keil and Walter Fuchs

ABSTRACT

Herpesviruses are highly successful pathogens infecting animals and man. Although there is a wide variety of different herpesviruses with different biological characteristics, they have in common basic properties such as morphology of the virion, highly regulated transcription and establishment of latency. In animal virology the most important herpesviruses belong to the *Alphaherpesvirinae*. Research on pseudorabies virus, the causative agent of Aujeszky's disease in pigs, has pioneered animal disease control with genetically modified vaccines. PrV is now extensively studied as a model for basic processes during lytic herpesvirus infection, and for unravelling molecular mechanisms of herpesvirus neurotropism, whereas bovine herpesvirus 1, the causative agent of bovine infectious rhinotracheitis and pustular vulvovaginitis, is analyzed to elucidate molecular mechanisms of latency. The avian infectious laryngotracheitis virus is phylogenetically distant from these two viruses and serves to underline similarity and diversity within the *Alphaherpesvirinae*.

INTRODUCTION

Herpesviruses have been grouped together primarily based on a common virion morphology (Fig. 1) and shared biological properties. The herpes virion consists of four distinguishable substructures: the core which contains the genomic linear double-stranded DNA, possibly wrapped around a proteinaceous support; the icosahedral capsid of T=16 symmetry; the envelope, a lipid bilayer derived from host cell membranes into which virally encoded (glyco)proteins are embedded; and the tegument which, similar to the matrix in RNA viruses, links envelope and capsid (Roizman and Pellet, 2001). Herpesvirus genomes vary in size between ca. 124 kbp (Varicella-Zoster-Virus; Davison and Scott, 1986) and close to 300 kbp (Koi-Herpesvirus; Aoki et al., 2007). Most of them contain extensive regions of repetitive sequences whose location in the viral genome has prompted the classification of herpesvirus genome structures A-E (Roizman and Pellet, 2001).

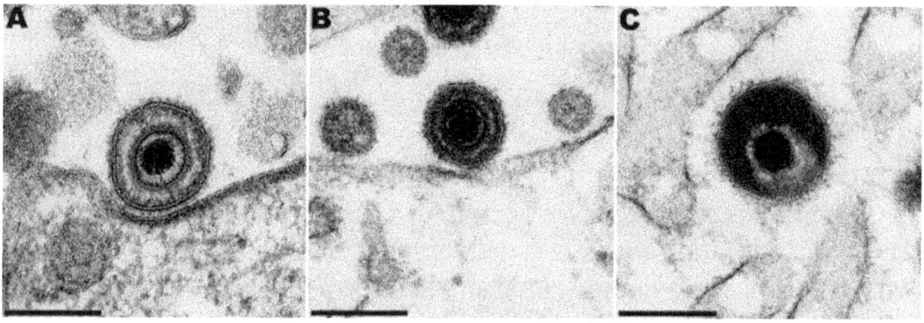

Fig. 1. Electron micrographs of thin sectioned PrV (A), BHV-1 (B) and ILTV (C) virions. Bar represents 200 nm. Pictures by courtesy of Dr. Harald Granzow, Friedrich-Loeffler-Institut, Insel Riems.

The family *Herpesviridae* comprises three subfamilies designated as *Alphaherpesvirinae*, *Betaherpesvirinae* and *Gamma-herpesvirinae* which contain herpesviruses of mammalian, avian and reptilian origin. Whereas there is a substantial level of sequence homology between many deduced proteins of these viruses, herpesvirus genomes of amphibians, fish and molluscs show only rudimentary similarity at the sequence level which prompted the proposal to establish a new taxonomic entity, the order 'Herpesvirales' containing the 'classical' *Herpesviridae* but also including two new families, the *Alloherpesviridae* which include fish and amphibian herpesviruses, and the *Malacoherpesviridae* containing the hitherto only identified invertebrate herpesvirus, ostreid herpesvirus 1 isolated from Pacific oyster (McGeoch et al., 2006).

Herpesviruses are important human and animal pathogens. In humans, eight different herpesviruses have so far been identified, designated as human herpesvirus 1 to 8 (HHV 1-8; Roizman and Pellet, 2001). However, most are widely known by their generic names such as herpes simplex viruses types 1 and 2 (HHV-1, HHV-2), varicella-zoster virus (HHV-3), Epstein-Barr virus (HHV-4), and human cytomegalovirus (HHV-5). In animals, herpesviruses cause serious, potentially life-threatening diseases but also milder illnesses which nevertheless may have a detrimental impact on economic success of animal husbandry. Of prime importance are the alphaherpesviruses pseudorabies virus (Suid herpesvirus 1), bovine herpesvirus 1, equid herpesviruses 1 and 4, infectious laryngotracheitis virus (gallid herpesvirus 1) and Marek's disease virus (gallid herpesvirus 2), and the gammaherpesviruses causing malignant catarrhal fever in ruminants, ovine herpesvirus 2 and alcelaphine herpesvirus.

In this review, selected aspects of the molecular biology of three animal alphaherpesviruses will be summarized. Pseudorabies virus (PrV) and bovine herpesvirus 1 (BHV-1) are closely related members of the genus *Varicellovirus*, whereas infectious laryngotracheitis virus (ILTV) is the type member of the genus *Iltovirus* (McGeoch et al., 2006). Phylogenetically, ILTV is the alphaherpesvirus most distant to the varicelloviruses (McGeoch et al., 2000) reflecting early separation from the evolutionary branch leading to mammalian herpesviruses and ongoing coevolution with its avian host, chicken. The recently determined DNA sequence of the parrot pathogen psittacid herpesvirus (PsHV-1) exhibits many similarities to ILTV, and, therefore, PsHV-1 should also be assigned to the same genus (Thureen and Keeler, 2006). Recent extensive reviews on each of these viruses are available (Pomeranz et al., 2005; Fuchs et al., 2007; Muylkens et al., 2007).

Although all herpesviruses share a basic virion structure, PrV and BHV-1 virions are morphologically largely identical with a centrally located nucleocapsid surrounded by an evenly distributed tegument (Fig. 1A, B). In contrast, ILTV virions contain variable amounts of tegument resulting in decentralized capsid location and nonsymmetrical tegument deposition (Fig. 1C). All three viruses share a common genome structure which has been designated as a herpesvirus type D genome (Fig. 2; Roizman and Pellet, 2001). It is characterized by two sets of unique sequences, designated as long and short unique regions (U_L, U_S), in which the latter is bracketed by inverted repeat sequences (internal repeat, IR_S; terminal repeat, TR_S). This organization results in the formation of two isomers of the genome during replication which are distinguished by the orientation of the U_S region. The biological relevance of the formation of two genome isomers is unclear. Compared to the prototypic alphaherpesvirus herpes simplex virus 1 (HSV-1), the genomes of PrV, BHV-1, ILTV and other alphaherpesviruses show an antiparallel orientation of the unique long region. Thus, gene UL1, which is located at the leftmost, 5'-end of the HSV-1 U_L region, is localized closest to the internal repeat in the PrV, BHV-1 and ILTV genomes (Schwyzer and Ackermann, 1996; Klupp et al., 2004; Thureen and Keeler, 2006). Since the HSV-1 type E genome contains inverted repeats bracketing both the U_L and the U_S regions, it is present in four isomers with different orientations of both the U_L and U_S regions (Roizman and Pellet, 2001). Thus, it can be deduced that the prototypic orientation of alphaherpesvirus U_L regions is antiparallel to that which has fortuitously been designated as prototypic in HSV-1. Nevertheless, gene designations which have been established for HSV-1 by sequential numbering on the basis of the location of the open reading frame on the different genome regions have, for clarity, also been adopted for several other alphaherpesviruses. This is reasonable, since most gene products exhibit a significant homology and are known or assumed to execute similar functions during the replicative cycle of the respective virus (Table 1).

gene PrV	BHV-1	ILTV	protein name	properties / functions	virion localization
-	Circ	-	Circ	α and γ protein[2]	yes (T)
ORF-1.2	-	-		unknown	yes
ORF-1	-	-		unknown	yes
UL54	UL54	UL54	ICP27	α or β[4] protein / gene regulation[4]	no
UL53	UL53	UL53	gK	membrane glycoprotein / viral egress[1]	yes (E)

gene PrV	BHV-1	ILTV	protein name	properties / functions	virion localization
UL52	UL52	UL52		helicase-primase subunit / DNA replication	no
UL51	UL51	UL51		unknown	yes (T)
UL50	UL50	UL50		dUTPase[7] / nucleotide metabolism	no
UL49.5	UL49.5	UL49.5	(gN)[5]	membrane (glyco)[5]protein / complex with gM[7]	yes (E)
UL49	UL49	UL49	VP22	interacts with gE and gM[1]	yes (T)
UL48	UL48	UL48	αTIF VP16	transactivator of α-gene expression[7] / interacts with UL3.5 protein[4] / viral egress[1]	yes (T)
UL47	UL47	UL47	VP13/14	viral egress[1], RNA transport[2]	yes (T)
UL46	UL46	UL46	VP11/12	unknown	yes (T)
-	-	UL45		putative membrane protein	?
-	-	ORF-A		unknown	?
-	-	ORF-B		unknown	?
-	-	ORF-C		unknown	?
-	-	ORF-D		unknown	?
-	-	ORF-E		unknown	?
UL44	UL44	UL44	gC	membrane glycoprotein / virion attachment to host cells[4]	yes (E)
UL43	UL43	UL43		membrane protein[1]	yes (E)
UL42	UL42	UL42		subunit of DNA-polymerase / DNA replication	no
UL41	UL41	UL41	VHS	virion host cell shutoff factor[4]	yes (T)

gene PrV	BHV-1	ILTV	protein name	properties / functions	virion localization
UL40	UL40	UL40	RR2	subunit of ribonucleotide reductase / nucleotide metabolism	?
UL39	UL39	UL39	RR1	subunit of ribonucleotide reductase / nucleotide metabolism	yes
UL38	UL38	UL38	VP19c	triplex protein / capsid formation	yes (C)
UL37	UL37	UL37		interacts with UL36 protein / viral egress[1]	yes (T)
UL36	UL36	UL36	VP1/2	interacts with UL36 protein / viral egress[1]	yes (T)
UL35	UL35	UL35	VP26	small capsid protein	yes (C)
UL34	UL34	UL34		interacts with UL31 protein / nuclear egress[1]	no
UL33	UL33	UL33		interacts with UL28 and UL15 proteins / DNA cleavage and encapsidation[1]	no
UL32	UL32	UL32		DNA cleavage and encapsidation[1]	?
UL31	UL31	UL31		interacts with UL34 protein / nuclear egress[1]	no
UL30	UL30	UL30		subunit of DNA-polymerase / DNA-replication	no
UL29	UL29	UL29	ICP8	single-stranded DNA-binding protein / DNA replication	no
UL28	UL28	UL28	ICP18.5	interacts with UL33 and UL15 proteins / DNA cleavage and encapsidation[1]	no

gene			protein		
PrV	BHV-1	ILTV	name	properties / functions	virion localization
UL27	UL27	UL27	gB	membrane glycoprotein[7] / virus entry and cell-to-cell spread[4]	yes (E)
UL26 UL26.5	UL26 UL26.5	UL26 UL26.5	VP24 VP22a	protease and scaffold protein / capsid formation[2]	yes no
UL25	UL25	UL25		DNA cleavage and encapsidation[4] nuclear egress[1]	yes
UL24	UL24	UL24		membrane protein	?
UL23	UL23	UL23	TK	thymidine kinase / nucleotide metabolism[7]	no
UL22	UL22	UL22	gH	membrane glycoprotein[7] / forms complex with gL / virus entry and cell-to-cell spread[4]	yes (E)
UL21	UL21	UL21		interacts with UL16 protein[1]	yes (T)
UL20	UL20	UL20		membrane protein / viral egress[1]	?
UL19	UL19	UL19	VP5	hexon and penton protein / capsid formation	yes (C)
UL18	UL18	UL18	VP23	triplex protein / capsid formation	yes (C)
UL15(e2)	UL15(e2)	UL15(e2)		interacts with UL33 and UL28 proteins / DNA cleavage and encapsidation[1]	no
UL17	UL17	UL17		DNA cleavage and encapsidation[1]	yes (C)
UL16	UL16	-		interacts with UL21 protein[1]	yes (T)
UL15(e1)	UL15(e1)	UL15(e1)		see above	no
UL14	UL14	UL14		unknown	?
UL13	UL13	UL13	VP18.8	protein kinase[1]	yes (T)

gene			protein		
PrV	BHV-1	ILTV	name	properties / functions	virion localization
UL12	UL12	UL12		alkaline exonuclease[2]	?
UL11	UL11	UL11		viral egress [1]	yes (T)
UL10	UL10	UL10	(gM)[4]	membrane (glyco)[4]protein / forms complex with gN[7]	yes (E)
UL9	UL9	UL9	OBP	origin binding protein / DNA replication	no
UL8	UL8	UL8		helicase-primase subunit / DNA replication	no
UL7	UL7	UL7		viral egress[1]	yes (T)
UL6	UL6	UL6		portal protein / DNA cleavage and encapsidation	yes (C)
UL5	UL5	UL5		helicase-primase subunit / DNA replication	no
UL4	UL4	UL4		unknown	no
UL3.5	UL3.5	UL3.5		interacts with UL48 protein[4] / viral egress[1]	yes (T)
UL3	UL3	UL3		nuclear protein[1]	no
UL2	UL2	UL2	UNG	uracil-DNA glycosylase[2] / DNA repair	no
UL1	UL1	UL1	gL	glycoprotein / forms complex with gH / virus entry and cell-to-cell spread[4]	yes (E)
-	UL0.7	-		putative ORF	?
-	UL0.5	-		putative ORF	?
?	LRORF2	-		latency related protein[2]	?
-	-	UL0		nuclear protein[3]	?
-	-	UL[-1]		nuclear protein[3]	?
EP0 gene	BICP0 gene	-	EP0 / BICP0	α[2] or β[1] protein / gene regulation[4]	yes (T)
IE180 gene	BICP4 gene	ICP4 gene	IE180 / BICP4	α-protein / gene regulation[4]	yes (T)

gene PrV	BHV-1	ILTV	protein name	properties / functions	virion localization
US1	US1	-	RSp40 / ICP22	α^2 protein / gene regulation	?
-	-	US10		unknown	?
-	-	SORF1		putative ORF	?
-	US1.67	-		unknown	?
US2	US2	US2		unknown	yes (T)
US3	US3	US3	PK	protein kinase[4]/ nuclear egress[1]	yes (T)
US4	US4	US4	gG	secreted glycoprotein[7]	yes (E)
-	-	US5	gJ	membrane glycoprotein[3]	yes (E)
US6	US6	US6	gD	membrane glycoprotein / virion attachment[4] and cell-to-cell spread[2]	yes (E)
US7	US7	US7	gI	membrane glycoprotein / forms complex with gE[4]	yes (E)
US8	US8	US8	gE	membrane glycoprotein / forms complex with gI[4]	yes (E)
US9	US9	US9		membrane protein / involved in axonal transport of virions[1] / induces apoptosis[2]	yes (E)

Table 1 – Boundaries between U_L, IR_S/TR_S, and U_S regions are marked by bold lines. Inversions and translocations of conserved genes in the individual virus genomes are indicated by shading. If determined, protein location in capsid (C), tegument (T) or envelope (E) of virions is given. [1]Demonstrated for PrV, [2]demonstrated for BHV-1, [3]demonstrated for ILTV, [4]demonstrated for PrV and BHV-1, [5]demonstrated for PrV and ILTV, [6]demonstrated for BHV-1 and ILTV, [7]demonstrated for PrV, BHV-1 and ILTV.

Interestingly, PrV and ILTV genomes contain a large inversion within their U_L regions (Fig. 2). In the PrV genome this inversion encompasses the genes from UL27 (gB) to UL44 (gC) (Ben-Porat et al., 1983), whereas in ILTV it contains genes from UL22 (gH) to UL44 (gC) (Ziemann et al., 1998a). A rearrangement identical to ILTV has also been found in the psittacid herpesvirus 1 genome (Thureen and Keeler, 2006). So far, the origin and biological relevance of these inversions is unclear. Despite having one of the breakpoints in common it is difficult to reconcile a common origin of these inversions with the phylogenetic distance of PrV and ILTV. Most likely these inversion events occurred independently during evolution of the respective viral genome.

Complete genome sequences are available for these three alphaherpesviruses. The PrV sequence, which has been assembled from genomic regions of several different strains, contains 143,461 nucleotides (Klupp et al., 2004; GenBank Acc. No. BK001744). 72 open reading frames predicted to encode 70 different proteins have been identified. The BHV-1 genomic sequence which is also a composite of different strains contains 135,301 nucleotides and encodes 76 open reading frames resulting in expression of 74 different proteins (Schwyzer and Ackermann, 1996; GenBank Acc. No. AJ004801). The ILTV genomic sequence has also been assembled from sequenced regions of different strains resulting in a total of 148,687 nucleotides with 76 open reading frames predicted to be translated into 73 different proteins (Thureen and Keeler, 2006; GenBank Acc. No. NC_006623). The G+C content varies considerably from 48 % in ILTV to 72 % in BHV-1 and 74 % in PrV. Most of the genes are conserved within these alphaherpesviruses (Table 1). However, virus-specific genes have been described in BHV-1 (UL0.5; Delhon et al., 2003) and ILTV (UL[0], UL[-1], ORF A to E; Ziemann et al., 1998a, b; Veits et al., 2003b), whereas no PrV-specific open reading frames have been detected. ORFs 1 and 1.2 of PrV are homologous to equine herpesvirus 1 and 4 gene 1. ILTV UL0 presumably resulted from an ancient duplication of UL[-1] (Ziemann et al., 1998b). Furthermore, compared to the genomes of BHV-1 and PrV, in ILTV the conserved UL47 gene is translocated from the U_L to the U_S genome region (Fig. 2), and ILTV lacks a UL16 homologue (Fuchs and Mettenleiter, 1999), Besides these particularities, the low sequence homologies between conserved genes demonstrate a considerable phylogenetic distance of ILTV or Iltoviruses not only to all mammalian, but also to other avian alphaherpesviruses like Marek's disease virus (McGeoch et al., 2000).

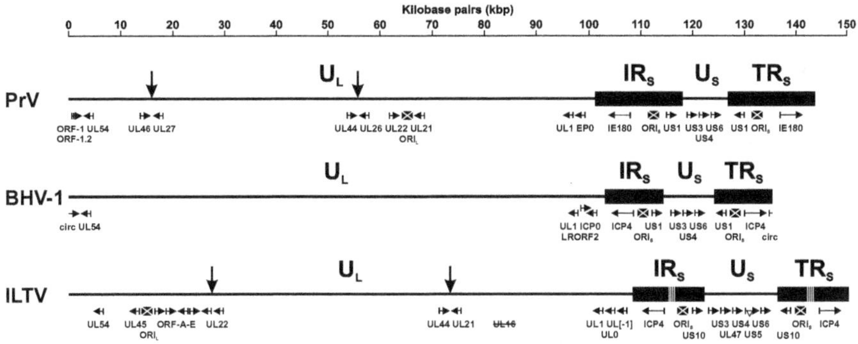

Fig. 2. Illustration of the genomes of PrV, BHV-1 and ILTV. Indicated are long and short unique regions (U_L, U_S) and inverted and terminal repeats (IR_S, TR_S). Location of origins of DNA replication (ORI_L, ORI_S) and of selected genes is depicted and transcriptional orientation is indicated by horizontal arrows. Vertical arrows denote inversions present in the U_L regions of PrV and ILTV genomes. Vertical lines indicate tandem repetitions of a 856 bp element in the IR_S and TR_S sequences of ILTV.

Although these data give a general picture of the respective genomes, genome size may vary dependent on strain and/or passage history. Thus, strains of PrV or ILTV have been described which contain smaller genomes due to deletions (Lomniczi et al., 1984; Fuchs et al., 2007). Size variation may also occur due to the presence of variable numbers of repeated sequence elements (Simon et al., 1989; Wild et al., 1996). In common, all three genomes contain two origins of viral DNA replication located within the IR regions (ORI_S). In PrV and ILTV genomes one origin is located in the middle of the U_L (ORI_L). Presence of an ORI_L has not yet been confirmed in BHV-1.

Besides their economic importance, all three viruses have increasingly been used to analyze fundamental processes in herpesvirus biology. PrV serves as a model to study the molecular basis for herpesvirus morphogenesis (Mettenleiter et al., 2006) and neurotropism (Enquist et al., 1999), whereas BHV-1 is a major focus on research on regulation of viral gene expression and latency (Jones et al., 2006). ILTV has a significant potential as a viral vector for live vaccination against other chicken pathogens (Fuchs et al., 2006a). These different aspects will be covered in this review.

PSEUDORABIES VIRUS (PrV)

PrV is the causative agent of Aujeszky's disease, a serious disease primarily affecting pigs (Pensaert and Kluge, 1989; Pejsak and Truszczynski, 2006). The infectious agent has first been isolated by the Hungarian physician Aladar Aujeszky (Aujeszky 1902), who also differentiated it from rabies. Although PrV is able to infect most mammals with the exception of higher primates including humans, infection in these species is almost always fatal except in pigs. Whereas piglets also frequently succumb to the disease, older animals may only present respiratory symptoms but survive to serve as the only animal reservoir for PrV. Domestic and feral pigs as well as wild boar are equally susceptible. Due to its wide host range which includes, besides the experimentally accessible natural host also small laboratory animals but omits humans, easy handling in the laboratory and excellent replication characteristics, PrV has become an important model to study herpesvirus infection in cell culture and in the animal. In particular, studies on basic aspects of viral replication and the molecular basis for herpesvirus neurotropism are being analyzed.

PrV VIRION COMPOSITION

Like other herpes virions PrV particles consist of core, capsid, tegument and envelope (Fig. 3). Approximately half of the PrV genes encode proteins which are located in the virion (Table 1; Klupp et al., 2004). Although ultrastructural studies of capsids have primarily been performed with human herpesviruses, general principles are probably valid for all herpesviruses. The herpesvirus capsid of T=16 symmetry contains 955 copies of the major capsid protein pUL19 which form 150 hexons and 11 pentons. Stability of the capsid is provided by a structure designated as triplex which consists of two copies of pUL18 and one of pUL38 and interconnects the capsomers. Whereas these proteins are required for formation of capsids, pUL35 which covers the top of the hexons, is nonessential for capsid formation. It has been used for fusion with green fluorescent protein to be able to intrinsically label and monitor capsids (Desai and Person, 1998; Smith et al., 2001), although this fusion results in a loss of function of pUL35 (Krautwald et al., in preparation). One pentonal site is occupied by the portal complex, a dodecameric ring composed of pUL6, through which DNA is incorporated and, probably, released (Chang et al., 2007). In the PrV envelope the ten glycoproteins B, C, D, E, H, I, K, L, M, and N have been identified (Mettenleiter, 2006). Whereas gB, gD, gH, and gL are considered essential for infection by wild-type viruses, gC, gE, gI, gK, gM, and gN are dispensable for a productive replication at least in cultured cells. gB is present in homooligomeric form and proteolytically processed into disulfide-linked subunits (Hampl et al., 1984; Lukacs et al., 1985; Wölfer et al., 1990). gE and gI (Zuckermann et al., 1988), gH and gL (Klupp et al., 1994), and gM and gN (Jöns et al., 1998) form heterodimeric complexes. The interaction between gE and gI as well as gH and gL is noncovalent. gN, the product of the UL49.5 gene, has first been identified as a glycoprotein (Jöns et al., 1996) and a disulfide-linked interaction partner for gM (Jöns et al., 1998) in PrV which has subsequently been confirmed for several other herpesviruses including BHV-1 (Wu et al., 1998), although the protein product of the UL49.5 reading frame is not glycosylated in BHV-1 (Liang et al., 1996). gB, gH, gL, gM and gN are conserved throughout the *Herpesviridae* (McGeoch et al., 2006). gH and gM are properly processed only in the presence of their complex partners gL and gN (Klupp et al., 1994; Jöns et al., 1998). However, whereas most herpesviral gM homologs require gN for incorporation into the virion, and most gH homologs require gL, this is not the case in PrV. gH and gM are present in the virion even in the absence of the complex partner, although they are not functional in this configuration. Another viral glycoprotein, gG, is not present in the virion but secreted from infected cells (Ben-Porat and Kaplan, 1970; Rea et al., 1985).

The viral tegument is probably the most complex structure of the virion. It contains numerous virally encoded proteins but also harbors cellular components. In PrV more than 15 viral proteins have been identified as part of the tegument (Table 1; Mettenleiter, 2004). They are assembled in a complex network of protein-protein interactions which will be described in detail below (Mettenleiter, 2006).

Besides virally encoded virion structural proteins, PrV particles also contain components of the host cell. Using proteomics analyses, the cellular proteins actin, heat shock proteins Hsp70 and Hsp90 as well as annexins I and II have been detected as part of purified virions, most likely located in the tegument (Wong and Chen, 1998; del Rio et al., 2005; Michael et al., 2006b). Although their function for viral replication, if any, is unclear, incorporation of cellular proteins, in particular actin, is enhanced in the absence of major viral tegument proteins such as pUL47 and pUL49 (del Rio et al., 2005b, 2006a). Whereas one function of the viral tegument is presumably to link envelope and nucleocapsid, it also contains proteins which, after release from the invading particle, prime the host cell for herpesvirus replication such as pUL41 or pUL48 (see below). Thus, the tegument may also be considered as the 'trunk' of the herpesvirus particle accepting 'cargo' to be delivered to the host cell upon infection.

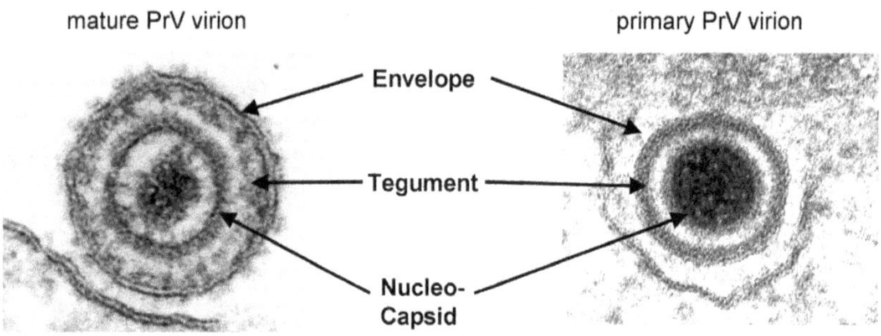

Envelope	Tegument		Nucleocapsid	Tegument		Envelope
pUL1 (gL)	IE180	pUL41	pUL19 (MCP)	pUL25	pUS3	pUL34
pUL10 (gM)	pUL3.5	pUL46	pUL18 (Triplex)	pUL31		
pUL20	pUL7	pUL47	pUL38 (Triplex)			
pUL22 (gH)	pUL11	pUL48	pUL35			
pUL27 (gB)	pUL13	pUL49	pUL6 (portal)			
pUL43	pUL16	pUS2	pUL26 (scaffold)			
pUL44 (gC)	pUL21	pUS3	pUL26.5 (protease)			
pUL49.5 (gN)	pUL25	Hsp70	pUL17			
pUL53 (gK)	pUL36	Hsp27				
pUS6 (gD)	pUL37	Actin				
pUS7 (gI)	pUL39	Annexin				
pUS8 (gE)	pUL51					
pUS9						

Fig. 3: Composition of primary and mature PrV virions

PrV REPLICATION

PrV replication in permissive cells in cell culture is characterized by a rapid progression. Infectious progeny can first be detected approx. 4h after infection and viral titers may reach over 10^8 plaque forming units per milliliter of supernatant. Thus, PrV can be grown at reasonable expense to high amounts allowing intricate biochemical and biological analyses (Michael et al., 2006a,b). For unravelling function of viral proteins, specific viral mutants have been isolated after cotransfection of plasmids containing mutated genomic regions and wild-type PrV genomic DNA followed by selection of recombinants which have arisen by homologous recombination between plasmid- and genome-borne sequences (Robbins et al., 1986; Mettenleiter et al., 1987a). Isolation of recombinant PrV has been facilitated by the insertion of marker genes expressing easily identifiable proteins such as ß-galactosidase (Mettenleiter and Rauh, 1990), firefly luciferase (Kovacs and Mettenleiter, 1991) or fluorescent proteins (e.g. GFP, Jöns and Mettenleiter, 1997; YFP, RFP, Banfield et al., 2003). In recent years, the cloning of infectious copies of the PrV genome in bacterial artifical chromosomes (Smith and Enquist, 1999) has allowed the use of bacterial genetic manipulation technology for faster construction and isolation of viral mutants (Smith and Enquist, 2000). Whereas mutants in nonessential genes, which make up approx. half of the PrV encoded genes (Klupp et al., 2004), can be propagated with ease in cell culture, mutants in genes encoding proteins either essential or very important for viral replication require trans-complementing cells which carry and express the respective gene mutated in the viral genome in its wild-type form and provide it *in trans* for viral propagation (Rauh and Mettenleiter, 1991; Peeters et al., 1992). In either case, careful controls including repair of the mutation which should result in the restoration of a wild-type phenotype are essential for the correct assignment of phenotype to genetic lesion. However, trans-effects of specific mutations on other viral (and cellular) proteins have also been demonstrated (Michael et al., 2006b) which have to be taken into account.

Infectious entry

Herpesvirus infection of a host cell starts by attachment of the extracellular virion to the cell surface (Fig. 4). In the initial interaction, binding of virion envelope components to heparan sulfate proteoglycans of the extracellular matrix occurs (WuDunn and Spear, 1989). The first identified heparan sulfate binding herpesvirus protein was glycoprotein C (gC) of pseudorabies virus (Mettenleiter et al., 1990). In HSV-1, gB also binds heparan sulfate in addition to gC (Herold et al., 1994). However, since PrV gC is not required for productive infection and neither is cellular heparan sulfate (Mettenleiter et al, 1988; Karger et al., 1995), other means of attachment have to be functional. It is now clear that specific binding to host cells is mediated via the interaction of the receptor-binding glycoprotein D (gD) with cell surface proteins. In the case of PrV, these comprise nectins 1 and 2 and poliovirus-receptor (Geraghty et al., 1998; Milne et al., 2001). Interaction between gD and its receptor triggers the fusion of virion envelope and cellular plasma membrane for which a fusion complex of glycoproteins B, H and L is required (Spear and Longnecker, 2003). Compared to other well described fusion processes e.g. in flavi- or orthomyxoviruses, it is still unclear how herpesviruses induce membrane fusion, although recent structural data on HSV-1 represent a significant step forward (Heldwein et al., 2006; Subramanian and Geraghty, 2007). Interestingly, gB, gH and gL are conserved throughout the mammalian and avian herpesviruses (McGeoch et al., 2006) indicating that similar mechanisms underlie infection initiation by these viruses. This is further supported by the demonstration that BHV-1 gB can substitute for PrV gB in a PrV background, although not vice versa (Rauh et al., 1991). A similar one-way complementation has been described for PrV gB in complementation of a lethal gB mutation in HSV-1 (Mettenleiter and Spear, 1994). On the other hand, the receptor-binding gD is only present in several alphaherpesviruses with the exception of varicella-zoster virus which lacks a gD homolog altogether (Davison and Scott, 1986) and Marek's disease virus in which the gD gene apparently is not expressed (Tan et al., 2001).

Most herpesviral gB homologs contain a polybasic cleavage site which is recognized by the trans-Golgi protease furin. The resulting proteolytic processing cleaves the glycosylated precursor into two polypeptides which remain linked by disulfide bonds. Activation of fusion-associated proteins by proteolytic cleavage is a common theme in virology. However, mutagenesis studies have shown that proteolytic cleavage of BHV-1 gB is not required for its function in either a BHV-1 or PrV background (Blewett and Misra, 1991; Kopp et al., 1994). Interestingly, HSV-1 gB is not cleaved (Sarmiento and Spear, 1979). Thus, the functional significance of this cleavage event remains unclear. However, proteolytic cleavage at this site can be used to release functionally active polypeptides from chimeric proteins containing insertions between reduplicated furin cleavage sites as has been shown for BHV-1 gB (Keil et al., 2005) effectively using gB as a transporter through the secretory pathway.

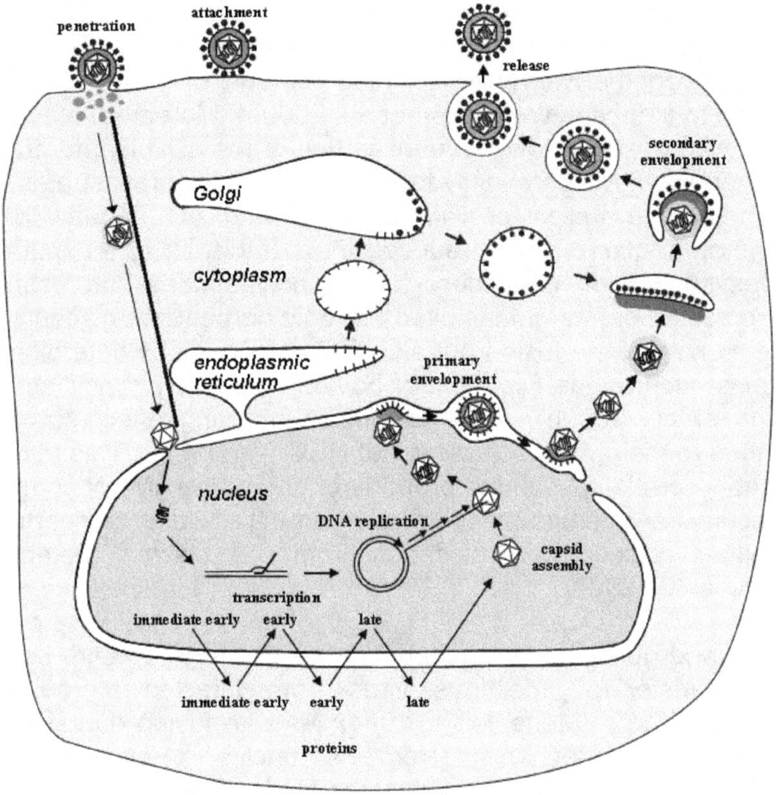

Fig. 4. The PrV replication cycle (in part reprinted from J. Virol. 76, 1537-1547 with permission from the American Society for Microbiology).

Although gL is considered an essential glycoprotein PrV mutants have been isolated which are infectious in the absence of gL. After repeated passaging of gL-deficient PrV by exploiting its limited capacity for direct cell-cell spread, a spontaneous mutant virus could be isolated which expressed a hybrid protein consisting of the amino terminal 2/3 of gD fused to the carboxyterminal 2/3 of gH (Klupp and Mettenleiter, 1999). This gDH protein apparently executes receptor-binding functions via its gD moiety, whereas the gH part participates in membrane fusion. Although gL is discussed as a 'chaperone' required for proper folding of gH, these data show that a functional conformation of gH can be obtained without gL.

In a transient expression-fusion assay, gB, gH and gL are sufficient to mediate fusion (Klupp et al., 2000b). However, fusogenic activity was enhanced when a carboxy-terminally truncated version of gB lacking the C-terminal 29 amino acids was used instead of wild-type gB. Together with this truncated gB, only gH was required in addition to induce fusion (Klupp et al., 2000b). Moreover, extensive fusion was observed after cotransfection of the gDH hybrid protein together with the C-terminally truncated gB. This demonstrates that for membrane fusion of PrV only gB and gH appear to be absolutely required which reduces the complexity of the putative herpesvirus fusion machinery to two polypeptides. Interestingly, coexpression of PrV gM inhibited not only herpesvirus-glycoprotein induced membrane fusion but also fusion induced by heterologous proteins such as the F protein of bovine respiratory syncytial virus (Klupp et al., 2000b). A similar fusion-inhibiting effect could be ascribed to the gM/gN complexes of HSV-1 and ILTV. This effect is probably correlated with the property of gM, or the gM/gN complex, to accumulate membrane proteins in an intracellular compartment by either retention or relocation from the plasma membrane (Crump et al., 2004).

gB, gH and gL are also required for direct cell-cell spread, i.e. the direct transfer of infectious virus from infected to adjacent noninfected cells (Mettenleiter, 1994). As with fusion during entry, the molecular mechanism remains unclear. However, it is interesting to note that HSV-1 requires gD for direct cell-cell spread (Ligas and Johnson, 1988), whereas in PrV gD is not needed (Rauh and Mettenleiter, 1991; Peeters et al., 1992). This demonstrates that interaction with a cellular gD receptor is not required for direct transer of infectivity from one infected cell to another which was supported by the finding that a single amino acid exchange in gH of gD-negative BHV-1 can complement the function of BHV-1 gD in direct spread (Schröder et al., 1997). In line with these observations, PrV mutants have been isolated that are able to initiate infection without having gD in their envelope (Schmidt et al., 1997). gD-independent infectivity has also been demonstrated in BHV-1 (Schröder et al., 1997). In PrV, mutations in gH and gB are associated with gD-independent entry (Schmidt et al., 2001) and it is presumed that these prime the fusion apparatus without the need for a gD-dependent trigger. Interestingly, gD-independent infectivity is sufficient for a lethal infection of mice, whereas in pigs the resulting mutant virus is avirulent. An infectious PrV mutant lacking both known attachment proteins, gC and gD, has also been isolated (Karger et al., 1998).

Whereas it is still generally assumed that herpesviruses initiate infection by fusion at the plasma membrane, reports are accumulating that, dependent on the host cell, infection via endocytosis and membrane fusion in the endocytic vesicle may also occur (Nicola et al., 2003). However, for PrV the primary mode of entry appears to be fusion of virion envelope and plasma membrane (Granzow et al., 1997).

Translocation to the nucleus

After infectious entry, envelope proteins remain in the plasma membrane, whereas part of the tegument proteins are released into the cytoplasm to prime the host cell for viral replication (Granzow et al., 2005; Luxton et al., 2005). These include the pUL48 immediate-early activator protein (see below) and the pUL41 virion host shutoff protein which degrades cellular mRNA. PrV pUL48 like its homologs in other alphaherpesviruses has been described as a potent transactivator of viral immediate-early genes (Fuchs et al., 2002b), though this function is not essential. However, in contrast to e.g. HSV-1 pUL41 (Taddeo et al., 2006), PrV pUL41 apparently does not induce an efficient host-cell shut off early after entry (Karger and Mettenleiter, unpublished), although it exhibits RNase activity (Lin et al., 2004). Whether the decrease in cellular protein expression observed during late times after infection is, at least in part, mediated by pUL41 remains to be analyzed.

In contrast to pUL41, pUL48 and other components of the outer tegument, an inner set of tegument proteins, notably homologs of the pUL36 and pUL37 proteins, remain attached to the virion during its retrograde transport via microtubuli to the nuclear pore (Granzow et al., 2005; Luxton et al., 2005). Although no specific PrV protein has been identified to interact functionally with the cellular dynein-kinesin motor system, pUL36 and pUL37 are good candidates. They form a physical complex (Klupp et al., 2002) which is important for recruitment of these proteins into the nascent virus particle (see below), and pUL36 has been shown to be required for intracellular capsid transport (Luxton et al., 2006). pUL36 homologs are the largest herpesvirus proteins encompassing, in PrV, 3084 amino acids (Klupp et al., 2004). Where analyzed, they have been found to be strictly essential for herpesvirus replication which also applies to PrV (Fuchs et al., 2004). However, it has recently been shown that a large internal proline/alanine-rich stretch of 709 amino acids is dispensable for viral replication which narrows the location of essential domains (Böttcher et al., 2006). The C-terminus of the protein is apparently essential (Böttcher et al., 2006, Lee et al., 2006), whereas an N-terminally located deubiquitinylation activity, which is also conserved in pUL36 homologs of all herpesvirus subfamilies (Kattenhorn et al., 2005), is apparently not required for its function (Lee et al., 2006).

pUL36 and pUL37 are also prime candidates for viral components interacting with the nuclear pore during docking of the capsid, although no direct interaction between either and any component of the nuclear pore complex has been described yet. Capsids invariably are located with one vertex towards the nuclear pore (Granzow et al., 1997) which opens for release of the viral genome through the nuclear pore into the nucleus for transcription and replication. It is unclear whether any vertex can serve as exit site or whether there is a selection for that pentonal site which contains the viral portal complex.

Transcription and genome replication

After release into the nucleus, the viral genome circularizes and transcription commences. The PrV genome contains 72 genes, 40 of which are conserved throughout all herpesvirus subfamilies (Klupp et al., 2004). Functions of the respective gene products as well as their location in the virus particle in the case of structural proteins is given in Table 1.

Gene transcription in herpesviruses occurs in a cascade-like, strictly regulated manner which also applies to PrV (Feldman et al., 1979). For transcription of their immediate-early, early and late genes herpesviruses utilize the cellular RNA polymerase II and further host cell factors for mRNA processing (Roizman and Knipe, 2001). Transcription of immediate-early genes occurs first, which is enhanced by the activity of the UL48 gene product, in HSV-1 designated as alpha-transinducing factor (α-TIF; Batterson and Roizman, 1983). pUL48 is brought into the host cell as part of the tegument and rapidly translocates to the nucleus. Although activity of α-TIF is clearly not required to start viral transcription since naked herpesviral genomes are infectious when transfected into appropriate host cells, it is beneficial for a kick-start. In PrV two genes expressed with immediate-early (IE) kinetics have been described. The major IE gene is located in the repeat sequences and, thus, is present in two copies in the viral genome (Feldman et al., 1979). It encodes a protein of 180 kDa molecular mass, designated as IE180 (Chlan et al., 1987), which is homologous to infected cell protein 4 (ICP4), the major immediate-early protein of HSV-1 encoded at a colinear position of the HSV-1 genome (Cheung, 1989b). A second PrV transcription unit regulated with immediate-early kinetics encompasses the gene encoding RSp40, a homolog of the HSV-1 immediate-early protein ICP22 (Fuchs et al., 2000b). The RSp40 gene is also located in the repeat regions and, thus, present in two copies per genome. It is translated from a 1.8kb mRNA from which two introns are spliced from the 5' noncoding region. Other genes with homology to immediate-early genes of HSV-1 such as UL54 or the gene encoding ICP0 are apparently not regulated with immediate-early kinetics in PrV.

IE180 acts as a transcriptional transactivator of several viral promotors (Workman et al., 1988), thereby initiating expression of early genes. However, it also binds to heterologous viral (Wu and Berk, 1988) and cellular promotors e.g. those governing expression of β-globin or Hsp70 (Cromlish et al., 1989). Interestingly, Hsp70 is also incorporated into mature PrV virions (Michael et al., 2006b).

Expression of the PrV homolog of HSV-1 ICP0 is regulated with early kinetics, and, consequently, the protein has been designated as early protein (EP) 0 (Cheung, 1991). Despite this difference in regulation, expression of the nonessential EP0 induces alterations in the cell nucleus similar to those observed after expression of HSV-1, VZV and BHV-1 homologs (Parkinson and Everett, 2000, 2001), and the respective HSV-1 and VZV proteins are able to rescue the replication defect of EP0-deleted PrV (Moriuchi et al., 1993). Besides HSV-1 ICP0, most information on function of this class of proteins has been derived from studies in BHV-1 which are reviewed below.

Early genes primarily encode proteins involved in nucleotide metabolism and genome replication. Homologs of the HSV-1 genome replication proteins pUL9 (origin binding protein), pUL5/8/52 (helicase-primase complex), pUL29 (single-strand DNA-binding protein), pUL30 (DNA polymerase) and pUL42 (DNA-polymerase associated processivity factor) are conserved throughout the *Alphaherpesvirinae* including PrV demonstrating common mechanisms of genome replication within the herpesviruses (Baines and Weller, 2005). dUTPase (pUL50), Uracil-DNA-Glycosylase (pUL2), ribonucleotide reductase (pUL39/40) and thymidine kinase (pUL23) are involved in nucleotide metabolism and genome replication, although they are nonessential for viral replication in cell culture. However, they may constitute important virulence factors since elimination of thymidine kinase (Kit et al., 1985; McGregor et al., 1985), ribonucleotide reductase (de Wind et al., 1993) or dUTPase (Jöns et al., 1997) activity strongly attenuates the virus in infection of animals. However, this may be virus and host species specific since deletion of UL50 reduces virulence of ILTV only modestly (Fuchs et al., 2000a). Herpesvirus genome replication occurs by a rolling circle mechanism in which long head-to-tail concatemers are produced which have to be processed prior to or during incorporation into capsids (Roizman and Pellet, 2001).

Capsid formation

Capsid proteins are translated in the cytoplasm and then recruited to the nucleus where self-assembly of the capsid occurs. Herpesvirus capsid formation has primarily been analyzed in HSV-1, but homology of viral gene products as well as ultrastructural studies on capsid formation and maturation indicate common mechanisms (Mettenleiter, 2006). The capsid shell forms around a scaffold consisting of pUL26 and its carboxy-terminal half pUL26.5. Interestingly, capsid formation during herpesvirus infection closely resembles head assembly of the large double-stranded DNA bacteriophages of the order Caudovirales (Baker et al., 2005) as does genome replication and packaging (Baines and Weller, 2005) which indicates common ancestry. In fact, the pUL15 terminase subunit of the presumed pUL15/pUL28 complex which translocates DNA into the virion and cleaves off unit-length molecules at specific recognition sites is the only protein with detectable homology throughout all identified herpesviruses including those of piscine and invertebrate origin, and also shows homology to the corresponding protein in dsDNA bacteriophages (Davison, 2002). Absence of PrV pUL15 inhibits genome packaging (Fuchs et al., unpublished; Mettenleiter et al., 1993). In cotransfection experiments, HSV-1 pUL15 was able to translocate PrV pUL28 into the nucleus (Koslowski et al., 1997) which strongly indicates direct complex formation between the two which, consequently, should also occur with PrV pUL15. In the nucleus three types of PrV capsids can be distinguished morphologically which, presumably, all arise from a procapsid precursor (Mettenleiter et al, 1993). C-capsids contain DNA and mature after translocation through the nuclear membrane in the cytoplasm to infectious virions. B-capsids are immature forms which contain internal scaffold which is proteolytically cleaved and, at least in part, extruded from the capsid during DNA packaging, although the N-terminal protease fragment of the scaffold protein is also detectable in purified extracellular virions (Michael et al., 2006b). A-capsids without any discernible internal structure are thought to have arisen by unsuccessful packaging events resulting in degradation and extrusion of scaffold but without concomitant insertion of genomic DNA. These three capsid forms have been found in all herpesviruses analyzed to date in this respect (Steven and Spear, 1997). Specific procapsids have not been described in PrV infected cells which may correlate with an efficient and rapid transformation of these primary capsids to the more stable A-, B- and C-forms.

Besides bona-fide capsid proteins recently other viral proteins associated with intranuclear capsids and involved in DNA cleavage and packaging have been described. They comprise pUL17 which is a virion component thought to enter the capsid with the newly introduced DNA (Klupp et al., 2005c), and pUL25 which resides at the outside of mature, DNA-filled capsids but is absent from immature capsid forms (Klupp et al., 2006). Lack of pUL17 abolishes DNA packaging and only immature capsids are present in the nucleus whereas lack of pUL25 resulted in continued genome incorporation into capsids, albeit at reduced levels, but inability of the capsids to leave the nucleus despite contacting the inner nuclear membrane. Moreover, recently deletion mutants of PrV pUL6, pUL32 and pUL33 have been isolated and shown to be impaired in genome cleavage and packaging similar to their HSV-1 homologs (Klupp, Fuchs, Mettenleiter, unpublished data).

Nuclear egress

Final morphogenesis of PrV virions occurs in the cytoplasm (Mettenleiter, 2006). To reach this cellular compartment, capsids have to cross the nuclear envelope. This is achieved by an envelopment-deenvelopment mechanism in which capsids first bud at the inner nuclear membrane thereby acquiring a primary envelope which then fuses with the outer nuclear membrane to release the capsids into the cytosol. Although this mechanism is not undisputed within the herpesvirus field, a multitude of biochemical, genetic and morphological data support this pathway (Mettenleiter and Minson, 2006). In contrast, transport of capsids through nuclear pores as hypothesized for BHV-1 and HSV-1 (Leuzinger et al., 2005; Wild et al., 2005) or transport through the ER-Golgi secretory pathway maintaining integrity of the primary enveloped virion (Roizman and Knipe, 2001) appear highly unlikely for reasons described below.

Two conserved herpesvirus proteins are required for nuclear egress of capsids, pUL31 and pUL34. Both form a complex (Fuchs et al., 2002a) and both are part of primary virions but absent from mature virus particles (Klupp et al., 2000a; Fuchs et al., 2002a; Michael et al., 2006a) which argues against the maintenance of integrity between primary and mature virions. Whereas pUL31 when expressed without other viral functions localizes diffusely to the nucleoplasm, pUL34 is lining the nuclear membrane. Coexpression of both proteins enhances pUL34 nuclear membrane localization and relocates pUL31 to the nuclear envelope (Fuchs et al., 2002a). In the absence of either pUL31 or pUL34 or both, capsids remain trapped in the nucleus. A similar picture is observed in the absence of the pUL25 protein (Klupp et al., 2006) indicating that pUL31/34 at the (inner) nuclear membrane has to interact directly or indirectly with pUL25 at the capsid to trigger primary envelopment. Unfortunately, no direct interaction between pUL25 and either pUL31 or pUL34 has so far been observed. Although HSV-1 and PrV homologs are reasonably conserved in sequence, functional homology by exchange of the respective proteins has so far not been shown, although similar properties can be assumed.

However, it has recently been demonstrated that coexpression of pUL31 and pUL34 in stably transfected cell lines induces alterations in the nuclear envelope resulting in invaginations of the inner nuclear membrane and vesicle formation from the nuclear envelope mimicking the formation of primary envelopes (Klupp et al., 2007). Thus, these two proteins suffice to form structures which are hallmarks of herpesvirus infection but have not been described in noninfected cells so far. Whether this is indicative of subversion of a hitherto unrecognized cellular pathway by herpesviruses or whether it is specific for herpesvirus infection has to be analyzed further.

pUL31 is a 271 aa phosphoprotein which is presumed to be a constituent of the primary tegument. pUL34 is a predicted 262 aa type II transmembrane protein which is supposedly located in the primary envelope. Moreover, the pUS3 kinase which is present only in alphaherpesviruses, is also part of the primary virion (Granzow et al., 2004). In its absence, primary enveloped virions accumulate in large invaginations of the inner nuclear membrane indicating a role for pUS3 in the fusion of primary envelope and outer nuclear membrane for deenvelopment (Wagenaar et al., 1995; Klupp et al., 2001a). An identical phenotype has also been demonstrated for US3 mutants of HSV-1 (Reynolds et al., 2002) and Marek's disease virus (Schumacher et al., 2005). Site-specific mutations demonstrated that the kinase activity is required for efficient nuclear egress of PrV (Gentner et al., unpublished). Expression of pUS3 also augments pUL34 localization to the nuclear membrane (Klupp et al., 2001a). Although in HSV-1 pUL34 has been demonstrated to be a substrate for pUS3 kinase (Purves et al., 1992), in PrV pUL34 is phosphorylated in the absence of pUS3 (Klupp et al., 2001a). Currently, it is unclear whether other viral (pUL13, pUL39) or cellular kinases recognize pUL34 in cells lacking pUS3.

Expression of herpesviral pUL31 and pUL34 homologs has been demonstrated to induce numerous changes in nuclear achitecture including recruitment of cellular kinases to phosphorylate and, probably, dissolve the nuclear lamina, redistribution of cellular chromatin, and direct interaction with cellular lamins (reviewed in Mettenleiter, 2006). Together, these effects should facilitate access of nascent capsids to the inner nuclear membrane for primary envelopment.

Whether the primary PrV virion also contains precursors or mature forms of the glycoproteins found in extracellular virions is unclear. These glycoproteins have been detected in primary HSV-1 virions (Jensen and Norrild, 1998; Baines et al., 2007) but, so far, not in PrV. In any case, it is evident that nuclear egress proceeds even in the absence of glycoproteins required for membrane fusion during entry demonstrating different molecular requirements for these two fusion processes.

Although coexpression of pUL31 and pUL34 is required for efficient nuclear egress, these proteins have no effect on the morphology of nuclear pores (Mettenleiter and Minson, 2006). Egress through grossly dilated nuclear pores has been postulated for HSV-1 and BHV-1 (Leuzinger et al., 2005; Wild et al., 2005). However, none of our numerous analyses supports this pathway. Moreover, nuclear pores appear unaffected during most of the replication cycle prior to total dissolution of the nucleus at very late times after infection resulting in cell death. Thus, it appears highly unlikely that the 110nm nucleocapsid can traverse through nuclear pores. In contrast, coexpression of pUL31 and pUL34 is sufficient for vesicle formation from the nuclear envelope as occurs during primary envelopment (Klupp et al., 2007). Whether these vesicles, which contain pUL31 and pUL34 as do primary virions are fusion competent remains to be established.

Thus, although the general outline of nuclear egress seems clear, questions on the molecular basis of the budding and fusion reaction remain (Minson and Mettenleiter, 2006).

Secondary envelopment

After release into the cytoplasm, a highly ordered network of protein-protein interactions drives tegumentation and secondary envelopment which includes incorporation into the nascent virion of more than 25 tegument and envelope proteins (Mettenleiter, 2006). Tegument formation initiates at two distinct sites, the capsid and the future envelope. Although, after their release from the nucleus, capsids appear morphologically 'naked' by conventional electron microscopy, immuno EM studies demonstrated that they carry the pUS3, pUL36 and pUL37 proteins (Fuchs et al., 2002b). Thus, these nascent capsids resemble in their composition the viral particles which enter the cell and are translocated to the nuclear pore (Granzow et al., 2005; Luxton et al., 2005). Immuno-EM data indicate that at least pUL36 and pUL37 must be recruited early after nuclear egress since they are absent from the primary PrV virion. In contrast, pUS3 is present in primary and mature virions and may remain associated with the capsid during nuclear egress (Granzow et al., 2004).

pUL36 presumably constitutes the innermost, capsid-apposed part of the tegument and it has been hypothesized that HSV-1 pUL36 contacts the capsid (Zhou et al., 1999). The interaction with pUL36 recruits pUL37 to the inner tegument (Klupp et al., 2002). pUL36 is the only tegument protein which has been shown to be indispensable for PrV replication. This essential character is separable from its interaction with the nonessential pUL37 (Klupp et al., 2000a; Fuchs et al., 2004) indicating that pUL36 performs functions during PrV replication beyond interacting with pUL37. So far only two functional regions have been mapped on PrV pUL36: the pUL37 interaction domain encompassing aa 312 to 398 of the 3084 aa pUL36, and a deubiquitinating activity with an active enzymatic site presumably at conserved cysteine at position 26. Neither is essential for pUL36 function (Fuchs et al., 2004; Luxton et al., 2006). Moreover, a 709-aa internal deletion removing a large proline/alanine-rich stretch within the C-terminal half of the molecule did not grossly impair pUL36 function, whereas deletion of the conserved extreme C-terminus leads to a lethal loss of pUL36 function (Böttcher et al., 2006; Luxton et al., 2006). In the absence of pUL36, capsids are still translocated to the cytoplasm but tegumentation and secondary envelopment is blocked. However, loss of pUL36 may also interfere with early events in PrV replication. In summary, pUL36, pUL37 and pUS3 are part of an inner, capsid-proximal layer of the tegument. It is important to note in this context that the inner tegument containing the pUL36 and pUL37 proteins is present in stoichiometrically conserved amounts in different mutants viruses whereas components of the outer tegument may vary considerably (Michael et al., 2006a, b).

At the future envelopment site in the trans-Golgi network glycoproteins gather which are retained from secretion or retrieved from the plasma membrane by interaction between each other and with viral proteins located at the cytoplasmic face of the vesicle or membrane. It has been shown that PrV gM and pUL43 are involved in *trans*-Golgi-retention and/or relocalization of envelope proteins, even of unrelated viruses (Klupp et al., 2000b; Crump et al., 2005; Klupp et al., 2005a). Several PrV membrane proteins have been shown to become efficiently endocytosed which include gB, gE, gI, and pUS9 (Tirabassi and Enquist, 1998; Brideau et al., 2000a; Nixdorf et al., 2001). However, the relevance of endocytosis for virion incorporation of these proteins is unclear since inhibition of endocytosis by mutation of endocytosis signals in the proteins does not interfere with virion localization of these proteins (Nixdorf et al. 2001, Brideau et al., 2000b). However, endocytosis may play a role during directional transneuronal infection in vivo (Brideau et al., 2000a, 2000b; see also below).

Major components of the outer tegument are the products of a gene cluster containing the UL46, UL47, UL48 and UL49 genes. Whereas deletion of UL48 showed a drastic effect on PrV virion formation (Fuchs et al., 2002b), deletion of UL46 (Kopp et al., 2002) or UL49 (del Rio et al., 2002) had no discernible consequences for viral replication in cell culture, whereas deletion of UL47 slightly impaired virion morphogenesis (Kopp et al., 2002). Even absence of all four tegument components does not abolish PrV replication (Fuchs et al., 2003) indicating that other proteins may substitute in redundant interactions within the tegument protein network. However, it has been shown that pUL49, a major tegument component, interacts with the cytoplasmic tails of gE and gM (Fuchs et al., 2002c). Whereas neither gE, nor gM, nor pUL49 are essential for viral replication, lack of both gE and gM, or of only the gE cytoplasmic tail and gM, drastically inhibits virion formation (Brack et al., 1999, 2000). In their absence large intracytoplasmic inclusions containing capsids and associated tegument accumulate. A similar, but even more drastic phenotype has been observed after deletion of gM and the membrane-associated tegument protein pUL11 (Kopp et al., 2004). In both cases, capsid-apposed tegumentation still proceeds but secondary envelopment is severely impaired demonstrating that these proteins are required for proper interaction between partially tegumented capsids and the envelopment site. Interestingly, in both cases simultaneous deletion of tegument and envelope components is required to induce this drastic phenotype. However, in the absence of either pUL11 (Kopp et al., 2003) or gM (Kopp et al., 2004), an effect on morphogenesis, albeit much less drastic, was already discernible.

HSV-1 pUL11 has been shown to have Golgi-targeting properties (Bowzard et al., 2000), whereas PrV gM is instrumental in collecting envelope proteins in the trans-Golgi network (Klupp et al., 2000a; Crump et al., 2004). Thus, simultaneous absence of both would inhibit glycoprotein accumulation at as well as localization of components of the outer tegument to the future envelopment site explaining the drastic phenotypes. Simultaneous absence of gE and gM may result in a decrease of glycoprotein-tegument interactions with the same detrimental consequences. A similar phenotype has also been described for HSV-1 lacking gE and gD (Farnsworth et al., 2003) but not for a gE/gM double mutant (Browne et al., 2004). The fact that very often only multiple deletions yield striking phenotypes gives testimony to the inbuilt redundancy in the complex network of protein-protein interactions (Mettenleiter, 2006).

HSV-1 pUL11 is able to interact with PrV pUL16 (Loomis et al., 2003) which, in turn, complexes with pUL21 (Klupp et al., 2005b). Although, in contrast to the situation in HSV-1 (Harper et al., 2006), a tripartite complex of PrV pUL11, pUL16 and pUL21 has not been demonstrated so far, available data indicate that it may exist (Klopfleisch et al., 2006). Since pUL11 is associated with Golgi-derived membranes and pUL21 is thought to be located close to the capsid (de Wind et al., 1992), the tripartite complex may contribute to linking envelope and capsid. However, it is clearly not required for this bridging since a triple mutant lacking all three proteins is replication competent and virion morphogenesis proceeds in the absence of all three proteins (Klupp et al., 2005b). Functional redundancy with other putative 'linkers' (see below) may explain this result.

The interaction between the partially tegumented capsid and tegument proteins assembling together with glycoproteins at the envelopment site may be mediated by pUL48 and pUL3.5. Besides its role as transcriptional activator pUL48 is an important structural component of the viral tegument. Deletion of pUL48 from the PrV genome results in the accumulation of unenveloped capsids in the cytoplasm (Fuchs et al., 2002b). A similar phenotype has been observed in the absence of the pUL3.5 protein (Fuchs et al., 1996). The corresponding UL3.5 gene is absent from the HSV-1 genome but present e.g. in BHV-1, ILTV and VZV genomes. It has first been shown for BHV-1 (Lam and Letchworth, 2000), and later confimed for PrV (Fuchs et al., in press), that pUL3.5 and pUL48 form a complex which, apparently, constitutes a functional unit. BHV-1 pUL3.5 can partially complement the defect in a pUL3.5 PrV deletion mutant indicating functional equivalence of these two proteins (Fuchs et al., 1997b). Interestingly, secondary envelopment of tegument but without a capsid can be observed frequently. Formation of capsidless enveloped particles designated, due to their decreased buoyant density in sucrose gradients, as 'light' or 'L-' particles occurs also during wild-type virus infection but at greatly increased levels in several virus mutants with defects in secondary envelopment (Fuchs et al., 2002b). However, L-particles are not formed in mutants deficient in the interaction between components of the outer tegument and envelope glycoproteins (Brack et al., 1999; Kopp et al., 2004).

Ascribing specific phenotypes to specific gene deletions is difficult due to the functional redundancy of the network and trans-effects of absence of tegument components on other structural proteins. Quantitative proteome analyses demonstrated that the inner tegument containing the pUL36 and pUL37 proteins is present in stoichiometrically conserved amounts in different mutant viruses, whereas components of the outer tegument may vary considerably. Thus, in the absence of pUS3, pUL47, pUL49 or gE, the relative amounts of the capsid protein pUL38, the capsid-associated pUL25 protein, tegument protein pUL36 and gH were unaffected, whereas the content of other tegument proteins such as pUL46, pUL48 and pUL49 varied significantly. Moreover, the cellular protein actin was enriched in virions of mutants deficient in tegument proteins pUL47, pUL49 or pUS3 (del Rio et al., 2005; Michael et al., 2006b). Deletion of pUL21 resulted in a drastic decrease in the incorporation of pUL46, pUL49 and pUS3, which is also observed in PrV-Ba (see below) which, among other defects, carries mutations in the UL21 gene (Michael et al., 2007). In the absence of pUL11 and pUL16, incorporation of full-length pUL36 is decreased and a C-terminally truncated pUL36 fragment is incorporated in increased amounts instead (Michael et al., 2006a).

Other proteins whose absence has been shown to impair virion morphogenesis to varying degrees include the structural pUL51 (Klupp et al., 2005) and pUL7 (Fuchs et al., 2005b) and the nonstructural pUL4 (Fuchs et al., 2006b).

Egress

The net result of secondary envelopment is a mature virus particle enshrined in a secretory vesicle. Fusion of the vesicle membrane with the plasma membrane releases the infectious particle into the extracellular environment. So far, it is unclear why the presumably fusion competent virion does not immediately fuse with the vesicle membrane. However, it has been shown that refusion of released virions with the cell they just exited is prevented by gK, a multiply membrane spanning viral glycoprotein (Klupp et al., 1998). Thus, presence of gK in viral and/or cellular membranes may help

prevent these undesired fusion events. gK requires another viral protein for proper processing, pUL20 (Dietz et al., 2000). Absence of pUL20 not only interferes with gK maturation but also inhibits proper virus release by resulting in the accumulation of virions in secretory vesicles (Fuchs et al., 1997a). Thus, both proteins, gK and pUL20, apparently are important for virus release.
Although not all details have yet been uncovered of this complex maturation process, studies on PrV have been important for our understanding of how herpesvirus particles are formed.

SUMMARY OF FUNCTION OF PrV GENES AND PROTEINS

Many PrV genes and proteins have been analyzed in detail so that PrV may qualify as the best analyzed animal herpesvirus. Summaries about each gene can be found under http://www.fli.bund.de/prv_gene_annotations.html.

VIRUS-HOST CELL INTERACTION

Infection by PrV exerts an overall effect on the host cell metabolism affecting cellular transcription. Both, up- and downregulation of transcripts have been observed in cell culture and infected rats which include RNAs encoding proteins involved in innate immunity, stress responses or signalling pathways (Ray and Enquist, 2004; Paulus et al., 2006). Moreover, PrV expresses proteins which modulate the host cell (Nauwynck et al., 2007). The pUS3 kinase induces the breakdown of actin stress fibers (van Minnebruggen et al., 2003) and further alters cell morphology by the induction of actin-containing cellular projections which may act as 'highways' for direct cell-to-cell spread of infectivity (Favoreel et al., 2005). Moreover, pUS3 possesses an activity which protects cells from apoptosis induced by infection or chemicals (Geenen et al., 2005). In contrast to HSV-1, pUS3 appears to be the only protein in PrV to protect infected cells from apoptosis. Interestingly, the BHV-1 homologue seems to be unable to prevent apoptosis (Takashima et al., 1999). pUS3 is present in two isoforms in infected cells which are derived from two differentially initiated transcripts (van Zijl et al., 1999). Thus, the longer isoform contains an N-terminal extension of 54 aa but is otherwise identical to the smaller isoform. Only the smaller isoform seems to be present in virions. Kinase activity is required for the anti-apoptotic effect and cellular rearrangement. However, pUS3 deleted PrV was only slightly attenuated in the rat eye (Olsen et al., 2006) or murine intranasal infection model

(Klopfleisch et al., 2006), whereas in pigs a more pronounced attenuation was observed (Kimman et al., 1992).

PrV LATENCY

Like all herpesviruses PrV also establishes latency for the lifetime of its host. Since only pigs can survive a productive PrV infection, they are also the only reservoir for PrV. During latency which, after natural oronasal infection, is established primarily in neurons of the trigeminal ganglion but also in tonsils, viral genome persists (Gutekunst, 1979; Rziha et al., 1986; Brown et al., 1995) but only a small segment of the viral genome is transcriptionally active which encompasses part of the IR and adjoining U_L region (Cheung, 1989a; Priola et al., 1990). The latency-associated transcript (LAT) gene is transcribed into three different RNAs of 8.4, 8.0 and 2.0 kb. The 8.4 and 2.0kb species are derived from a larger precursor by splicing (Cheung, 1991). During latency only the 8.4kb large latency transcript (LLT) is produced, whereas the 8.0 and 2.0 species are transcribed during lytic infection (Jin and Scherba, 1999). Consequently, two different promotors, latency-active promotor (LAP) 1 and 2, have been identified (Cheung and Smith, 1999; Jin et al., 2000). LAP1 is characterized by a TATA box located 34 nucleotides upstream from the LLT initiation site, whereas the LAP2 TATA region is situated 143 bp downstream. LAP1 may be neuron specific since it is not required for LLT transcription in cell culture (Taharaguchi et al., 2002, 2003), whereas LAP2 is active in neuronal and nonneuronal cells (Cheung and Smith, 1999). Due to the overlapping antiparallel transcription of IE180 mRNA and LLT, it has been hypothesized that hybrid formation of the corresponding transcripts may modulate establishment and maintenance of latency or reactivation. IE180 has been shown to bind LAP1 resulting in downregulation of LAP-1 driven transcription (Ou et al., 2002). Whether this is a preparatory step for reactivation remains to be analyzed.

PrV IMMUNOLOGY

Major immunogens of PrV in eliciting antibody-dependent and cell-mediated immunity have been identified (Mettenleiter, 1996). Most are envelope glycoproteins, although the major immediate-early protein IE180 also induces an antibody response (Cheung, 1990). The attachment protein gC is a major target of neutralizing antibodies (Hampl et al., 1984; Lukacs et al., 1985) and T-cell mediated immune responses (Zuckermann et al., 1990; Ober et al.,

1998; Ober et al., 2000). Surprisingly, immunity induced by vaccinating immunodeficient mice with parapoxvirus recombinants expressing PrV gC resulted in protection even in the absence of either $CD4^+$ or $CD8^+$ T-cells, B cells or an intact perforin pathway (Fischer et al., 2003). Complement-independently neutralizing antibodies have also been isolated that are directed against gB and gD (Hampl et al., 1984; Lukacs et al., 1985). Interestingly, binding of antibodies to viral glycoproteins presented at the cell surface leads to their internalization which may hide virus-infected cells from immune destruction (Favoreel et al., 1997, 1999; van de Walle et al., 2003). This effect is mediated by endocytosis motifs in the C-terminal intracellular domains of gD and gB (Ficinska et al., 2005). The gE-gI complex exhibits Fc receptor activity (Favoreel et al., 1997; van de Walle et al., 2003) which may also contribute to immune evasion as does binding of complement factor 3 by PrV gC (Huemer et al., 1992).

A second mechanism of immune evasion includes inhibition of the MHC class I-dependent pathway of antigen presentation. Several members of the varicellovirus genus including PrV and BHV-1 downregulate MHC-I presentation (Mellencamp et al., 1991; Sparks-Thissen and Enquist, 1999; Ambagala et al., 2000) by inhibiting the transporter associated with antigen processing, TAP (Koppers-Lalic et al., 2005). The responsible viral protein is the product of the UL49.5 gene, gN. Although gN is also found complexed with gM, free gN, which is present at high levels in cells early after infection, is required to mediate this effect. BHV-1 gN directly binds to TAP, arrests it in a translocation-incompetent conformation and induces its proteasomal degradation (Koppers-Lalic et al., 2005). The exact mode of action of PrV gN is apparently different but awaits further elucidation (Wiertz, personal comm.).

PrV NEUROTROPISM

Like other alphaherpesviruses, PrV exhibits a distinct neurotropism. After replication in peripheral epithelial cells PrV enters nerve endings and is transported to the neuronal cell bodies where productive viral replication or establishment of latency ensues (Enquist, 1994). PrV may then spread either laterally to adjacent neurons or across synapses into higher areas of the nervous system (Enquist et al., 1999). Predominant transsynaptic spread is apparently secured by the relative nonpermissiveness of cells lining the neuron which may become infected by virus exiting the axons at nonsynaptic sites such as varicosities (Tomishima and Enquist, 2002; de Regge et al., 2006). Glycoproteins gE and gI are important determinants for PrV neurovirulence and neuronal spread in rodent infection models and in the pig (Mettenleiter et al., 1987b; Card et al., 1992; Mulder et al., 1994; Babic et al., 1996). Absence of either glycoprotein abolished anterograde spread in synaptically connected neurons after infection of rats in the retinal circuit without affecting retrograde transport, a property shared by the gE-deleted PrV-strain Bartha (Whealy et al., 1993; see below). Interestingly, the homologous proteins of BHV-1 were able to complement absence of gE and gI in PrV in neuronal infection (Knapp and Enquist, 1997; Knapp et al., 1997). Since they are transported rather faithfully along connected neuronal circuits and allow longer labeling times, attenuated, gE-deleted strains are increasingly being used in neurobiology as self-amplifying tracers to uncover neural networks (Loewy, 1995).

gE and gI affect neurotropism as well as neurovirulence (Tirabassi et al., 1997). Neurotropism is defined as the extent of infection within the nervous system, whereas neurovirulence describes pathogenic consequences of neuronal infection. Besides the nonessential gE and gI, the essential gB has also a profound effect on neurovirulence. Whereas gB deletion mutants of PrV are completely innocuous due to the essential character of this protein (Babic et al., 1993; Heffner et al., 1993), substitution of PrV gB by the homologous protein of BHV-1 results in a replication-competent chimeric virus. Surprisingly, this hybrid virus exhibits a drastically increased neurovirulence in pigs which parallels a change in CNS entry from primarily the trigeminal towards the olfactory pathway (Gerdts et al., 2000). Apparently, this phenotype is species-specific since it has not been observed after intranasal infection of mice (Damann et al., 2006a).

Besides glycoproteins, other structural virion proteins also have a distinct effect on PrV neurotropism which include components of the viral tegument (Klopfleisch et al., 2006). To analyze the effects of specific gene deletions on neurotropism and neurovirulence in a highly standardized assay, PrV infection in small animal models such as mice and rats has been established. In most cases, there was a strict correlation between extent of neuronal infection and survival times. Moreover, generally survival times correlated with the relative defects of the mutant viruses in cell culture. However, differences were observed which indicated specific effects of deletions on neuroinvasion. For example, deletion of the nonessential pUs9 membrane protein impaired anterograde spread between neurons, presumably by inhibition of the entry of envelope glycoproteins into the axon (Tomishima and Enquist, 2001). Absence of the tegument proteins pUL37 and pUL48 produced similar overall replication defects in cell culture (Klupp et al., 2001b; Fuchs et al., 2002b). However, only absence of pUL37 abolished neuroinvasion in vivo, whereas animals infected with pUL48 deletion mutants still succumbed to the infection, albeit at significantly later time points compared to wild-type virus infections (Klopfleisch et al., 2004). Thus, these data show a different contribution of tegument proteins to neuronal infection of PrV.

The extent of viral spread in the CNS does not correlate with increased neurovirulence but only with the survival time of the animal. Thus, PrV-Ba produces a more widespread infection in the CNS since animals survive twice as long as after a wild-type virus infection (Card et al., 1992; Klopfleisch et al., 2006). However, despite the restricted wild-type viral spread, animals succumb to the disease early. The molecular basis for this phenomenon is still unclear.

PrV virions reach the neuronal cell body and the synapse by fast axonal transport (Smith et al., 2001). Direct labelling of capsids via the fusion of green fluorescent protein with the nonessential capsid component pUL35 allowed in vivo observations in real time to establish the velocity and the mode of virion transport (Smith et al., 2001). After entry and loss of most tegument proteins except pUL36 and pUL37 (Luxton et al., 2005), which is similar to entry in cultured nonpolarized cells, capsids are retrogradely transported along microtubuli to the neuronal nucleus. After intranuclear assembly of capsids followed by nuclear egress, anterograde transport to the synapse occurs. It has been shown that in fact there is dynein- and kinesin-driven bidirectional transport at both stages (Smith et al., 2001). However, during entry the dynein-mediated minus-end directed transport is more efficient, whereas the opposite is true for egress. It has been hypothesized that these distinct transport efficiencies are due to differences in the composition of the cargo: incoming capsids carry only the inner tegument including pUL36 and pUL37, whereas capsids during exit are also covered by outer tegument proteins (Luxton et al., 2005). Moreover, enveloped virions move only in anterograde direction (Antinone and Smith, 2006). Thus, local modulation of kinesin-mediated plus-end transport targets PrV capsids during entry towards the cell body and towards the synapse during exit (Smith et al., 2004), whereas enveloped particles move strictly towards the synapse (Antinone and Smith, 2006).

The physical nature of the virus particle which is transported in the axon is still under debate. Earlier studies were congruent with a separate transport of capsids and envelopes with final assembly only occurring at the site where virions leave the axon (Tomishima et al., 2001). However, recent data using live-cell imaging microscopy with mutant viruses which express differentially labelled capsid and envelope proteins indicate that in most instances fully enveloped virions are transported down the axon (Antinone and Smith, 2006). These results are not only in conflict with earlier data on PrV but also with current results on HSV-1 transport (Miranda-Saksena et al., 2006; Synder et al., 2006) and, thus, must await further confirmation.

As concerns exit sites, PrV apparently leaves axons not only at the synapse but also at 'varicosities', i.e. bulges in the membrane along the axon (de Regge et al., 2006). It has previously been established that PrV can exit from axons at nonsynaptic sites and infect surrounding astrocytes and brain macrophages (Tomishima and Enquist, 2002). However, these cells appear to be unable to sustain a complete round of viral replication, and, thus, 'insulate' the axon resulting in directional virus transmission via the synapse (Rinaman et al., 1993).

Due to this specific neurotropism, PrV has become a widely applied biological tracer to elucidate and map synaptic connections in neuronal networks (Loewy, 1995; Song et al., 2005). Compared to chemical compounds, PrV acts as a self-amplifying tracer due to its replication in the transsynaptically infected neurons. Owing to the wide host range of PrV, neuronal networks of numerous species of animals including mice and rats can successfully be traced, which, however, depends on the viral strain used, and the time after experimental infection. More virulent strains kill infected animals before any meaningful tracing can be observed. They also tend to produce lateral infections more readily than attenuated strains. In contrast, attenuated strains allow longer survival times of the animal and faithfully propagate along interconnected chains of neurons.

In particular, the live vaccine strain PrV-Bartha (Bartha, 1961; see below) has become a widely-appreciated tracer. Tracing is facilitated by the use of different variants of PrV-Bartha which express marker proteins such as ß-galactosidase or autofluorescent proteins (Loewy et al., 1991; Banfield et al., 2003). These differentially marked but otherwise isogenic strains also allow double-labelling experiments in which central nervous system neurons connected to different peripheral sites can simultaneously be identified (Kerman et al., 2003; Cano et al., 2004). PrV tracing has for example been used to demonstrate remodeling of intraspinal circuitry after injury (Bareyre et al., 2004) or to identify central command neurons in the brain directing the fight-or-flight response (Jansen et al., 1995).

Surprisingly, infection by PrV does not grossly interfere with electrophysiological recordings from neurons. Thus, neurons can first be identified by PrV tracing and, subsequently, used for electrophysiological experiments (Smith et al., 2000; Irnaten et al., 2001; Damann et al., 2006b).

PrV CONTROL: APPLIED MOLECULAR BIOLOGY

Aujeszky's disease is an important economical factor in pig husbandry. Thus, early on inactivated and attenuated live vaccines have been used to limit losses and spread of the disease. Classical attenuated live vaccines have been isolated after repeated passaging of field-virus in tissue culture or embryonated chicken or duck eggs. Although these strains are for the most part genetically ill-defined, one vaccine strain, designated strain Bartha (PrV-Ba) after Adorjan Bartha, an Hungarian veterinarian who isolated this virus (Bartha, 1961), has gained major attention. Restriction fragment length polymorphism analyses (Lomniczi et al., 1984), marker rescue (Lomniczi et al., 1987) and sequencing studies showed that PrV-Ba harbors several independent defects which jointly are responsible for its attenuated phenotype. Most prominent is a large deletion in the U_S region which encompasses the genes US7 (encoding gI), US8 (encoding gE), US9 (encoding the 11kDa pUS9, a small membrane-associated protein) and part of US2 (encoding a 28K protein of unknown function; Mettenleiter et al.,

1985; Petrovskis et al., 1986). Moreover, it contains a mutation in the signal sequence for gC (Robbins et al., 1990), specifies point mutations in the UL21 gene (Klupp et al., 1995), and expresses a UL10 gene product (gM) which is not glycosylated due to mutation of the consensus N-glycosylation signal (Dijkstra et al., 1998). The U_S deletion and the UL21 mutations have been shown to be most important for the observed attenuation since repair of these defects restores the virulent phenotype (Klupp et al., 1995). The observation that PrV-Ba lacks gE yielded the first proof that this glycoprotein is dispensable for viral replication and set the basis for a new strategy in animal disease control: the 'marker' or 'DIVA' (differentiation between infected and vaccinated animals; van Oirschot, 1999) concept. Vaccination is the most cost-efficient means of disease prevention. However, vaccine-induced antibodies could not be differentiated from infection-induced immunity and, thus, vaccinated animals which may still harbor latent field-virus continued to cause concern. This could be overcome using vaccines which induce an immune response that can be differentiated from that elicited by natural infection (van Oirschot et al., 1986). Vaccine strains lacking proteins which are nonessential for viral replication and dispensable for inducing solid immunity despite being immunogenic themselves induce characteristic antibody patterns with specificity for all immunogenic polypeptides except the missing protein. After wild-type infection, a full scale immunological response is produced. Thus, antibodies against the so-called 'marker' protein are distinctive of a wild-type virus infection. The natural absence of gE in PrV-Ba provides such a marker. PrV-Ba lost its gE gene 'naturally' during passaging in chicken embryo fibroblasts since absence of gE yields a selective advantage in this cell culture system (Mettenleiter et al., 1988). However, appropriate deletions were also genetically engineered into wild-type strains to yield mutants which lack gE (Quint et al., 1987), gC (Kit et al., 1987) or gG (Marchioli et al., 1987) to provide the appropriate markers. In these genetically modified strains, besides the deletion of the marker gene attenuating mutations/deletions of the UL23 gene encoding the virulence factor thymidine kinase have also been introduced for safe and stable attenuation. Using either 'classical' or genetically modified, preferentially gE-deleted, vaccines, the marker concept has been used to successfully control and eliminate Aujesky's Disease from domestic pig populations of several heavily infected countries including Germany (Müller et al., 2003) and the USA (Anderson and Leafstedt, 2004). Within these eradication campaigns, genetically modified live vaccines have for the first time been used on a large scale which gives testimony to direct impact of basic molecular biology to applied disease control. A similar approach is now being pursued to eliminate BHV-1 infections using gE-deleted vaccine strains and accompanying marker diagnostics (van Engelenburg et al., 1994; Kaashoek et al., 1995).

BOVINE HERPESVIRUS 1

Bovine herpesvirus 1 is a neurotropic virus of cattle which causes respiratory (infectious bovine rhinotracheitis (IBR)) and genital tract infections (infectious bovine vulvovaginitis (IPV) or infectious bovine balanopothitis (IPB)), abortions, and, rarely, central nervous disorders. In contrast to adult animals in which infection often results in only mild clinical symptoms, BHV-1 inflicts serious problems in younger calves and newborns. BHV-1 is one of the pathogens involved in the bovine respiratory disease complex or shipping fever (Tikoo et al., 1995; Muylkens et al., 2007). Due to the high financial losses entailed by BHV-1 infections, cattle are routinely vaccinated in most parts of the world. To date, several countries in Europe are BHV-1 free and several others have started eradication programs (Ackermann and Engels, 2006) which involve vaccination with marker vaccines that allow differentiation between vaccinated and field virus infected animals. This so called DIVA approach (van Oirschot, 1999) has successfully been used in several countries for the eradication of Aujeszky´s disease in pigs caused by the alphaherpesvirus PrV (see above).

BHV-1 GENE EXPRESSION

Like in all herpesviruses, BHV-1 gene expression in infected permissive cells is temporally regulated in a cascade-like fashion. The first transcripts are the immediate-early (IE) RNAs which originate from IE transcription units 1 and 2 (Wirth et al., 1991). IE transcription is enhanced by the UL48 encoded protein named α-TIF, a structural protein of BHV-1 virions (Misra, 1995). Since transfected naked genomic DNA is infectious, at least in cell culture, the transactivating function of the UL48 gene product is not essential for BHV-1 replication. IE transcription unit 1 codes for bICP0 and bICP4, IE transcription unit 2 encodes bICP22 (Fig. 5). Presence of immediate early proteins is essential to initiate early gene expression which then leads to viral genome replication and late gene expression, and finally to virion assembly and release.

Among the IE proteins, bICP0 is of special significance for BHV-1 replication. bICP0 activates *in trans* all viral promotors and is present in high abundance in infected cells during the entire replication cycle (Wirth et al., 1991, 1992; Fraefel et al., 1994b). Despite its importance for BHV-1 propagation, analysis of cell culture characteristics of a bICP0 null mutant has shown that bICP0 is dispensable. It grows, however, to at least 100-fold lower titers which underlines the significance of bICP0 for BHV-1 replication (Geiser et al., 2005). Within bICP0 are several important functional domains: a zinc RING finger motif near the aminoterminal end, two separate transcriptional activation domains and a nuclear localization signal that is required for efficient trans-activation (Fraefel et al., 1994b; Inman et al., 2001b; Zhang and Jones, 2005). Functional studies have shown that bICP0 represses the interferon-β promoter and a promoter with consensus interferon stimulated response elements which requires the zinc RING finger motif (Henderson et al., 2005). It is believed that repression of interferon induced transcription is of importance for BHV-1 productive infection and pathogenesis (Jones et al., 2006). Besides the effects on transcription, bICP0 colocalizes with and disrupts the proto-oncogene promyelotic leukemia (PML) protein-containing nuclear domains (Jones et al., 2006). Expression of bICP0 in murine neuroblastoma cells and bovine turbinate cells leads to activation of caspase 3, a key regulatory protein for the induction of apoptosis (Henderson et al., 2004), which is also activated during productive infection (Devireddy and Jones, 1999). Furthermore, bICP0 interacts with and reduces the steady state levels of histone deacetylase1 (HDAC1), which results in increased transcriptional activity (Zhang and Jones, 2001) and, thus, may lead to an improved viral replication.

Analysis of the immediate-early gene expression of BHV-1 revealed a so far unique feature among the alphaherpesviruses. A spliced IE RNA of 1.5 kb (IER1.5), composed of two exons, arises by transcription over the terminal inverted repeat sequence (TR_S) - U_L junction which is formed after circularization of the genomic DNA in infected cells. Exon 1 originates from the TRs, whereas exon 2 of IER1.5 is transcribed from sequences located within the U_L segment. Sequence analysis showed that the U_L part of IER1.5 is 3' coterminal with an unspliced late transcript of 1.1 kb (LR1.1), whose synthesis is directed by an alternative promoter, and that both RNAs contain an open reading frame (ORF) with the same

247 codons (Fraefel et al.,1993). The predicted protein has some homology with proteins encoded by varicella-zoster virus ORF2 and the equine herpesvirus 1 gene 3. Due to its origin from a circular BHV-1 genome, the gene specifying IER1.5 was designated *circ*. The *circ*-encoded protein is a 34 kDa apparent molecular weight component of BHV-1 virions (Fraefel et al., 1994a) which is myristoylated as is the equine herpesvirus 1 gene 3 encoded protein which, however, is translated from an early expressed mRNA (Harty et al., 1993). In BHV-1 infected cells, the circ protein is located in the cytoplasm where it associates with membranes (Schwyzer et al., 2002). The function of the circ protein is largely unclear. Deletion of the *circ* gene from the BHV-1 genome did not affect viral growth in cell culture (Fraefel et al., 1994a). Infection of bovine macrophages with the *circ*-deleted mutant and transduction of bovine monocytes with circ protein expressing HSV-1 based amplicons suggested that it may be involved in downregulation of MHC-II protein expression and may cause a specific cytopathogenic effect in monocytes (Schwyzer et al., 2002).

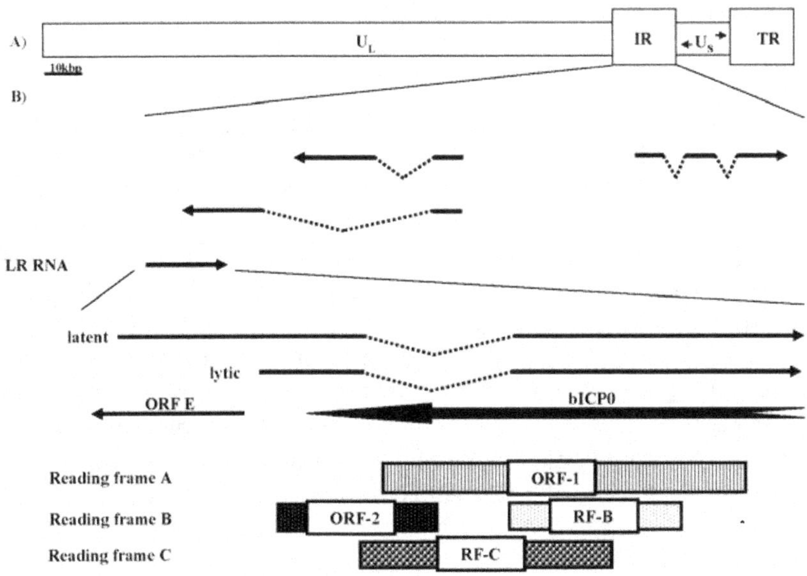

Fig. 5: Organization of the LAT region of BHV-1

BHV-1 LATENCY

In infected cattle, BHV-1 initially replicates in the mucosal epithelium of the respiratory or genital tract, spreads to peripheral sensory neurons and establishes lifelong latency in the appendant ganglia (Ackermann et al., 1982). Establishment of latency comprises penetration of virions into ganglionic neurons where initial lytic viral gene expression and DNA replication is turned off, and abundant viral RNA transcription is restricted to the latency-related (LR) gene and the ORF-E gene (Rock et al., 1987, 1992; Kutish et al., 1990; Inman et al., 2004). ORF E is located within the promoter of the LR gene downstream of the bICP0 ORF (Fig. 5). Orientation of the 402 nt ORF-E is antisense to the LR gene. Its transcription may be controlled by a long A-T rich motif within the LR gene promoter (Clinton et al., 2006). Transcripts encompassing ORF-E were found in productively infected cultured bovine cells and in the trigeminal ganglia of latently infected calves (Inman et al., 2004). Fusion proteins of ORF-E with GFP localize to the nucleus of neuroblastoma cells but are found throughout rabbit skin cells. The significance and the function of the ORF-E protein during establishment and maintenance of latency or for reactivation of BHV-1 from the latent state remains to be elucidated (Jones et al., 2006).

The LR gene is abundantly transcribed in latently infected neurons (Rock et al., 1987, 1992; Kutish et al., 1990). Direction of transcription is opposite to the mRNAs which code for bICP0. The LR gene contains two open reading frames named ORF-1 and ORF-2, and two reading frames, RF-B and RF-C, that lack a translation initiation codon ATG. Monospecific antibodies directed against the amino terminus of the ORF-2 encoded protein reacted with a protein which interacts with cyclin-dependent kinase 2 during productive infection (Jiang, 1998). In trigeminal ganglia of infected cattle a portion of the LR-RNA is polyadenylated and alternative splicing of LR-RNA yields a family of transcripts which led to the assumption that more than one LR protein might be expressed at specific times during establishment and maintenance of latency and reactivation (Hossain et al., 1995; Devireddy and Jones, 1999).

In vitro analyses showed that expression of the LR gene products inhibits entry of cells into the S phase, bICP0 expression (Bratanic et al., 1992; Geiser et al., 2002; Schang et al., 1996) and apoptosis (Ciacci-Zanella et al., 1999). Consistent with the requirement of LR gene directed protein synthesis to inhibit apoptosis, Meyer et al. (2007) recently described a 40 kDa protein (LR fusion protein) which is encoded by a transcript that represents an ORF-2/ORF-1 fusion (Devireddy and Jones 1998; Devireddy et al., 2003). This mRNA is expressed in trigeminal ganglia at 7 days after infection of cattle which is an important time point since the majority of BHV-1 genes are quiescent thereafter (Schang and Jones, 1997). Bacterial two-hybrid assays identified interaction of the LR fusion protein with Bid (a BH3-only pro-death Bcl-2 family protein), CDC42 (cell division cycle 42, aRho GTPase) and CCAAT enhancer binding protein alpha (C/EBP-α). Bid links the extrinsic and intrinsic pathways of apoptosis (Li et al., 1998; Wang, 2001) and activation of CDC42 results in apoptosis in primary sympathetic neurons deprived of nerve growth factor (Kanamoto et al., 2000; Xu et al., 2001). These findings suggest that the LR fusion protein interferes which induction of apoptosis which promotes the establishment of latency. Overexpression of C/EBP-α enhances productive BHV-1 expression which led to the assumption that interactions of the LR fusion protein with C/EBP-α represses productive infection in trigeminal neurons and, thus, supports the establishment of latency (Meyer et al., 2007).

To analyse the significance of LR protein expression *in vivo*, a BHV-1 mutant (LR mutant) with three stop codons at the 5′-end of LR ORF-2 has been used in animal experiments. In comparison to wild type BHV-1 or the LR rescued virus, the LR mutant had similar growth characteristics in productively infected cultured cells and in the nasal cavities of calves during acute infection, whereas clinical symptoms and virus shedding from the eye or trigeminal ganglia was reduced. In addition, the LR mutant could not be reactivated by treatment of latently infected calves with dexamethasone, whereas in calves latently infected with wild type BHV-1 and the LR rescued virus this treatment resulted in reactivation and shedding of infectious virus which provides strong evidence that expression of the authentic LR gene and protein is indispensable for the complete latency-reactivation cycle in BHV-1 infected cattle (Inman et al., 2001a, 2002; Jiang et al., 2004; Jones et al., 2006). Trigeminal ganglia of calves infected with the LR mutant contained LR RNA, and harboured genomic BHV-1 DNA but at reduced levels in comparison to calves latently infected with wild-type BHV-1 (Inman et al., 2002; Lovato et al., 2003). Comparison of the frequency of apoptosis in calves infected with the LR mutant compared to wild type BHV-1 infection demonstrated that the levels of apoptosis were higher in calves infected with the LR mutant than in calves infected with wild type BHV-1 or in mock-infected calves which is in good agreement with the antiapoptotic function of LR protein expression in cell culture. Taken together, these results point to an important role of LR gene encoded antiapoptotic properties for the establishment of latency (Lovato et al., 2003).

As mentioned above, expression of proteins from the LR gene appeared to be not required for inhibiting cell proliferation and bICP0 expression in cultured cells. These functions of the non-coding LR RNAs are probably also of importance in the latency-reactivation cycle. Inhibition of bICP0 protein synthesis interferes with productive infection (Geiser et al., 2005) and, thus, could contribute to the survival of infected neurons. Inhibition of cell proliferation may also be beneficial for neuronal survival since proteins involved in cell cycle regulation can induce apoptosis. The mechanism by which the non-coding LR RNAs exert their function is, however, unclear. The gene encoding the herpes simplex virus type 1 (HSV-1) latency associated transcript (LAT), which shares structural and functional properties with the BHV-1 LR gene, encodes a microRNA (miRNA) which confers resistance to apoptosis by downregulation of transforming growth factor ß-1 and SMAD3 expression (Fraser, 2006). It should be noted that protein expression from the LAT transcripts has not been reliably detected in HSV-1 so far. Potential stem-loop structures which are characteristic for miRNAs have also been found in BHV-1 LR transcripts (Geiser and Jones, 2005) which might be indicative for a miRNA mediated silencing mechanism in inhibition of bICP0 expression and cell growth. Further studies are needed to prove this hypothesis.

INFECTIOUS LARYNGOTRACHEITIS VIRUS

Infectious laryngotracheitis virus causes an economically important disease of chicken which results in delayed growth, reduced egg production, and also frequently leads to death of the animals. After acute infection of the upper respiratory tract, the virus enters the nervous system and establishes latency in the central nervous system. Subsequent reactivation and virus shedding can then result in infection of naive chickens. For prevention of ILTV, conventionally attenuated live vaccines are available which, however, are genetically noncharacterized and occasionally revert to a virulent phenotype. ILTV has a very narrow host range naturally infecting only chicken and pheasant (Guy and Bagust, 2003).

ILTV TRANSCRIPTS AND PROTEINS

Like other herpesviruses, ILTV exhibits a cascade-like regulation of gene expression in infected cells (Prideux et al., 1992; Roizman and Knipe, 2001). At least one viral gene is abundantly transcribed in the absence of de novo protein synthesis. It encodes the homologue of the major immediate-early protein ICP4 of HSV-1, which is required for subsequent transcription of other virus genes (Johnson et al., 1995; Ziemann et al., 1998b). The virion protein encoded by the UL48 homologue of ILTV has been shown to stimulate viral immediate-early gene expression (Helferich et al., 2007a), which also parallels the situation found with HSV-1 and other alphaherpesviruses (Roizman and Knipe, 2001). Due to their transcription activating functions, both the ICP4 and the UL48 gene products of ILTV are able to enhance the infectivity of genomic virion DNA in transfected cells (Fuchs et al., 2000a). Only few ILTV genes have been shown to contain introns. Whereas pUL0, pUL[-1] and pUL15 seem to be exclusively expressed from spliced transcripts (Ziemann et al., 1998b; Fuchs and Mettenleiter, 1999), distinct proteins are translated from spliced and unspliced US5 mRNAs (Fuchs et al., 2005a).

The ILTV genome contains numerous conserved ORFs which are predicted to encode proteins involved in DNA replication, nucleotide metabolism, or protein modification (Table 1), and the enzymatic activities of the viral dUTPase (UL50) and thymidine-kinase (UL23) have already been demonstrated (Schnitzlein et al., 1995; Fuchs et al., 2000a). Other viral gene products, including tegument proteins, envelope glycoproteins, and nonstructural proteins could be directly identified by the use of monospecific antisera raised against respective bacterial fusion proteins, or by monoclonal antibodies. The protein products of UL37, UL46, UL47, UL48 and UL49 have been detected in purified ILTV particles, and are considered to represent major components of their pronounced tegument layer (Helferich et al., 2007a).

Two envelope glycoproteins have been shown to be the preferred targets of mouse monoclonal antibodies prepared against ILT virions: the UL44 gene product gC (Fig. 6), and the US5 gene product gJ (Veits et al., 2003a). Antibodies against these two proteins are also detectable in most sera of ILTV-infected chickens (Fuchs et al., 2005a). Glycoprotein J is modified by addition of N- and O-linked carbohydrates, and multiple forms of the protein result from translation of spliced and unspliced mRNAs, and presumably also from proteolytic cleavage of the primary translation products (Fuchs et al., 2005a). For gC only N-glycosylation could be demonstrated (Kingsley et al., 1994; Fuchs et al., 2005a). The gC homologues of many alphaherpesviruses are involved in primary attachment to their host cells by binding to heparan sulfate chains of proteoglycans at the plasma membrane (Roizman and Knipe, 2001). In contrast, neither gC, nor other envelope glycoproteins of ILTV possess similar binding activities, indicating attachment to cells by a different, yet unknown mechanism (Kingsley et al., 1999). The ILTV homologues of two herpesvirus glycoproteins which are essential for penetration of host cells, gB (UL27) and gH (UL22), have been identified using monospecific antisera (Fuchs et al., unpublished results; Poulsen and Keeler, 1997). Like the corresponding proteins of most other herpesviruses gB of ILTV is processed by N-glycosylation and by proteolytic cleavage into two subunits, which are linked in a homooligomeric protein complex (Poulsen and Keeler, 1997). Another disulfide-linked, but heterodimeric complex is formed by glycoproteins gM (UL10) and gN (UL49.5) of many herpesviruses including ILTV (Fuchs and Mettenleiter, 2005). However, unlike all known gM homologues, the UL10 protein of ILTV is apparently not glycosylated at all, whereas gN is modified by O-glycosylation (Fuchs and Mettenleiter, 2005). The ILTV genome also contains ORFs encoding homologues of the alphaherpesvirus envelope glycoproteins gD, gE, gI and gK, but the respective gene products have not been detected up to now (Table 1). The conserved N-glycosylated gG (US4) of ILTV and many other alphaherpesviruses is not incorporated into mature virus particles, but secreted from infected cells (Kongsuwan et al., 1993; Helferich et al., 2007a). Whereas in most cases secretion of gG results from proteolytic cleavage of an originally membrane-anchored protein, this step is presumably not required for the ILTV homologue, since the deduced gene product (Wild et al., 1996) does not contain hydrophobic amino acid stretches which could serve as membrane anchor domains.

Another nonstructural protein of ILTV is encoded by the UL31 gene. This protein could be localized along the nuclear membranes of infected host cells (Helferich et al., 2007a), which indicates a function during egress of nucleocapsids from the nucleus, as it has been described for the homologous UL31 gene products of other herpesviruses (Mettenleiter, 2002). The products of the ILTV- or Iltovirus-specific genes UL0, UL[-1], and ORF A to E were also not detectable in purified virions, but all seven genes are transcribed and translated in infected cells (Ziemann et al., 1998a, b; Veits et al., 2003b). Whereas the low abundances of the ORF A to E proteins did not permit their precise localization, the UL0 and UL[-1] gene products were preferentially found in the cell nuclei. Thus, these two proteins might be involved in regulation of viral genes like the ICP0 homologues encoded at the corresponding genome positions of HSV-1 and other alphaherpesviruses (Roizman and Knipe, 2001). However, UL0 and UL[-1] are not expressed under immediate-early conditions, and do not contain zinc-finger domains which are characteristic for ICP0 homologues (Ziemann et al., 1998b).

ILTV DNA REPLICATION AND VIRUS MORPHOGENESIS

Whereas primary enveloped ILTV particles in the perinuclear space are indistinguishable from those of other herpesviruses, finally enveloped virions frequently contain huge amounts of tegument leading to particle diameters of up to 350 nm (Fig. 1C; Granzow et al., 2001). Besides complete virions, exceptionally high proportions of capsidless light particles, which only consist of tegument and envelope are released from ILTV-infected cells. Within the cytoplasm, tubular aggregations of presumably viral proteins, and large vacuoles containing numerous virus particles are frequently observed (Guo et al., 1993). First infectious progeny viruses are detectable in cultured chicken cells between 8 and 12 h, and maximum titers are achieved 24 to 30 h after synchronized infection (Fuchs et al., 2000a). However, besides the chorioallantoic membrane of embryonated chicken eggs and primary chicken kidney and liver cells, only one permanent chicken hepatoma cell line has been shown to permit productive replication of ILTV (Schnitzlein et al., 1994).

IN VIVO REPLICATION, PATHOGENICITY AND CONTROL

ILTV exhibits a very narrow host range, not only in vitro, but also in vivo. Up to now, signs of disease have been observed only in domestic chickens and pheasants, and, besides from these species, ILTV could be isolated only from peafowl and from experimentally infected turkeys (Guy and Bagust, 2003). ILTV infections predominantly affect mucosa and submucosa of the upper respiratory tract, but conjunctivitis is also a frequently observed symptom. Further symptoms of milder forms of infectious laryngotracheitis (ILT) are nasal discharge and reduced egg production, whereas severe forms are additionally characterized by gasping, coughing, expectoration of bloody mucus, and marked dyspnoea which may lead to suffocation. Depending on virus dose and strain virulence incubation periods can vary from 2 to 12 days, and mortality rates in infected flocks are ranging between 0 and 70 %. In chickens surviving the acute phase of disease lasting ca. 7 days, ILTV, like most other alphaherpesviruses, can establish lifelong latent infections in the central nervous system (Bagust, 1986; Williams et al., 1992). The ILTV genome could be detected predominantly within the trigeminal ganglia, whereas latency associated transcripts or proteins have not been described up to now. Sporadic reactivations of latent ILTV infections are in most cases clinically inapparent, but lead to productive replication in the respiratory tract and virus shedding, which can result in infection of susceptible contact animals (Hughes et al., 1989).

ILT occurs worldwide and causes considerable economic losses. For prevention chickens can be immunized with live-virus vaccines, which had been attenuated by passage in embryonated chicken eggs or cell cultures (Guy and Bagust, 2003). Vaccination, like natural infection, induces both humoral and cellular immune responses (Fahey and York, 1990), and confers long-lasting protection against subsequent infections with pathogenic ILTV strains. However, the vaccine strains used up to now are not characterized genetically, and many of them possess residual virulence which can further increase during animal passage (Guy et al., 1990, 1991; Kotiw et al., 1995). Therefore, inactivated whole-virus and preparations of immunogenic envelope glycoproteins were tested as alternative vaccines (Barboom et al., 1986; York and Fahey, 1991). In laboratory experiments these vaccines proved to be protective, but, due to the high costs of production and administration, they are not suitable for immunization of large chicken flocks.

GENETICALLY ENGINEERED ILTV MUTANTS

Genetic engineering is a promising tool to obtain safe, reliable, and affordable ILT vaccines. In several approaches the apathogenic herpesvirus of turkeys or attenuated fowlpox virus strains were used as expression vectors for immunogenic envelope proteins of ILTV, and the obtained virus recombinants were shown to protect chickens against subsequent challenge infections with virulent ILTV (Saif et al., 1994; Tong et al., 2001; Davison et al., 2006).

It was also attempted to generate stably attenuated ILTV mutants by identification and directed deletion of virulence-determining genes. These experiments are still hampered by the lack of infectious full length clones of the ILTV genome, and of suitable cell lines for efficient in vitro replication and manipulation. Nevertheless, up to now ILTV recombinants with deletions of 14 different genes could be isolated after cotransfection of permissive chicken cells with virion DNA and transfer plasmids. In many of these approaches, virus replication in transfected cells was enhanced by expression plasmids for viral trans-activators (Fuchs et al., 2000a), and identification of the desired virus mutants was facilitated by concomitant insertion of reporter genes encoding β-galactosidase or enhanced green fluorescent protein.

The ILTV genes deleted so far include two enzymes of the nucleotide metabolism, the UL23 gene product thymidine kinase (TK) (Okamura et al., 1994; Schnitzlein et al., 1995; Han et al., 2002), and the UL50 protein dUTPase (Fuchs et al., 2000a). Since both enzymes possess almost ubiquitous cellular homologues, in vitro replication of the respective ILTV mutants was not significantly affected. Deletion of the US4 gene encoding the secreted glycoprotein gG did also not affect virus growth in cultured chicken cells (Devlin et al., 2006b; Helferich et al., 2007b). In contrast, ILTV mutants lacking the envelope proteins encoded by UL10 (gM homologue), UL49.5 (gN), or US5 (gJ), or the tegument protein products of UL47 or UL21 exhibited detectable replication defects with respect to maximum virus titers and/or plaque formation in cell monolayers (Fuchs and Mettenleiter, 1999, 2005; Fuchs et al., 2005a, 2007; Helferich et al., 2007b). This also applies to deletions of the ILTV- or Iltovirus-specific genes ORF-A, ORF-B, ORF-C, ORF-D, ORF-E, and UL0 (Veits et al., 2003b, 2003c). Thus, the protein products of these additional genes are not essential, but might be beneficial for replication in avian host cells. The precise reasons for the in vitro growth defects of the described ILTV recombinants remain to be elucidated. For these studies deletion mutants without foreign gene insertions are required, since reporter gene expression was shown to affect ILTV replication directly (Fuchs et al., 2000a; Veits et al., 2003c). Furthermore, rescue mutants have to be generated to rule out possible effects of unwanted second site mutations. Up to now, these prerequisites are given for several, but not for all of the analyzed genes.

All genes which had been hitherto deleted from the ILTV genome are dispensable for propagation in cell culture, and, as far as present and investigated, their homologues are also nonessential in other alphaherpesviruses (Roizman and Knipe, 2001). Repeated attempts to delete several other ILTV genes were not successful, indicating that the encoded proteins might be essential for productive virus replication. Among them are the ILTV homologues of the conserved UL31, UL37, and UL48 genes which were also shown to be essential in HSV-1 and very important in PrV (Roizman and Knipe, 2001; Mettenleiter, 2002; Helferich et al., 2007a). In contrast, the tegument proteins encoded by UL46 and UL49, as well as the envelope glycoproteins gI and gE (US7 and US8 gene products), proved to be dispensable for mammalian alphaherpesviruses, but could not be deleted from the ILTV genome (Devlin et al., 2006a; Helferich et al., 2007a). Remarkably, UL49, gE and gI are also essential for another chicken pathogen, Marek's disease virus (MDV) (Schumacher et al., 2001; Dorange et al., 2002). These parallels indicate that the mentioned proteins might possess functions which are specifically required for replication of alphaherpesviruses in avian cells. One of the Iltovirus-specific genes, UL[-1], seems to be also indispensable for ILTV. Thus, although structurally related to the adjoining nonessential UL0 gene, UL[-1] might have different or additional functions, possibly in regulation of viral gene expression (Veits et al., 2003c). However, it should be kept in mind that up to now the essential roles of none of the proposed ILTV genes could be verified by purification of the respective deletion mutants in trans-complementing cells.

Several of the described gene deletion mutants of ILTV have already been used for animal trials to test their suitability as live-virus vaccines. As in other herpesviruses deletion of the TK gene led to a marked attenuation (Okamura et al., 1994; Schnitzlein et al., 1995; Han et al., 2002), whereas deletion of the dUTPase gene reduced virulence only moderately (Fuchs et al., 2000). Attenuation was also observed after deletion of gG, or the tegument protein

UL47 (Devlin et al., 2006b; Helferich et al., 2007b), and deletion of gJ or the UL0 gene rendered ILTV almost completely apathogenic for chickens (Veits et al., 2003c; Fuchs et al., 2005a). A single live-virus vaccination of chickens with any of the tested ILTV recombinants conferred solid protection against challenge infections with virulent ILTV. Furthermore, ILTV recombinants with deletions of immunogenic glycoproteins like gJ seem to be suitable as marker vaccines, which permit serological differentiation of field-virus infected from vaccinated animals (DIVA strategy; see above) on the basis of presence or absence of antibodies against the respective viral gene product (Fuchs et al., 2005a). Thus, in the near future genetically engineered ILTV mutants might replace the genetically undefined vaccine strains which are currently in use (Fig. 6).

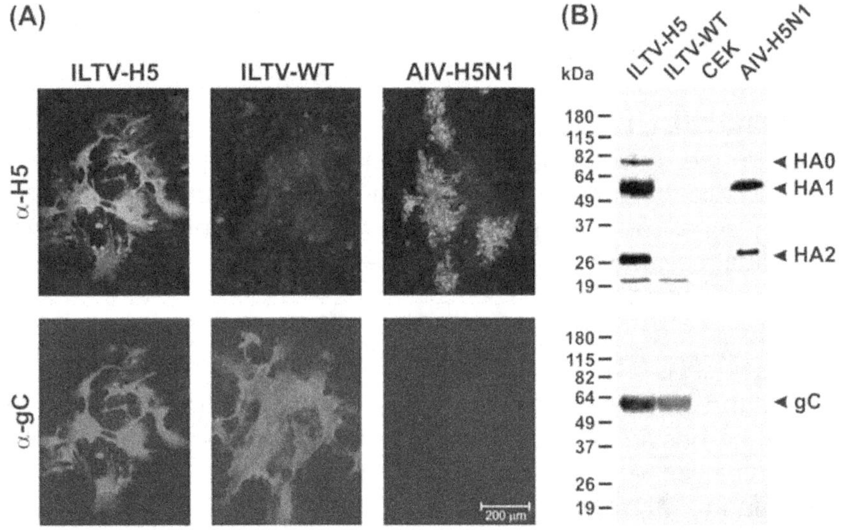

Fig. 6. Influenza virus hemagglutinin expression by recombinant ILTV. Chicken kidney cells infected with mutant or wild-type ILTV (ILTV-H5, ILTV-WT), or avian influenza virus (AIV-H5N1) and uninfected cells (CEK) were analyzed by indirect immunfluorescence tests (A) and Western blotting (B). Reactions of antibodies specific for H5-hemagglutinin (upper panels), and ILTV glycoprotein gC (lower panels) are shown. Molecular masses of marker proteins and the processing products of hemagglutinin (HA0, HA1, HA2) are indicated.

Since ILT live-virus vaccines permit economical mass application to large flocks via aerosol or drinking water (Guy and Bagust, 2003), they might be suitable as vectors for immunogenic proteins of other important chicken viruses, like MDV, Newcastle disease virus, infectious bronchitis virus, infectious bursal disease virus or fowlpox virus, as well as of pathogenic bacteria or parasites. In first attempts of this kind the H5 and H7 hemagglutinin (HA) genes of highly pathogenic avian influenza viruses (HPAIV) were inserted into ILTV recombinants lacking the dUTPase gene or UL0 (Lüschow et al., 2001; Veits et al., 2003c). Expression of the foreign proteins could be detected in infected cells (Fig. 6), and HA-specific antibodies were induced in experimentally infected chickens. After eye-drop immunization the animals were well protected against challenge infections with virulent ILTV, as well as with HPAIV of the respective serotypes. Protection against fowl plague was comparable to that obtained with inactivated influenza virus preparations which, however, require subcutaneous or intramuscular application. These results clearly demonstrate that ILTV recombinants might be useful viral vectors for the development of polyvalent vaccines.

ACKNOWLEDGEMENTS

Work in the authors laboratories has been supported by the Federal Ministry for Food, Agriculture and Consumer Protection, the Deutsche Forschungsgemeinschaft and Intervet Intl. We thank our colleagues within and beyond the FLI for their encouragement and cooperation.

REFERENCES

Ackermann, M., Peterhans, E., and Wyler, R. (1982). DNA of bovine herpesvirus type 1 in the trigeminal ganglia of latently infected calves. Am. J. Vet. Res. *43*, 36-40.

Ackermann, M., and Engels, M. (2006). Pro and contra IBR-eradication. Vet. Mic. *113*, 293-302.

Ambagala, A., Hinkley, S., and Srikumaran, S. (2000). An early pseudorabies virus protein down-regulates porcine MHC class I expression by inhibition of transporter associated with antigen processing (TAP). J. Immunol. *164*, 93-99.

Anderson, P.L., and Leafstedt, J.W. (2004). Report of the Committee on Pseudorabies. Proc. 108th Annual Meeting of the United States Animal Health Assoc. Greensboro, NC, Oct. 21-27, 2004, pp. 480-489.

Antinone, S.E., and Smith, G.E. (2006). Two modes of herpesvirus trafficking in neurons: membrane acquisition directs motion. J. Virol. *80*, 11235-11240.

Aoki, T., Hirono, I., Kurokawa, K., Fukuda, H., Nahary, R., Eldar, A., Davison, A.J., Waltzek, T.B., Bercovier, H., and Hedrick, R.P. (2007). Genome sequences of three koi herpesvirus isolates representing the expanding distribution of an emerging disease threatening koi and common carp worldwide. J. Virol. (in press).

Aujeszky, A. (1902). Über eine neue Infektionskrankheit bei Haustieren. Zbl. Bakteriol. Infektionskrankh., Parasitenkde. 1. Abt. Orig. *32*, 353-357 (in German).

Babic, N., Mettenleiter, T.C., Flamand, A., and Ugolini, G. (1993). Role of the essential glycoproteins gII and gp50 in transneuronal transfer of pseudorabies virus from the hypoglossal nerve of mice. J. Virol. *67*, 4421-4426.

Babic, N., Klupp, B.G., Brack, A., Mettenleiter, T.C., and Flamand, A. (1996). Deletion of glycoprotein gE reduces the propagation of pseudorabies virus in the nervous system of the mouse after intranasal inoculation. Virology *219*, 279-284.

Bagust, T.J. (1986). Laryngotracheitis (gallid-1) herpesvirus infection in the chicken. 4. Latency establishment by wild and vaccine strains of ILT virus. Avian Pathol. *15*, 581-595.

Baines, J., and Weller, S.K. (2005). Cleavage and packaging of herpes simplex virus 1 DNA. In: Viral genome packaging machines: genetics, structures and mechanisms. Edited by C.E. Catalano. Landes Biosciences, pp. 135-150.

Baines, J.D., Wills, E., Jacob, R.J., Pennington, J., and Roizman, B. (2007). Glycoprotein M of herpes simplex virus 1 is incorporated into virions during budding at the inner nuclear membrane. J. Virol. *81*, 800-812.

Baker, M.L., Jiang, W., Rixon, F.J., and Chiu, W. (2005). Common ancestry of herpesviruses and tailed DNA bacteriophages. J. Virol. *79*, 14967-14970.

Banfield, B.W., Kaufman, J.D., Randall, J.A., and Pickard, G.E. (2003). Development of pseudorabies virus strains expressing red fluorescent proteins: new tools for multisynaptic labeling applications. J. Virol. *77*, 10106-10112.

Barboom, S.A., Forgacs, A., and Solyom, F. (1986). Development of an inactivated vaccine against laryngotracheitis (ILT) – serological and protection studies. Avian Pathol. *15*, 213-221.

Bareyre, F., Kerschensteiner, M., Raineteau, O., Mettenleiter, T.C., and Schwab, M.E. (2004). Recovery in absence of regeneration: spontaneous corticospinal remodelling is involved in the formation of a new intraspinal circuit mediating physiological recovery after spinal cord injury in rats. Nature Neurosci. *7*, 269-277.

Bartha, A. (1961). Experimental reduction of virulence of Aujeszky's disease virus. Mag. Allatorv. Lapja *16*, 42-45 (in Hungarian).

Batterson, W., and Roizman, B. (1983). Characterization of the herpes simplex virion-associated factor responsible for the induction of α genes. J. Virol. *46*, 371-377.

Ben-Porat, T., and Kaplan, A.S. (1970). Synthesis of proteins in cells infected with herpesvirus. V. Viral glycoproteins. Virology *41*, 265-273.

Ben-Porat, T., Veach, R.A., and Ihara, S. (1983). Localization of the regions of homology between the genomes of herpes simplex virus type 1 and pseudorabies virus. Virology *127*, 194-204.

Blewett, E., and Misra, V. (1991). Cleavage of the bovine herpesvirus glycoprotein B is not essential for its function. J. Gen. Virol. *72*, 2083-2090.

Böttcher, S., Klupp, B.G., Granzow, H., and Mettenleiter, T.C. (2006). Identification of a 709-amino-acid internal nonessential region within the essential conserved tegument protein (p)UL36 of pseudorabies virus. J. Virol. *80*, 9910-9915.

Bowzard, J.B., Visalli, R.J., Wilson, C.B., Loomis, J.S., Callahan, E.M., Courtney, R.J., and Wills, J.W. (2000). Membrane targeting properties of a herpesvirus tegument protein-retrovirus Gag chimera. J. Virol. *74*, 8692-8699.

Brack, A.R., Dijkstra, J., Granzow, H., Klupp, B.G., and Mettenleiter, T.C. (1999). Inhibition of virion maturation by simultaneous deletion of glycoproteins E, I and M of pseudorabies virus. J. Virol. *73*, 5364-5372.

Brack, A.R., Klupp, B.G., Granzow, H., Tirabassi, R., Enquist, L.W., and Mettenleiter, T.C. (2000). Role of the cytoplasmic tail of pseudorabies virus glycoprotein E in virion formation. J. Virol. *74*, 4004-4016.

Bratanic, A. C., and Jones, C. J. (1992). Localization of cis-acting sequences in the latency-related promoter of bovine herpesvirus 1 which are regulated by neuronal cell type factors and immediate-early genes. J. Virol. *66*, 6099-6106.

Brideau, A., Eldridge, M.G., and Enquist, L.W. (2000a). Directional transneuronal infection by pseudorabies virus is dependent on an acidic internalization motif in the Us9 cytoplasmic tail. J. Virol. *74*, 4549-4561.

Brideau, A., Enquist, L.W., and Tirabassi, R.S. (2000b). The role of virion membrane protein endocytosis in the herpesvirus life cycle. J. Clin. Virol. *17*, 69-82.

Brown, T.T., Shin, K.O., and Fuller, F.J. (1995). Detection of pseudorabies viral DNA in tonsillar epithelial cells of latently infected pigs. Am. J. Vet. Res. *56*, 587-594.
Browne, H., Bell, S., and Minson, T. (2004). Analysis of the requirements for glycoprotein M in herpes simplex virus type 1 morphogenesis. J. Virol. *78*, 1039-1041.
Cano, G., Card, J.P., and Sved, A.F. (2004). Dual viral transneuronal tracing of central autonomic circuits involved in the innervation of the two kidneys in rat. J. Comp. Neurol. *471*, 462-481.
Card, J.P., Whealy, M.E., Robbins, A.K., and Enquist, L.W. (1992). Pseudorabies virus envelope glycoprotein gI influences both neurotropism and virulence during infection of the rat visual system. J. Virol. *66*, 3032-3041.
Chang, J.T., Schmid, M.F., Frazer, J., Rixon, F.J., and Chiu, W. (2007). Electron cryotomography reveals the portal in the herpesvirus capsid. J. Virol. *81*, 2065-2068.
Cheung, A.K. (1989a). Detection of pseudorabies virus transcripts in trigeminal ganglia of latently infected swine. J. Virol. *63*, 2908-2913.
Cheung, A.K. (1989b). DNA nucleotide sequence analysis of the immediate-early gene of pseudorabies virus. Nucl. Acids Res. *17*, 4637-4646.
Cheung, A.K. (1990). Humoral immune response to immediate-early protein of pseudorabies virus in swine with induced or naturally acquired infection. Am. J. Vet. Res. *51*, 222-226.
Cheung, A.K. (1991). Cloning of the latency gene and the early protein 0 gene of pseudorabies virus. J. Virol. *65*, 5260-5271.
Cheung, A.K., and Smith, T.A. (1999). Analysis of the latency-associated transcript/UL1-3.5 gene cluster promoter complex of pseudorabies virus. Arch. Virol. *144*, 381-391.
Chlan, C.A., Coulter, C., and Feldman, L.T. (1987). Binding of the pseudorabies virus immediate-early protein to single-stranded DNA. J. Virol. *61*, 1855-1860.
Ciacci-Zanella, J., Stone, M., Henderson, G., and Jones, C. (1999). The latency-related gene of bovine herpesvirus 1 inhibits programmed cell death. J. Virol. *73*, 9734-9740.
Cromlish, W.A., Abmayr, S.M., Workman, J.L., Horikoshi, M., and Roeder, R.G. (1989). Transcriptionally active immediate-early protein of pseudorabies virus binds to specific sites on class II gene promoters. J. Virol. *63*, 1869-1876.

Crump, C., Bruun, B., Bell, S., Pomeranz, L.E., Minson, T., and Browne, H.M. (2004). Alphaherpesvirus glycoprotein M causes the relocalization of plasma membrane proteins. J. Gen. Virol. *85*, 3517-3527.

Damman, N., Klopfleisch, R., Rothermel, M., Mettenleiter, T.C., Hatt, H., and Wetzel, C. (2006a). Neuronal pathways of viral invasion in mice after intranasal inoculation of pseudorabies virus PrV-9112C2 expressing bovine herpesvirus 1 glycoprotein B. J. Neurovirol. *12*, 60-64.

Damann, N., Rothermel, M., Klupp, B.G., Mettenleiter, T.C., Hatt, H., and Wetzel, C.H. (2006b). Pseudorabies virus mediated live-cell tracing of primary afferent neurons allows functional analysis of identified neurons in cell culture. BMC Neurosc. *7*, 46.

Davison, A.J., and Scott, J.E. (1986). The complete DNA sequence of varicella-zoster virus. J. Gen. Virol. *67*, 1759-1816.

Davison, A.J. (2002). Evolution of the herpesviruses. Vet. Microbiol. *86*, 69-88.

Davison, S., Gingerich, E.N., Casavant, S., and Eckroade, R.J. (2006). Evaluation of the efficacy of a live fowlpox-vectored infectious laryngotracheitis/avian encephalomyelitis virus against ILT viral challenge. Avian Dis. *50*, 50-54.

de Regge, N., Nauwynck, H.J., Geenen, K., Krummenacher, C., Cohen, G.H., Eisenberg, R.J., Mettenleiter, T.C., and Favoreel, H.W. (2006). α-Herpesvirus glycoprotein D interaction with sensory neurons triggers formation of varicosities that serve as virus exit sites. J. Cell Biol. *174*, 267-275.

de Wind, N., Wagenaar, F., Pol, J., Kimman, T., and Berns, A. (1992). The pseudorabies virus homolog of the herpes simplex virus UL21 gene product is a capsid protein which is involved in capsid maturation. J. Virol. *66*, 7096-7103.

de Wind, N., Berns, A., Gielkens, A., and Kimman, T. (1993). Ribonucleotide reductase-deficient mutants of pseudorabies virus are avirulent for pigs and induce partial protective immunity. J. Gen. Virol. *74*, 351-359.

del Rio, T., Werner, H.C., and Enquist, L.W. (2002). The pseudorabies virus VP22 homologue (UL49) is dispensable for virus growth in vitro and has no effect on virulence and neuronal spread in rodents. J. Virol. *76*, 774-782.

del Rio, T., Decoste, C.J., and Enquist, L.W. (2005). Actin is a component of the compensation mechanism in pseudorabies virus virions lacking the major tegument protein VP22. J. Virol. *79*, 8614-8619.

Delhon, G., Moraes, M.P., Lu, Z., Alfonso, C.L., Flores, E.F., Weiblen, R., Kutish, G.F., and Rock, D.L. (2003). Genome of bovine herpesvirus 5. J. Virol. *77*, 10339-10347.

Desai, S., and Person, S. (1998). Incorporation of the green fluorescent protein into the herpes simplex virus type 1 capsid. J. Virol. *72*, 7563-7568.

Devireddy, L., and Jones, C. (1998). Alternative splicing of the latency-related transcript of bovine herpesvirus 1 yields RNAs containing unique open reading frames. J. Virol. 72, 7294-7301.

Devireddy, L. R., and Jones, C. (1999). Activation of caspases and p53 by bovine herpesvirus 1 infection results in programmed cell death and efficient virus release. J.Virol. *73*, 3778-3788.

Devireddy, L. R., Zhang. Y. and Jones, C. J. (2003). Cloning and initial characterization of an alternatively spliced transcript encoded by the bovine herpesvirus 1 latency-related gene. J. Neurovirol. *9*, 612-622.

Devlin, J.M., Browning, G.F., and Gilkerson, J.R. (2006a). A glycoprotein I- and glycoprotein E-deficient mutant of infectious laryngotracheitis virus exhibits impaired cell-to-cell spread in cultured cells. Arch. Virol. *151*, 1281-1291.

Devlin, J.M., Browning, G.F., Hartley, C.A., Kirkpatrick, N.C., Mahmoudian, A., Noormohammadi, A.H., and Gilkerson, J.R. (2006b). Glycoprotein G is a virulence factor in infectious laryngotracheitis virus. J. Gen. Virol. *87*, 2839-2847.

Dietz, P., Klupp, B.G., Fuchs, W., Köllner, B., Weiland, E., and Mettenleiter, T.C. (2000). Pseudorabies virus glycoprotein K requires the UL20 gene product for processing. J. Virol. *74*, 5083-5090.

Dijkstra, J.M., Brack, A., Jöns, A., Klupp, B.G., and Mettenleiter, T.C. (1998). Different point mutations within the conserved N-glycosylation motif of pseudorabies virus glycoprotein M result in expression of a nonglycosylated form of the protein. J. Gen. Virol. *79*, 851-854.

Dorange, F., Tischer, B.K., Vautherot, J.F., and Osterrieder, N. (2002). Characterization of Marek's disease virus serotype 1 (MDV-1) deletion mutants that lack UL46 to UL49 genes: MDV-1 UL49, encoding VP22, is indispensable for virus growth. J. Virol. *76*, 1959-1970.

Enquist, L.W. (1994). Infection of the mammalian nervous system by pseudorabies virus (PRV). Semin. Virol. *5*, 221-231.

Enquist, L.W., Husak, P.J., Banfield, B.W., and Smith, G.A. (1999). Infection and spread of alphaherpesviruses in the nervous system. Adv. Virus Res. *51*, 237-347.
Fahey, K.J., and York, J.J. (1990). The role of mucosal antibody in immunity to infectious laryngotracheitis virus in chickens. J. Gen. Virol. *71*, 2401-2405.
Farnsworth, A., Goldsmith, K., and Johnson, D.C. (2003). Herpes simplex virus glycoproteins gD and gE/gI serve essential but redundant functions during acquisition of the virion envelope in the cytoplasm. J. Virol. *77*, 8481-8494.
Favoreel, H.W., Nauwynck, H.J., van Oostveldt, P., Mettenleiter, T.C., and Pensaert, M. (1997). Antibody-induced and cytoskeleton-mediated redistribution and shedding of viral glycoproteins, expressed on pseudorabies virus-infected cells. J. Virol. *71*, 8254-8261.
Favoreel, H.W., Nauwynck, H.J., Halewyck, H.M., Van Oostveldt, P., Mettenleiter, T.C., and Pensaert, M.B. (1999). Antibody-induced endocytosis of viral glycoproteins and major histocompatibility complex class I on pseudorabies virus-infected monocytes. J. Gen. Virol. *80*, 1283-1291.
Favoreel, H.W., van Minnebruggen, G., Adriaensen, D., and Nauwynck, H.-J. (2005). Cytoskeletal rearrangements and cell extensions induced by the US3 kinase of an alphaherpesvirus are associated with enhanced spread. Proc. Natl. Acad. Sci. USA *102*, 8990-8995.
Feldman, L., Rixon, F.J., Jean, J.-H., Ben-Porat, T., and Kaplan, A.S. (1979). Transcription of the genome of pseudorabies virus (a herpesvirus) is strictly controlled. Virology *97*, 316-327.
Ficinska, J., van Minnebruggen, G., Nauwynck, H.J., Bienkowska-Szewczyk, K., and Favoreel, H.W. (2005). Pseudorabies virus glycoprotein gD contains a functional endocytosis motif that acts in concert with an endocytosis motif in gB to drive internalization of antibody-antigen complexes from the surface of infected monocytes. J. Virol. *79*, 7248-7254.
Fischer, T., Planz, O., Stitz, L., and Rziha, H.-J. (2003). Novel recombinant parapoxvirus vectors induce protective humoral and cellular immunity against lethal herpesvirus challenge infection in mice. J. Virol. *77*, 9312-9323.
Fraefel, C., Wirth, U.V., Vogt, B., and Schwyzer, M. (1993). Immediate-early transcription over covalently joined genome ends of bovine herpesvirus 1: the circ gene. J. Virol. *67*, 1328-1333.

Fraefel, C., Ackermann, M., and Schwyzer, M. (1994a). Identification of the bovine herpesvirus 1 circ protein, a myristylated and virion-associated polypeptide which is not essential for virus replication in cell culture. J. Virol. *68*, 8082-8088.

Fraefel, C., Zeng, J., Choffat, Y., Engels, M., Schwyzer, M., and Ackermann, M. (1994b). Identification and zinc dependence of the bovine herpesvirus 1 transactivator protein BICP0. J. Virol. *68*, 3154-3162.

Fuchs, W., Klupp, B.G., Granzow, H., Rziha, H.-J., and Mettenleiter, T.C. (1996). Identification and characterization of the pseudorabies virus UL3.5 protein, which is involved in virus egress. J. Virol. *70*, 3517-3527.

Fuchs, W., Klupp, B.G., Granzow, H., and Mettenleiter, T.C. (1997a). The UL20 gene product of pseudorabies virus functions in virus egress. J. Virol. *71*, 5639-5646.

Fuchs, W., Granzow, H., and Mettenleiter, T.C. (1997b). Functional complementation of UL3.5-negative pseudorabies virus by the bovine herpesvirus 1 UL3.5 homolog. J. Virol. *71*, 8886-8892.

Fuchs, W., and Mettenleiter, T.C. (1999). DNA sequence of the UL6 to UL20 genes of infectious laryngotracheitis virus and characterization of the UL10 gene product as a nonglycosylated and nonessential virion protein. J. Gen. Virol. *80*, 2173-2182.

Fuchs, W., Ziemann, K., Teifke, J.P., Werner, O., and Mettenleiter, T.C. (2000a). The non-essential UL50 gene of avian infectious laryngotracheitis virus encodes a functional dUTPase which is not a virulence factor. J. Gen. Virol. *81*, 627-638.

Fuchs, W., Ehrlich, C., Klupp, B.G., and Mettenleiter, T.C. (2000b). Characterization of the replication origin (Ori_S) and adjoining parts of the inverted repeat sequences of the pseudorabies virus genome. J. Gen. Virol. *81*, 1539-1541.

Fuchs, W., Klupp, B.G., Granzow, H., Osterrieder, N., and Mettenleiter, T.C. (2002a). The interacting UL31 and UL34 gene products of pseudorabies virus are involved in egress from the host-cell nucleus and represent components of primary enveloped but not of mature virions. J. Virol. *76*, 364-378.

Fuchs, W., Granzow, H., Klupp, B.G., Kopp, M., and Mettenleiter, T.C. (2002b). The multifunctional UL48 tegument protein of pseudorabies virus is critical for intracytoplasmic assembly of infectious virions. J. Virol. *76*, 6729-6742.

Fuchs, W., Klupp, B.G., Granzow, H., Hengartner, C., Brack, A., Mundt, A., Enquist, L.W., and Mettenleiter, T.C. (2002c). Physical interaction between envelope glycoproteins E and M of pseudorabies virus and the major tegument protein UL49. J. Virol. *76*, 8208-8217.

Fuchs, W., Granzow, H., and Mettenleiter, T.C. (2003). A pseudorabies virus recombinant simultaneously lacking the major tegument proteins encoded by the UL46, UL47, UL48 and UL49 genes is viable in cultured cells. J. Virol. *77*, 12891-12900.

Fuchs, W., Klupp, B.G., Granzow, H., and Mettenleiter, T.C. (2004). The essential function of the PrV UL36 gene product is independent of its interaction with the UL37 protein. J. Virol. *78*, 11879-11889.

Fuchs, W., and Mettenleiter, T.C. (2005). The nonessential UL49.5 gene of infectious laryngotracheitis virus encodes an O-glycosylated protein which forms a complex with the non-glycosylated UL10 gene product. Virus Res. *112*, 108-114.

Fuchs, W., Wiesner, D., Veits, J., Teifke, J.P., and Mettenleiter, T.C. (2005a). In vitro and in vivo relevance of infectious laryngotracheitis virus gJ proteins that are expressed from spliced and nonspliced mRNAs. J. Virol. *79*, 705-716.

Fuchs, W., Granzow, H., Klopfleisch, R., Klupp, B.G., Rosenkranz, D., and Mettenleiter, T.C. (2005b). The UL7 gene of pseudorabies virus encodes a nonessential structural protein which is involved in virion formation. J. Virol. *79*, 13442-13453.

Fuchs, W., Veits, J., and Mettenleiter, T.C. (2006a). Recombinant viruses of poultry as vector vaccines against fowl plague. Berl. Münch. Tierärztl. Wochenschr. *119*, 160-166.

Fuchs, W., Granzow, H., Klopfleisch, R., Klupp, B.G., and Mettenleiter, T.C. (2006b). The UL4 gene of pseudorabies virus encodes a minor infected cell protein which is dispensable for virus replication. J. Gen. Virol. *87*, 2517-2525.

Fuchs, W., Veits, J., Helferich, D., Granzow, H., Teifke, J.P., and Mettenleiter, T.C. (2007). Molecular biology of avian infectious laryngotracheitis virus. Vet. Res. *38*, 261-279.

Geenen, K., Favoreel, H.W., Olsen, L., Enquist, L.W., and Nauwynck, H.J. (2005). The pseudorabies virus US3 protein kinase possesses anti-apoptotic activity that protects cells from apoptosis during infection and after treatment with sorbitol or staurosporine. Virology *331*, 114-150.

Geiser, V., Inman, M., Zhang, Y., and Jones, C. (2002). The latency-related gene of bovine herpesvirus 1 can inhibit the ability of bICP0 to activate productive infection. J. Gen. Virol. *83*, 2965-2971.

Geiser, V., and Jones (2005). Localization of sequences within the latency-related gene of bovine herpesvirus 1 that inhibit mammalian cell growth. J. Neurovirol. *11*, 563-570.
Geiser, V., Zhang, Y., and Jones, C. (2005). Analysis of a bovine herpesvirus 1 recombinant virus that does not express the bICP0 protein. J. Gen. Virol *86*, 1987-1996.
Gelenczei, E.F., and Marty, E.W. (1965). Strain stability and immunologic characteristics of tissue-culture modified infectious laryngotracheitis virus. Avian Dis. *9*, 44-56.
Geraghty, R.J., Krummenacher, C., Eisenberg, R.J., Cohen, G.H., and Spear, P.G. (1998). Entry of alphaherpesviruses mediated by poliovirus receptor related protein 1 and poliovirus receptor. Science *280*, 1618-1620.
Gerdts, V., Beyer, J., Lomniczi, B., and Mettenleiter, T.C. (2000). Pseudorabies virus expressing the bovine herpesvirus 1 glycoprotein B exhibits altered neurotropism and increased neurovirulence. J. Virol. *74*, 817-827.
Granzow, H., Weiland, F., Jöns, A., Klupp, B.G., Karger, A., and Mettenleiter, T.C. (1997). Ultrastructural analysis of the replication cycle of pseudorabies virus in cell culture: a reassessment. J. Virol. *71*, 2072-2082.
Granzow, H., Klupp, B.G., Fuchs, W., Veits, J., Osterrieder, N., and Mettenleiter, T.C. (2001). Egress of alphaherpesviruses: comparative ultrastructural study. J. Virol. *75*, 3675-3684.
Granzow, H., Klupp, B.G., and Mettenleiter, T.C. (2004). The pseudorabies virus US3 protein is a component of primary and of mature virions. J. Virol. *78*, 1314-1323.
Granzow, H., Klupp, B.G., and Mettenleiter, T.C. (2005). Entry of pseudorabies virus: an immunogold labelling study. J. Virol. *79*, 3831-3840.
Guo, P., Scholz, E., Turek, J., Nodgreen, R., and Maloney, B. (1993). Assembly pathway of avian infectious laryngotracheitis virus. Am. J. Vet. Res. *54*, 2031-2039.
Gupta, A., Gartner, J.J., Sethupathy, P., Hatzigeorgiou, A.G., and Fraser, N.W. (2006). Anti-apoptotic function of a microRNA encoded by the HSV-1 latency-associated transcript. Nature *442*, 82-85.
Gutekunst, D.E. (1979). Latent pseudorabies virus infection in swine detected by RNA-DNA hybridization. Am. J. Vet. Res. *40*, 1568-1572.
Guy, J.S., Barnes, H.J., and Morgan, L.M. (1990). Virulence of infectious laryngotracheitis viruses: comparison of modified-live

vaccine viruses and North Carolina field isolates. Avian Dis. *34*, 106-113.
Guy, J.S., Barnes, H.J., and Smith, L.G. (1991). Increased virulence of modified-live infectious laryngotracheitis vaccine virus following bird-to-bird passage. Avian Dis. *35*, 348-355.
Guy, J.S., and Bagust, T.J. (2003). Laryngotracheitis, in: Saif, Y.M., Barnes, H.J., Glisson, J.R., Fadly, A.M., McDougald, L.R., Swayne, D.E. (Eds.), Diseases of Poultry, Iowa State Press, Ames, 2003, pp.121-134.
Hampl, H., Ben-Porat, T., Ehrlicher, L., Habermehl, K.O., and Kaplan, A.S. (1984). Characterization of the envelope proteins of pseudorabies virus. J. Virol. *52*, 583-590.
Han, M.G., Kweon, C.H., Mo, I.P., and Kim, S.J. (2002). Pathogenicity and vaccine efficacy of a thymidine kinase gene deleted infectious laryngotracheitis virus expressing the green fluorescent protein gene. Arch. Virol. *147*, 1017-1031.
Harper, A.L., Ward, M.D., Semmes, J.O., Courtney, R.J., and Wills, J.W. (2006). Characterization of the UL11-UL16-UL 21 interactions during HSV-1 asssembly. 31st Interntl. Herpesvirus Workshop, Seattle, Washington, Abstract 11.45.
Harty, R.N., Caughman, G.B., Holden, V.R., and O'Callaghan, D.J. (1993). Characterization of the myristylated polypeptide encoded by the UL1 gene that is conserved in the genome of defective interfering particles of equine herpesvirus 1. J. Virol. *67*, 4122-413.
Heffner, S., Kovacs, F., Klupp, B.G., and Mettenleiter, T.C. (1993). Glycoprotein gp50-negative pseudorabies virus: a novel approach toward a nonspreading live herpesvirus vaccine. J. Virol. *67*, 1529-1537.
Heldwein, E., Lou, H., Bender, F., Cohen, G.H., Eisenberg, R.J., and Harrison, S.C. (2006). Crystal structure of glycoprotein B from herpes simplex virus. Science *313*, 217-220.
Helferich, D., Veits, J., Mettenleiter, T.C., and Fuchs, W. (2007a). Identification of transcripts and protein products of the UL31, UL37, UL46, UL47, UL48, UL49 and US4 gene homologues of avian infectious laryngotracheitis virus. J. Gen. Virol. *88*, 719-731.
Helferich, D., Veits, J., Teifke, J.P., Mettenleiter, T.C., and Fuchs, W. (2007b). The UL47 gene of avian infectious laryngotracheitis virus is not essential for in vitro replication but relevant for virulence in chickens. J. Gen. Virol. *88*, 732-742.
Henderson, G., Zhang, Y., Inman, M., Jones, D., and Jones, C. (2004). Infected cell protein 0 encoded by bovine herpesvirus 1 can

activate caspase 3 when overexpressed in transfected cells. J. Gen. Virol. *85*, 3511-3516.

Henderson, G., Zhang, Y., and Jones, C. (2005). The bovine herpesvirus 1 gene encoding infected cell protein 0 (bICP0) can inhibit interferon-dependent transcription in the absence of other viral genes. J. Gen. Virol *86*, 2697-2702.

Herold, B.C., Visalli, R., Susmarski, N., Brandt, C., and Spear, P.G. (1994). Glycoprotein C-independent binding of herpes simplex virus to cells requires cell surface heparan sulphate and glycoprotein B. J. Gen. Virol. *75*, 1211-1222.

Hossain, A., Schang, L. M., and Jones, C. (1995). Identification of gene products encoded by the latency-related gene of bovine herpesvirus 1. J. Virol. *69*, 5345-52.

Huemer, H.P., Larcher, C., and Coe, N.E. (1992). Pseudorabies virus glycoprotein III derived from virions and infected cells binds to the third component of complement. Virus Res. *23*, 271-280.

Hughes, C.S., Gaskell, R.M., Jones, R.C., Bradbury, J.M., and Jordan, F.T.W. (1989). Effects of certain stress factors on the re-excretion of infectious laryngotracheitis virus from latently infected carrier birds. Res. Vet. Sci. *46*, 274-276.

Inman, M., Lovato, L., Doster, A., and Jones, C. (2001a). A mutation in the latency-related gene of bovine herpesvirus 1 leads to impaired ocular shedding in acutely infected calves. J. Virol. *75*, 8507-8515.

Inman, M., Zhang, Y., Geiser, V., and Jones, C. (2001b). The zinc ring finger in the bICP0 protein encoded by bovine herpesvirus-1 mediates toxicity and activates productive infection. J. Gen. Virol. *82*, 483-492.

Inman, M., Lovato, L., Doster, A., and Jones, C. (2002). A mutation in the latency-related gene of bovine herpesvirus 1 disrupts the latency-reactivation cycle in calves. J. Virol. *76,* 6771-6779.

Inman, M., Zhou, J., Webb, H., and Jones, C. (2004). Identification of a novel bovine herpesvirus 1 transcript containing a small open reading frame that is expressed in trigeminal ganglia of latently infected cattle. J. Virol. *78*, 5438-5447.

Irnaten, M., Neff, R.A., Wang, J., Loewy, A.D., Mettenleiter, T.C., and Mendelowitz, D. (2001). Activity of cardiorespiratory networks revealed by transsynaptic virus expressing GFP. J. Neurophysiol. *85*, 435-438.

Jansen, A.S., Nguyen, X.V., Karpitskyi, V., Mettenleiter, T.C., and Loewy, A.D. (1995). Central command neurons of the sympathetic

nervous system: basis of the fight-or-flight response. Science *270*, 644-646.
Jensen, H.L., and Norrild, B. (1998). Herpes simplex virus type 1-infected human embryonic lung cells studied by optimized immunogold cryosection electron microscopcy. J. Histochem. Cytochem. *46*, 487-496.
Jiang, Y., Hossain, A., Winkler, M.T., Holt, T., Doster, A., and Jones, C. (1998). A protein encoded by the latency-related gene of bovine herpesvirus 1 is expressed in trigeminal ganglionic neurons of latently infected cattle and interacts with cyclin-dependent kinase 2 during productive infection. J. Virol. *72*, 8133-8142.
Jiang, Y., Inman, M., Zhang, Y., Posadas, N.A., and Jones, C. (2004). A mutation in the latency-related gene of bovine herpesvirus 1 inhibits protein expression from open reading frame 2 and an adjacent reading frame during productive infection. J. Virol. *78,* 3184-3189.
Jin, L., and Scherba, G. (1999). Expression of the pseudorabies virus latency-associated transcript gene during productive infection of cultured cells. J. Virol. *73*, 9781-9788.
Jin, L., Schnitzlein, W., and Scherba, G. (2000). Identification of the pseudorabies virus promoter required for latency-associated transcript gene expression in the natural host. J. Virol. *74*, 6333-6338.
Johnson, M.A., Tyack, S.G., Prideaux, C.T., Kongsuwan, K., and Sheppard, M. (1995). Nucleotide sequence of infectious laryngotracheitis virus (gallid herpesvirus 1) ICP4 gene. Virus Res. *35*, 193-204.
Jones, C. (2003). Herpes simplex virus type 1 and bovine herpesvirus 1 latency. Clin. Microbiol. Rev. *16*, 79-95.
Jones, C., Geiser, V., Henderson, G., Jiang, Y., Meyer, F., Perez, S., and Zhan, Y. (2006). Functional analysis of bovine herpesvirus 1 (BHV-1) genes expressed during latency. Vet. Microbiol. *113*, 199-210.
Jöns, A., Granzow, H., Kuchling, R., and Mettenleiter, T.C. (1996). The UL49.5 gene of pseudorabies virus codes for an O-glycosylated structural protein. J. Virol. *70*, 1237-1241.
Jöns, A., and Mettenleiter, T.C. (1997). Green fluorescent protein expressed by recombinant pseudorabies virus as an in vivo marker for viral replication. J. Virol. Meth. *66*, 283-292.
Jöns, A., Gerdts, V., Lange, E., Kaden, V., and Mettenleiter, T.C. (1997). Attenuation of dUTPase-deficient pseudorabies virus for the natural host. Vet. Microbiol. *56*, 47-54.

Jöns, A., Dijkstra, J., and Mettenleiter, T.C. (1998). Glycoproteins M and N of pseudorabies virus form a disulfide-linked complex. J. Virol. *72*, 550-557.

Kaashoek, M., Moerman, J.A., Madic, J., Weerdmeester, K., Maris-Veldhuis, M., Rijsewijk, F.A., and van Oirschot, J.T. (1995). An inactivated vaccine based on a glycoprotein E-negative strain of bovine herpesvirus 1 induces protective immunity and allows serological differentiation. Vaccine *13*, 342-346.

Kanamoto, T., Mota, M., Takeda, K.L., Rubin, L.L., Miyazono, K., Ichijo, H., and Bazenet, C.E. (2000). Role of apoptosis signal-regulating kinase in regulation of the c-Jun N-terminal kinase pathway and apoptosis in sympathetic neurons. Mol. Cell. Biol. *20*, 196-204.

Karger, A., Saalmüller, A., Tufaro, F., Banfield, B.W., and Mettenleiter, T.C. (1995). Cell surface proteoglycans are not essential for infection by pseudorabies virus. J. Virol. *69*, 3482-3489.

Karger, A., Schmidt, J. and Mettenleiter, T.C. (1998). Infectivity of a pseudorabies virus mutant lacking attachment glycoproteins C and D. J. Virol. *72*, 7341-7348.

Kattenhorn, L.M., Korbel, G.A., Kessler, B.M., Spooner, E., and Ploegh, H.L. (2005). A deubiquitinating enzyme encoded by HSV-1 belongs to a family of cysteine proteases that is conserved across the family herpesviridae. Mol. Cell *19*, 547-557.

Keil, G.M., Höhle, C., Giesow, K., and König, P. (2005). Engineering glycoprotein B of bovine herpesvirus 1 to function as transporter for secreted proteins: a new protein expression approach. J. Virol. *79*, 791-799.

Kerman, I.A., Enquist, L.W., Watson, S.J., and Yates, B.J. (2003). Brainstem substrates of sympatho-motor circuitry identified using trans-synaptic tracing with pseudorabies virus recombinants. J. Neurosc. *23*, 4657-4666.

Kimman, T.G., de Wind, N., Oei-Lie, N., Pol, J.M.A., Berns, A.J.M., and Gielkens, A.L.J. (1992). Contribution of single genes within the unique short region of Aujeszky's disease virus (suid herpesvirus type 1) to virulence, pathogenesis and immunogenicity. J. Gen. Virol. *73*, 243-251.

Kingsley, D.H., Hazel, J.W., and Keeler, C.L. (1994). Identification and characterization of the infectious laryngotracheitis virus glycoprotein C gene. Virology *203*, 336-343.

Kingsley, D.H., and Keeler, C.L. (1999). Infectious laryngotracheitis virus, an alpha herpesvirus that does not interact with cell surface heparan sulfate. Virology *256*, 213-219.

Kit, S., Kitt, M., and Pirtle, E.C. (1985). Attenuated properties of thymidine kinase-negative deletion mutant of pseudorabies virus. Am. J. Vet. Res. *46*, 1359-1367.

Kit, S., Sheppard, M., Ichimura, H., and Kit, M. (1987). Second-generation pseudorabies virus vaccine with deletions in thymidine kinase and glycoprotein genes. Am. J. Vet. Res. *48*, 780-793.

Klopfleisch, R., Teifke, J.P., Fuchs, W., Klupp, B.G., Kopp, M., and Mettenleiter, T.C. (2004). Role of herpesvirus tegument proteins UL11, UL37, U46, UL47 and UL48 in neuroinvasion. J. Virol. *78*, 2956-2966.

Klopfleisch, R., Klupp, B.G., Fuchs, W., Kopp, M., Teifke, J.P., and Mettenleiter, T.C. (2006). Influence of pseudorabies virus proteins on neuroinvasion and neurovirulence in mice. J. Virol. *80*, 5571-5576.

Klupp, B.G., Baumeister, J., Karger, A., Visser, N., and Mettenleiter, T.C. (1994). Identification and characterization of a novel structural glycoprotein in pseudorabies virus, PrV gL. J. Virol. *68*, 3868-3878.

Klupp, B.G., Lomniczi, B., Visser, N., Fuchs, W., and Mettenleiter, T.C. (1995). Mutations affecting the UL21 gene contribute to avirulence of pseudorabies virus vaccine strain Bartha. Virology *212*, 466-473.

Klupp, B.G., Baumeister, J., Dietz, P., Granzow, H., and Mettenleiter, T.C. (1998). Pseudorabies virus glycoprotein gK is a structural component of virions involved in virus release but not required for entry. J. Virol. *72*, 1949-1958.

Klupp, B.G., and Mettenleiter, T.C. (1999). Glycoprotein gL independent infectivity of pseudorabies virus is mediated by a gD-gH fusion protein. J. Virol. *73*, 3014-3022.

Klupp, B.G., Granzow, H., and Mettenleiter, T.C. (2000a). Primary envelopment of pseudorabies virus at the nuclear membrane requires the UL34 gene product. J. Virol. *74*, 10063-10071.

Klupp, B.G., Nixdorf, R., and Mettenleiter, T.C. (2000b). Pseudorabies virus glycoprotein M inhibits membrane fusion. J. Virol. *74*, 7137-7145.

Klupp, B.G., Granzow, H., and Mettenleiter, T.C. (2001a). Effect of the pseudorabies virus US3 protein on nuclear membrane localization of the UL34 protein and virus egress from the nucleus. J. Gen. Virol. *82*, 2363-2371.

Klupp, B.G., Granzow, H., Mundt, E., and Mettenleiter, T.C. (2001b). Pseudorabies virus UL37 gene product is involved in secondary envelopment. J. Virol. *75*, 8927-8936.
Klupp, B.G., Fuchs, W., Granzow, H., Nixdorf, R., and Mettenleiter, T.C. (2002). The pseudorabies virus UL36 tegument protein physically interacts with the UL37 gene product. J. Virol. *76*, 3065-3071.
Klupp, B.G., Hengartner, C.J., Mettenleiter, T.C., and Enquist, L.W. (2004). Complete, annotated sequence of the pseudorabies virus genome. J. Virol. *78*, 424-440.
Klupp, B.G., Altenschmidt, J., Granzow, H., Fuchs, W., and Mettenleiter, T.C. (2005a). Identification and characterization of the pseudorabies virus UL43 protein. Virology *334*, 224-233.
Klupp, B.G., Böttcher, S., Granzow, H., Kopp, M., and Mettenleiter, T.C. (2005b). Complex formation between the UL16 and UL21 tegument proteins of pseudorabies virus. J. Virol. *79*, 1510-1522.
Klupp, B.G., Granzow, H., Karger, A., and Mettenleiter, T.C. (2005c). Identification, subviral localization and functional characterization of the pseudorabies virus UL17 protein. J. Virol. *79*, 13442-13453.
Klupp, B.G., Mundt, E., and Mettenleiter, T.C. (2005d). Functional analysis of the pseudorabies virus UL51 protein. J. Virol. *79*, 3831-3840.
Klupp, B.G., Granzow, H., and Mettenleiter, T.C. (2006). The capsid-associated UL25 protein of the alphaherpesvirus pseudorabies virus is nonessential for cleavage and packaging of genomic DNA but is required for nuclear egress of capsids. J. Virol. *80*, 6235-6246.
Klupp, B.G., Granzow, H., Fuchs, W., Keil, G.M., Finke, S., and Mettenleiter, T.C. (2007). Vesicle formation at the nuclear membrane induced by coexpression of two conserved herpesvirus proteins. Proc. Natl. Acad. Sci. USA *104*, 7241-7246.
Knapp, A.C., and Enquist, L.W. (1997). Pseudorabies virus recombinants expressing functional virulence determinants gE and gI from bovine herpesvirus 1.1 J. Virol. *71*, 2731-2739.
Knapp, A.C., Husak, P.J., and Enquist, L.W. (1997). The gE and gI homologs from two alphaherpesviruses have conserved and divergent neuroinvasive properties. J. Virol. *71*, 5820-5827.
Kongsuwan, K., Johnson, M.A., Prideaux, C.T., and Sheppard, M. (1993). Identification of an infectious laryngotracheitis virus gene encoding an immunogenic protein with a predicted M_r of 32 kilodaltons. Virus Res. *29*, 125-140.

Kopp, A., Blewett, E., Misra, V., and Mettenleiter, T.C. (1994). Proteolytic cleavage of glycoprotein gB of bovine herpesvirus 1 (BHV-1) is not necessary for its function in BHV-1 or pseudorabies virus. J. Virol. *68*, 1667-1674.

Kopp, M., Klupp, B.G., Granzow, H., Fuchs, W., and Mettenleiter, T.C. (2002). Identification and characterization of the pseudorabies virus tegument proteins UL46 and UL47: role for UL47 in virion morphogenesis in the cytoplasm. J. Virol. *76*, 8820-8833.

Kopp, M., Granzow, H., Fuchs, W., Klupp, B.G., Mundt, E., Karger, A., and Mettenleiter, T.C. (2003). The pseudorabies virus UL11 protein is a virion component involved in secondary envelopment in the cytoplasm. J. Virol. *77*, 5339-5351.

Kopp, M., Granzow, H., Fuchs, W., Klupp, B.G., and Mettenleiter, T.C. (2004). Simultaneous deletion of pseudorabies virus tegument protein UL11 and envelope glycoprotein M severely impairs secondary envelopment. J. Virol. *78*, 3024-3034.

Koppers-Lalic, D., Reits, E., Ressing, M., Lipinska, A., Abele, R., Koch, J., Rezende, M., Admiraal, P., van Leeuwen, D., Bienkowska-Szeyczyk, K., Mettenleiter, T.C., Rijsewijk, F., Tampe, R., Neffjes, J., and Wiertz, E. (2005). Varicelloviruses avoid T cell recognition by UL49.5-mediated inactivation of the transporter associated with antigen processing. Proc. Natl. Acad. Sci. USA *102*, 5144-5149.

Koslowski, K., Shaver, P., Wang, X.-Y., Tenney, D., and Pedersen, N.E. (1997). The pseudorabies virus UL28 protein enters the nucleus after coexpression with the herpes simplex virus UL15 protein. J. Virol. *71*, 9118-9123.

Kotiw, M., Wilks, C.R., and May, J.T. (1995). The effect of serial in vivo passage on the expression of virulence and DNA stability of an infectious laryngotracheitis virus strain of low virulence. Vet. Microbiol. *45*, 71-80.

Kovacs, F., and Mettenleiter, T.C. (1991). Firefly luciferase as a marker for herpesvirus (pseudorabies virus) replication in vitro and in vivo. J. Gen. Virol. *72*, 2999-3008.

Kutish, G., Mainprize, T., and Rock, D. (1990). Characterization of the latency-related transcriptionally active region of the bovine herpesvirus 1 genome. J. Virol. *64*, 5730-5737.

Lam, N.G., and Letchworth, G. (2000). Bovine herpesvirus 1 U_L3.5 interacts with bovine herpesvirus 1 α-transinducing factor. J. Virol. *74*, 2876-2884.

Lee, J.I., Luxton, G.W., and Smith, G.A. (2006). Identification of an essential domain in the herpesvirus VP1/2 tegument protein: the

carboxy terminus directs incorporation into capsid assemblons. J. Virol. *80*, 12086-12094.
Leuzinger, H., Ziegler, U., Schraner, E.M., Fraefel, C., Glauser, D.L., Heid, I., Ackermann, M., Mueller, M., and Wild, P. (2005). Herpes simplex virus 1 envelopment follows two diverse pathways. J. Virol. *79*, 13047-13059.
Li, H., Zhu, H., Xu, C.J., and Yuan, J. (1998). Cleavage of BID by caspase 8 mediates the mitochondrial damage in the Fas pathway of apoptosis. Cell *94*, 491-501.
Liang, X., Chow, B., Raggo, C., and Babiuk, L.A. (1996). Bovine herpesvirus 1 UL49.5 homolog gene encodes a noval viral envelope protein that forms a disulfide-linked complex with a second virion structural protein. J. Virol. *70*, 1448-1454.
Ligas, M.W., and Johnson, D.C. (1988). A herpes simplex virus mutant in which glycoprotein D sequences are replaced by ß-galactosidase sequences binds to but is unable to penetrate into cells. J. Virol. *62*, 1486-1494.
Lin, H.-W., Chang, Y.-Y., Wong, M.-L., Lin, J.-W., and Chang, T.-J. (2004). Functional analysis of virion host shutoff protein of pseudorabies virus. Virology *324*, 412-418.
Loewy, A.D., Bridgman, P.C., and Mettenleiter, T.C. (1991). ß-galactosidase expressing recombinant pseudorabies virus for light and electron microscopic study of transneuronally labeled CNS neurons. Brain Res. *555*, 346-352.
Loewy, A. (1995). Pseudorabies Virus: a transneuronal tracer for neuroanatomical studies. In: M. Kaplitt and A.D. Loewy (eds.). Viral vectors. Academic Press, San Diego, p. 349-366.
Lomniczi, B., Watanabe, S., Ben-Porat, T., and Kaplan, A.S. (1984). Genetic basis of the neurovirulence of pseudorabies virus. J. Virol. *52*, 198-205.
Lomniczi, B., Watanabe, S., Ben-Porat, T., and Kaplan, A.S. (1987). Genome location and identification of functions defective in the Bartha vaccine strain of pseudorabies virus. J. Virol. *61*, 796-801.
Loomis, J.S., Courtney, R.J., and Wills, J.W. (2003). Identification of binding partners for the UL11 tegument protein of herpes simplex virus type 1. J. Virol. *77*, 11417-11424.
Lovato, L., Inman, M., Henderson, G., Doster, A., and Jones, C. (2003). Infection of cattle with a bovine herpesvirus 1 strain that contains a mutation in the latency related gene leads to increased apoptosis in trigeminal ganglia during the transition from acute infection to latency. J. Virol. *77*, 4848-4857.

Lukacs, N., Thiel, H.-J., Mettenleiter, T.C., and Rziha, H.-J. (1985). Demonstration of three major species of pseudorabies virus glycoproteins and identification of a disulfide-linked glycoprotein complex. J. Virol. *56*, 166-173.

Lüschow, D., Werner, O., Mettenleiter, T.C., and Fuchs, W. (2001). Protection of chickens from lethal avian influenza A virus infection by live-virus vaccination with infectious laryngotracheitis virus recombinants expressing the hemagglutinin (H5) gene. Vaccine *19*, 4249-4259.

Luxton, G.W., Haverlock, S., Coller, K.E., Antinone, S.E., Pincetic, A., and Smith, G.A. (2005). Targeting of herpesvirus capsid transport in axons is coupled to association with specific sets of tegument proteins. Proc. Natl. Acad. Sci. USA *102*, 5832-5837.

Luxton, G.W., Lee, J.I.-H., Haverlock-Moyns, S., Schober, J.M., and Smith, G.A. (2006). The pseudorabies virus VP1/2 tegument protein is required for intracellular capsid transport. J. Virol. *80*, 201-209.

Marchioli, C.C., Yancey, R.J., Wardley, R.C., Thomsen, D.R., and Post, L.E. (1987). A vaccine strain of pseudorabies virus with deletions in the thymidine kinase and glycoprotein X genes. Am. J. Vet. Res. *48*, 1577-1583.

McGeoch, D.J., Dolan, A., and Ralph, A.C. (2000). Toward a comprehensive phylogeny for mammalian and avian herpesviruses. J. Virol. *74*, 10401-10406.

McGeoch, D.J., Rixon, F.J., and Davison, A.J. (2006). Topics in herpesvirus genomics and evolution. Virus Res. *117*, 90-104.

McGregor, S., Easterday, B., Kaplan, A.S., and Ben-Porat, T. (1985). Vaccination of swine with thymidine kinase-deficient mutants of pseudorabies virus. Am. J. Vet. Res. *46*, 1494-1497.

Mellencamp, M.W., O'Brien, P.C.M., and Stevenson, J.R. (1991). Pseudorabies virus-induced suppression of majoar histocompatibility complex class I antigen expression. J. Virol. *65*, 3365-3368.

Mettenleiter, T.C., Lukacs, N., and Rziha, H.-J. (1985). Pseudorabies virus avirulent strains fail to express a major glycoprotein. J. Virol. *56*, 307-311.

Mettenleiter, T.C., Schreurs, C., Zuckermann, F., and Ben-Porat, T. (1987a). Role of pseudorabies virus glycoprotein gI in virus release from infected cells. J. Virol. *61*, 2764-2769.

Mettenleiter, T.C., Zsak, L., Kaplan, A., Ben-Porat, T., and Lomniczi, B. (1987b). Role of a structural protein of pseudorabies in virus virulence. J. Virol. *61*, 4030-4032.

Mettenleiter, T.C., Lomniczi, B., Sugg, N., Schreurs, C., and Ben-Porat, T. (1988). Host cell-specific growth advantage of pseudorabies virus with a deletion in the genome sequences encoding a structural glycoprotein. J. Virol. *62*, 12-19.

Mettenleiter, T.C., and Rauh, I. (1990). A glycoprotein gX-ßgalactosidase fusion gene as insertional marker for rapid identification of pseudorabies virus mutants. J. Virol. Meth. *30*, 55-66.

Mettenleiter, T.C., Zsak, L., Zuckermann, F., Sugg, N., Kern, H., and Ben-Porat, T. (1990). Interaction of glycoprotein gIII with a cellular heparinlike substance mediates adsorption of pseudorabies virus. J. Virol. *64*, 278-286.

Mettenleiter, T.C., Saalmüller, A., and Weiland, F. (1993). Pseudorabies virus protein homologous to herpes simplex virus type 1 ICP18.5 is necessary for capsid maturation. J. Virol. *67*, 1236-1245.

Mettenleiter, T.C. (1994). Initiation and spread of α-herpesvirus infections. Trends Microbiol. *2*, 2-4.

Mettenleiter, T.C., and Spear, P.G. (1994). Glycoprotein gB (gII) of pseudorabies virus can functionally substitute for glycoprotein gB in herpes simplex virus type 1. J. Virol. *68*, 500-504.

Mettenleiter, T.C. (1996). Immunobiology of pseudorabies (Aujeszky's disease). Vet. Immunol. Immunopathol. *54*, 221-229.

Mettenleiter, T.C. (2002). Herpesvirus assembly and egress. J. Virol. *76*, 1537-1547.

Mettenleiter, T.C. (2004). Budding events in herpesvirus morphogenesis. Virus Res. *106*, 167-180.

Mettenleiter, T.C. (2006). Intriguing interplay between viral proteins during herpesvirus assembly - the herpesvirus assembly puzzle. Vet. Microbiol. *113*, 163-169.

Mettenleiter, T.C., and Minson, T. (2006). Egress of alphaherpesviruses. J. Virol. *80*, 1610-1612.

Mettenleiter, T.C., Klupp, B.G., and Granzow, H. (2006). Herpesvirus assembly: a tale of two membranes. Curr. Opin. Microbiol. *9*, 423-429.

Meyer, F., Perez, S., Geiser, V., Sintek, M., Inman, M., and Jones, C. (2007). A protein encoded by the bovine herpesvirus 1 latency-related gene interacts with specific cellular regulatory proteins, including CCAAT enhancer binding protein alpha. J. Virol. *81,* 59-67.

Michael, K., Böttcher, S., Klupp, B.G., Karger, A., and Mettenleiter, T.C. (2006a). Pseudorabies virus particles lacking tegument proteins pUL11 or pUL16 incorporate less full-length pUL36 than wild-type virus, but specifically accumulate a pUL36 N-terminal fragment. J. Gen. Virol. *87*, 3503-3507.

Michael, K., Klupp, B.G., Mettenleiter, T.C., and Karger, A. (2006b). Composition of pseudorabies virus particles lacking tegument proteins US3, UL47, UL49 or glycoprotein E. J. Virol. *80*, 1332-1339.

Michael, K., Klupp, B.G., Karger, A., and Mettenleiter, T.C. (2007). Efficient incorporation of tegument proteins pUL46, pUL49, and pUS3 into pseudorabies virus particles depends on the presence of pUL21. J. Virol. *81*, 1048-1051.

Milne, R.S., Connolly, S.A., Krummenacher, C., Eisenberg, R.J., and Cohen, G.H. (2001). Porcine HveC, a member of the highly conserved HveC/Nectin 1 family, is a functional alphaherpesvirus receptor. Virology *281*, 315-328.

Minson, T., and Mettenleiter, T.C. (2006). Herpesvirus egress: the unanswered questions. J. Virol. *80*, 6718-6719.

Miranda-Saksena, M., Wakisaka, H., Tijono, B., Boadle, R., Rixon, F., Takahashi, H., and Cunningham, A.L. (2006). Herpes simplex virus type 1 accumulation, envelopment, and exit in growth cones and varicosiites in mid-distal regions of axons. J. Virol. *80*, 3592-3606.

Misra, V., Walker, S., Hayes, S., and O'Hare, P. (1995). The bovine herpesvirus alpha gene trans-inducing factor activates transcription by mechanisms different from those of its herpes simplex virus type 1 counterpart VP16. J. Virol. *69*, 5209-5216.

Moriuchi, H., Moriuchi, M., Dean, H., Cheung, A.K., and Cohen, J.I. (1995). Pseudorabies virus EP0 is functionally homologous to varicella-zoster virus ORF 61 protein and herpes simplex virus type 1 ICP0. Virology *209*, 281-283.

Mulder, W., Jacobs, L., Priem, J., Kok, G., Wagenaar, F., Kimman, T.G., and Pol, J.M.A. (1994). Glycoprotein gE-negative pseudorabies virus has a reduced capability to infect second- and third-order neurons of the olfactory and trigeminal routes in the porcine central nervous system. J. Gen. Virol. *75*, 3095-3106.

Muylkens, B., Thiry, J., Kirten, P., Schynts, F. and Thiry, E. (2007). Bovine herpesvirus 1 infection and infectious rhinotracheitis. Vet. Res. *38*, 181-209.

Müller, T., Bätza, H.-J., Schlüter, H., Conraths, F.J., and Mettenleiter, T.C. (2003). Eradication of Aujeszky's disease in Germany. J. Vet. Med. B *50*, 207-213.

Nauwynck, H.J., Glorieux, S., Favoreel, H., and Pensaert, M. (2007). Cell biological and molecular characteristics of pseudorabies virus infections in cell cultures and in pigs with emphasis on the respiratory tract. Vet. Res. *38*, 229-241.
Nicola, A.V., McEvoy, A.M., and Straus, S.E. (2003). Roles for endocytosis and low pH in herpes simplex virus entry into HeLa and Chinese hamster ovary cells. J. Virol. *77*, 5324-5332.
Nixdorf, R., Klupp, B.G., and Mettenleiter, T.C. (2001). Role of the cytoplasmic tails of pseudorabies virus glycoproteins B, E and M in intracellular localization and virion incorporation. J. Gen. Virol. *82*, 215-226.
Ober, B., Summerfield, A., Mattlinger, C., Wiesmüller, K.-H., Jung, G., Pfaff, E., Saalmüller, A., and Rziha, H.-J. (1998). Vaccine-induced, pseudorabies virus-specific, extrathymic $CD4^+/CD8^+$ memory T-helper cells in swine. J. Virol. *72*, 4866-4873.
Ober, B., Teufel, B., Wiesmüller, K.-H., Jung, G., Pfaff, E., Saalmüller, A., and Rziha, H.-J. (2000). The porcine humoral immune response against pseudorabies virus specifically targets attachment sites on glycoprotein gC. J. Virol. *74*, 17552-1760.
Okamura, H., Sakaguchi, M., Honda, T., Taneno, A., Matsuo, K., and Yamada, S. (1994). Construction of recombinant infectious laryngotracheitis virus expressing the LacZ gene of *E. coli* with thymidine kinase gene. J. Vet. Med. Sci. *56*, 799-801.
Olsen, L., Ch'ng, T., Card, J.P., and Enquist, L.W. (2006). Role of pseudorabies virus Us3 protein kinase during neuronal infection. J. Virol. *80*, 6387-6398.
Ou, C.J., Wong, M.L., Huang, C., and Chang, T.J. (2002). Suppression of promoter activity of the LAT gene by IE180 of pseudorabies virus. Virus Genes *25*, 227-239.
Parkinson, J., and Everett, R.D. (2000). Alphaherpesvirus proteins related to herpes simplex virus type 1 ICP0 affect cellular structures and proteins. J. Virol. *74*, 10006-10017.
Parkinson, J., and Everett, R.D. (2001). Alphaherpesvirus proteins related to herpes simplex virus type 1 ICP0 induce the formation of colocalizing, conjugated ubiquitin. J. Virol. *75*, 5357-5362.
Paulus, C., Sollars, P.J., Pickard, G.E., and Enquist, L.W. (2006). Transcriptome signature of virulent and attenuated pseudorabies virus-infected rodent brain. J. Virol. *80*, 1773-1786.
Peeters, B., de Wind, N., Hooisma, M., Wagenaar, F., Gielkens, A., and Moormann, R. (1992). Pseudorabies virus envelope glycoproteins gp50 and gII are essential for virus penetration, but only gII is involved in membrane fusion. J. Virol. *66*, 894-905.

Pejsak, Z.K., and Truszczynski, M.J. (2006). Aujeszky's disease (pseudorabies), p. 419-433. In: A. D. Lehman (ed.), Diseases of swine.

Pensaert, M.B., and Kluge, J.P. (1989). Pseudorabies virus (Aujeszky's disease), p.39-64. In: M.B. Pensaert (ed.), Virus infections of porcines. Elsevier Science Publishers, Amsterdam, The Netherlands.

Petrovskis, E.A., Timins, J.G., Gierman, T.M., and Post, L.E. (1986). Deletions in vaccine strains of pseudorabies virus and their effect on synthesis of glycoprotein gp63. J. Virol. *60*, 1166-1169.

Pomeranz, L.E., Reynolds, A.E., and Hengartner, C.J. (2005). Molecular biology of pseudorabies virus: impact on neurovirology and veterinary medicine. Microbiol. Mol. Biol. Rev. *69*, 462-500.

Poulsen, D.J., and Keeler, C.L.Jr. (1997). Characterization of the assembly and processing of infectious laryngotracheitis virus glycoprotein B. J. Gen. Virol. *78*, 2945-2951.

Prideaux, C.T., Kongsuwan, K., Johnson, M.A., Sheppard, M., and Fahey, K.J. (1992). Infectious laryngotracheitis virus growth, DNA replication, and protein synthesis. Arch. Virol. *123*, 181-192.

Priola, S., Gustafson, D.P., Wagner, E.K., and Stevens, J.G. (1990). A major portion of the latent pseudorabies virus genome is transcribed in trigeminal ganglia of pigs. J. Virol. *64*, 4755-4760.

Purves, F., Spector, D., and Roizman, B. (1992). UL34, the target of the herpes simplex virus Us3 protein kinase, is a membrane protein which in its unphosphorylated state associates with novel phosphoproteins. J. Virol. *66*, 4295-4303.

Quint, W., Gielkens, A.L.J., van Oirschot, J.T., Berns, A., and Cuypers, H.T. (1987). Construction and characterization of deletion mutants of pseudorabies virus: a new generation of 'live' vaccines. J. Gen. Virol. *68*, 523-534.

Rauh, I., and Mettenleiter, T.C. (1991). Pseudorabies virus glycoproteins gII and gp50 are essential for virus penetration. J. Virol. *65*, 5348-5356.

Rauh, I., Weiland, F., Fehler, F., Keil, G., and Mettenleiter, T.C. (1991). Pseudorabies virus mutants lacking the essential glycoprotein gII can be complemented by glycoprotein gI of bovine herpesvirus 1. J. Virol. *65*, 621-631.

Ray, N., and Enquist, L.W. (2004). Transcriptional response of a common permissive cell type to infection by two diverse alphaherpesviruses. J. Virol. *78*, 3489-3501.

Rea, T.J., Timmins, J.G., Long, G.W., and Post, L.E. (1985). Mapping and sequence of the gene for the pseudorabies virus glycoprotein which accumulates in the medium of infected cells. J. Virol. *54*, 21-29.

Reynolds, A., Wils, E.G., Roller, R.J., Ryckman, B., and Baines, J.D. (2002). Ultrastructural localization of the herpes simplex virus type 1 UL31, UL34 and US3 proteins suggests specific roles in primary envelopment and egress of nucleocapsids. J. Virol. *76*, 8939-8952.

Rinaman, L., Card, J.P., and Enquist, L.W. (1993). Spatiotemporal responses of astrocytes, ramified microglia, and brain macrophages to central neuronal infection with pseudorabies virus. J. Neurosci. *13*, 685-702.

Robbins, A.K., Whealy, M.E., Watson, R.J., and Enquist, L.W. (1986). Pseudorabies virus gene encoding glycoprotein gIII is not essential for growth in tissue culture. J. Virol. *59*, 635-645.

Robbins, A.K., Ryan, J.P., Whealy, M.E., and Enquist L.W. (1990). The gene encoding the gIII envelope protein of pseudorabies virus vaccine strain Bartha contains a mutation affecting protein localization. J. Virol. *63*, 250-258.

Rock, D.L., Beam, S.L., and Mayfield, J.E. (1987). Mapping bovine herpesvirus type 1 latency-related RNA in trigeminal ganglia of latently infected rabbits. J. Virol. *61*, 3827-3831.

Rock, D., Lokensgard, J., Lewis, T., and Kutish, G. (1992). Characterization of dexamethasone-induced reactivation of latent bovine herpesvirus 1. J. Virol. *66*, 2484-2890.

Roizman, B., and Knipe, D.M. (2001). Herpes simplex viruses and their replication, in: Knipe, D.M., Howley, P.M. (Eds.), Fields Virology, Lippincott Williams and Wilkins, Philadelphia, 2001, pp. 2399-2459.

Roizman, B., and Pellett, P.E. (2001). The family *Herpesviridae*: A brief introduction, in: Knipe, D.M., Howley P.M. (Eds.), Fields Virology, Lippincott Williams and Wilkins, Philadelphia, 2001, pp. 2381-2397.

Rziha, H.-J., Mettenleiter, T.C., Ohlinger, V., and Wittmann, G. (1986). Herpesvirus (pseudorabies virus) latency in swine: occurrence and physical state of viral DNA in neural tissues. Virology *155*, 600-613.

Saif, Y.M., Rosenberger, J.K., Cloud, S.S., Wild, M.A., McMillen, J.K., and Schwartz, R.D. (1994). Efficacy and safety of a recombinant herpesvirus of turkeys containing genes from infectious laryngotracheitis virus, in: Proc. Am. Vet. Med. Assoc., Minneapolis, 1994, p. 154.

Sarmiento, M., and Spear, P.G. (1979). Membrane proteins specified by herpes simplex viruses. IV: Conformation of the virion glycoprotein designated VP7(B_2). J. Virol. *29*, 1159-1167.
Schang L. M., Hossain, A., and Jones, C. (1996). The latency-related gene of bovine herpesvirus 1 encodes a product which inhibits cell cycle progression.J. Virol. *70*, 3807-3814.
Schang L. M., and Jones, C. (1997). Analysis of bovine herpesvirus 1 transcripts during a primary infection of trigeminal ganglia of cattle. J. Virol. *71*, 6786-6795.
Schmidt, J., Klupp, B.G., Karger, A., and Mettenleiter, T.C. (1997). Adaptability in herpesviruses: glycoprotein D-independent infectivity of pseudorabies virus. J. Virol. *71*, 17-24.
Schmidt, J., Gerdts, V., Beyer, J., Klupp, B.G., and Mettenleiter, T.C. (2001). Glycoprotein D-independent infectivity of pseudorabies virus results in an alteration of in vivo host range and correlates with mutations in glycoproteins B and H. J. Virol. *75*, 10054-10064.
Schnitzlein, W.M., Radzevicius, J., and Tripathy, D.N. (1994). Propagation of infectious laryngotracheitis virus in an avian liver cell line. Avian Dis. *38*, 211-217.
Schnitzlein, W.M., Winans, R., Ellsworth, S., and Tripathy, D.N. (1995). Generation of thymidine kinase-deficient mutants of infectious laryngotracheitis virus. Virology *209*, 304-314.
Schröder, C., Linde, G., Fehler, F., and Keil, G.M. (1997). From essential to beneficial: glycoprotein D loses importance for replication of bovine herpesvirus 1 in cell culture. J. Virol. *71*, 25-33.
Schumacher, D., Tischer, B.K., Reddy, S.M., and Osterrieder, N. (2001). Glycoproteins E and I of Marek's disease virus serotype 1 are essential for virus growth in cultured cells. J Virol *75*, 11307-11318.
Schumacher, D., Tischer, K., Trapp, S., and Osterrieder, N. (2005). The protein encoded by the Us3 orthologue of Marek's disease virus is required for efficient de-envelopment of perinuclear virions and involved in actin stress fiber breakdown. J. Virol. *79*, 3987-3997.
Schwyzer, M., and Ackermann, M. (1996). Molecular virology of ruminant herpesviruses. Vet. Microbiol. *53*, 17-29.
Schwyzer, M., Fischer-Bracher, C., Fraefel, C., Bächi, T., Nuñez, R., Engels, M., and Ackermann, M. (2002). Transduction of Vero cells and bovine monocytes with a herpes simplex virus-1 based amplicon carrying the gene for the bovine herpesvirus-1 circ protein. Vet. Microbiol. *86*, 165-174.

Simon, A., Mettenleiter, T.C., and Rziha, H.-J. (1989). Pseudorabies virus displays variable numbers of a repeat unit adjacent to the 3'-end of the glycoprotein gII gene. J. Gen. Virol. *70*, 1239-1246.
Smith, B.E., Banfield, B.W., Smeraski, C.A., Wilcox, C.L., Dudek, F.E., Enquist, L.W., and Pickard, G.E. (2000). Pseudorabies virus expressing enhanced green fluorescent protein: a tool for in vitro electrophysiological analysis of transsynaptically labeled neurons in identified central nervous system circuits. Proc. Natl. Acad. Sci. USA *97*, 9264-9269.
Smith, G.W., and Enquist, L.W. (1999). Construction and transposon mutagenesis in Escherichia coli of a full-length infectious clone of pseudorabies virus, an alphaherpesvirus. J. Virol. *68*, 2339-2346.
Smith, G.W., and Enquist, L.W. (2000). A self-recombining bacterial artificial chromosome and its application for analysis of herpesvirus pathogenesis. Proc. Natl. Acad. Sci. USA *97*, 4873-4878.
Smith, G.A., Gross, S.P., and Enquist, L.W. (2001). Herpesviruses use bidirectional fast-axonal transport to spread in sensory neurons. Proc. Natl. Acad. Sci. USA *98*, 3466-3470.
Smith, B.E., Pomeranz, L., Gross, S.P., and Enquist, L.W. (2004). Local modulation of plus-end transport targets herpesvirus entry and egress in sensory axons. Proc. Natl. Acad. Sci. USA 101, 16034-16039.
Snyder, A., Wisner, T.W., and Johnson, D.W. (2006). Herpes simplex virus capsids are transported in neuronal axons without an envelope containing viral glycoproteins. J. Virol. *80*, 11165-11177.
Song, C.K., Enquist, L.W., and Bartness, T.J. (2005). New developments in tracing neural circuits with herpesviruses. Virus Res. *111*, 235-249.
Sparks-Thissen, R., and Enquist, L.W. (1999). Differential regulation of D^k and K^k major histocompatibility complex class I proteins on the cell surface after infection of murine cells by pseudorabies virus. J. Virol. *73*, 5748-5756.
Spear, P.G., and Longnecker, R. (2003). Herpesvirus entry: an update. J. Virol. *77*, 10179-10185.
Steven, A.C., and Spear, P.G. (1997). Herpesvirus capsid assembly and envelopment. In: Chiu, W., Burnett, R.M., Garcea, R. (eds)., Structural biology of viruses. Oxford University Press, new York, pp. 312-351.

Subramanian, R.P., and Geraghty, R.J. (2007). Herpes simplex virus type 1 mediates fusion through a hemifusion intermediate by sequential activity of glycoproteins D, H, L, and B. Proc. Natl. Acad. Sci. USA *104*, 2903-2908.

Taddeo, B., Zhang, W., and Roizman, B. (2006). The UL41 protein of herpes simplex virus 1 degrades RNA by endonucleolytic cleavage in absence of other cellular or viral proteins. Proc. Natl. Acad. Sci. USA *103*, 2827-2832.

Taharaguchi, S., Kobayashi, T., Yoshino, S., and Ono, E. (2002). Analysis of regulatory functions for the region located upstream from the latency-associated transcript (LAT) promoter of pseudorabies virus in cultured cells. Vet. Microbiol. *85*, 197-208.

Taharaguchi, S., Yoshino, S., Amagai, K., and Ono, E. (2003). The latency-associated transcript promoter of pseudorabies virus directs neuron-specific expression in trigeminal ganglia of transgenic mice. J. Gen. Virol. *84*, 2015-2022.

Takashima, Y., Tamura, H., Xuan, X., and Otsuka, H. (1999). Identification of the US3 gene product of BHV-1 as a protein kinase and characterization of BHV-1 mutants of the US3 gene. Virus Res. *59*, 23-34.

Tan, X., Brunovskis, P., and Velicer, L.F. (2001). Transcriptional analysis of Marek's disease virus glycoprotein D, I and E genes: expression is undetectable in cell culture. J. Virol. *75*, 2067-2075.

Thureen, D.R., and Keeler, C.L. (2006). Psittacid herpesvirus 1 and infectious laryngotracheitis virus: comparative genome sequence analysis of two avian alphaherpesviruses. J. Virol. *80*, 7863-7872.

Tikoo, S.K., Campos, M., and Babiuk, L.A. (1995). Bovine herpesvirus 1 (BHV-1): biology, pathogenesis, and control. Adv. Virus Res. *45*, 191-223.

Tirabassi, R., Townley, R.A., Eldridge, M.G., and Enquist, L.W. (1997). Characterization of pseudorabies virus mutants expressing carboxy-terminal truncations of gE: evidence for envelope incorporation, virulence, and neurotropism domains. J. Virol. *71*, 6455-6464.

Tirabassi, R., and Enquist, L.W. (1998). Role of envelope protein gE endocytosis in the pseudorabies virus life cycle. J. Virol. *72*, 4571-4579.

Tomishima, M.J., and Enquist, L.W. (2001). A conserved alpha-herpesvirus protein necessary for axonal localization of viral membrane proteins. J. Cell Biol. *154*, 741-752.

Tomishima, M.J., Smith, G.A., and Enquist, L.W. (2001). Sorting and transport of alpha herpesviruses in axons. Traffic *2*, 429-436.

Tomishima, M.J., and Enquist, L.W. (2002). In vivo egress of an alphaherpesvirus from axons. J. Virol. *76*, 8310-8317.
Tong, G.Z., Zhang, S.J., Meng, S.S., Wang, L., Qiu, H.J., Wang, Y.F., Yu, L., and Wang, M. (2001). Protection of chickens from infectious laryngotracheitis with a recombinant fowlpox virus expressing glycoprotein B of infectious laryngotracheitis virus. Avian Pathol. *30*, 143-148.
van de Walle, G., Favoreel, H.W., Nauwynck, H.J., and Pensaert, M.B. (2003). Antibody-induced internalization of viral glycoproteins and gE-gI Fc receptor activity protect pseudorabies virus-infected monocytes from efficient complement-mediated lysis. J. Gen. Virol. *84*, 939-948.
van Engelenburg, F., Kaashoek, M., Rijsewijk, F., van den Burg, L., Moerman, A., Gielkens, L., and van Oirschot, J.T. (1994). A glycoprotein E deletion mutant of bovine herpesvirus 1 is avirulent in calves. J. Gen. Virol. *75*, 2311-2318.
van Minnebruggen, G., Favoreel, H.W., Jacobs, L., and Nauwynck, H.-J. (2003). Pseudorabies virus US3 protein kinase mediates actin stress fiber breakdown. J. Virol. *77*, 9074-9080.
van Oirschot, J.T., Rziha, H.-J., Moonen, P.J.L.M., Pol, J.M.A., and van Zaane, D. (1986). Differentiation of serum antibodies from pigs vaccinated or infected with Aujeszky's disease virus by a competitive enzyme immunoassay. J. Gen. Virol. *67*, 1179-1182.
van Oirschot, J.T. (1999). Diva vaccines that reduce virus transmission. J. Biotechnol. *73*, 195-205.
van Zijl, M., van der Gulden, H., de Wind, N., Gielkens, A., and Berns, A. (1990). Identification of two genes in the unique short region of pseudorabies virus: comparison with herpes simplex and varicella-zoster virus. J. Gen. Virol. *76*, 1851-1859.
Veits, J., Köllner, B., Teifke, J.P., Granzow, H., Mettenleiter, T.C., and Fuchs, W. (2003a). Isolation and characterization of monoclonal antibodies against structural proteins of infectious laryngotracheitis virus. Avian Dis. *47*, 330-342.
Veits, J., Mettenleiter, T.C., and Fuchs, W. (2003b). Five unique open reading frames of infectious laryngotracheitis virus are expressed during infection but are dispensable for virus replication in cell culture. J. Gen. Virol. *84*, 1415-1425.
Veits, J., Lüschow, D., Kindermann, K., Werner, O., Teifke, J.P., Mettenleiter, T.C., and Fuchs, W. (2003c). Deletion of the non-essential UL0 gene of infectious laryngotracheitis (ILT) virus leads to attenuation in chickens, and UL0 mutants expressing influenza

virus haemagglutinin (H7) protect against ILT and fowl plague. J. Gen. Virol. *84*, 3343-3352.
Wagenaar, F., Pol, J.M., Peeters, B., Gielkens, A.L.J., de Wind, N., and Kimman, T.G. (1995). The US3-encoded protein kinase from pseudorabies virus affects egress of virions from the nucleus. J. Gen. Virol. *76*, 1851-1859.
Wang, X. (2001). The expanding role of mitochondria in apoptosis. Genes Dev. *15*, 2922-2933.
Whealy, M.E., Card, J.P., Robbins, A.K., Dubin, J.R., Rziha, H.-J., and Enquist, L.W. (1993). Specific pseudorabies virus infection of the rat visual system requires both gI and gp63 glycoproteins. J. Virol. *67*, 3786-3797.
Wild, M.A., Cook, S., and Cochran, M. (1996). A genomic map of infectious laryngotracheitis virus and the sequence and organization of genes present in the unique short and flanking regions. Virus Genes *12*, 107-116.
Wild, P., Engels, M., Senn, C., Tobler, K., Ziegler, U., Schraner, E.M., Loepfe, E., Ackermann, M., Mueller, M., and Walther, P. (2005). Impairment of nuclear pores in bovine herpesvirus 1-infected MDBK cells. J. Virol. *79*, 1071-1083.
Williams, R.A., Benett, M., Bradbury, J.M., Gaskell, R.M., Jones, R.C., and Jordan, F.T.W. (1992). Demonstration of sites of latency of infectious laryngotracheitis virus using the polymerase chain reaction. J. Gen. Virol. *73*, 2415-2420.
Wirth, U.V., Vogt, B., and Schwyzer, M. (1991). The three major immediate-early transcripts of bovine herpesvirus 1 arise from two divergent and spliced transcription units. J. Virol. *65*, 195-205.
Wirth, U.V., Fraefel, C., Vogt, B., Vlcek, C., Paces, V., and Schwyzer, M. (1992). Immediate-early RNA 2.9 and early RNA 2.6 of bovine herpesvirus 1 are 3' coterminal and encode a putative zinc finger transactivator protein. J. Virol. *66*, 2763-2772.
Wölfer, U., Kruft, V., Sawitzky, D., Hampl, H., Wittmann-Liebold, B., and Habermehl, K.-O. (1990). Processing of pseudorabies virus glycoprotein gII. J. Virol. *64*, 3122-3125.
Wong, M.-L., and Chen, C.-H. (1998). Evidence for the internal location of actin in the pseudorabies virion. Virus Res. *56*, 191-197.
Workman, J.L., Abmayr, S.M., Cromlish, W.A., and Roeder, R.G. (1988). Transcriptional regulation by the immediate-early protein of pseudorabies virus during in vitro nucleosome assembly. Cell *55*, 211-219.

Wu, L., and Berk, A.J. (1988). Transcriptional activation by the pseudorabies virus immediate-early protein requires the TATA box element in the adenovirus 2 E1B promoter. Virology *167*, 318-322.

Wu, S.X., Zhu, X.P., and Letchworth, G.J. (1998). Bovine herpesvirus 1 glycoprotein M forms a disulfide-linked heterodimer with the U(L)49.5 protein. J. Virol. *72*, 3029-3036.

WuDunn, D., and Spear, P.G. (1989). Initial interaction of herpes simplex virus with cells is binding to heparan sulfate. J. Virol. *63*, 52-58.

Xu, Z., Maroney, A.C., Dobrzanski, P., Kukekov, N.V., and Greene, L.A. (2001). The MLK family mediates c-Jun N-terminal kinase activation in neuronal apoptosis. Mol. Cell. Biol. *21*, 4713-4724.

York, J.J., and Fahey, K.J. (1991). Vaccination with affinity-purified glycoproteins protects chickens against infectious laryngotracheitis herpesvirus. Avian Pathol. *20*, 693-704.

Zhang, Y., and Jones, C. (2001). The bovine herpesvirus 1 immediate-early protein (bICP0) associates with histone deacetylase 1 to activate transcription. J. Virol. *75*, 9571-9578.

Zhang, Y., Zhou, J., and Jones ,C. (2005). Identification of functional domains within the bICP0 protein encoded by bovine herpesvirus 1. J. Gen. Virol. *86*, 879-886.

Zhou, Z., Chen, D., Jakana, J., Rixon, F.J., and Chiu, W. (1999). Visualization of tegument-capsid interactions and DNA in intact herpes simplex virus type 1 virions. J. Virol. *73*, 3210-3218.

Ziemann, K., Mettenleiter, T.C., and Fuchs, W. (1998a). Gene arrangement within the unique long genome region of infectious laryngotracheitis virus is distinct from that of other alphaherpesviruses. J. Virol. *72*, 847-852.

Ziemann, K., Mettenleiter, T.C., and Fuchs, W. (1998b). Infectious laryngotracheitis herpesvirus expresses a related pair of unique nuclear proteins which are encoded by split genes located at the right end of the U_L genome region. J. Virol. *72*, 6867-6874.

Zuckermann, F., Mettenleiter, T.C., Schreurs, C., Sugg, N., and Ben-Porat, T. (1988). Complex between glycoproteins gI and gp63 of pseudorabies virus: its effect on virus replication. J. Virol. *62*, 4622-4626.

Zuckermann, F., Zsak, L., Mettenleiter, T.C., and Ben-Porat, T. (1990). Pseudorabies virus glycoprotein gIII is a major target antigen for murine and swine virus-specific cytotoxic T-lymphocytes. J. Virol. *64*, 802-812.

Chapter 10

African Swine Fever Virus

Linda K. Dixon, Charles C. Abrams, Dave G. Chapman and Fuquan Zhang

---

ABSTRACT

African swine fever virus (ASFV) is a large double-stranded DNA virus which replicates in the cytoplasm of infected cells and is the only member of the *Asfarviridae* family. In common with other viral haemorrhagic fevers, the main target cells for replication are those of monocyte, macrophage lineage. The virus causes a haemorrhagic fever with high mortality rates in pigs, but persistently infects its natural hosts, warthogs, bushpigs and soft ticks of the *Ornithodoros* species with no disease signs.

The virus encodes enzymes required for replication and transcription of the genome, including elements of a base excision repair system, structural proteins and many proteins that are not essential for replication in cells but have roles in virus survival and transmission in its hosts. Virus replication takes place in perinuclear factory areas. Assembly of the icosahedral capsid occurs on modified membranes from the endoplasmic reticulum. Products from proteolytically processed polyproteins form the core shell between the internal membrane and the nucleoprotein core. An additional outer membrane is gained as particles bud from the plasma membrane. The virus encodes proteins that inhibit signalling pathways in infected macrophages and thus modulate transcriptional activation of immune response genes. In addition the virus encodes proteins which inhibit apoptosis of infected cells to facilitate production of progeny virions. Viral membrane proteins with similarity to cellular adhesion proteins modulate interaction of virus-infected cells and extracellular virions with host components.

INTRODUCTION

African swine fever virus (ASFV) is a large DNA virus which replicates in the cytoplasm of infected cells. ASFV is the only member of the *Asfarviridae* family and is the type member of the only genus, the *Asfivirus* genus (Dixon, 2005). The virus shares similarities in its gene complement with other large DNA virus families, which replicate at least partially in the cytoplasm. These virus families may share a common ancestor. This virus superfamily, named the nucleo-cytoplasmic large DNA virus family (NCLDV), also includes the *Poxviridae*, *Iridoviridae, Phycodnaviridae* and *Mimiviridae* (Iyer et al., 2006). ASFV shares most in common, in terms of its replication strategy, with the *Poxviridae.* Phylogenetic analysis based on the DNA polymerase gene shows the *Asfarviridae* as a separate clade between the *Poxviridae* and *Iridoviridae* (Raoult et al., 2004).

ASFV causes an acute haemorrhagic fever in domestic pigs with mortality rates approaching 100%. The disease was first identified in Kenya in the 1920s following spread from infected wildlife (Montgomery RE, 1921). Subsequently it was established that ASFV infection has been present over a very long period in Eastern and Southern Africa in warthogs (*Phacochoerus aethiopicus)*, bushpigs (*Potamochoerus porcus)* and a species of soft tick (*Ornithodoros porcinus)*. These wildlife hosts can be persistently infected over long time periods without showing disease signs. Ticks become infected by feeding on warthogs and young warthogs develop a sufficiently high viraemia to infect ticks. Virus can be transmitted between ticks transadially, transovarially and transexually from male to female. Infected ticks transmit virus to warthogs on which they feed but direct transmission between warthogs is thought to be infrequent and the tick vector thus plays an important role in the sylvatic cycle (Plowright W, 1969; Thomson, 1985; Thomson et al., 1980). The virus is very well-adapted to these hosts, probably reflecting long term evolution in this transmission cycle. Once introduced into the domestic pig population, virus can be transmitted directly between pigs without the requirement of the tick vector, although these may provide an important reservoir of virus if present in pig houses.

ASF spread to most sub-Saharan African countries and was introduced to Portugal in 1957 and 1960. The disease remained endemic in Spain and Portugal until the mid-1990s. In southern Spain and Portugal the tick species *Ornithodoros erraticus* acted as a vector for the virus contributing to the difficulties in disease eradication. ASF is thought to have spread to other European countries and to Brazil and the Carribean from Spain and Portugal. However, outside Africa ASF has now been eradicated except for in Sardinia.

Although most ASFV isolates cause an acute disease with high mortality, less virulent isolates have been described in pigs. These include moderately virulent isolates from which 30 to 50% of pigs can recover and low virulence isolates which can cause a chronic form of disease but very low mortality (Boinas et al., 2004; Leitao et al., 2001; Wilkinson et al., 1981). In Africa ASF causes devastating economic losses and limits pig production, thus restricting the supply of a cheap source of high quality protein. The lack of a vaccine limits the strategies for disease control and means that the disease is a constant threat to pig farming worldwide.

The main target cells for virus replication are cells of monocyte/macrophage lineage and those expressing cell surface markers characteristic of intermediate/late stages of differentiation are permissive for virus replication (McCullough et al., 1999). In the early stages of infection virus replication is seen primarily in macrophages and this is most likely the primary event leading to the haemorrhagic pathology (del Moral et al., 1999; Gomezvillamandos et al., 1995a; Oura et al., 1998). In common with other viral haemorrhagic fevers, ASF is characterised by damage to vascular endothelial cells and this contributes to vascular permeability. This is thought to be caused by factors released from virus infected macrophages, although direct infection of endothelial cells may be involved. *In vitro* infection of endothelial cells by ASFV leads to induction of apoptosis which could contribute to the destruction of vascular endothelial cells in infected pigs (Vallee et al., 2001). In the later stages of disease disseminated intravascular coagulation (DIC) develops as demonstrated by the appearance of fibrin degradation products and presence of numerous fibrin thrombi in blood (Villeda et al., 1995;

Villeda et al., 1993a; Villeda et al., 1993b). Elevated levels of TNF-α and IL1-β are observed in serum from pigs experimentally infected with ASFV (del Moral et al., 1999; Salguero et al., 2002). TNF-α induces vasodilation, an increase in vascular permeability and activation of the vascular endothelium. All of these may tip the balance in favour of pro-coagulation activities favouring the generation of microthrombi. In addition TNF-α can lead to induction of apoptosis and this may contribute to the destruction of endothelial cells (del Moral et al., 1999).

A striking feature of ASFV infection is the massive apoptosis of lymphocytes which occurs in infiltrate lymphocytes and lymphoid structure in pigs infected with virulent isolates (Carrasco et al., 1996; Gomezvillamandos et al., 1995a; Oura et al., 1998; Salguero et al., 2004). Neither T nor B lymphocytes are directly infected by the virus. The proximity of apoptotic lymphocytes to virus infected macrophages and the correlation between the number of virus infected macrophages and extent of lymphocyte apoptosis, suggests that factors produced by infected macrophages are responsible for the induction of lymphocyte apoptosis (Oura et al., 1998).

VIRUS GENOME

The structure of the ASFV genomes resembles that of the poxviruses since it consists of a single molecule of linear, double-stranded (ds) DNA that is cross-linked at the termini. The end sequences are present in two flip-flop forms that are inverted and complimentary with respect to each other. The sequence of these hairpin loops has been determined in the BA71V isolate. They are 37 nucleotides long and are incompletely base-paired (Gonzalez et al., 1986). Adjacent to both termini are inverted repeats which consist of several different tandem repeat arrays and vary in length between 2.1 kilobase pairs (kbp) and 2.5 kbp. In the BA71V isolate the terminal inverted repetition (TIR) consists of three different types of tandem direct repeats and two types of direct repeats. These are interspersed with unique sequences (Yanez et al., 1995). Variation is observed in the sequence and number of direct repeats from different isolates.

The number of open reading frames (ORFs) encoded varies between 160 and 175 depending on the isolate. These are closely spaced and read from both DNA strands. Short sequences upstream from ORFs contain virus specific promoters that temporally regulate transcription of three classes of genes, early, intermediate and late (Almazan et al., 1992; Almazan et al., 1993; Rodriguez et al., 1996c). The genome is AT rich (~ 61%)(Yanez et al., 1995). The nomenclature of ORFs varies between different isolates. Those ORFs on the BA71V isolate were named with the *Hind* III restriction fragment followed by the number of amino acids encoded and a letter to indicate if they are read towards the left or right end of the genome (Yanez et al., 1995). Those from the right end of the Malawi isolate genome have been named with the *Sal* I fragment, a number to indicate the position from the left end of that fragment and a letter to indicate whether read towards the left or right end of the genome (Dixon et al., 1994). In addition some other nomenclatures refer to the function of the protein encoded. For clarity in this review the nomenclature of the BA71V ORF will be used or included along with more commonly cited names from the literature if appropriate. For those ORFs not present on the BA71V genome nomenclature from other isolates (usually Malawi LIL20/1) is included.

GENOME VARIATION

Variability of the ASFV genome has been determined by comparing isolates obtained from outbreaks in domestic pigs and from wildlife sources (warthogs and soft ticks) collected over a 45 year time span from different geographical locations and from isolates passaged in tissue culture. These comparisons involved sequencing of selected genes and analysis of restriction fragment length polymorphisms. In addition, the complete genome sequences of one tissue-culture adapted Spanish isolate, BA71V, and 9 field isolates have been determined. Eight of these field isolates are from domestic pigs or wildlife sources in Africa and one is a low virulence isolate from Portugal. Annotated genome sequences are available on the website www.virology.ca. The length of the genomes varies between 170 and 192 kbp due mainly to the gain or loss of sequences within 40 kbp from the left end of the genome and within 25 kbp from the right end of the genome. As discussed below these genome regions encode copies of six different multigene families as well as several other genes. The length variations are largely due to gain or loss of members of these families. Comparison of the genome of the tissue-culture adapted ASFV strain BA71V with those of field isolates identified a sequence of about 3 kbp close to the right end that was not present in the BA71V genome but in all the other genomes analyzed. This region encodes 5 ORFs (I7L, I8L, I9R, I10L, I11L) including one or two additional copies of the gene encoding the p22 (KP177R) structural protein (Chapman DG., Tschereparov V., Upton C., and Dixon LK unpublished results). The three other ORFs are of unknown function, although deletion of one, I11L, was shown not to reduce virus replication in macrophages or virulence in pigs (Kleiboeker et al., 1998). In addition both the BA71V isolate and the non-pathogenic isolates OURT88/3 and NHP68 have a deletion of 8 or 9 kbp from the same region at the left end of the genome. This region encodes 6 copies of MGF 360 and 1 or 2 copies of MGF 530.

Partial sequencing of the gene encoding the major capsid protein P72 (B646L) revealed that isolates obtained from domestic pig outbreaks in West and Central Africa, Europe and the Caribbean over a forty year period were all closely related and could not be distinguished by sequencing this gene fragment alone. However, isolates from ticks inhabiting warthog burrows in East and Southern Africa were shown to be genetically very diverse and this is consistent with the long term evolution of the virus in these hosts. Isolates from domestic pigs in these regions were also more diverse probably reflecting several introductions from wildlife reservoirs into domestic pigs (Bastos et al., 2003; Dixon and Wilkinson, 1988; Lubisi et al., 2005). One virus lineage was detected from domestic pig isolates obtained over a 40 year period in Malawi, Zambia, Mozambique and Zimbabwe. This suggests that even where the virus is endemic in the sylvatic cycle, introduction from the sylvatic cycle into domestic pigs is a relatively rare event (Lubisi et al., 2005; Sumption et al., 1990).

GENES ENCODED

Analysis of the 10 complete genome sequences available shows that the number of ORFs encoded varies between 160 and 175 in different isolates. Of these 110 are present as a single copy on the genomes of all isolates. Other ORFs conserved between the different isolates are present in six different multigene families (MGF) which are located in regions close to the genome termini. The number of genes in these MGFs varies between 11 and 19 for MGF 360, between 5 and 13 for MGF 110, between 8 and 10 for MGF 505/530, between 3 and 4 for MGF 300, between 2 and 3 for MGF 100, and between 1 and 3 for the P22 (KP177R) family (Chapman DG., Tschereparov V., Upton C., and Dixon LK unpublished results). The organization of these genes on the genome indicates that they have evolved by a process of gene duplication and sequence divergence. There is also evidence for transposition of some genes from one end of the genome to the other. For example copies of MGF360 and P22 (KP177R) are present close to both ends of the genome whereas other MGFs families are present at one end or the other; MGF110, 505/530 and 300 are located close to the left end and MGF100 close to the right end. There is considerable sequence divergence between individual MGF family members encoded by the same isolate. For example MGF 360 family members encoded by a single genome share between about 30 and 90% amino acid identity (Almendral et al., 1990; Delavega et al., 1990; Gonzalez et al., 1990; Rodriguez et al., 1994; Yozawa et al., 1994). In contrast orthologous copies of MGF family members encoded by different isolates are more closely related. The level of identity varies but for individual members of MGF360 the amino acid identity range is between 75 and 100%. This suggests that gene duplication events are relatively rare.

Of the 110 conserved ORFs, 39 encode proteins of known function, a further 42 contain motifs contained in other proteins and 28 are of unknown function. As is predicted from the cytoplasmic site of virus replication a number of the proteins encoded are involved in replication and transcription of the virus genome. In addition proteins of defined functions include virion structural proteins and proteins involved in evading the host's defences. Many virus encoded-proteins are not essential for replication in cells but have important roles in interactions with hosts which are important for virus survival and transmission. More detailed description of the proteins encoded is given below.

Proteins involved in DNA replication and repair and mRNA transcription and processing

Due to the limiting pools of dNTPs present in cells many large DNA viruses encode enzymes involved in nucleotide metabolism. These function to increase the precursor pools of dNTPs required for virus DNA replication. For ASFV this is particularly important since the main target cells for replication are mature macrophages which are non-dividing and therefore have restricted pools of dNTPs. Indeed it has been shown that although the virus-encoded thymidine kinase (K196R) and dUTPase (E165R) genes are non-essential for virus replication in dividing tissue culture cells, their deletion dramatically reduces virus replication in macrophages (Moore et al., 1998; Oliveros et al., 1999). The virus also encodes thymidylate kinase (A240L) and ribonucleotide reductase (F134L, F778R) enzymes that are involved in nucleotide metabolism.

Enzymes and proteins, that are likely to be directly involved in replicating the virus DNA, include a DNA polymerase type B (G1211R) and a PCNA-like DNA clamp (E301R).

Table 1. Known functions of ASFV encoded proteins. The left column indicates the protein functional class and predicted function. Experimentally demonstrated functions are indicated with an asterisk. The right column gives the ORF name from the BA71V isolate and other nomenclature used in brackets.

Nucleotide metabolism, transcription, replication and repair	
Thymidylate kinase	A240L
Thymidine kinase	K196R
dUTPase*	E165R (k1R)
Ribonucleotide reductase (small subunit)	F334L
Ribonucleotide reductase (large subunit)	F778R
DNA polymerase β	G1211R
DNA topoisomerase type II*	P1192R (i7R)
Proliferating cell nuclear antigen (PCNA) like	E301R (j15R)
DNA polymerase family X*	O174L
DNA ligase*	NP419L (g3L)
Putative DNA primase	C962R
AP endonuclease class II*	E296R (k4R)
RNA polymerase subunit 2	EP1242L (g2L)
RNA polymerase subunit 6	C147L (g6L)
RNA polymerase subunit 1	NP1450L
RNA polymerase subunit 3	H359L (j1L)
RNA polymerase subunit 5	D205R (i2R)
RNA polymerase subunit 10	CP80R
TFIIB like	C315R
Helicase superfamily II	A859L
Helicase superfamily II similar to origin binding protein	F1055L
Helicase superfamily II	B962L
Helicase superfamily II VV D6/D11 like involved in transcription termination	D1133L (g10L)
Helicase superfamily II VV D5 like	Q706L (j10L)
Helicase superfamily II VV A18 like	QP509L
Transcription factor SII	I243L (k9L)
Guanyl transferase*	NP868R (g4R)
Poly A polymerase large subunit	C475L
FTS J like methyl transferase domain	EP424R
ERCC4 nuclease domain	EP364R
Lambda-like exonuclease	D345L
VV A2L like transcription factor	B385R
VV A7L like transcription factor	G1340L
VV VLTF2 like late transcription factor, FCS like finger	B175L
VV D5 like ATPase involved in replication	C962R

Other enzymes with unknown roles	
Prenyl transferase*	B318L
Serine protein kinase*	R298L (j8L)
Ubiquitin conjugating enzyme*	I215L (k13L)
Nudix hydrolase*	D250R (g5R)
Host cell interactions	
IAP apoptosis inhibitor*	A224L (4CL)
Bcl 2 apoptosis inhibitor*	A179L
IkB homolog and inhibitor of calcineurin phosphatase*	A238L (5EL)
C type lectin like*	EP153R
CD2 like. Causes haemadsorption to infected cells*	EP402R (CD2v, Mw8R)
Similar to HSV ICP34.5 neurovirulence factor and GADD34 DNA damage response protein, protein phosphatise 1 docking motif	DP71L (I14L, NL)
Nif S like PLP dependent transferase	QP383R (j11R)
Mn dependent superoxide dismutase	C129R
Structural proteins and proteins involved in morphogenesis	
P22 Transmembrane domain. Present in 1 to 3 copies on genomes of different virus isolates.	KP177R (p22)
A104R Histone-like structural protein. HF-like DNA binding protein	A104R
P11.5	A137R
P10	A78R (p10)
P72 Major capsid protein. Involved in virus entry	B646L (p72, p73)
B438L. Required for formation of vertices in icosahedral capsid.	B438L (p49)
B602L Chaperone. Involved in folding of p72 capsid protein. Not incorporated into virus particles.	B602L (9RL)
B119L ERV 1 like. Involved in redox metabolism*. Not incorporated into virus particles.	B119L (9GL)
Sumo 1 like protease. Involved in polyprotein cleavage	S273R
P220 Polyprotein precursor of p150, p37, p14, p34 core shell components. Required for packaging of nucleoprotein core.	CP2475L
P30 Phosphoprotein. Involved in virus entry.	CP204L (p30, p32)
P60 Polyprotein precursor of p35 and p15. Present in core shell.	CP530R (p60, p62)
P12 Attachment protein involved in virus entry. Transmembrane domain.	O61R (p12)
P17 Transmembrane domain.	D117L (i11L)
J5R Transmembrane domain.	H108R (j5R)
P54 Binds to LC8 chain of dynein, involved in virus entry. Transmembrane domain. Required for recruitment of membranes to virus factories.	E183L (j13L, p54)
J18L Transmembrane domain. VV J5 like membrane protein	E199L (j18L)
E248R Possible component of redox pathway required disulfide bond formation. Structural protein.	E248R (k2R)

A151R Contains CXXC motif similar to that in thioredoxins. Binds to E248R protein. Possible component of redox pathway.	A151R
P14.5 DNA binding. Required for movement of virions to plasma membrane	E120R (k3R)
XP124L Multigene family 110 member. Contains KDEL ER retrieval sequence and transmembrane domain. Involved in modification of ER in preparation for virus assembly.	XP124L
EP402R Similar to host CD2 protein. Required for binding red blood cells to infected cells and extracellular virus particles. Glycoprotein inserted into external virus envelope.	EP402R (CD2v, MW8R)

ASFV also encodes a DNA topoisomerase type II (P1192R) and putative DNA primase (C962R) and helicase (F1055L) related to enzymes involved in binding to origins of replication and these are probably involved in initiating DNA replication. An ERCC4-like nuclease (EP364R) is related to the principle Holliday junction resolvase, Mus81, of eucaryotes and the virus also encodes a lambda type exonuclease (D345L). These enzymes may be involved in resolution of the concatemers formed during virus replication.

Another adaptation which may have evolved to facilitate ASFV replication in macrophages is the presence of genes involved in DNA repair. The production of reactive oxygen species by macrophages induce a wide spectrum of microbial damage including DNA lesions which have either miscoding properties or are blocks for DNA and RNA polymerases. This would lead to a block in viral replication. The majority of lesions induced by reactive oxygen species are repaired by a base excision repair pathway (BER). Mammalian BER pathways involve replacement of either a single nucleotide (short-patch) or up to 10 nucleotides (long patch). Three enzymes, a DNA polymerase type X (O174L), which is the smallest known, together with an ATP-dependent -DNA ligase (NP419L) and a class II apurinic/apyrimidinic (AP) endonuclease (E296R) comprise likely components of an ASFV minimalist DNA BER mechanism. The *in vitro* properties of the ASFV DNA polymerase X indicate that this enzyme acts during a short patch BER of damaged DNA (Beard and Wilson, 2001; Garcia-Escudero et al., 2003; Jezewska et al., 2006; Showalter et al., 2001). The repair of oxidative DNA lesions is often initiated with a bifunctional DNA glycosylase that has an associated ligase activity.This glycosylase is capable of removing a chemically modified base to generate an abasic site. A gene encoding a DNA glycosylase has not been identified in the ASFV genome so possibly spontaneously generated abasic sites might serve as the initiating lesions in ASFV repair. These could be incised by the putative ASFV AP endonuclease and the resulting nucleotide gap filled by the DNA polymerase X. Prior to nick ligation by the DNA ligase, the 5' deoxyribose phosphate generated by the AP endonuclease incision of the sugar phosphate backbone would need to be removed (Garcia-Escudero et al., 2003). This activity has not yet been identified in the ASFV BER. These "missing" activities for the BER pathway might be provided by novel genes or additional functions of the enzymes already identified. The requirement of the AP endonuclease for efficient replication in macrophages, but not tissue culture cells, supports the hypothesis that this BER repair system is an adaptation to virus replication in the macrophage cytoplasm (Lamarche and Tsai, 2006; Redrejo-Rodriguez et al., 2006).

The ASFV DNA polymerase X is the smallest DNA polymerase known (174 residues) and was the first DNA polymerase solution structure determined by nuclear magnetic resonance. The structure shows that the enzyme does not resemble a "hand" as has been described for other DNA polymerases since the "fingers" and "thumb" are missing and the topology of the "palm" is unique. Thus it was predicted to lack the N-terminal lyase domain and duplex DNA binding subdomain of DNA polymerase B but to retain the catalytic and C-terminal subdomains involved in dNTP selection (Beard and Wilson, 2001; Showalter et al., 2001). Biochemical analysis of the ASFV DNA polymerase X suggests that it exhibits low fidelity (Showalter et al., 2001), although there are conflicting reports about this (Garcia-Escudero et al., 2003). In particular the enzyme showed a high propensity to form G:G mispairs. It was suggested that the error prone nature of this repair polymerase might contribute to increase mutation frequency in the viral genome. This could help to generate antigenic variants and facilitate virus survival (Showalter et al., 2001). For this hypothesis to be relevant the ASFV DNA ligase should be low fidelity and capable of sealing nicks at 3' mismatched base pairs. In support of this analysis of the catalytic efficiency of nick sealing, of all possible matched and mismatched base pair combinations, showed the ASFV DNA ligase is the lowest fidelity DNA ligase ever reported. Comparison of the mismatch specificity of the ASFV DNA polymerase X with that of ASFV DNA ligase suggests that the ligase may have evolved towards low fidelity for the purpose of generating the broadest spectrum of sealed mismatches (Lamarche et al., 2005). However, the *in vivo* relevance of this model remains to be determined.

As expected for a virus which replicates in the cytoplasm, transcription of virus genes does not require the host RNA polymerase II. It therefore seems likely that the virus encodes all of the enzymes and factors required for transcription of its own genes (Salas et al., 1988). Genes with similarity to six subunits of host RNA polymerase have been identified (EP1242L, C147L, NP1450L, H359L, D205R, CP80R) as well as an othologue of the basal eukaryotic transcription factor TFIIB (C315R) (Iyer et al., 2006). The vaccinia virus RNA polymerase contains 9 subunits therefore additional ASFV encoded subunits may remain to be identified. It is predicted that ASFV should encode specific transcription factors that are required to transcribe different temporally regulated classes of genes. Although there is no experimental data to identify these factors, comparison with poxvirus encoded proteins has identified three ORFs encoding proteins with similarity to known vaccinia virus transcription factors, A2L (B385R), A7L (G1340L) and VLTF2 (B175L) (Iyer et al., 2006). ASFV also encodes six members of the helicase superfamily (A859L, F105L, B92L, D1133L,Q706L, QP509L). Comparison with helicases encoded by vaccinia virus suggests that three of these may be components of the transcription apparatus. ASFV transcripts are capped at the 5' end and polyadenylated at the 3' end. The gene encoding the mRNA capping enzyme ((NP868R) contains all three domains, namely a triphosphatase, a guanyl transferase and a methyltransferase, required for this function. A putative polyA polymerase (C475L) with similarity to the Poxvirus enzymes is also encoded. Since transcription takes place in the cytoplasm no introns are present in the viral genes, and enzymes involved in mRNA splicing are not encoded.

Genes encoding other enzymes

Apart from enzymes with defined roles in the virus replication cycle, several other enzymes are encoded. These could have roles in virus replication or regulating host cell function. One of these is a predicted FTS-J-like RNA methyltransferase (EP424R). The role of this enzyme is unknown but possibly it could play a role in stabilizing rRNA in infected cells this inhibiting ribosome degradation. This could help to prevent shut-off of protein synthesis and thus facilitate virus replication (Iyer et al., 2006).

The virus also encodes two enzymes involved in post-translational modification of proteins, a serine/threonine protein kinase (R298L) and a ubiquitin conjugating enzyme (I215L). The substrates for these enzymes are not known and may be either virus or host proteins. The protein kinase is packaged into virus particles suggesting it may have a role early in infection (Baylis et al., 1993). The ubiquitin conjugating enzyme shuttles between the nucleus and cytoplasm and binds to a host ARID DNA binding domain containing protein. This suggests it may have a role in regulating a host nuclear function (Bulimo et al., 2000; Hingamp et al., 1992).

The virus encodes a trans-prenyl transferase (B318L) that catalyzes *in vitro* the condensation of farnesyl diphosphate and isopentenyl diphosphate to synthesize geranylgeranyl diphosphate and longer chain prenyl diphosphates (Alejo et al., 1997). The localisation of this enzyme was determined in ASFV infected cells and in cells infected with a recombinant pseudo-Sindbis virus carrying the complete B318L gene. In this latter system, the trans-prenyltransferase was found to colocalize with the endoplasmic reticulum marker protein-disulfide isomerase, whereas in cells infected with ASFV, the viral enzyme was present in cytoplasmic viral assembly sites, associated with precursor viral membranes derived from the endoplasmic reticulum (ER). These results indicate that the viral B318L protein may have a role in factory and/or virion assembly (Alejo et al., 1999).

A nudix hydrolase encoded by the virus (D250R) showed a much wider range of nucleotide substrate specificity compared to similar eukaryotic hydrolases, and had highest activity with GTP, followed by adenosine 5'-pentaphosphate (p(5)A) and dGTP. Diadenosine and diguanosine nucleotides were substrates, but the enzyme showed no activity with cap analogues such as 7mGp(3)A indicating that it is not involved in synthesis of precursors for 5' cap structures of the mRNAs. In addition to nucleotide substrates the ASFV enzyme also hydrolyzed the diphosphoinositol polyphosphates PP-InsP(5), and [PP](2)-InsP (Cartwright et al., 2002). ASFV-infection led to a reduction in the levels of PP-InsP(5), ATP and GTP by ca. 50% at late times postinfection, suggesting these may be substrates for the virus enzyme. The association of the enzyme with the rough endoplasmic reticulum suggests a possible role for the ASFV enzyme in regulating a stage of viral morphogenesis involving diphosphoinositol polyphosphate-mediated membrane trafficking (Cartwright et al., 2002).

The virus encodes two enzymes involved in redox metabolism, a homologue of the bacterial NifS protein (QP383R) and an ERV1 homologue (B119L). The later is likely part of a redox chain involved in virus assembly (Lewis et al., 2000; Rodriguez et al., 2006).

Genes encoding structural proteins

The complex virion contains around 50 proteins as estimated by two-dimensional gel electrophoresis (Tabares et al., 1980). These include enzymes and factors that are packaged into core particles to initiate the next round of infection as well as proteins that form part of the virus structure. To date 17 ORFs encoding virion structural proteins have been identified (Dixon, 2005). These include two ORFs encoding polyproteins, pp220 (CP2475L) and pp60 (CP530R), which are proteolytically cleaved by a virus-encoded SUMO-like protease (S273R) to yield the mature virion proteins p150, p37, p14, p34 and p35 and p15 respectively (Andres et al., 1997; Simonmateo et al., 1997; Simonmateo et al., 1993). As is predicted from the presence of both internal and external membranes within the virus particles, several of the virion structural proteins contain transmembrane domains. In addition a chaperone (B602L) required for assembly of the major capsid protein has been identified but this is not incorporated into virus particles (Cobbold et al., 2001). In vaccinia virus 55 proteins have been identified that are essential for virus morphogenesis and about 70 proteins have been identified in virus particles (Condit et al., 2006). It is therefore possible that additional ASFV structural proteins will be identified.

Genes encoding proteins involved in evading host defences

Since ASFV shares a similar replication strategy with poxviruses, a prediction is that the number of proteins involved in virus replication in cells may be the same. A core of 90 genes is conserved between different *genera* of the *Chordopoxvirinae* subfamily, suggesting that this represents the minimal number of genes required for replication (Upton et al., 2003). Thus, 70 to 85 ASFV ORFs may encode proteins that are not essential for replication but play an important role in interactions with the host which facilitate virus survival and transmission. These include a number of proteins which help the virus to evade host defences and persist in the host.

By interfering with macrophage function the virus could disrupt both the initial innate response to infection as well as the later acquired immune response. The identified virus proteins with roles in evading host defences can be classified into three different types (Dixon et al., 2004). Firstly, the virus encodes proteins which interfere with transcriptional activation of host factors involved in transcription of genes encoding immunomodulatory proteins. A second strategy is to encode adhesion proteins which can modulate interactions of infected cells or extracellular virions with the extracellular environment. Thirdly, the virus encodes proteins that inhibit apoptosis and can thus prolong survival of infected cells and facilitate virus replication. These proteins are discussed in more detail in a later section.

Multigene families

As described above, a large proportion of the genome (~ 30% of the coding information) encodes related ORFs that are present in multiple copies (MGF100, 110, 300, 360, 505/530 and p22) and these vary in number between different isolates (Aguero et al., 1990; Pires et al., 1997; Yanez et al., 1995; Yozawa et al., 1994). This large investment in coding for multiple copies of several MGFs implies that these offer a selective advantage to the virus. Although their role is largely unknown, the current knowledge is described in the following sections. One possibility is that they may represent a mechanism for antigenic variation thus helping the virus to escape from the host's immune response. Transcript analysis showed that all copies of one MGF, MGF110, were transcribed during infection arguing against the selected transcription of genes encoding antigenic variants (Almazan et al., 1992). It is also unclear if individual members of each multigene family have the same or different functions. The large sequence divergence between individual genes and presence of additional functional domains in some copies of MGF families suggests that at least some of these genes have different functions. For example amongst the MGF110 members two genes encode proteins which contain a KDEL endoplasmic reticulum (ER) retrieval sequence (XP124L, V118L) (Netherton et al., 2004b) and several encode proteins with predicted signal peptides and cleavage sites suggesting that they are secreted from infected cells. A subset of the MGF360 genes, encode ankyrin protein interaction domains.

STRUCTURE

ASFV particles have a complex multi-layered structure (Carrascosa et al., 1984). The nucleoprotein core structure is 70 to 100 nm in diameter and contains the virus genome as well as enzymes and other proteins that are packaged into virus particles and used in the early stages of infection following virus entry. These include a DNA dependent RNA polymerase, mRNA capping and polyadenylation enzymes as well as other factors required for early gene transcription. The nucleoprotein core is surrounded by a core shell and an internal envelope onto which the icosahedral capsid is assembled. The capsid is 170 to 190 nm in diameter. Although earlier reports suggested that this internal membrane consisted of a collapsed double membrane layer (Andres et al., 1998; Cobbold et al., 1996), it has also been suggested that only one membrane layer is present. The capsid exhibits icosahedral symmetry (T=189-217) corresponding to 1892 to 2172 capsomers (each capsomer is 13 nm in diameter and appears as a hexagonal prism with a central hole.; intracapsomer distance is 7.4 to 8.1 nm). Extracellular virions which have a diameter of 175 to 215 nm (Carrascosa et al., 1984) contain a loosely fitting external envelope which is derived by budding through the plasma membrane.

The locations of some of the structural proteins in the virus particle have been defined by immunogold electron microscopy, confocal microscopy and using biochemical and genetic methods. Thus, the products of the p220 (CP2475L) and pp60 (CP530R) polyproteins (Andres et al., 2002a; Andres et al., 1997; Carrascosa et al., 1986) have been localized in the core shell. The major virus capsid protein is p72 (B646L), which is assembled on the internal viral membrane but is not an integral membrane protein (Cobbold and Wileman, 1998). A candidate protein for forming the vertices of the icosahedral capsid is protein B438L. Components of the internal viral membrane include one of the MGF110 (XP124L) members which has an ER targeting sequence and transmembrane domain (Netherton et al., 2004a). One of the most important integral membrane proteins incorporated into the internal envelope is encoded by ORF E183L. This protein has been designated as p54, j13L or E183L. Several other virion proteins contain transmembrane domains and are incorporated into the internal envelope. These include j18L (E199L), j5R (H108R), p12 (O61R), p17 or i1L (D117L) (Alcami et al., 1992; Brookes et al., 1998b; Sun et al., 1996). Components of the external envelope of the virus particle include the protein encoded by ORF EP402R, also named

CD2v (RuizGonzalvo et al., 1996). This is the only virus encoded glycoprotein incorporated into the particles.

REPLICATION CYCLE

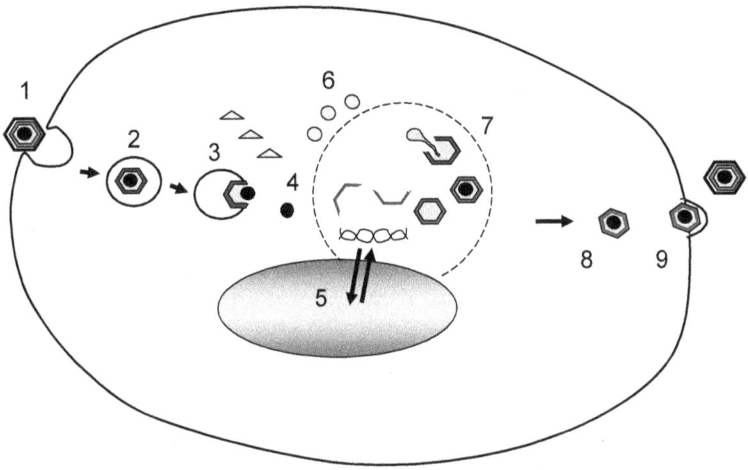

Fig. 1. African swine fever virus replication cycle.
1. Virus enters cells by receptor mediated endocytosis. The loose fitting outer membrane is lost. 2. Virus is internalised in endosomes. 3. A low pH induced fusion between the viral membrane and vesicle membrane releases virus cores into the cytoplasm. 4. Early mRNA synthesis begins in the cytoplasm using enzymes and factors packaged in the virus core. Early virus proteins encode enzymes required for DNA replication. 5. An early stage of DNA replication of subgenomic fragments is detected in the nucleus. 6. The microtubule network is required for formation of perinuclear factory areas. Following the onset of DNA replication in the factories, a shift occurs in the pattern of gene transcription to late virus genes. These encode structural proteins and enzymes and factors which are packaged into virus particles for use during the next round of infection. 7. Virus assembly takes place in factories. 8. Mature virions move on microtubules, using the conventional kinesin motor, to the plasma membrane. 9. Virus particles bud through the plasma membrane gaining a loose fitting external envelope.

ENTRY

The main target cells for virus replication *in vitro* and *in vivo* are macrophages. Interdigitating dendritic cells have also been reported to be infected at early times post-infection *in vivo* and endothelial cells can also be productively infected *in vitro* although *in vivo* infection of these cells is rarely observed (Gregg et al., 1995; Rodriguez et al., 1996a; Vallee et al., 2001). Only at later stages of disease has infection has been reported in a variety of cell types including endothelial cells, megakaryocytes, platelets, neutrophils and hepatocytes (Carrasco et al., 1992; Fernandez et al., 1992a; Fernandez et al., 1992b; Fernandez et al., 1992c; GomezVillamandos et al., 1996; Gomezvillamandos et al., 1995b; Gomezvillamandos et al., 1995c; Sierra et al., 1991). In addition some virus isolates have been adapted to grow in tissue culture cell lines such as Vero cells.

Macrophages which express cell surface markers characteristic of intermediate and late stages of differentiation are permissive for infection. Separation of porcine macrophages populations using antibodies specific for either SWC9 or CD163 cell surface markers yielded a population of cells carrying these cell surface markers that were susceptible to infection and a population without these markers which were not susceptible (McCullough et al., 1999; Sanchez-Torres et al., 2003). It was shown that purified virus particles bind to saturable binding sites on permissive cells. This binding is sensitive to proteolytic treatment of the cells, indicating that binding to a protein receptor is involved. Drugs which increase the pH of lysosomes, inhibit entry of ASFV into both macrophages and tissue culture cells indicating that entry is via receptor mediated endocytosis (Alcami et al., 1989; Valdeira et al., 1998; Valdeira and Geraldes, 1985). It remains uncertain from which type of endocytic vesicle virus cores are released into the cytosol. This is likely to involve a process of fusion of the internal viral envelope with that of the endocytic vesicle and subsequent release of cores. Evidence from electron micrographs supports this model of virus entry. The exact mechanism will require further definition of the structure of virus particles and investigation of the vesicles from which virus is released.

The number of internal membranes in ASFV particles has not yet been resolved. This question has also been controversial in the poxvius field for many years. Earlier reports had indicated that the ASFV internal envelope consists of a double membrane layer derived from the endoplasmic reticulum which collapsed by extrusion of cellular proteins. This has been called into question following use of more advanced electron microscopy techniques. Freeze fracture microscopy has e.g. indicated that vaccinia virus particles contain a single internal membrane. The particles are assembled on a scaffold and this had previously been interpreted to be an additional membrane layer (Heuser, 2005; Szajner et al., 2005). However, the mechanism by which the virus acquires a single internal membrane layer is not known (Condit et al., 2006). During entry vaccinia virus particles shed their external membrane at the cell surface and enter by fusion of a single internal membrane at the cell surface (Law et al., 2006). Certainly the entry of ASFV would also be easier to explain if the loose fitting external envelope were lost at the surface and entry of cores into the cytoplasm was achieved by fusion of a single internal envelope with those of endocytic vesicles.

Recombinant scavenger receptor CD163 has been shown to bind to virus particles and to inhibit virus infection, suggesting that it may act as a virus receptor (Sanchez-Torres et al., 2003). Antibodies against or recombinant virus proteins p12 (O61R), p72 (B646L) and p54 (E183L) inhibit virus binding to cells and those against p30 (CP204L) inhibit virus internalisation (Angulo et al., 1993; Borca et al., 1994a; Gomez-Puertas et al., 1998). This suggests that these proteins have a role in this process. Experiments have been carried out to determine if ASFV particles bound to antibodies might enter macrophages via the Fc receptor but evidence was not obtained for antibody enhancement of infection (Alcami and Vinuela, 1991). These data suggest that one determinant of host cell tropism could be the presence of a cell surface receptor to which virus particles can bind. In one study cell lines from different species were tested for their sensitivity to virus infection. This showed that the most restricted cells showed a reduced efficiency of virus binding and virus internalization. Other resistant cells were restricted only after translation of early virus proteins or viral DNA replication, proving that the presence of virus-specific receptors may be necessary, but not sufficient, to guarantee the cell permissiveness to the virus, and that the ASFV infection can be arrested at different steps of the replication cycle (Carrascosa et al., 1999). Following entry, virus cores are transported to perinuclear assembly sites using the microtubule network. These assembly sites are adjacent to the microtubule organising center as determined by confocal microscopy (Heath et al., 2001).

mRNA TRANSCRIPTION

Early virus gene transcription begins in the cytoplasm immediately following virus entry, using virus-encoded RNA polymerase, mRNA capping and polyadenylation enzymes and factors packaged in the nucleoprotein core. Amongst these are transcription factors required for early gene transcription (Salas, 1999). As yet there is experimental evidence to identify these transcription factors although three ORFs share sequence similarity with the poxvirus transcription factors A2L (B385R), A7L (G1340L) and VLTF2

(B175L) (Iyer et al., 2006). Early gene transcription begins at promoter sequences located a short distance upstream of the translation start codon of individual ORFs. These sequences are AT rich and relatively short (40 to 50 bp). Early gene transcription terminates at a sequence consisting of at least 7 thymidine residues (T) located downstream from the translation termination signal. Some readthrough of the 7 T signal can occur and a sequence of at least 10 Ts is required to terminate all transcription. Transcription termination signals may be located at some distance from the translation termination signal resulting in the production of transcripts which contain sequences from more than one ORF(Almazan et al., 1992; Almazan et al., 1993; Garcia et al., 1995; Goatley et al., 2002). There is no evidence that these transcripts can function as polycistronic mRNAs and translation of only the first ORF in each mRNA is thought to occur. Transcripts are polyadenylated at the 3' end at the position of transcript termination and capped at the 5' end. Virus-encoded enzymes packaged into the virus particle carry out these modifications (Kuznar et al., 1980; Pena et al., 1993; Salas et al., 1988; Salas, 1999). Early genes encode proteins that include enzymes and factors needed for later in the replication cycle. For example enzymes involved in nucleotide metabolism and DNA replication are expressed early post-infection as well as transcription factors needed for expression of later gene classes.

A shift in the pattern of transcription occurs after the onset of DNA replication in the cytoplasm. Intermediate and late gene classes have also been identified and transcription of the latter is dependent on the onset of DNA replication. The similarity in the temporal pattern of gene expression to that of the poxviruses suggests that ASFV transcription follows a similar regulation cascade. In this model, transcription factors required for the next phase of transcription are synthesised in the previous phase. Thus, intermediate transcription factors are synthesised early in infection, late transcription factors at intermediate times and early transcription factors are synthesised at late times and packaged into virus particles. Transcription of intermediate and late ASFV mRNAs also begins at short AT rich promoter sequences immediately upstream of translation initiation codons and terminates at the same termination signals of 7 to 10 Ts as early genes. These transcripts are also polyadenylated and capped. There are several differences in transcription of ASFV genes compared to transcription of vaccinia virus genes. For example, late vaccinia virus transcripts do not terminate at specific sequences and are heterogeneous in length. In addition, vaccinia virus late transcripts have a 5' poly A sequence, added as a consequence of slippage of the RNA polymerase (Ahn and Moss, 1989; Ink and Pickup, 1990), and these are not found in ASFV transcripts. Although both ASFV and vaccinia virus promoters are short AT rich sequences, the ASFV promoters are not generally recognised by vaccinia virus RNA polymerase as assayed by transfection of ASFV genome fragments including promoter elements into vaccinia virus infected cells and vice versa (Hammond and Dixon, 1991). Late genes encode virion structural proteins and enzymes and factors such as RNA polymerase and early transcription factors that are packaged into virus particles for use during the next round of infection. Transcription of several early genes continues throughout the replication cycle.

Although ASFV transcription occurs independently of the host RNA polymerase, productive infection requires the presence of the cell nucleus. The reasons for this are unknown and might reflect a requirement for nuclear gene transcripts or nuclear proteins. In addition, at early stages of ASFV infection synthesis DNA fragments of sub-genomic length has been detected in the nucleus. However it is not known if this is an essential part of the replication cycle. One suggestion is that a host nuclear primase may be necessary to initiate virus DNA replication. Synthesis of complete genomes from these subgenomic fragments may then take place in the cytoplasm (Garciabeato et al., 1992; Rojo et al., 1999). Another requirement of this model would be a mechanism to transport the viral genome into the nucleus and subgenomic fragments from the nucleus into the cytoplasm. There is no information about how this might be achieved although, the demonstration that the structural protein p35 contains two active nuclear localisation signals has led to suggestions that this protein could be involved in transporting the genome into the nucleus (Eulalio et al., 2004). It is interesting to note that members of the *Iridoviridae* have an early stage of DNA replication in the nucleus and later replicatie in the cytoplasm. In this respect ASFV may resemble the *Iridoviridae*. However, entry of the Iridovirus genome into the nucleus is essential since RNA polymerase is not packaged into Iridovirus particles. Obviously, this is not a requirement for ASFV.

Replication of full length ASFV genomes occurs in the cytoplasmic factory areas via head to head concatemers, which are resolved to unit length genomes and packaged into virus particles in the virus factories (Gonzalez et al., 1986; Meireles and Costa, 1994; Rojo et al., 1999). The mechanism of DNA replication and transcription is similar to that of poxviruses, although an early phase of poxvirus DNA replication in the nucleus has not been detected.

ASSEMBLY

Virus morphogenesis takes place in perinuclear factory areas that are located adjacent to the microtubule organising centre (Heath et al., 2001). The microtubule network is required for formation of factories since depolymerisation of microtubules using nocodazole inhibits this process. A dominant negative form of dynamin inhibits factory formation, demonstrating that the minus-end directed motor dynein is required. The virus protein p54 encoded by the E183L gene binds to the LC8 chain of the dynein motor via a motif close to the C-terminus of the protein. This may provide a mechanism for transport of virions within the cytoplasm to factory sites (Alonso et al., 2001; Valdeira et al., 1998). However, as this is an integral membrane protein, it may be lost from the virus structure following fusion with the endosomal membrane. Thus, it remains possible that another virus protein, perhaps exposed on virus core particles may be required for transport to sites of assembly.

Virus factories resemble aggresomes since they are surrounded by a vimentin cage and an increased number of mitochondria. Aggresomes are formed in response to cell stress and function to remove misfolded proteins. Thus, the virus may take advantage of this cellular stress response to form its assembly sites (Heath et al., 2001; Rojo et al., 1998). The first stage in vimentin rearrangement during ASFV infection involves a microtubule-dependent concentration of vimentin into an "aster" within virus assembly sites located close to the microtubule organizing center. Conversion of this aster into a cage requires ASFV DNA replication and this also results in the activation of calcium calmodulin-dependent protein kinase II (CaM kinase II) and phosphorylation of the N-terminal domain of vimentin on serine 82. Possibly this is necessary for cage formation or may simply be a consequence of activation of CaM kinase II by ASFV. The vimentin cage may serve a cytoprotective function and prevent movement of viral components into the cytoplasm and at the same time concentrate late structural proteins at sites of virus assembly (Stefanovic et al., 2005).

The assembly of virions at the virus factories commences with the formation of precursor membranous structures, which are recruited from the surrounding endoplasmic reticulum (Andres et al., 1998; Cobbold et al., 1996). The p54 (E183L) protein is targeted to the ER when it is transiently expressed in transfected cells. Using a conditional lethal recombinant ASFV virus, repression of p54 (E183L) synthesis was shown to arrest virus morphogenesis at a very early stage, even prior to the formation of the precursor membranes. Under restrictive conditions, the virus factories appeared as discrete electron-lucent areas essentially free of viral structures. In contrast, outside the assembly sites, large amounts of aberrant zipper-like structures are formed. These results indicate that the p54 (E183L) protein is critical for the recruitment and transformation of the ER membranes into the precursors of the viral envelope (Rodriguez et al., 2004). An attractive model is that the p54 (E183L) protein is inserted into the ER and that the exposed C-terminal LC8 dynein binding site recruits these membranes to the virus factories as minus-end directed cargo on the microtubule network. Another vrion protein predicted to play an important role in preparing the ER for its role in viral morphogenesis is the XP124L member of MGF110. Its expression, in the absence of other ASFV genes, had a dramatic effect on the contents of the ER that was dependent on its C-terminal ER retrieval sequence KEDL. The normal subcellular distribution of a number of ER resident proteins was drastically altered in cells expressing wild-type XP124L protein (Netherton et al., 2004a). As discussed previously, a number of other structural proteins contain predicted transmembrane domains and are likely to be incorporated into the internal virion membranes.

These modified ER membranes are incorporated as an inner envelope into virus particles and become an icosahedral structure by the progressive assembly of the viral capsid protein p72 (B646L) (Cobbold and Wileman, 1998; Garcia-Escudero et al., 1998). Although earlier analysis had indicated that the membranes incorporated into virions consisted of a collapsed double membrane derived from the ER this data is open to re-interpretation and it remains possible that a single membrane layer is incorporated.

Assembly of the p72 (B646L) protein into ASF virions requires a chaperone function which is encoded by the B602L protein. Biochemical analysis indicated that the p72 (B646L) and B602L proteins interact (Cobbold et al., 2001). Construction of a conditional lethal mutant in which the B602L gene was placed under control of an inducible promoter confirmed that this protein is essential for virion morphogenesis, although it is not incorporated into the virion structure. In the absence of B602L expression, p72 (B646L) was rapidly turned over. These data are consistent with the proposed role of B602L as a chaperone required for p72 (B646L) assembly. Surprisingly, confocal microscopy showed that the B602L protein was excluded from the virus factory and instead was localised in other parts of the cytoplasm (Epifano C, 2006). This apparent paradox has still to be resolved but one possibility is that B602L protein may have a function in addition to acting as a chaperone.

The B438L protein is suggested to have a role in assembly of the capsid by formation of the vertices of the icosahedral structure. Placing the gene encoding the B438L protein under control of an inducible promoter showed that the protein was essential for the formation of infectious virus particles. Under repressive conditions the virus assembly sites contained, instead of icosahedrons, filamentous particles. These contained an aberrant core-shell domain, and an inner envelope and were covered by a layer which contained the major capsid protein p72, but was not icosahedral. These filamentous particles could move from the virus assembly sites to the plasma membrane and exit the cell by budding. These data support a role for this protein in the construction or stabilization of the icosahedral vertices of the virus particle (Epifano et al., 2006).

The two polyprotein precursors, pp220 (CP2475L) and pp60 (CP530R) play a crucial role in assembly of the core shell. These are proteolytically processed by a Sumo-like virus encoded protease (S273R) into six major components of the virus particle which together account for about 30% of the virion protein mass and represent the major components of the core shell (Andres et al., 2001a). The pp220 (CP2475L) polypeptide is N-myristoylated and linked to the virion internal membranes by this modification. This polyprotein gives rise to the major structural proteins p150, p37, p34, and p14. Polyprotein pp60 (CP530R) is processed into p35 and p15. The mature products of both polyproteins localize at the core shell, a matrix-like domain placed between the DNA-containing nucleoprotein core and the inner envelope. They are present in equimolecular amounts in the virus particle. Pulse-chase experiments showed that the processing of both polyprotein precursors occurs concomitantly with virus assembly. By using viruses in which the genes encoding pp220 or p72 were placed under control of inducible promoters, it was shown that pp60 processing requires the expression of the pp220 core precursor, whereas the processing of both precursors pp220 and pp60 is dependent on expression of the major capsid protein p72. Interestingly, when p72 expression is blocked, unprocessed pp220 and pp60 polyproteins assemble into aberrant zipper-like elements consisting of an elongated membrane-bound protein structure reminiscent of the core shell (Alejo et al., 2003; Andres et al., 2002a; Andres et al., 2002b; Andres et al., 1997; Heath et al., 2003).

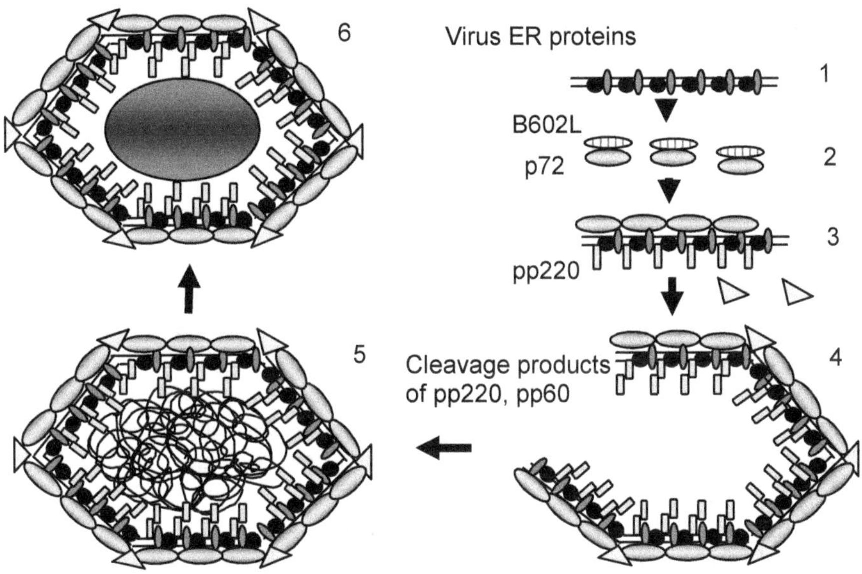

Fig. 2. Assembly of African swine fever virus particles. 1. The endoplasmic reticulum is modified by insertion of virus proteins. 2. The major capsid protein p72 (B646L) is associated with a virus-encoded chaperone B602L. 3. The p72 protein associates with the modified endoplasmic reticulum membranes. The polyprotein pp220 is linked to the viral internal membranes by myristoylation. 4. Expression of the B438L virus protein induces sequential folding of the linear structures to form the sides of the icosahedral structure. Processing of the pp220 and pp60 polyproteins occurs to form the core shell. 5. The viral DNA is inserted and the structure is closed. 6. The nucleoprotein core condenses to mature particles which are transported on the microtubule network to the plasma membrane. Virus particles gain an extra membrane as they bud through the plasma membrane.

It has also been shown that repression of pp220 expression leads to the assembly of icosahedral particles virtually devoid of the core structure. These mutant particles did not contain the pp220-derived products, a number of major core proteins nor the viral DNA but they could move to the plasma membrane and exit from cells (Andres et al., 2002b). It was also confirmed, by using a recombinant virus with the S273R gene under control of an inducible promoter, that the virus-encoded protease S273R is essential for correct processing of the polyprotein precursors and assembly of infectious virions. Inhibition of proteolytic processing leads to the assembly of defective icosahedral particles containing an abnormal nucleoid and core shell. These defective particles were shown to contain the unprocessed pp220 and pp60 precursors, as well as the major DNA-binding nucleoid proteins p10 (A78R) and A104R. This suggests that polyprotein processing is essential for proper core assembly and infectivity, although not for incorporation of the polyproteins or the DNA-containing nucleoid into the assembling particles. Instead, the ASFV proteinase is involved in a late maturational step (Alejo et al., 2003).

The final step in virion maturation involves the condensation of the nucleoid core of the particle to produce mature virus particles with the appearance of being “full” rather than “empty”. The morphological data indicate that the viral DNA begins to condense into a pronucleoid and is then inserted, at a single vertex, into an "empty" particle. Further maturation of the viral particle, including closure of the narrow opening in the icosahedron, gives rise to "intermediate" particles, where the nucleoprotein core undergoes additional consolidation to produce the characteristic mature or "full" virions. The site of particle closure may represent a "weak point" at one vertex, but the mechanisms and structures involved in the packaging and release of the virus genome via such a port are yet to be determined (Brookes et al., 1996; Brookes et al., 1998a). The virus protein E120R binds to both DNA and to the capsid protein p72 (B646L), suggesting a possible role in packaging of the genome (Andres et al., 2001b).

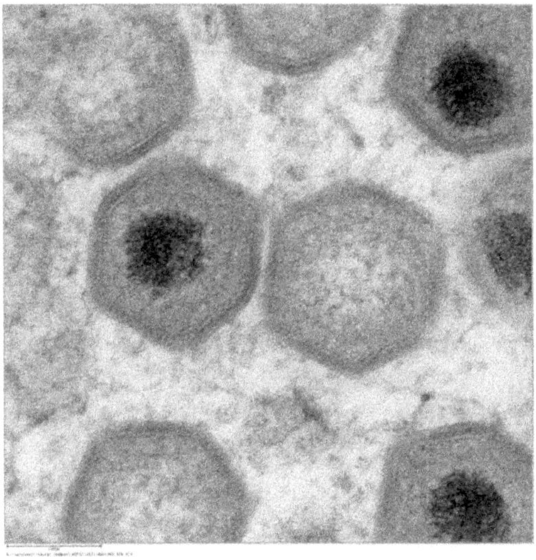

Fig. 3. Immature and mature intracellular virus particles. The mature particles contain a dense nucleoprotein core. Immature particles are visible in which the nucleoprotein core has not condensed to a dense structure. In addition immature particles in which the icosahedral structure is not complete are visible. Electron micrographs were kindly provided by Dr Paul Monaghan and Pippa Hawes Institute for Animal Health Pirbright Lab.

Both, ASFV and poxviruses assemble in the reducing environment of the cell cytosol and encode proteins involved in a redox pathway which is involved in formation of disulphide bonds in some proteins in the virus factory. The ASFV B119L protein has been demonstrated to be a flavine adenine dinucleotide-linked sulfhydrl oxidase which is not incorporated into virus particles but is required for efficient virion maturation. The B119L protein interacts with the A151R protein, which contains a CXXC motif similar to that found in thioredoxins and this binds to a virion structural protein E248R. Possibly these proteins are all components of a system for formation of disulfide bonds in virions that is required for their correct assembly (Rodriguez et al., 2006).

VIRUS EXIT

In order for the virus particles to exit from cells, mature virions must be transported from the virus factories to the plasma membrane. Some intracellular virus particles may be directly released as a result of cell death by necrosis or apoptosis. Movement of virus particles from the factory sites to the plasma membrane involves transport on microtubules since virus movement is inhibited by depolymerisation of the microtubule network using nocodazole. Mature virus particles can also be observed aligned along the microtubule network by confocal microscopy at late times post-infection. Immunofluorescence microscopy showed that conventional kinesin was recruited to virus factories and to a large fraction of virus particles in the cytoplasm. Overexpression of the cargo-binding domain of the kinesin light chain severely inhibited the movement of particles to the plasma membrane consistent with the use of conventional kinesin as a plus-end microtubule motor to move virus from perinuclear assembly sites to the plasma membrane (Jouvenet et al., 2004). Construction of a recombinant virus expressing the p54 structural protein fused to green fluorescent protein (GFP) has provided a tool to study kinetics of virus factory formation and movement of virus particles. These studies have confirmed that the trajectory and speed of particle movement is consistent with their transport by microtubules (Hernaez et al., 2006). The mechanism and virus protein involved in binding to kinesin have not yet been identified but a strong candidate is the E120R protein. By placing the E120R gene under control of an inducible promoter this protein was shown to be required for transport of virus particles from assembly sites to the plasma membrane. Although movement of virus from the factories to the plasma membrane was inhibited in the absence of E120R, recovery of total infectious virus was not reduced. This shows that the E120R protein is not required for the infection process and that intracellular mature virions are infectious (Andres et al., 2001b). The mechanism by which movement to the cell periphery is initiated is also unknown. The demonstration that E120R protein is post-translationally modified by acetylation raises the possibility that a conformational change induced by this modification may expose a kinesin binding site on the protein (Alfonso et al., 2004). Also of interest is the observation that virus infection induces increased acetylation of microtubules as well as their protection against depolymerization by nocodazole (Jouvenet et al., 2004). Possibly, acetylation of virus and host proteins could play a role in regulating virus transport. Interestingly, the p54 protein is not detected by immunogold or confocal microscopy on the surface of mature virions, although the protein is detected as a major component by Western blotting of purified virions, and antibodies against the protein inhibit virus entry (Brookes et al., 1998b; Gomez-Puertas et al., 1998; Rodriguez et al., 1996b; Sun et al., 1995). As this protein contains an LC8 dynein binding motif, exposure of this on the surface of virions could result in minus-end directed transport thus inhibiting virus release. Possibly, the LC8 dynein binding motif on the p54 protein may be altered by a conformational change or by processing at late times post-infection after formation of the mature virions. Alternatively, another protein may block the LC8 binding site on p54 to facilitate plus-end directed movement of mature virions to the plasma membrane.

At the cell periphery virus particles are observed budding through the plasma membrane thereby gaining an additional loose fitting external lipid envelope. As discussed above this membrane is not essential for infectivity since intracellular mature virions are infectious. In the presence of red blood cells intracellular virus particles can be seen below the membranes of infected macrophages in close association with red blood cells. Virus particles are also observed budding into pits on the surface of red blood cells (Dixon et al., 2004). A virus encoded-protein, EP402R (or CD2v), is required for the binding of red blood cells to virus infected cells and to extracellular virus particles. These observations suggest a possible role for this protein in the budding process. Although the EP402R protein is not required for virus infectivity detailed studies on its potential role in virus release have not been carried out (Borca et al., 1994b; Rodriguez et al., 1993). In pigs infected with haemadsorbing viruses, which have a functional EP402R protein, a large proportion of the total virus (>90%) in blood is associated with red blood cells (Quintero et al., 1986).

Vaccinia virus induces actin polymerization and formation of a characteristic actin tail on the cytosolic face of the plasma membrane, directly beneath the extracellular particle. The actin tail acts to propel the virus away from the cell surface to enhance its cell-to-cell spread (Cudmore et al., 1995; Rietdorf et al., 2001; Sanderson et al., 1998). ASFV has also been shown to stimulate the polymerization of actin at the cell surface. Intracellular ASFV particles project out at the tip of long filopodia-like protrusions. However, projection of virus particles out of cells on these protrusions has not been demonstrated (Hernaez et al., 2006; Jouvenet et al., 2006).

Another mechanism for virus release and spread is suggested by the observation that ASFV infection induces apoptosis at late stages of infection. The spread of virus particles in apoptotic bodies may provide a mechanism to evade the immune system by avoiding exposure of extracellular virions to components of the immune system. Moreover, macrophages, the target cells for virus replication, recognise scavenger signals such as phosphatidyl serine on the surface of apoptotic bodies and cells might thus actively take up apoptotic bodies containing virus particles. This speculative model has attractions but has not been investigated experimentally.

EVASION OF HOST DEFENCES

In common with other large DNA viruses ASFV encodes many proteins which help the virus to evade host defences. Approximately 90 proteins are predicted to be required for virus replication (Upton et al., 2003). Of the remaining 70 to 85 encoded proteins many may interfere with host defence systems. Viruses tend to show some specialisation in strategies they use to evade host defences although certain pathways are targeted by most viruses (Alcami and Koszinowski, 2000; Gale and Foy, 2005; Lybarger et al., 2005; Pinto and Hill, 2005; Seet et al., 2003; Vischer et al., 2006). The replication of ASFV in macrophages provides an opportunity for the virus to interfere with both, the initial innate and later acquired immune response to infection by manipulating macrophage function. Three different classes of ASFV encoded proteins which interfere with host defences have been identified (Dixon et al., 2004). A major strategy used by ASFV is to interfere with signalling pathways in infected macrophages and thereby inhibit transcriptional activation of important immunomodulatory genes. Secondly, ASFV has several strategies to inhibit apotosis and facilitate virus replication. Thirdly, several membrane proteins are encoded by ASFV which can alter the adhesion properties of infected cells and virus particles. Our current knowledge of these different strategies is summarised below.

MODULATION OF SIGNALLING PATHWAYS

Cells respond to infection in a number of ways including the reprogramming of gene transcription to activate host defences. This activation of signalling cascades can be triggered by extra or intracellular stimuli. Knowledge of these pathways has been rapidly advancing and their relative importance in anti-viral defences is evident from the number and variety of viral strategies to block these pathways.

The NFκB pathway was the first host pathway shown to be inhibited by ASFV (Powell et al., 1996; Revilla et al., 1998). NFκB complexes act as transcriptional transactivators (in some cases suppressors) of a wide range of immunomodulatory genes including those encoding proinflammatory cytokines and chemokines, anti-apoptotic genes such as IAP proteins, Bcl-2 and Bcl-Xl, acute phase proteins and adhesion molecules (Ghosh et al., 1998). It is thus a key pathway for virus blockade. Sequence analysis identified a region in the centre of the ASFV A238L protein which contains ankyrin repeats similar to those found in the IκB inhibitor of the host NFκB transcription factor (Miskin et al., 2000; Miskin et al., 1998; Powell et al., 1996; Revilla et al., 1998). The suggestion that A238L acts as a functional homologue of IκB was confirmed by experiments showing that A238L expression inhibits activation of an NFkB dependent reporter construct (Powell et al., 1996; Revilla et al., 1998). Members of the IκB protein family bind to the NFκB transcription factor retaining it in an inactive state (Baeuerle and Henkel, 1994; Ghosh and Karin, 2002). A variety of stimuli, including TNF-α and IL-1β, result in activation of an IκB kinase complex, IKK which phosphorylates IκB on two serine residues near the N-terminus. This leads to poly-ubiquitination of IκB and its degradation by the proteasome. A nuclear localisation signal on NFκB is exposed and results in nuclear localisation of the complex and transcriptional activation of genes containing the appropriate binding sites in their promoter regions. Further levels of regulation are provided by phosphorylation and acetylation of NFκB. A feed back mechanism to shut-off NFkB dependent transcription is provided by newly synthesised IkB which is one of the genes activated by NFkB.

Several lines of evidence suggest that A238L acts to inhibit the p65/ p50 heterodimeric form of NFκB. First, A238L has been shown to co-precipitate with the p65 subunit of NFκB suggesting that it may bind directly to p65. Secondly, electrophoretic mobility shift assays (EMSA) suggested that A238L inhibits binding of p65/p50 heterodimers to target DNA sequences. A238L does not displace IκB from NFκB and is thought to bind to NFκB following signal induced degradation of IκB (Revilla et al., 1998; Tait et al., 2000). At earlier times post-infection most A238L is in the cytoplasm but at later times it accumulates in the nucleus (Bowick G., Abrams C.C and Dixon L.K. unpublished results). Expression of A238L does not inhibit nuclear shuttling of NFkB suggesting that A238L can act within the nucleus to inhibit NFkB. Recombinant

A238L protein added to nuclear extracts inhibited binding of NFkB to target DNA sequences and displaced NFkB from preformed complexes with DNA indicating that A238L could act by preventing NFkB binding to DNA (Revilla et al., 1998).

An unexpected additional function for A238L was defined following the identification of the cellular phosphatase calcineurin (protein phosphatase 2B) as a binding partner of A238L using the yeast two-hybrid system. Subsequently, A238L was shown to inhibit calcineurin phosphatase activity, and thus, inhibit calcineurin activated pathways such as activation of the NFAT transcription factor. This activity of A238L is similar to that of the immunosuppressive drug cyclosporin A which binds in a complex with cyclophilin A to calcineurin inhibiting its phosphatase activity (Clipstone and Crabtree, 1992; Huai et al., 2002; Jin and Harrison, 2002). A critical motif PxIxIT close to the A238L C-terminus is required for A238L binding to the catalytic subunit of calcineurin. This motif is conserved in the NFAT substrate and several cellular inhibitors of calcineurin (Miskin et al., 2000; Miskin et al., 1998). The A238L C-terminal 82 amino acids, which includes this motif, acts as a potent inhibitor of calcineurin (Abrams C.C, Najjam S., and Dixon L.K unpublished results).

The mechanism by which these two functions of A238L are regulated is likely to involve post-translational modification since two forms of the A238L protein are present in infected cells. These are expressed from the same cDNA and it is assumed that they differ by post-translational modification, although the nature of the modification is not known. The higher molecular weight form of A238L co-precipitates with the p65 subunit of NFκB (Tait et al., 2000) and this form of the protein accumulates in the nucleus late in infection (Silk et al., 2007). However, a pool of A238L, comprising both forms of the protein, is present in the cytoplasm, and nuclear A238L is exported by a CRM-1 mediated pathway suggesting that the protein shuttles between nucleus and cytoplasm.

Calcineurin has a wide range of functions which include activation of transcription factors including NFAT and Elk 1, modulation of the activity of receptors and regulation of apoptosis by activating the pro-apoptotic protein Bad (Clipstone and Crabtree, 1992; Crabtree and Olson, 2002; Graef et al., 2001; Graef et al., 2003; Hamalainen et al., 2002; Lopez-Rodriguez et al., 1999; Macian et al., 2000;

Macian et al., 2001; Tian and Karin, 1999; Wang et al., 1999). Thus, A238L is predicted to act as a potent immunomodulatory protein with a broad inhibitory effect on transcription of host genes regulated by both NFκB, NFAT and possibly other factors.

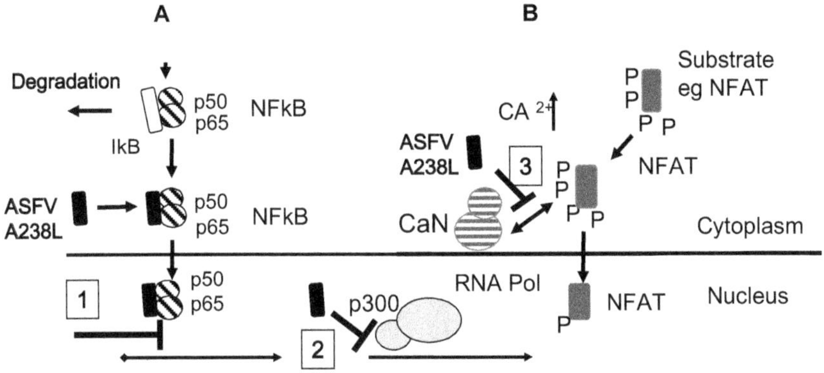

Fig. 4. Functions of the A238L protein. The signalling pathways involved in activation of the A NF-κB and B NFAT transcription factors are shown. 1. A238L inhibits activation of NF-κB transcription factor. A238L has been shown to be in a complex with the p65 subunit of NF-κB at early times of infection in the cytoplasm. This complex forms after signal induced degradation of the inhibitor IκB. At later times of infection a modified form of A238L accumulates in the nucleus. Recombinant A238L prevents binding of NF-κB to DNA and displaces NF-κB from preformed complexes with DNA. 2. Expression of A238L has been shown to inhibit the transcriptional co-activation mediated by the p300 protein. 3. A238L protein binds directly to the calcineurin phosphatase via a PxIxIT motif. As a consequence the phosphatise activity is inhibited and dephosphrylation of substrates including the NFAT transcription factor are inhibited.

Studies on the effects of A238L protein on specific promoters have revealed further complexities in its mode of action. In one example the effect of A238L on expression from the cyclooxygenase-2 promoter was studied using a Jurkat T cell line stably expressing A238L. These cells were transfected with cyclooxygenase-2 promoter-luciferase constructs containing deletion or point mutations. Measurement of luciferase activity in cell extracts indicated the level of promoter activation. This showed that A238L inhibited PMA and ionophore induced expression from the cyclooxygenase promoter and this was dependent on NFAT elements within the promoter. The inhibition was reversed by co-transfection of constitutively active calcineurin or NFAT but not by the p65 subunit of NFκB. The evidence suggested that A238L acted by inhibiting the ability of NFAT to transactivate transcription rather than by inhibiting its binding to DNA. Cyclooxygenase-2 is transiently induced upon cell activation or viral infections, and is the key enzyme involved in prostaglandin biosynthesis. Prostaglandins are important mediators of inflammation, and thus, are a target for virus inhibition. By using wild type ASFV and a deletion mutant lacking the A238L gene it was shown that A238L expression causes a reduction in COX2 expression in infected cells and production of prostaglandin E2 (Granja et al., 2004b).

A similar approach was used to analyse the effect of A238L on the TNF-α promoter. In Jurkat T cells stably expressing A238L, the cAMP-responsive element and kappa 3 sites on the TNF-α promoter were responsible for promoter activation, and A238L inhibited TNF-α expression through these DNA binding sites. This inhibition was partially reverted by overexpression of the transcriptional factors NF-AT, NF-κB, and c-jun. Again, the data indicated that A238L inhibited NF-κB, NF-AT, and c-jun transactivation through a mechanism that involves CREB binding protein/p300 function. A238L was shown to bind to the cyclic AMP-responsive element/kappa 3 complex, thus displacing the CREB binding protein/p300 coactivators (Granja et al., 2006a).

Using a virus deletion mutant and a macrophage cell line (RAW264.7) expressing A238L, this protein was also shown to inhibit expression of inducible nitric oxide synthase (iNOS) another key regulator of inflammation. The detailed mechanism by which activation of the iNOS promoter by LPS and interferon-γ was inhibited by A238L has been studied in the RAW264.7 cell line. This showed that overexpression of p300 co-activator reversed the A238L-mediated inhibition of iNOS promoter activity suggesting that A238L may target p300. Furthermore, A238L expression inhibited interaction of p300 and p65 and prevented their binding to NFkB sites in the promoter. A238L also abrogated p300 mediated transactivation (Granja et al., 2006b).

Together these results suggest that A238L can act at the level of the transcriptional co-activator to inhibit transcription dependent on different factors. The exact mechanism remains to be elucidated. Other virus proteins, such as adenovirus E1A, are known to modulate interaction of several transcription factors with co-activators, so possibly a similar mechanism is involved. As discussed earlier, A238L is also present in the cytoplasm where it was shown to in a complex with p65 at early times post-infection.

Thus, A238L may act on more than one target and in more than one location to inhibit NFkB activation and calcineurin phosphatase activity.

ASFV encodes other proteins which have been demonstrated or are predicted to inhibit host signalling pathways. Comparison between host gene transcription in macrophages infected with wild-type ASFV or in a deletion mutant lacking 5 members of multigene family 360 and 530 demonstrated that these genes are involved in transcriptional activation of type I interferon (IFN), and possibly, also of IFN-activated pathways (Afonso et al., 2004). It is not known whether all or a subset of MGF 530 and 360 genes is necessary for this activity and how they act. A key part of the innate response to virus infections is the induction of type I IFN and activation of the IFN-induced anti-viral state. This can limit virus replication and dissemination and is a key factor in controlling virus infections. Many viruses encode proteins which interfere with induction of IFN and IFN-activated pathways, and these can have key roles in virus pathogenesis and persistence (Goodbourn et al., 2000).

Defining host proteins that bind to ASFV proteins using the yeast two-hybrid system has identified other virus proteins which may inhibit host signalling pathways. The ASFV j4R protein binds to the α-chain of nascent polypeptide associated complex (α-NAC) (Goatley et al., 2002). α-NAC was first suggested to have a role in translation by preventing non-specific targeting of proteins lacking signal peptides to the secretory pathway (Wiedmann et al., 1994). α-NAC can also act as a transcriptional co-activator potentiating transcription dependent on c-Jun (Moreau et al., 1998; Yotov et al., 1998). By binding to α-NAC, j4R may interfere with the ability of α-NAC to act as a transcriptional co-activator. The transcription factor c-Jun is activated in response to stress and is involved in transcription of many immunomodulatory genes. α-NAC has also been shown to bind to FADD (Fas Associated Death Domain). FADD is involved in assembly of the death inducing signalling complex (DISC) at the TNF-α receptor cytoplasmic tail following TNF-α binding, and thus, in induction of apoptosis. Binding of α-NAC to FADD is suggested to prevent FADD oligomerization and assembly of the DISC complex in the absence of TNF-α (Stilo et al., 2003). Possibly, interaction of j4R with α-NAC in some way modulates assembly of the DISC complex and induction of apoptosis by TNF-α.

The ASFV ubiquitin-conjugating enzyme, UBCv, has been implicated in having a role in modulating host gene transcription since it was shown to bind to a host nuclear protein SMCy, which contains an AT rich interaction (ARID) DNA binding domain and is involved in transcriptional regulation (Bulimo et al., 2000).

The ASFV DP71L gene (also named I14L and NLS) encodes a protein with similarity over a conserved C-terminal domain to the herpes simplex virus encoded neurovirulence factor ICP34.5. The ICP34.5 protein is multifunctional and its roles include preventing double-stranded RNA activated, PKR-mediated shut-off of host protein synthesis (Brown et al., 1994; He et al., 1997; He et al., 1998; Mao and Rosenthal, 2003). The DP71L protein is expressed as either a short or long form in different ASFV isolates. Both forms are present predominantly in the nucleus and the long form is targeted to the nucleolus by an N-terminal basic domain (Goatley et al., 1999). The ICP34.5 protein contains a nuclear export signal and amino acid repeats of sequence PAT which anchor the protein in the cytoplasm (Cheng et al., 2002). The absence of these from both forms of the DP71L protein explains their nuclear localization. Both ICP34.5 and DP71L proteins contain a protein phosphatase 1 (PP1) docking motif and may act as regulatory subunits to target PP1 to dephosphorylate specific substrates. ICP34.5 targets PP1 to dephosphorylate translation initiation factor eiF2-α thus returning it to a form active in translation (He et al., 1998). The nuclear location of DP71L suggests it may target a nuclear or nucleolar substrate rather than eiF2-α. By using a virus deletion mutant it has been demonstrated that DP71L is required for the activation of PP1 phosphatase activity that is induced by ASFV infection (Riviera J., Abrams CC., Dixon LK. and Alonso C. in press J. Virology).

APOPTOSIS

Induction of apoptosis is a host defence response which limits virus replication by inducing cell death before production of progeny virions. Consequently many viruses have strategies to block apoptotic pathways (Benedict et al., 2002). The balance between pro and anti-apoptotic effectors encoded by the host and by the virus may limit or enhance virus replication, and thus, play a role in cell tropism and pathogenesis. ASFV infection triggers apoptosis in the absence of protein synthesis or virus DNA replication demonstrating that it occurs at an early stage in the infection process, possibly during virus uncoating (Carrascosa et al., 2002).

One possible mechanism by which infection may trigger apoptosis involves the interaction of the ASFV structural protein, p54 (E183L), with the LC8 chain of the microtubule motor protein dynein. This interaction requires a motif similar to that by which the pro-apoptotic Bcl-2 family member Bim binds to LC8. Interaction of p54 with LC8 may displace Bim from dynein resulting in its translocation to mitochondria and activation of apoptosis. Transient expression of p54 resulted in caspase-3 activation and induction of apoptosis. Interestingly, p54 mutants lacking the 13 amino acid dynein-binding motif, failed to activate caspase 3 supporting the hypothesis that apoptosis induced by p54 results from its interaction with LC8 (Hernaez et al., 2004). The relevance of this mechanism in infected cells remains to be determined but it may be involved either at early stages of infection during entry and uncoating or at late replication stages.

ASFV infection induces transcription of p53 and hence activates transcription of p53 target genes such as those encoding the pro-apoptotic proteins Bax and p21. Interestingly, mdm2 expression is also induced. Although mdm2 normally functions to inactivate p53 in the nucleus, in ASFV infected cells mdm2 is retained in virus factories (Granja et al., 2004a). This may have the effect of prolonging the pro-apoptotic effect of p53. Stress signaling from mitochondria and the ER leads to the induction of the proapoptotic transcription factor CHOP/GADD153. Since ASFV is assembled on and enveloped by ER this may lead to activation of CHOP/GADD153 and induction of apoptosis. Unlike other viruses which assemble on the ER, ASFV was shown not to activate CHOP and was able to inhibit the induction of CHOP/GADD153 by a number of exogenous stimuli. The mechanism of this inhibition is unknown (Netherton et al., 2004c).

Despite the pro-apoptotic stimuli that are triggered during infection, extensive apoptosis in infected cells is observed only at later times. The probable reason is that ASFV encodes three proteins with roles in apoptosis inhibition and these presumably help to delay the onset of apoptosis. Two of these proteins are similar to known apoptosis regulators. The A224L gene encodes a protein with similarity to the IAP family of apoptosis inhibitors. Transient expression of the A224L protein was shown to substantially inhibit caspase 3 activity and cell death (Chacon et al., 1995; Nogal et al., 2001). In cells infected with an ASFV deletion mutant lacking the A224L gene, caspase 3 activation was also substantially increased.

A224L protein acts by direct interaction with the proteolytic fragment of caspase-3 and inhibition of the protease activity (Chacon et al., 1995; Nogal et al., 2001). Since the A224L protein is expressed late during virus infection and packaged into virions, it may also have a role early during the next round of infection following virus entry. Cellular IAP proteins can interact with the cytoplasmic tail of the TNF-α receptor to induce NFκB activation. Expression of the A224L protein was also shown to increase activation of NFκB which includes amongst its gene targets the anti-apoptotic cellular IAP genes (Rodriguez et al., 2002). Although this could provide another mechanism for the anti-apoptotic effect of A224L the virus encoded inhibitor of NFκB activation, A238L, is likely to counteract this anti-apoptotic effect in infected cells.

The ASFV bcl-2 homologue, A179L, was shown to protect the human myeloid cell line K562 from apoptosis induced by cycloheximide and actinomycin D or by treatment with cytosine arabinoside. A179L contains a single BH1 domain and mutations within this domain abrogated the anti-apoptotic effect of A179L. The A179L protein is expressed both early and late during virus infection and is thought to be required for virus infection (Afonso et al., 1996; Neilan et al., 1993; Revilla et al., 1997). Anti-apoptotic activity of the A179L protein in insect cells was shown to be dependent on cell anchorage. This suggested that A179L may play different roles in infected circulating compared to adherent macrophages (Brun et al., 1998). A third ASFV protein, EP153R, has also been shown to inhibit apoptosis (Hurtado et al., 2004). This protein is a C-type lectin that is non-essential for infectivity (Galindo et al., 2000). In cells infected with a deletion mutant lacking the EP153R gene, increased levels of caspase-3 activity, and of cell death, were observed compared to cells infected with wild type virus. Expression of the EP153R gene resulted in a partial protection of transfected lines from apoptosis induced by virus infection or external stimuli. Although the mechanism by which the EP153R protein acts is not clear, its expression was shown to reduce the transactivating activity of the cellular protein p53 (Hurtado et al., 2004).

The ASFV A238L protein is predicted to have both pro-apoptotic and anti-apoptotic effects. By inhibiting NFκB activation A238L is predicted to have an pro-apoptotic effect, since NFκB target genes encode anti-apoptotic proteins such as IAP proteins (Wang et al., 1998). However, by inhibiting calcineurin phopshatase activity,

A238L is predicted to prevent dephosphorylation and activation of the pro-apoptotic protein BAD and hence inhibit apoptosis induced by BAD (Wang et al., 1999). However, as yet there has been no clear demonstration of either a pro- or anti-apoptotic effect of A238L.

ADHESION PROTEINS

Another class of ASFV proteins which modulate host responses to infection are transmembrane proteins with similarity to cell adhesion proteins. The protein encoded by the EP402R gene shares low sequence similarity over its extracellular domain with the T cell adhesion molecule CD2 (Borca et al., 1994b; Rodriguez et al., 1993). Both are members of the Ig superfamily and contain two immunoglobulin-like domains. The interaction between host CD2 and its ligand (CD58 in man and CD49 in rodents) stabilizes the contact between T cells and antigen presenting cells. The cytoplasmic tail of CD2 interacts with an actin binding adaptor protein, CD2AP, which is involved in reorganization of the actin cytoskeleton during formation of the immunological synapse (Dustin et al., 1998; Lee et al., 2003).

The ASFV CD2 like protein (named CD2v or EP402R) is expressed at the surface of infected cells and causes the characteristic haemadsorbtion of red blood cells around virus infected cells. As the virus buds from infected cells it acquires an external envelope containing the EP402R protein which causes binding of virus particles to the surface of red blood cells (RuizGonzalvo et al., 1996). In pigs infected with virus isolates that cause HAD, most virus is associated with red blood cells (Sierra et al., 1991). When this gene is deleted from the virus, onset of viraemia and virus dissemination to lymph nodes is delayed. Possibly, virus "hidden" on red blood cells may be cleared by components of the immune system more slowly.

Macrophages infected *in vitro* with ASFV inhibit proliferation of bystander, non-infected lymphocytes in response to mitogens, and this inhibition is abrogated when the EP402R gene is deleted (Borca et al., 1998). The mechanism involved has not been elucidated but may require direct interaction of the extracellular domain of the EP402R protein with lymphocytes. Alternatively, an indirect mechanism may be involved in which EP402R may modulate the pattern of cell surface or secreted proteins in infected macrophages, thus altering co-stimulatory signals delivered to lymphocytes. Interestingly, the cytoplasmic tail of the EP402R protein binds to an actin binding adaptor protein SH3P7 (also named mabp1 or HIP55) (Kay-Jackson et al., 2004). SH3P7 has been suggested to have an essential role in endocytosis and in protein transport through the Golgi (Fucini et al., 2002; Kessels et al., 2001; Mise-Omata et al., 2003; Warren et al., 2002) and modulates the JNK1 signalling pathway (Ensenat et al., 1999). Interaction of the cytoplasmic tail of the EP402R protein with SH3P7 may modulate protein trafficking or signaling pathways in infected macrophages and this is one hypothetical mechanism by which co-stimulatory signals provided to bystander lymphocytes could be modulated. Within infected cells most EP402R protein is present in membrane areas that surround virus factories rather than at the cell surface. This provides some support for a model whereby EP402R affects protein trafficking in infected cells (Kay-Jackson et al., 2004).

A second ASFV protein, EP153R, is a type II transmembrane protein containing a C-type lectin domain. This protein is non-essential for virus replication and has low sequence similarity with several receptors expressed on the surface of natural killer (NK) cells (Galindo et al., 2000). The extracellular ligands for the CD2v and EP153R proteins are not yet known.

Viral interference with cell surface expression of proteins is a common immune evasion mechanism. For example, cell surface expression of major histocompatibility complex (MHC) class I proteins is inhibited by several virus proteins including many from herpes viruses. Although this can help the virus to avoid an immune response by inhibiting antigen presentation to T cells in conjunction with class I molecules, the absence of class I molecules can trigger activation of NK cells (Alcami and Koszinowski, 2000).

Cell surface transport assays using a temperature-sensitive mutant of the G protein of vesicular stomatitis virus showed that ASFV significantly reduces the rate at which the protein is delivered to the plasma membrane. This was linked to a general reorganization of the secretory pathway during infection and a specific, microtubule-dependent disruption of structural components of the trans-Golgi network. This may be functionally important because infection of macrophages with virulent isolates of ASFV increased the expression of MHC class I genes, but there was no parallel increase in MHC class I delivery to the plasma membrane (Netherton et al., 2006). Virus proteins might also target specifically MHC class I molecules to reduce their cell surface expression, in addition to the more general effects of virus infection on the secretory pathway.

GENES THAT AFFECT VIRUS VIRULENCE OR TROPISM

The targeted deletion of genes from the ASFV genome has identified genes which cause a reduction in virulence in domestic pigs, altered cell tropism or reduced the ability of virus to replicate in ticks (Tulman and Rock, 2001). The ability of the virus to replicate in macrophage cultures *in vitro* is reduced by deleting genes encoding enzymes involved in nucleotide metabolism, such as thymidine kinase gene and dUTPase. Ability of these deletion mutants to replicate in cell lines was not impaired suggesting that the limiting pool of nucleotides in non-dividing macrophage cultures may restrict replication of viruses lacking these enzymes. Deletion of the thymidine kinase gene was shown to reduce virulence in pigs (Moore et al., 1998; Oliveros et al., 1999).

Deletion of a group of two MGF530 genes and six MGF MGF360 genes from the Pr4 isolate reduced virus growth in macrophages by about 100-fold possibly by promoting survival of infected cells. A deletion mutant lacking these genes was also attenuated in domestic pigs. However, deletions of subsets of these genes including from one to three genes did not reduce virus replication in macrophages This suggested a possible gene dosage effect with increased copy numbers of these genes promoting virus replication in macrophages. By using a porcine microarray to compare transcription profiles from macrophages infected with wild type virus compared to the deletion mutant, it was shown that these MGF360 and 530 genes control induction of IFN-α and type I IFN-activated pathways. Deletion of this same group of genes reduced the ability of virus to replicate in ticks by 100- to 1000-fold, although the mechanism by which this is mediated is unknown. Deletion of a subset of three of the MGF360 genes was sufficient to reduce virus replication in ticks by 100 to 1000 fold. Virus replication was not detected in the midgut, hemolymph, salivary glands, coxal gland, or reproductive organs showing that the genes are required for efficient virus replication in the midgut and for generalization of infection. Since arthropods do not have an interferon system this reduced replication in the absence of these MGF360 genes must involve a different mechanism. This same group of MGF360 and

530 genes has been deleted from the genome of non-pathogenic ASFV isolates (NHP/68 and OURT88/3) which may represent the molecular basis for attenuation of these isolates (Chapman D, Tscherepov V., Upton C., Dixon LK unpublished results).

Deletion of the DP71L gene (also named NL, I14L), which shares similarity with the HSV ICP34.5 neurovirulence factor, from the genome of a virulent Spanish ASFV isolate, E70, reduced virulence of the virus for domestic pigs (Zsak et al., 1996), whereas deletion of the gene from the virulent Malawi LIL 20/1 or Pr4 isolates did not reduce virulence. This suggested that the Malawi and Pr4 isolates may encode other virulence determinants which can compensate for the loss of DP71L (Afonso et al., 1998).

To map this putative virulence determinant, *in vivo* marker rescue experiments were performed by inoculating swine with infection-transfection lysates containing E70 NL deletion mutant virus and DNA clones from the Malawi NL gene deletion mutant. A 20 kbp fragment, encoding members of multigene families (MGF) 360 and 530, was capable of fully restoring virulence to the E70 NL mutant. Nucleotide sequence analysis identified an 8-kb deletion in the E70 NL deletion mutant which resulted in the deletion and/or truncation of three MGF 360 genes and four MGF 530 genes. This suggests that these genes are virulence factors for domestic swine and that their presence may compensate for the loss of the NL gene. A virus mutant lacking a further three members of MGF360 and MGF530 genes from the Malawi NL deletion mutant replicated normally in macrophage cell culture but was avirulent in swine (Zsak et al., 2001). An 11 kbp region encoding members of MGF 360 and 530 was also able to restore replication of tissue-culture adapted virus isolates in macrophages (Neilan et al., 2002; Zsak et al., 2001).

Deletion of the UK ORF from the E70 isolate did not affect the growth characteristics of the virus in macrophage cell cultures. However, the deletion mutant, but not a revertant virus, exhibited reduced virulence and a 100- to 1000-fold reduction in viraemia in infected pigs (Zsak et al., 1998).

Deletion of another gene from close to the right hand end of the genome, I11L, which is not present in the cell culture-adapted BA71V isolate, had no effect on virulence. The I11L gene is highly conserved among pathogenic African, European and Caribbean ASFV field isolates, and it exists either in a short form, encoding a protein of 77-78 amino acids (9.1 kDa) or in a longer form of 93-94 amino acids (11.1 kDa), The presence of two predicted membrane-spanning segments suggests that I11L is an integral membrane protein (Kleiboeker et al., 1998). Deletion of the 9GL (B119L) gene, which is part of a redox chain required for efficient virion morphogenesis, reduced virus replication in macrophages 100-fold and dramatically reduced virulence in pigs (Lewis et al., 2000).

Surprisingly, deletion of several genes encoding host defence evasion proteins including the A238L, A224L (IAP homologue), and EP153R gene (C-type lectin) gene products, did not have any apparent effect on pathogenesis of a virulent isolate in pigs (Neilan et al., 1999; Neilan et al., 1997a; Neilan et al., 1997b). A possible explanation is that the virus encodes other genes with similar functions which can compensate for the loss of these genes. Other large DNA viruses, including both pox and herpes viruses, encode several proteins with complementary functions in evasion of host defences. As discussed in the previous section, deletion of the EP402R gene delayed the onset of viraemia and virus dissemination but did not reduce the mortality rate of a virulent isolate in pigs (Borca et al., 1998). The EP402R protein was also indicated to have an important role in enhancing virus replication in the tick vector. Following feeding of Ornithodoros ticks across a membrane, replication of a parental ASFV strain in which the EP402R gene was disrupted was reduced by 10^3 or 10^4 compared to recombinant viruses in which the gene was restored. The block in replication was shown to be at the level of virus uptake across the midgut since similar high levels of virus replication were observed following inoculation of both parental and recombinant viruses directly into the haemocel bypassing the gut barrier (Nix RJ, Boinas F., Dixon LK unpublished results). Possibly, binding of virus particles to red blood cells can enhance their uptake across the gut since electron microscopy identified virus particles bound to red blood cells in phagolysosomes of midgut epithelium (Kleiboeker et al., 1999).

The studies with virus deletion mutants have shown that reducing the ability of the virus to replicate in macrophages dramatically reduces the virulence of the virus in pigs. Presumably, the reduced levels of replication enable the defence mechanisms of the host to clear virus infection. It is also possible that the processes which lead to the haemorrhagic pathology are not induced. However, deleting some genes did not reduce virus replication in macrophages but still reduced virulence in pigs. In these examples the reasons for the reduced virulence in pigs is not clear. One hypothesis is that the proteins encoded block the innate immune response and that in their absence virus infection is more rapidly cleared. In addition, as discussed above, possibly host responses which induce the haemorrhagic pathology are not initiated.

HOST CELL RESPONSES TO ASFV INFECTION

ASFV targets signalling pathways in infected macrophages to interfere with host immunomodulatory gene transcription. Analysis of gene transcription in ASFV infected macrophages supports this hypothesis. Two reports showed that ASFV infection of macrophages in *vitro* reduces transcript levels, at late times post-infection, for proinflammatory cytokines including TNF-α, IL-8 and IFN-β (del Moral et al., 1999; Powell et al., 1996). In one of these studies an early increase in TNF-α transcript levels was detected as well as secretion of TNF-α (del Moral et al., 1999). More recently a porcine cDNA microarray has been used to monitor changes in macrophage gene expression following infection with a virulent ASFV isolate (Zhang et al., 2006). This study showed an early increase in expression level of about 100 genes following infection but expression of most of these returned to the base level observed in mock-infected cells at later times post-infection. These genes included proinflammatory cytokines (TNF-α, IL-6 and IFN-β), chemokines, cell surface receptors, adhesion molecules and vesicle sorting proteins. These results are consistent with an early activation of host transcription by virus infection and a later suppression of this response due to expression of virus-encoded proteins, such as A238L and MGF 360/530 that inhibit pathways leading to the transcriptional activation. A smaller number of genes were increased in expression level later during infection and these may encode some proteins that are required by the virus for replication (Zhang et al., 2006). By expressing a library of antisense cDNAs in ASFV infected Hela cells, host genes that are required for infection, were identified (Chang et al., 2006). These included one gene, BAT3, involved in apoptotic pathways. Inhibiting BAT3 expression altered transcription of apoptosis-related genes, as determined by cDNA microarray analysis, and increased cellular sensitivity to staurosporine-induced apoptosis (Chang et al., 2006).

A comparison of cytokine gene expression using real time PCR showed increased transcript levels for IL6, TNF-α, IL-12 and IL-15 in macrophages infected with a low virulence isolate compared to a high virulence isolate. This suggests that virulence of ASFV isolates may depend on the capacity to regulate the expression of macrophage-derived cytokines relevant for the development of host protective responses (Gil et al., 2003).

In complementary studies a proteomic approach has been used to monitor host proteins that are altered following ASFV infection (Alfonso et al., 2004). This identified increased levels of host proteins including those induced by stress such as heat shock proteins, thioredoxin as well as superoxide dismutase and apoliprotein E and A. Increased levels of proteins secreted from infected cells included a soluble form of galectin 3. These proteins play important roles in regulating apoptosis in the immune system.

FUTURE PROSPECTS

Recent research on the molecular aspects of ASFV replication, on virus manipulation of host defence pathways, and on virus genes involved in virulence and cell tropism has identified several novel molecular and cellular mechanisms of general interest for the understanding of host-pathogen interactions. This information has also provided a solid basis for development of improved disease control strategies, in particular for development of an effective vaccine. Thus, identification of molecular determinants of virus virulence and of mechanisms used by the virus to evade the immune response of the host, provides the basic information needed for the rational design of attenuated virus vaccines by sequential gene deletion. Analysis of the role of individual virus proteins in assembly has suggested methods for constructing non-replicating viruses as vaccines. Knowledge of the virus structure and mechanism of replication has helped to identify critical antigens which are being incorporated into subunit vaccines. Finally, knowledge of host pathways and virus proteins important for virus replication has suggested some potential targets for antiviral development of antivirals.

ACKNOWLEDGEMENTS

We would like to thank colleagues for helpful discussions including: Paul Monagahan, Pippa Hawes, Haru Takamatsu, Rhiannon Silk, Rebecca Nix, Becky Herbert, Lynnette Goatley, Chris Netherton. We thank BBSRC, DEFRA and Wellcome Trust for funding.

REFERENCES

Afonso, C.L., Neilan, J.G., Kutish, G.F., and Rock, D.L. (1996). An African swine fever virus Bcl-2 homolog, 5-HL, suppresses apoptotic cell death. J. Virol. *70*, 4858-4863.

Afonso, C.L., Piccone, M.E., Zaffuto, K.M., Neilan, J., Kutish, G.F., Lu, Z., Balinsky, C.A., Gibb, T.R., Bean, T.J., Zsak, L., and Rock, D.L. (2004). African swine fever virus multigene family 360 and 530 genes affect host interferon response. J. Virol. *78*, 1858-1864.

Afonso, C.L., Zsak, L., Carrillo, C., Borca, M.V., and Rock, D.L. (1998). African swine fever virus NL gene is not required for virus virulence. Journal of General Virology *79*, 2543-2547.

Aguero, M., Blasco, R., Wilkinson, P., and Vinuela, E. (1990). Analysis of naturally-occurring deletion variants of African swine fever virus - multigene Family-110 is not essential for infectivity or virulence in pigs. Virology *176*, 195-204.

Ahn, B.Y., and Moss, B. (1989). Capped Poly(a) Leaders of variable lengths at the 5' ends of vaccinia virus late messenger-RNAs. J. Virol. *63*, 226-232.

Alcami, A., Angulo, A., Lopezotin, C., Munoz, M., Freije, J.M.P., Carrascosa, A.L., and Vinuela, E. (1992). Amino-acid-sequence and structural-properties of protein-P12, an African swine fever virus attachment protein. J. Virol. *66*, 3860-3868.

Alcami, A., Carrascosa, A.L., and Vinuela, E. (1989). Saturable binding-sites mediate the entry of African swine fever virus into vero cells. Virology *168*, 393-398.

Alcami, A., and Koszinowski, U.H. (2000). Viral mechanisms of immune evasion. Trends Microbiol. *8*, 410-418.

Alcami, A., and Vinuela, E. (1991). Fc-receptors do not mediate African swine fever virus- replication in macrophages. Virology *181*, 756-759.

Alejo, A., Andres, G., and Salas, M.L. (2003). African swine fever virus proteinase is essential for core maturation and infectivity. J. Virol. *77*, 5571-5577.

Alejo, A., Andres, G., Vinuela, E., and Salas, M.L. (1999). The African swine fever virus prenyltransferase is an integral membrane trans-geranylgeranyl-diphosphate synthase. J. Biol. Chem. *274*, 18033-18039.

Alejo, A., Yanez, R.J., Rodriguez, J.M., Vinuela, E., and Salas, M.L. (1997). African swine fever virus trans-prenyltransferase. J. Biol. Chem. *272*, 9417-9423.

Alfonso, P., Rivera, J., Hernaez, B., Alonso, C., and Escribano, J.M. (2004). Identification of cellular proteins modified in response to African swine fever virus infection by proteomics. Proteomics *4*, 2037-2046.

Almazan, F., Rodriguez, J.M., Andres, G., Perez, R., Vinuela, E., and Rodriguez, J.F. (1992). Transcriptional analysis of multigene family 110 of African swine fever virus. J. Virol. *66*, 6655-6667.

Almazan, F., Rodriguez, J.M., Angulo, A., Vinuela, E., and Rodriguez, J.F. (1993). Transcriptional mapping of a late gene coding for the P12 attachment protein of African swine fever virus. J. Virol. *67*, 553-556.

Almendral, J.M., Almazan, F., Blasco, R., and Vinuela, E. (1990). Multigene Families in African swine fever virus - family-110. J. Virol. *64*, 2064-2072.

Alonso, C., Miskin, J., Hernaez, B., Fernandez-Zapatero, P., Soto, L., Canto, C., Rodriguez-Crespo, I., Dixon, L., and Escribano, J.M. (2001). African swine fever virus protein p54 interacts with the microtubular motor complex through direct binding to light-chain dynein. J. Virol. *75*, 9819-9827.

Andres, G., Alejo, A., Salas, J., and Salas, M.L. (2002a). African swine fever virus polyproteins pp220 and pp62 assemble into the core shell. J. Virol. *76*, 12473-12482.

Andres, G., Alejo, A., Simon-Mateo, C., and Salas, M.L. (2001a). African swine fever virus protease, a new viral member of the SUMO-1-specific protease family. J. Biol. Chem. *276*, 780-787.

Andres, G., Garcia-Escudero, R., Salas, M.L., and Rodriguez, J.M. (2002b). Repression of African swine fever virus polyprotein pp220-encoding gene leads to the assembly of icosahedral core-less particles. J. Virol. *76*, 2654-2666.

Andres, G., Garcia-Escudero, R., Simon-Mateo, C., and Vinuela, E. (1998). African swine fever virus is enveloped by a two-membraned collapsed cisterna derived from the endoplasmic reticulum. J. Virol. *72*, 8988-9001.

Andres, G., Garcia-Escudero, R., Vinuela, E., Salas, M.L., and Rodriguez, J.M. (2001b). African swine fever virus structural protein pE120R is essential for virus transport from assembly sites to plasma membrane but not for infectivity. J. Virol. *75*, 6758-6768.

Andres, G., SimonMateo, C., and Vinuela, E. (1997). Assembly of African swine fever virus: Role of polyprotein pp220. J. Virol. *71*, 2331-2341.

Angulo, A., Vinuela, E., and Alcami, A. (1993). Inhibition of African swine fever virus binding and infectivity by purified recombinant virus attachment protein P12. J. Virol. *67*, 5463-5471.

Baeuerle, P.A., and Henkel, T. (1994). Function and activation of Nf-kappa-B in the immune-system. Annu. Rev. Immunol. *12*, 141-179.
Bastos, A.D.S., Penrith, M.L., Cruciere, C., Edrich, J.L., Hutchings, G., Roger, F., Couacy-Hymann, E., and Thomson, G.R. (2003). Genotyping field strains of African swine fever virus by partial p72 gene characterisation. Arch. Virol. *148*, 693-706.
Baylis, S.A., Banham, A.H., Vydelingum, S., Dixon, L.K., and Smith, G.L. (1993). African swine fever virus encodes a serine protein-kinase which Is packaged into virions. J. Virol. *67*, 4549-4556.
Beard, W.A., and Wilson, S.H. (2001). DNA polymerases lose their grip. Nat. Struct. Biol. *8*, 915-917.
Benedict, C.A., Norris, P.S., and Ware, C.F. (2002). To kill or be killed: viral evasion of apoptosis. Nat. Immunol. *3*, 1013-1018.
Boinas, F.S., Hutchings, G.H., Dixon, L.K., and Wilkinson, P.J. (2004). Characterization of pathogenic and non-pathogenic African swine fever virus isolates from Ornithodoros erraticus inhabiting pig premises in Portugal. J. Gen. Virol. *85*, 2177-2187.
Borca, M.V., Carrillo, C., Zsak, L., Laegreid, W.W., Kutish, G.F., Neilan, J.G., Burrage, T.G., and Rock, D.L. (1998). Deletion of a CD2-like gene, 8-DR, from African swine fever virus affects viral infection in domestic swine. J. Virol. *72*, 2881-2889.
Borca, M.V., Irusta, P., Carrillo, C., Afonso, C.L., Burrage, T., and Rock, D.L. (1994a). African swine fever virus structural protein P72 contains a conformational neutralizing epitope. Virology *201*, 413-418.
Borca, M.V., Kutish, G.F., Afonso, C.L., Irusta, P., Carrillo, C., Brun, A., Sussman, M., and Rock, D.L. (1994b). An African swine fever virus gene with similarity to the T- lymphocyte surface-antigen Cd2 mediates hemadsorption. Virology *199*, 463-468.
Brookes, S.M., Dixon, L.K., and Parkhouse, R.M.E. (1996). Assembly of African swine fever virus: Quantitative ultrastructural analysis in vitro and in vivo. Virology *224*, 84-92.
Brookes, S.M., Hyatt, A.D., Wise, T., and Parkhouse, R.M.E. (1998a). Intracellular virus DNA distribution and the acquisition of the nucleoprotein core during African swine fever virus particle assembly: Ultrastructural in situ hybridisation and DNase-gold labelling. Virology *249*, 175-188.
Brookes, S.M., Sun, H., Dixon, L.K., and Parkhouse, R.M.E. (1998b). Characterization of African swine fever virion proteins j5R and j13L: immuno-localization in virus particles and assembly sites. J. Gen. Virol. *79*, 1179-1188.

Brown, S.M., Maclean, A.R., Aitken, J.D., and Harland, J. (1994). Icp34.5 influences Herpes-Simplex Virus Type-1 Maturation and egress from infected-cells in-vitro. J. Gen. Virol. *75*, 3679-3686.
Brun, A., Rodriguez, F., Escribano, J.M., and Alonso, C. (1998). Functionality and cell anchorage dependence of the African swine fever virus gene A179L, a viral bcl-2 homolog, in insect cells. J. Virol. *72*, 10227-10233.
Bulimo, W.D., Miskin, J.E., and Dixon, L.K. (2000). An ARID family protein binds to the African swine fever virus encoded ubiquitin conjugating enzyme, UBCv1. FEBS Lett. *471*, 17-22.
Carrasco, L., deLara, F.C.M., delasMulas, J.M., Gomez Villamandos, J.C., Perez, J., Wilkinson, P.J., and Sierra, M.A. (1996). Apoptosis in lymph nodes in acute African swine fever. J. Comp. Pathol. *115*, 415-428.
Carrasco, L., Fernandez, A., Villamandos, J.C.G., Mozos, E., Mendez, A., and Jover, A. (1992). Kupffer cells and pims in acute experimental African swine fever. Histol. Histopathol. *7*, 421-425.
Carrascosa, A.L., Bustos, M.J., Galindo, I., and Vinuela, E. (1999). Virus-specific cell receptors are necessary, but not sufficient, to confer cell susceptibility to African swine fever virus. Arch. Virol. *144*, 1309-1321.
Carrascosa, A.L., Bustos, M.J., Nogal, M.L., de Buitrago, G.G., and Revilla, Y. (2002). Apoptosis induced in an early step of African swine fever virus entry into Vero cells does not require virus replication. Virology *294*, 372-382.
Carrascosa, J.L., Carazo, J.M., Carrascosa, A.L., Garcia, N., Santisteban, A., and Vinuela, E. (1984). General morphology and capsid fine-structure of African swine fever virus-particles. Virology *132*, 160-172.
Carrascosa, J.L., Gonzalez, P., Carrascosa, A.L., Garciabarreno, B., Enjuanes, L., and Vinuela, E. (1986). Localization of structural proteins in African swine fever virus-particles by immunoelectron microscopy. J. Virol. *58*, 377-384.
Cartwright, J.L., Safrany, S.T., Dixon, L.K., Darzynkiewicz, E., Stepinski, J., Burke, R., and McLennan, A.G. (2002). The g5R (D250) gene of African swine fever virus encodes a nudix hydrolase that preferentially degrades diphosphoinositol polyphosphates. J. Virol. *76*, 1415-1421.
Chacon, M.R., Almazan, F., Nogal, M.L., Vinuela, E., and Rodriguez, J.F. (1995). The African swine fever virus IAP homolog is a late structural polypeptide. Virology *214*, 670-674.
Chang, A.C.Y., Zsak, L., Feng, Y.A., Mosseri, R., Lu, Q., Kowalski, P., Zsak, A., Burrage, T.G., Neilan, J.G., Kutish, G.F., *et al.* (2006).

Phenotype-based identification of host genes required for replication of African swine fever virus. J. Virol. *80*, 8705-8717.
Cheng, G.F., Brett, M.E., and He, B. (2002). Signals that dictate nuclear, nucleolar, and cytoplasmic shuttling of the gamma(1)34.5 protein of herpes simplex virus type 1. J. Virol. *76*, 9434-9445.
Clipstone, N.A., and Crabtree, G.R. (1992). Identification of calcineurin as a key signaling enzyme in lymphocyte-T activation. Nature *357*, 695-697.
Cobbold, C., Whittle, J.T., and Wileman, T. (1996). Involvement of the endoplasmic reticulum in the assembly and envelopment of African swine fever virus. J. Virol. *70*, 8382-8390.
Cobbold, C., and Wileman, T. (1998). The major structural protein of African swine fever virus, p73, is packaged into large structures, indicative of viral capsid or matrix precursors, on the endoplasmic reticulum. J. Virol. *72*, 5215-5223.
Cobbold, C., Windsor, M., and Wileman, T. (2001). A virally encoded chaperone specialized for folding of the major capsid protein of African swine fever virus. J. Virol. *75*, 7221-7229.
Condit, R.C., Moussatche, N., and Traktman, P. (2006). In a nutshell: Structure and assembly of the vaccinia virion. Adv. Virus Res. *66*, 31-124.
Crabtree, G.R., and Olson, E.N. (2002). NFAT signaling: Choreographing the social lives of cells. Cell *109* Suppl., S67-79.
Cudmore, S., Cossart, P., Griffiths, G., and Way, M. (1995). Actin-Based Motility of Vaccinia Virus. Nature *378*, 636-638.
del Moral, M.G., Ortuno, E., Fernandez-Zapatero, P., Alonso, F., Alonso, C., Ezquerra, A., and Dominguez, J. (1999). African swine fever virus infection induces tumor necrosis factor alpha production: Implications in pathogenesis. J. Virol. *73*, 2173-2180.
Delavega, I., Vinuela, E., and Blasco, R. (1990). Genetic-variation and multigene families in African swine fever virus. Virology *179*, 234-246.
Dixon, L.K., Abrams, C.C., Bowick, G., Goatley, L.C., Kay-Jackson, P.C., Chapman, D., Liverani, E., Nix, R., Silk, R., and Zhang, F.Q. (2004). African swine fever virus proteins involved in evading host defence systems. Vet. Immunol. Immunopathol. *100*, 117-134.
Dixon, L.K., Escribano, J.M., Martins, C., Rock, D.L., Salas, M.L. and Wilkinson, P.J. (2005). Asfarviridae. In Virus Taxonomy, VIIIth Report of the ICTV, C.M. Fauquet, M.A. Mayo, J. Maniloff, U. Desselberger, and L.A. Ball, eds. (London: Elsevier/Academic Press), pp. 135-143.
Dixon, L.K., Twigg, S.R.F., Baylis, S.A., Vydelingum, S., Bristow, C., Hammond, J.M., and Smith, G.L. (1994). Nucleotide-Sequence

of a 55 Kbp Region from the Right End of the Genome of a Pathogenic African Swine Fever Virus Isolate (Malawi Lil20/1). J. Gen. Virol. *75*, 1655-1684.

Dixon, L.K., and Wilkinson, P.J. (1988). Genetic Diversity of African Swine Fever Virus Isolates from Soft Ticks (Ornithodoros-Moubata) Inhabiting Warthog Burrows in Zambia. J. Gen. Virol. *69*, 2981-2993.

Dustin, M.L., Olszowy, M.W., Holdorf, A.D., Li, J., Bromley, S., Desai, N., Widder, P., Rosenberger, F., van der Merwe, P.A., Allen, P.M., and Shaw, A.S. (1998). A novel adaptor protein orchestrates receptor patterning and cytoskeletal polarity in T-cell contacts. Cell *94*, 667-677.

Ensenat, D., Yao, Z.B., Wang, X.H.S., Kori, R., Zhou, G.S., Lee, S.C., and Tan, T.H. (1999). A novel Src homology 3 domain-containing adaptor protein, HIP-55, that interacts with hematopoietic progenitor kinase 1. J. Biol. Chem. *274*, 33945-33950.

Epifano C., Krijnse-Locker, J., Salas M.L., Rodriguez J.M., and Salas J. (2006). The African swine fever virus nonstructural protein pB602L Is required for formation of the icosahedral capsid of the virus particle. J. Virol. *80*, 12260-12270.

Epifano, C., Krijnse-Locker, J., Salas, M.L., Salas, J., and Rodriguez, J.M. (2006). Generation of filamentous instead of icosahedral particles by repression of African swine fever virus structural protein pB438L. J. Virol. *80*, 11456-11466.

Eulalio, A., Nunes-Correia, I., Carvalho, A.L., Faro, C., Citovsky, V., Simoes, S., and de Lima, M.C.P. (2004). Two African swine fever virus proteins derived from a common precursor exhibit different nucleocytoplasmic transport activities. J. Virol. *78*, 9731-9739.

Fernandez, A., Perez, J., Carrasco, L., Bautista, M.J., Sanchez Vizcaino, J.M., and Sierra, M.A. (1992a). Distribution of Asfv Antigens in Pig-Tissues Experimentally Infected with 2 Different Spanish Virus Isolates. J. Vet. Med. B *39*, 393-402.

Fernandez, A., Perez, J., Carrasco, L., Sierra, M.A., Sanchez Vizcaino, M., and Jover, A. (1992b). Detection of African swine fever viral-antigens in paraffin-embedded tissues by use of immunohistologic methods and polyclonal antibodies. Am. J. Vet. Res. *53*, 1462-1467.

Fernandez, A., Perez, J., Delasmulas, J.M., Carrasco, L., Dominguez, J., and Sierra, M.A. (1992c). Localization of African Swine Fever Viral-Antigen, Swine Igm, Igg and C1q in Lung and Liver-Tissues of Experimentally Infected-Pigs. J. Comp. Pathol. *107*, 81-90.

Fucini, R.V., Chen, J.L., Sharma, C., Kessels, M.M., and Stamnes, M. (2002). Golgi vesicle proteins are linked to the assembly of an actin complex defined by mAbp1. Mol. Biol. Cell *13*, 621-631.

Gale, M., and Foy, E.M. (2005). Evasion of intracellular host defence by hepatitis C virus. Nature *436*, 939-945.

Galindo, I., Almazan, F., Bustos, M.J., Vinuela, E., and Carrascosa, A.L. (2000). African swine fever virus EP153R open reading frame encodes a glycoprotein involved in the hemadsorption of infected cells. Virology *266*, 340-351.

Garcia-Escudero, R., Andres, G., Almazan, F., and Vinuela, E. (1998). Inducible gene expression from African swine fever virus recombinants: Analysis of the major capsid protein p72. J. Virol. *72*, 3185-3195.

Garcia-Escudero, R., Garcia-Diaz, M., Salas, M.L., Blanco, L., and Salas, J. (2003). DNA polymerase x of African swine fever virus: Insertion fidelity on gapped DNA substrates and AP lyase activity support a role in base excision repair of viral DNA. J. Mol. Biol. *326*, 1403-1412.

Garcia, R., Almazan, F., Rodriguez, J.M., Alonso, M., Vinuela, E., and Rodriguez, J.F. (1995). Vectors for the Genetic Manipulation of African Swine Fever Virus. J. Biotechnol. *40*, 121-131.

Garciabeato, R., Salas, M.L., Vinuela, E., and Salas, J. (1992). Role of the host-cell nucleus in the replication of African swine fever virus-DNA. Virology *188*, 637-649.

Ghosh, S., and Karin, M. (2002). Missing pieces in the NF-kappa B puzzle. Cell *109*, S81-S96.

Ghosh, S., May, M.J., and Kopp, E.B. (1998). NF-kappa B and rel proteins: Evolutionarily conserved mediators of immune responses. Annu. Rev. Immunol. *16*, 225-260.

Gil, S., Spagnuolo-Weaver, M., Canals, A., Sepulveda, N., Oliveira, J., Aleixo, A., Allan, G., Leitao, A., and Martins, C.L.V. (2003). Expression at mRNA level of cytokines and A238L gene in porcine blood-derived macrophages infected in vitro with African swine fever virus (ASFV) isolates of different virulence. Arch. Virol. *148*, 2077-2097.

Goatley, L.C., Marron, M.B., Jacobs, S.C., Hammond, J.M., Miskin, J.E., Abrams, C.C., Smith, G.L., and Dixon, L.K. (1999). Nuclear and nucleolar localization of an African swine fever virus protein, I14L, that is similar to the herpes simplex virus-encoded virulence factor ICP34.5. J. Gen. Virol. *80*, 525-535.

Goatley, L.C., Twigg, S.R.F., Miskin, J.E., Monaghan, P., St-Arnaud, R., Smith, G.L., and Dixon, L.K. (2002). The African swine fever virus protein j4R binds to the alpha chain of nascent polypeptide-associated complex. J. Virol. *76*, 9991-9999.

Gomez-Puertas, P., Rodriguez, F., Oviedo, J.M., Brun, A., Alonso, C., and Escribano, J.M. (1998). The African swine fever virus proteins p54 and p30 are involved in two distinct steps of virus attachment and both contribute to the antibody-mediated protective immune response. Virology *243*, 461-471.

Gomez Villamandos, J.C., Bautista, M. J., Hervas, J., Carrasco, L., deLara, F.C.M., Perez, J., Wilkinson, P.J., and Sierra, M.A. (1996). Subcellular changes in platelets in acute and subacute African swine fever. J. Comp. Pathol. *115*, 327-341.

Gomez Villamandos, J.C., Hervas, J., Mendez, A., Carrasco, L., Delasmulas, J.M., Villeda, C.J., Wilkinson, P.J., and Sierra, M.A. (1995a). Experimental African swine fever - Apoptosis of lymphocytes and virus-replication in other cells. J. Gen. Virol. *76*, 2399-2405.

Gomez Villamandos, J.C., Hervas, J., Mendez, A., Carrasco, L., Villeda, C.J., Wilkinson, P.J., and Sierra, M.A. (1995b). Pathological-changes in the renal interstitial capillaries of pigs inoculated with 2 different strains of African swine fever virus. J. Comp. Pathol. *112*, 283-298.

Gomez Villamandos, J.C., Hervas, J., Mendez, A., Carrasco, L., Villeda, C.J., Wilkinson, P.J., and Sierra, M.A. (1995c). Ultrastructural-study of the renal tubular system in acute experimental African swine fever - Virus-replication in glomerular mesangial cells and in the collecting ducts. Arch. Virol. *140*, 581-589.

Gonzalez, A., Calvo, V., Almazan, F., Almendral, J.M., Ramirez, J.C., Delavega, I., Blasco, R., and Vinuela, E. (1990). Multigene families in African swine fever virus - Family-360. J. Virol. *64*, 2073-2081.

Gonzalez, A., Talavera, A., Almendral, J.M., and Vinuela, E. (1986). Hairpin loop structure of African swine fever virus-DNA. Nucleic Acids Res. *14*, 6835-6844.

Goodbourn, S., Didcock, L., and Randall, R.E. (2000). Interferons: cell signalling, immune modulation, antiviral responses and virus countermeasures. J. Gen. Virol. *81*, 2341-2364.

Graef, I.A., Chen, F., Chen, L., Kuo, A., and Crabtree, G.R. (2001). Signals transduced by Ca2+/calcineurin and NFATc3/c4 pattern the developing vasculature. Cell *105*, 863-875.

Graef, I.A., Wang, F., Charron, F., Chen, L., Neilson, J., Tessier-Lavigne, M., and Crabtree, G.R. (2003). Neurotrophins and netrins require calcineurin/NFAT signaling to stimulate outgrowth of embryonic axons. Cell *113*, 657-670.

Granja, A.G., Nogal, M.L., Hurtado, C., del Aguila, C., Carrascosa, A.L., Salas, M.L., Fresno, M., and Revilla, Y. (2006a). The viral protein A238L inhibits TNF-alpha expression through a CBP/p300 transcriptional coactivators pathway. J. Immunol. *176*, 451-462.

Granja, A.G., Nogal, M.L., Hurtado, C., Salas, J., Salas, M.L., Carrascosa, A.L., and Revilla, Y. (2004a). Modulation of p53 cellular function and cell death by African swine fever virus. J. Virol. *78*, 7165-7174.

Granja, A.G., Nogal, M.L., Hurtado, C., Vila, V., Carrascosa, A.L., Salas, M.L., Fresno, M., and Revilla, Y. (2004b). The viral protein A238L inhibits cyclooxygenase-2 expression through a nuclear factor of activated T cell-dependent transactivation pathway. J. Biol. Chem. *279*, 53736-53746.

Granja, A.G., Sabina, P., Salas, M.L., Fresno, M., and Revilla, Y. (2006b). Regulation of inducible nitric oxide synthase expression by viral A238L-mediated inhibition of p65/RelA acetylation and p300 transactivation. J. Virol. *80*, 10487-10496.

Gregg, D.A., Mebus, C.A., and Schlafer, D.H. (1995). Early infection of Interdigitating Dendritic Cells in the pig lymph-node with African swine fever viruses of high and low virulence - Immunohistochemical and ultrastructural studies. J. Vet. Diagn. Invest. *7*, 23-30.

Hamalainen, M., Lahti, A., and Moilanen, E. (2002). Calcineurin inhibitors, cyclosporin A and tacrolimus inhibit expression of inducible nitric oxide synthase in colon epithelial and macrophage cell lines. Eur. J. Pharmacol. *448*, 239-244.

Hammond, J.M., and Dixon, L.K. (1991). Vaccinia virus-mediated expression of African swine fever virus genes. Virology *181*, 778-782.

He, B., Gross, M., and Roizman, B. (1997). The gamma(1)34.5 protein of herpes simplex virus I complexes with protein phosphatase 1 alpha to dephosphorylate the alpha subunit of the eukaryotic translation initiation factor 2 and preclude the shutoff of protein synthesis by double-stranded RNA-activated protein kinase. Proc. Natl. Acad. Sci. USA *94*, 843-848.

He, B., Gross, M., and Roizman, B. (1998). The gamma(1)34.5 protein of herpes simplex virus 1 has the structural and functional attributes of a protein phosphatase I regulatory subunit and is present in a high molecular weight complex with the enzyme in infected cells. J. Biol. Chem. *273*, 20737-20743.

Heath, C.M., Windsor, M., and Wileman, T. (2001). Aggresomes resemble sites specialized for virus assembly. J. Cell Biol. *153*, 449-455.

Heath, C.M., Windsor, M., and Wileman, T. (2003). Membrane association facilitates the correct processing of pp220 during production of the major matrix proteins of African swine fever virus. J. Virol. *77*, 1682-1690.

Hernaez, B., Diaz-Gil, G., Garcia-Gallo, M., Quetglas, J.I., Rodriguez-Crespo, I., Dixon, L., Escribano, J.M., and Alonso, C. (2004). The African swine fever virus dynein-binding protein p54 induces infected cell apoptosis. FEBS Lett. *569*, 224-228.

Hernaez, B., Escribano, J.M., and Alonso, C. (2006). Visualization of the African swine fever virus infection in living cells by incorporation into the virus particle of green fluorescent protein-p54 membrane protein chimera. Virology *350*, 1-14.

Heuser, J. (2005). Deep-etch EM reveals that the early poxvirus envelope is a single membrane bilayer stabilized by a geodetic "honeycomb" surface coat. J. Cell Biol. *169*, 269-283.

Hingamp, P.M., Arnold, J.E., Mayer, R.J., and Dixon, L.K. (1992). A Ubiquitin Conjugating Enzyme Encoded by African Swine Fever Virus. EMBO J. *11*, 361-366.

Huai, Q., Kim, H.Y., Liu, Y.D., Zhao, Y.D., Mondragon, A., Liu, J.O., and Ke, H.M. (2002). Crystal structure of calcineurin-cyclophilin-cyclosporin shows common but distinct recognition of immunophilin-drug complexes. Proc. Natl. Acad. Sci. USA *99*, 12037-12042.

Hurtado, C., Granja, A.G., Bustos, M.J., Nogal, M.L., de Buitrago, G.G., de Yebenes, V.G., Salas, M.L., Revilla, Y., and Carrascosa, A.L. (2004). The C-type lectin homologue gene (EP153R) of African swine fever virus inhibits apoptosis both in virus infection and in heterologous expression. Virology *326*, 160-170.

Ink, B.S., and Pickup, D.J. (1990). Vaccinia virus directs the synthesis of early messenger-RNAs containing 5' poly(a) sequences. Proc. Natl. Acad. Sci. USA *87*, 1536-1540.

Iyer, L.A., Balaji, S., Koonin, E.V., and Aravind, L. (2006). Evolutionary genomics of nucleo-cytoplasmic large DNA viruses. Virus Res. *117*, 156-184.

Jezewska, M.J., Marcinowicz, A., Lucius, A.L., and Bujalowski, W. (2006). DNA polymerase X from African swine fever virus: Quantitative analysis of the enzyme-ssDNA interactions and the functional structure of the complex. J. Mol. Biol. *356*, 121-141.

Jin, L., and Harrison, S.C. (2002). Crystal structure of human calcineurin complexed with cyclosporin A and human cyclophilin. Proc. Natl. Acad. Sci. USA *99*, 13522-13526.
Jouvenet, N., Monaghan, P., Way, M., and Wileman, T. (2004). Transport of African swine fever virus from assembly sites to the plasma membrane is dependent on microtubules and conventional kinesin. J. Virol. *78*, 7990-8001.
Jouvenet, N., Windsor, M., Rietdorf, J., Hawes, P., Monaghan, P., Way, M., and Wileman, T. (2006). African swine fever virus induces filopodia-like projections at the plasma membrane. Cell. Microbiol. *8*, 1803-1811.
Kay-Jackson, P.C., Goatley, L.C., Cox, L., Miskin, J.E., Parkhouse, R.M.E., Wienands, J., and Dixon, L.K. (2004). The CD2v protein of African swine fever virus interacts with the actin-binding adaptor protein SH3P7. J. Gen. Virol. *85*, 119-130.
Kessels, M.M., Engqvist-Goldstein, A.E.Y., Drubin, D.G., and Qualmann, B. (2001). Mammalian Abp1, a signal-responsive F-actin-binding protein, links the actin cytoskeleton to endocytosis via the GTPase dynamin. J. Cell Biol. *153*, 351-366.
Kleiboeker, S.B., Kutish, G.F., Neilan, J.G., Lu, Z., Zsak, L., and Rock, D.L. (1998). A consented African swine fever virus right variable region gene, I1 1L, is non-essential for growth in vitro and virulence in domestic swine. J. Gen. Virol. *79*, 1189-1195.
Kleiboeker, S.B., Scoles, G.A., Burrage, T.G., and Sur, J.H. (1999). African swine fever virus replication in the midgut epithelium is required for infection of Ornithodoros ticks. J. Virol. *73*, 8587-8598.
Kuznar, J., Salas, M.L., and Vinuela, E. (1980). DNA-dependent RNA-polymerase in African swine fever virus. Virology *101*, 169-175.
Lamarche, B.J., Showalter, A.K., and Tsai, M.D. (2005). An error-prone viral DNA ligase. Biochemistry *44*, 8408-8417.
Lamarche, B.J., and Tsai, M.D. (2006). Contributions of an endonuclease IV homologue to DNA repair in the African swine fever virus. Biochemistry *45*, 2790-2803.
Law, M., Carter, G.C., Roberts, K.L., Hollinshead, M., and Smith, G.L. (2006). Ligand-induced and nonfusogenic dissolution of a viral membrane. Proc. Natl. Acad. Sci. USA *103*, 5989-5994.
Lee, K.H., Dinner, A.R., Tu, C., Campi, G., Raychaudhuri, S., Varma, R., Sims, T.N., Burack, W.R., Wu, H., Kanagawa, O., *et al.* (2003). The immunological synapse balances T cell receptor signaling and degradation. Science *302*, 1218-1222.
Leitao, A., Cartaxeiro, C., Coelho, R., Cruz, B., Parkhouse, R.M.E., Portugal, F.C., Vigario, J.D., and Martins, C.L.V. (2001). The non-

haemadsorbing African swine fever virus isolate ASFV/NH/P68 provides a model for defining the protective anti-virus immune response. J. Gen. Virol. *82*, 513-523.

Lewis, T., Zsak, L., Burrage, T.G., Lu, Z., Kutish, G.F., Neilan, J. G., and Rock, D.L. (2000). An African swine fever virus ERV1-ALR homologue, 9GL, affects virion maturation and viral growth in macrophages and viral virulence in swine. J. Virol. *74*, 1275-1285.

Lopez-Rodriguez, C., Aramburu, J., Rakeman, A.S., and Rao, A. (1999). NFAT5, a constitutively nuclear NFAT protein that does not cooperate with Fos and Jun. Proc. Natl. Acad. Sci. USA *96*, 7214-7219.

Lubisi, B.A., Bastos, A.D.S., Dwarka, R.M., and Vosloo, W. (2005). Molecular epidemiology of African swine fever in East Africa. Arch. Virol. *150*, 2439-2452.

Lybarger, L., Wang, X.L., Harris, M., and Hansen, T.H. (2005). Viral immune evasion molecules attack the ER peptide-loading complex and exploit ER-associated degradation pathways. Curr. Opin. Immunol. *17*, 71-78.

Macian, F., Garcia-Rodriguez, C., and Rao, A.J.N. (2000). Gene expression elicited by NFAT in the presence or absence of cooperative recruitment of Fos and Jun. EMBO J. *19*, 4783-4795.

Macian, F., Lopez-Rodriguez, C., and Rao, A. (2001). Partners in transcription: NFAT and AP-1. Oncogene *20*, 2476-2489.

Mao, H.W., and Rosenthal, K.S. (2003). Strain-dependent structural variants of herpes simplex virus type 1 ICP34.5 determine viral plaque size, efficiency of glycoprotein processing, and viral release and neuroinvasive disease potential. J. Virol. *77*, 3409-3417.

McCullough, K.C., Basta, S., Knotig, S., Gerber, H., Schaffner, R., Kim, Y.B., and Saalmuller, A. (1999). Intermediate stages in monocyte-macrophage differentiation modulate phenotype and susceptibility to virus infection. Immunology *98*, 203-212.

Meireles, M., and Costa, J.V. (1994). Nucleotide-sequence of the telomeric region of the African swine fever virus genome. Virology *203*, 193-196.

Mise-Omata, S., Montagne, B., Deckert, M., Wienands, J., and Acuto, O. (2003). Mammalian actin binding protein 1 is essential for endocytosis but not lamellipodia formation: functional analysis by RNA interference. Biochem. Biophys. Res. Commun. *301*, 704-710.

Miskin, J.E., Abrams, C.C., and Dixon, L.K. (2000). African swine fever virus protein A238L interacts with the cellular phosphatase

calcineurin via a binding domain similar to that of NFAT. J. Virol. *74*, 9412-9420.
Miskin, J.E., Abrams, C.C., Goatley, L.C., and Dixon, L.K. (1998). A viral mechanism for inhibition of the cellular phosphatase calcineurin. Science *281*, 562-565.
Montgomery, R.E. (1921). On a form of swine fever occurring in British East Africa (Kenya Colony). J. Comp. Pathol. *34*, 159-191, 243-262.
Moore, D.M., Zsak, L., Neilan, J.G., Lu, Z., and Rock, D.L. (1998). The African swine fever virus thymidine kinase gene is required for efficient replication in swine macrophages and for virulence in swine. J. Virol. *72*, 10310-10315.
Moreau, A., Yotov, W.V., Glorieux, F.H., and St-Arnaud, R. (1998). Bone-specific expression of the alpha chain of the nascent polypeptide-associated complex, a coactivator potentiating c- Jun-mediated transcription. Mol. Cell. Biol. *18*, 1312-1321.
Neilan, J.G., Borca, M.V., Lu, Z., Kutish, G.F., Kleiboeker, S.B., Carrillo, C., Zsak, L., and Rock, D.L. (1999). An African swine fever virus ORF with similarity to C-type lectins is non-essential for growth in swine macrophages in vitro and for virus virulence in domestic swine. J. Gen. Virol. *80*, 2693-2697.
Neilan, J.G., Lu, Z., Afonso, C.L., Kutish, G.F., Sussman, M.D., and Rock, D.L. (1993). An African Swine Fever Virus Gene with Similarity to the Protooncogene Bcl-2 and the Epstein-Barr-Virus Gene Bhrf1. J. Virol. *67*, 4391-4394.
Neilan, J.G., Lu, Z., Kutish, G.F., Zsak, L., Burrage, T.G., Borca, M.V., Carrillo, C., and Rock, D.L. (1997a). A BIR motif containing gene of African swine fever virus, 4CL, is nonessential for growth in vitro and viral virulence. Virology *230*, 252-264.
Neilan, J.G., Lu, Z., Kutish, G.F., Zsak, L., Lewis, T.L., and Rock, D.L. (1997b). A conserved African swine fever virus I kappa B homolog, 5EL, is nonessential for growth in vitro and virulence in domestic swine. Virology *235*, 377-385.
Neilan, J.G., Zsak, L., Lu, Z., Kutish, G.F., Afonso, C.L., and Rock, D.L. (2002). Novel swine virulence determinant in the left variable region of the African swine fever virus genome. J. Virol. *76*, 3095-3104.
Netherton, C., Rouiller, I., and Wileman, T. (2004a). The subcellular distribution of multigene family 110 proteins of African swine fever virus is determined by differences in C-terminal KDEL endoplasmic reticulum retention motifs. J. Virol. *78*, 3710-3721.
Netherton, C., Rouiller, I., and Wileman, T. (2004b). The subcellular distribution of multigene family 110 proteins of African

swine fever virus is determined by differences in C-terminal KDEL endoplasmic reticulum retention motifs. J. Virol. *78*, 3710-3721.

Netherton, C.L., McCrossan, M.C., Denyer, M., Ponnambalam, S., Armstrong, J., Takamatsu, H.H., and Wileman, T.E. (2006). African swine fever virus causes microtubule-dependent dispersal of the trans-golgi network and slows delivery of membrane protein to the plasma membrane. J. Virol. *80*, 11385-11392.

Netherton, C.L., Parsley, J.C., and Wileman, T. (2004c). African swine fever virus inhibits induction of the tress-induced proapoptotic transcription factor CHOP/GADD153. J. Virol. *78*, 10825-10828.

Nogal, M.L., de Buitrago, G.G., Rodriguez, C., Cubelos, B., Carrascosa, A.L., Salas, M.L., and Revilla, Y. (2001). African swine fever virus IAP homologue inhibits caspase activation and promotes cell survival in mammalian cells. J. Virol. *75*, 2535-2543.

Oliveros, M., Garcia-Escudero, R., Alejo, A., Vinuela, E., Salas, M.L., and Salas, J. (1999). African swine fever virus dUTPase is a highly specific enzyme required for efficient replication in swine macrophages. J. Virol. *73*, 8934-8943.

Oura, C.A.L., Powell, P.P., and Parkhouse, R.M.E. (1998). African swine fever: a disease characterized by apoptosis. J. Gen. Virol. *79*, 1427-1438.

Pena, L., Yanez, R.J., Revilla, Y., Vinuela, E., and Salas, M.L. (1993). African Swine Fever Virus Guanylyltransferase. Virology *193*, 319-328.

Pinto, A.K., and Hill, A.B. (2005). Viral interference with antigen presentation to CD8(+) T cells: Lessons from cytomegalovirus. Viral Immunol. *18*, 434-444.

Pires, S., Ribeiro, G., and Costa, J.V. (1997). Sequence and organization of the left multigene family 110 region of the Vero-adapted L60V strain of African swine fever virus. Virus Genes *15*, 271-274.

Plowright, W.P.J., and Pierce, M.A. (1969). African swine fever in ticks (Ornithodoros moubata, Murray) collected from animal burrows in Tanzania. Nature *221*, 1071-1073.

Powell, P.P., Dixon, L.K., and Parkhouse, R.M.E. (1996). An I kappa B homolog encoded by African swine fever virus provides a novel mechanism for downregulation of proinflammatory cytokine responses in host macrophages. J. Virol. *70*, 8527-8533.

Quintero, J.C., Wesley, R.D., Whyard, T.C., Gregg, D., and Mebus, C.A. (1986). In vitro and in vivo association of African swine fever virus with swine erythrocytes. Am. J. Vet. Res. *47*, 1125-1131.

Raoult, D., Audic, S., Robert, C., Abergel, C., Renesto, P., Ogata, H., La Scola, B., Suzan, M., and Claverie, J.M. (2004). The 1.2-megabase genome sequence of mimivirus. Science *306*, 1344-1350.

Redrejo-Rodriguez, M., Garcia-Escudero, R., Yanez-Munoz, R.J., Salas, M.L., and Salas, J. (2006). African swine fever virus protein pE296R is a DNA repair apurinic/apyrimidinic endonuclease required for virus growth in swine macrophages. J. Virol. *80*, 4847-4857.

Revilla, Y., Callejo, M., Rodriguez, J.M., Culebras, E., Nogal, M.L., Salas, M.L., Vinuela, E., and Fresno, M. (1998). Inhibition of nuclear factor kappa B activation by a virus- encoded I kappa B-like protein. J. Biol. Chem. *273*, 5405-5411.

Revilla, Y., Cebrian, A., Baixeras, E., Martinez, C., Vinuela, E., and Salas, M.L. (1997). Inhibition of apoptosis by the African swine fever virus bcl-2 homologue: Role of the BH1 domain. Virology *228*, 400-404.

Rietdorf, J., Ploubidou, A., Reckmann, I., Holmstrom, A., Frischknecht, F., Zettl, M., Zimmermann, T., and Way, M. (2001). Kinesin-dependent movement on microtubules precedes actin-based motility of vaccinia virus. Nat. Cell. Biol. *3*, 992-1000.

Rodriguez, C.I., Nogal, M.L., Carrascosa, A.L., Salas, M.L., Fresno, M., and Revilla, Y. (2002). African swine fever virus IAP-like protein induces the activation of nuclear factor kappa B. J. Virol. *76*, 3936-3942.

Rodriguez, F., Fernandez, A., Perez, J., delasMulas, J.M., Sierra, M.A., and Jover, A. (1996a). African swine fever: Morphopathology of a viral haemorrhagic disease. Vet. Rec. *139*, 249-254.

Rodriguez, F., Ley, V., GomezPuertas, P., Garcia, R., Rodriguez, J.F., and Escribano, J.M. (1996b). The structural protein p54 is essential for African swine fever virus viability. Virus Res. *40*, 161-167.

Rodriguez, I., Redrejo-Rodriguez, M., Rodriguez, J.M., Alejo, A., Salas, J., and Salas, M.L. (2006). African swine fever virus pB119L protein is a flavin adenine dinucleotide-linked sulfhydryl oxidase. J. Virol. *80*, 3157-3166.

Rodriguez, J.M., Garcia-Escudero, R., Salas, M.L., and Andres, G. (2004). African swine fever virus structural protein p54 is essential for the recruitment of envelope precursors to assembly sites. J. Virol. *78*, 4299-4313.

Rodriguez, J.M., Salas, M.L., and Vinuela, E. (1996c). Intermediate class of mRNAs in African swine fever virus. J. Virol. *70*, 8584-8589.

Rodriguez, J.M., Yanez, R.J., Almazan, F., Vinuela, E., and Rodriguez, J.F. (1993). African swine fever virus encodes a Cd2 homolog responsible for the adhesion of erythrocytes to infected-cells. J. Virol. *67*, 5312-5320.

Rodriguez, J.M., Yanez, R.J., Pan, R., Rodriguez, J.F., Salas, M.L., and Vinuela, E. (1994). Multigene families in African swine fever virus - family-505. J. Virol. *68*, 2746-2751.

Rojo, G., Chamorro, M., Salas, M.L., Vinuela, E., Cuezva, J.M., and Salas, J. (1998). Migration of mitochondria to viral assembly sites in African swine fever virus-infected cells. J. Virol. *72*, 7583-7588.

Rojo, G., Garcia-Beato, R., Vinuela, E., Salas, M.L., and Salas, J. (1999). Replication of African swine fever virus DNA in infected cells. Virology *257*, 524-536.

Ruiz Gonzalvo, F., Rodriguez, F., and Escribano, J.M. (1996). Functional and immunological properties of the baculovirus-expressed hemagglutinin of African swine fever virus. Virology *218*, 285-289.

Salas, J., Salas, M.L., and Vinuela, E. (1988). Effect of inhibitors of the host-cell RNA polymerase-II on African swine fever virus multiplication. Virology *164*, 280-283.

Salas, M., ed. (1999). African swine fever virus (Asfarviridae) (London: Academic press).

Salguero, F.J., Ruiz-Villamor, E., Bautista, M.J., Sanchez-Cordon, P.J., Carrasco, L., and Gomez-Villamandos, J.C. (2002). Changes in macrophages in spleen and lymph nodes during acute African swine fever: expression of cytokines. Vet. Immunol. Immunopathol. *90*, 11-22.

Salguero, F.J., Sanchez-Cordon, P.J., Sierra, M.A., Jover, A., Nunez, A., and Gomez-Villamandos, J.C. (2004). Apoptosis of thymocytes in experimental African swine fever virus infection. Histol. Histopathol. *19*, 77-84.

Sanchez-Torres, C., Gomez-Puertas, P., Gomez-del-Moral, M., Alonso, F., Escribano, J.M., Ezquerra, A., and Dominguez, J. (2003). Expression of porcine CD163 on monocytes/macrophages correlates with permissiveness to African swine fever infection. Arch. Virol. *148*, 2307-2323.

Sanderson, C.M., Way, M., and Smith, G.L. (1998). Virus-induced cell motility. J. Virol. *72*, 1235-1243.

Seet, B.T., Johnston, J.B., Brunetti, C.R., Barrett, J.W., Everett, H., Cameron, C., Sypula, J., Nazarian, S.H., Lucas, A., and McFadden, G. (2003). Poxviruses and immune evasion. Annu. Rev. Immunol. *21*, 377-423.

Showalter, A.K., Byeon, I.J.L., Su, M.I., and Tsai, M.D. (2001). Solution structure of a viral DNA polymerase X and evidence for a mutagenic function. Nat. Struct. Biol. *8*, 942-946.
Sierra, M.A., Gomez Villamandos, J.C., Carrasco, L., Fernandez, A., Mozos, E., and Jover, A. (1991). In vivo study of hemadsorption in African swine fever virus-infected cells. Vet. Pathol. *28*, 178-181.
Simon Mateo, C., Andres, G., Almazan, F., and Vinuela, E. (1997). Proteolytic processing in African swine fever virus: Evidence for a new structural polyprotein, pp62. J. Virol. *71*, 5799-5804.
Simon Mateo, C., Andres, G., and Vinuela, E. (1993). Polyprotein processing in African swine fever virus - a novel gene-expression strategy for a DNA virus. EMBO J. *12*, 2977-2987.
Silk, R.N., Bowick, G.C., Abrams, C.C., and Dixon L.K. (2007). African swine fever virus A238L inhibitor of NF-κB and of calcineurin phosphatase is imported actively into the nucleus and exported by a CRM1-mediated pathway. J. Gen. Virol. *88*, 411-419.
Stefanovic, S., Windsor, M., Nagata, K., Inagaki, M., and Wileman, T. (2005). Vimentin rearrangement during African swine fever virus infection involves retrograde transport along microtubules and phosphorylation of vimentin by calcium calmodulin kinase II. J. Virol. *79*, 11766-11775.
Stilo, R., Liguoro, D., di Jeso, B., Leonardi, A., and Vito, P. (2003). The a-chain of the nascent polypeptide-associated complex binds to and regulates FADD function. Biochem. Biophys. Res. Commun. *303*, 1034-1041.
Sumption, K.J., Hutchings, G.H., Wilkinson, P.J., and Dixon, L.K. (1990). Variable regions on the genome of Malawi isolates of African swine fever virus. J. Gen. Virol. *71*, 2331-2340.
Sun, H.C., Jacobs, S.C., Smith, G.L., Dixon, L.K., and Parkhouse, R.M.E. (1995). African swine fever virus gene J13l encodes a 25-27 Kda virion protein with variable numbers of amino-acid repeats. J. Gen. Virol. *76*, 1117-1127.
Sun, H.C., Jenson, J., Dixon, L.K., and Parkhouse, R.M.E. (1996). Characterization of the African swine fever virion protein j18L. J. Gen. Virol. *77*, 941-946.
Szajner, P., Weisberg, A.S., Lebowitz, J., Heuser, J., and Moss, B. (2005). External scaffold of spherical immature poxvirus particles is made of protein trimers, forming a honeycomb lattice. J. Cell Biol. *170*, 971-981.
Tabares, E., Marcotegui, M.A., Fernandez, M., and Sanchez-Botija, C. (1980). Proteins specified by African swine fever virus. 1. Analysis of viral structural proteins and antigenic properties. Arch. Virol. *66*, 107-117.

Tait, S.W.G., Reid, E.B., Greaves, D.R., Wileman, T.E., and Powell, P.P. (2000). Mechanism of inactivation of NF-kappa B by a viral homologue of I kappa B alpha - Signal-induced release of I kappa B alpha results in binding of the viral homologue to NF-kappa B. J. Biol. Chem. *275*, 34656-34664.
Thomson, G.R. (1985). The epidemiology of African swine fever - the role of free-living hosts in Africa. Onderstepoort J. Vet. Res. *52*, 201-209.
Thomson, G.R., Gainaru, M.D., and Vandellen, A.F. (1980). Experimental-infection of warthog (phacochoerus-aethiopicus) with African swine fever virus. Onderstepoort J. Vet. Res. *47*, 19-22.
Tian, J.M., and Karin, M. (1999). Stimulation of Elk1 transcriptional activity by mitogen- activated protein kinases is negatively regulated by protein phosphatase 2B (calcineurin). J. Biol. Chem. *274*, 15173-15180.
Tulman, E.R., and Rock, D.L. (2001). Novel virulence and host range genes of African swine fever virus. Curr. Opin. Microbiol. *4*, 456-461.
Upton, C., Slack, S., Hunter, A.L., Ehlers, A., and Roper, R.L. (2003). Poxvirus orthologous clusters: toward defining the minimum essential poxvirus genome. J. Virol. *77*, 7590-7600.
Valdeira, M.L., Bernardes, C., Cruz, B., and Geraldes, A. (1998). Entry of African swine fever virus into Vero cells and uncoating. Vet. Microbiol. *60*, 131-140.
Valdeira, M.L., and Geraldes, A. (1985). Morphological-study on the entry of African swine fever virus into cells. Biol. Cell *55*, 35-40.
Vallee, I., Tait, S.W.G., and Powell, P.P. (2001). African swine fever virus infection of porcine aortic endothelial cells leads to inhibition of inflammatory responses, activation of the thrombotic state, and apoptosis. J. Virol. *75*, 10372-10382.
Villeda, C.J., Gomez Villamandos, J.C., Williams, S.M., Hervas, J., Wilkinson, P.J., and Vinuela, E. (1995). The role of fibrinolysis in the pathogenesis of the hemorrhagic syndrome produced by virulent isolates of African swine fever virus. Thromb. Haemost. *73*, 112-117.
Villeda, C.J., Williams, S.M., Wilkinson, P.J., and Vinuela, E. (1993a). Consumption coagulopathy associated with shock in acute African swine fever. Arch. Virol. *133*, 467-475.
Villeda, C.J., Williams, S.M., Wilkinson, P.J., and Vinuela, E. (1993b). Hemostatic abnormalities in African swine fever - a comparison of 2 virus-strains of different virulence (Dominican-Republic 78 and Malta 78). Arch. Virol. *130*, 71-83.

Vischer, H.F., Vink, C., and Smit, M.J. (2006). A viral conspiracy: Hijacking the chemokine system through virally encoded pirated chemokine receptors. In Chemokines and Viral Infection, pp. 121-154.

Wang, C.Y., Mayo, M.W., Korneluk, R.G., Goeddel, D.V., and Baldwin, A.S. (1998). NF-kappa B antiapoptosis: Induction of TRAF1 and TRAF2 and c- IAP1 and c-IAP2 to suppress caspase-8 activation. Science *281*, 1680-1683.

Wang, H.G., Pathan, N., Ethell, I.M., Krajewski, S., Yamaguchi, Y., Shibasaki, F., McKeon, F., Bobo, T., Franke, T.F., and Reed, J.C. (1999). Ca2+-induced apoptosis through calcineurin dephosphorylation of BAD. Science *284*, 339-343.

Warren, D.T., Andrews, P.D., Gourlay, C.W., and Ayscough, K.R. (2002). Sla1p couples the yeast endocytic machinery to proteins regulating actin dynamics. J. Cell Sci. *115*, 1703-1715.

Wiedmann, B., Sakai, H., Davis, T.A., and Wiedmann, M. (1994). A Protein Complex Required for Signal-Sequence-Specific Sorting and Translocation. Nature *370*, 434-440.

Wilkinson, P.J., Wardley, R.C., and Williams, S.M. (1981). African swine fever virus (Malta-78) in pigs. J. Comp. Pathol. *91*, 277-284.

Yanez, R.J., Rodriguez, J.M., Nogal, M.L., Yuste, L., Enriquez, C., Rodriguez, J.F., and Vinuela, E. (1995). Analysis of the complete nucleotide-sequence of African swine fever virus. Virology *208*, 249-278.

Yotov, W.V., Moreau, A., and St-Arnaud, R. (1998). The alpha chain of the nascent polypeptide-associated complex functions as a transcriptional coactivator. Mol. Cell. Biol. *18*, 1303-1311.

Yozawa, T., Kutish, G.F., Afonso, C.L., Lu, Z., and Rock, D.L. (1994). 2 Novel multigene families, 530 and 300, in the terminal variable regions of African-swine-fever virus genome. Virology *202*, 997-1002.

Zhang, F.Q., Hopwood, P., Abrams, C.C., Downing, A., Murray, F., Talbot, R., Archibald, A., Lowden, S., and Dixon, L.K. (2006). Macrophage transcriptional responses following in vitro infection with a highly virulent African swine fever virus isolate. J. Virol. *80*, 10514-10521.

Zsak, L., Caler, E., Lu, Z., Kutish, G.F., Neiland, J.G., and Rock, D.L. (1998). A nonessential African swine fever virus gene UK is a significant virulence determinant in domestic swine. J. Virol. *72*, 1028-1035.

Zsak, L., Lu, Z., Burrage, T.G., Neilan, J.G., Kutish, G.F., Moore, D.M., and Rock, D.L. (2001). African swine fever virus multigene family 360 and 530 genes are novel macrophage host range determinants. J. Virol. *75*, 3066-3076.

Zsak, L., Lu, Z., Kutish, G.F., Neilan, J.G., and Rock, D.L. (1996). An African swine fever virus virulence-associated gene NL-S with similarity to the herpes simplex virus ICP34.5 gene. J. Virol. *70*, 8865-8871.

Epilogue

Animal Virology: A Showcase of Evolution

Esteban Domingo and Marian C. Horzinek

Darwin's modern "watchdog" Theodosius Dobzhansky (1900-1975) coined the adage "Nothing in Biology Makes Sense Except in the Light of Evolution", which – due to their rapid replication and gargantuan progeny - applies to viruses *a forteriori*. Genetic variation is guided by chance, which certainly did not guide the editors of the present volume to select the authors of this epilogue: all along our main interest has been viral evolution, and "in this light" we shall comment - and enlarge on pertinent remarks made in the Preface. The impressive compilation of review articles on major virus groups with an impact in the veterinary environment leads to considerations regarding their evolution and their long-term survival as pathogens in animal populations.

In co-evolving with their hosts, DNA and RNA viruses have developed very different adaptive strategies, based on requirements to evade the host's defense systems. However, some common trends can be discerned. Irrespective of its genetic material and mechanism of replication, each virus encodes multiple proteins, which are often multifunctional, and which may form oligo-

or multimeric complexes either with other virus-coded proteins or with those of the host. This is reflected by the possibility both of complementation in the case of trans-acting functions, and of modulation of viral functions by mutant versions of trans-acting proteins.

The modulation of viral functions by variant forms of viral proteins is well documented for RNA viruses. The mutant spectra that constitute replicating quasispecies can modulate the expression of specific mutants contained in that quasispecies, resulting in the selective suppression of mutants of superior fitness. A key contribution of the mutant spectrum to the replicative behaviour of a specific variant was predicted by quasispecies theory, and documented experimentally for the first time by de la Torre and Holland (1990). Recent results with human poliovirus and hepatitis C virus, as well as with the animal foot-and mouth disease virus have provided solid proof of these modulating activities (Pfeiffer and Kirkegaard, 2005; Vignuzzi et. al. 2006; Perales et. al. 2007; other cases *in vivo* and in cell culture are reviewed in different chapters of Domingo, 2006). In the case of complex DNA viruses, remarkably similar conclusions must be drawn: from the characteristics of virus-coded proteins it can be assumed that these also may play a modulatory role in complex mixtures of viral mutants, as e.g. documented for cytomegalovirus (Cicin-Sain et. al. 2005).

Interactions among mutants can be further facilitated by a tendency - documented for some virus-host systems - of double infections of the same subsets of cells *in vivo*, occurring at higher frequencies than predicted by random double hits (reviewed in Domingo, 2007). The present volume quotes several examples of structural and non-structural proteins that form complexes with other viral proteins or with host proteins, likely resulting in complexes with altered biological activities when mutated versions of a given protein are captured into abnormal assemblies. These examples include the triskelions of bluetongue virions, each composed of three VP2 molecules (and each, in turn, interacting with other structural proteins), or the VP5 trimers that conform globular structures interspersed between the triskelions; the interactions among proteins in viral capsids which are universally essential for particle stability, or the heterodimeric complexes between surface glycoproteins of herpes- and arteriviruses; the multiple subunits of the DNA-dependent RNA polymerase encoded by African swine fever virus, to name but a few. In fact, several key proteins for the viral life cycle are encoded by highly variable DNA genomic regions.

Altered viral proteins may be encoded by genomes displaying different degrees of defectiveness, with different potential to modulate the replicative capacity of a virus population as a whole (Grande-Pérez et al. 2005; Perales et. al. 2007). Time and again defective genomes have been detected in viral populations, often in excess over their standard predecessors. However, except for the defective-interfering particles and some symptom-modulating, defective satellite RNAs of plant viruses, the biological relevance of defective genomes has yet to be appreciated.

It is unlikely that complex DNA viruses will generally exploit high mutation rates the same way as RNA viruses do, because of the need to preserve a complex genome (complex in the sense of amount of encoded information). Nevertheless, the data summarized in this book suggest that at least some critical aspects of quasispecies behaviour - such as internal interactions among gene products of a replicative ensemble - can operate in principle as efficiently for DNA as for RNA viruses.

One consequence of the modulatory effects of mutant spectra is the observation that high mutation rates may turn out to the detriment of the virus. Indeed, enhanced mutagenesis can lead to virus extinction, as has been shown for several RNA viruses, both in animals and in cell culture (overviews in Domingo, 2006 and Domingo, 2007). This possibility had been mathematically predicted from quasispecies theory: the enzymatic machinery responsible for copying the genetic message must display a minimum copying accuracy for the maintenance of genetic information in any replicating system. As accuracy decreases and as the number of copying mistakes exceeds an “error threshold”, information is irretrievably lost and the message no longer maintained. This phenomenon is currently being explored as a new antiviral strategy termed “virus entry into error catastrophe” or “lethal mutagenesis”. A remarkable conclusion from the comparative accounts in this book is that lethal mutagenesis might also be applicable to DNA viruses, should this strategy prove efficient for RNA viruses.

Intra-population modulatory effects are by no means the only aspect relevant to virus evolution that distills from reading this book. Genetic flexibility has been associated with RNA viruses in the first place (due to their high rates of genetic variation, rapid replication, and generally high fecundity), but given sufficient time, complex DNA viruses can diversify into multiple types, subtypes, biotypes and so on, rendering each isolate virtually unique. Again, the diversity of herpesviruses and of different isolates of African swine fever virus serve to illustrate this point. Significantly, a proposal to establish the order "Herpesvirales" as a new taxonomical entity matches a similar need to create the order "Picornavirales" for this genetically simpler and seemingly more flexible group of RNA viruses. The need to periodically update the classification schemes testifies to the dynamic nature of the "virosphere".

African Swine Fever Virus (ASFV) is a very complex DNA virus indeed, the only member of the monogeneric family Asfaviridae, and until recently classified as an iridovirus (family *Iridoviridae,* represented by invertebrate, fish, amphibian, and reptile viruses). Its unique host spectrum (*Suidae* only) and epidemiological features (vector transmission) beg evolutionary questions. Thus adaptation to the feral warthog host and the tick vector suggests a long-term evolution in the arthropod-vertebrate transmission cycle. Virus isolates from ticks infesting warthog burrows were genetically very diverse, whereas the major capsid protein gene sequences obtained from domestic pig isolates from Africa, Europe and Latin America over a forty year period were closely related. Once introduced into domestic pigs, direct virus transmission (without the tick requirement) will occur; it may be speculated that the difference between the narrow porcine/acarid bottleneck (due to the small transmitted sample size and the founder effect in a heterologous host) versus the less restricted contact infection have influenced this evolutionary trend.

The genetic plasticity of a complex DNA virus is illustrated by the 10% difference in the number of ORFs among different isolates of ASFV; its genetics provide evidence of transposition, an event typical of cellular DNA variation. The presence of a multigene family in complex DNA viruses may favor tolerance (or robustness) to insertions and deletions in the genome. Also, ASFV encodes two genes for enzymes that have bacterial homologues, one of them likely part of a redox chain involved in virus assembly. Comparative chladistic analysis of the protein sequences encoded by poxviruses, asfarviruses, iridoviruses and phycodnaviruses (infecting algae) supports their monophyletic evolution and has allowed a tentative reconstruction of the ancestral viral genome. This hypothetical common predecessor was a nucleocytoplasmic virus with an icosahedral capsid, which encoded complex enzyme systems for DNA replication and transcription, probably inhibitors of apoptosis, and the mentioned redox protein. Conservation of the latter and of two virion membrane proteins indicates that the odd-shaped virions of poxviruses have evolved from the common icosahedral capsid architecture in asfarviruses, iridoviruses, and phycodnaviruses (Iyer et. al. 2001).

Upon reading this book, some striking examples of viral “behaviour” catch the eye that are a corollary of viral evolutionary history. One such issue is virulence. Being a complex phenotype, virulence usually resists disclosing its molecular roots. Take the genetically most humble of viruses reviewed in this book, with a genome of less than 1800 nucleotides! Whether circoviruses are pathogenic or not may depend on minimal and subtle genetic change. What we concisely term “pathogenesis” may hide a complex cascade of effects on the host. A triggering event may debilitate the immune system and may favor secondary infections by known, but also by as yet unrecognized environmental microbes. Does this bear on evolution in any way? Obviously it does. It relates to evolution of both the virus and the host, with alternative alleles presenting different selective advantages to the dynamic swarms of virus variants. The availability of different cellular receptors offers a playground for the selection of virus variants displaying different cell tropism and even host range. The identification of key receptor(s) and an understanding of the potential shifts in receptor usage remain a fertile area of research; advances in this field should help to understand why viruses cause disease, one of the main reasons why we study them. Veterinary virology offers a great potential to approach this question thanks to the availability of multiple model systems to analyze infections *in vivo*.

Infections of animals by their authentic pathogens permit a detailed scrutiny of a viral invasion, including its effects on host functions and a dissection of microevolutionary events of the pathogen in the host. A deeper knowledge of evolutionary mechanisms in response to the host defences should help in the design of preventive and therapeutic strategies. In fact, the genetic flexibility of most viral pathogens has a positive side: the possibility to engineer vaccine strains, which are replication-competent, attenuated, and endowed with markers, the latter with the objective of distinguishing infected from vaccinated animals, a critical issue in animal health (the so called DIVA products). Thus evolution is at the roots of the difficulties we encounter when trying to keep viruses at bay, but it may also unlock possibilities to design virus variants, which can be marshaled to prevent infection and disease.

Would the arguments summarized in this epilogue not suffice to grant its due weight to veterinary virology, the emergence of zoonotic diseases provides additional compelling evidence. “Eleven out of twelve human emerging infectious diseases in the world have arisen from animal sources. So what we really need to work on is the relationship between the human health surveillance system and the animal health surveillance system. One very important point of intersection is the laboratories. We have to do more to share our laboratory capacity.....” These are the words of Julie Gerberding, Director General of CDC in Atlanta, at a news conference in January 2004. We like to quote her words to emphasize the need for transdisciplinarity, to meet major challenges in science, be it e.g. the application of general theories to the understanding of biological systems (think of the quasispecies concept as a connection between theoretical biophysics and virology) or the bridging of animal and human health. Society would benefit if more collaborative comparative medicine research projects were conducted by physicians and veterinarians to investigate zoonotic agent-host interactions (Kahn, 2006). The “one medicine” concept is gaining acceptance in the world community of biomedical researchers (Zinsstag et al., 2005).

For reasons of zoonotic threat, pandemic timeliness and their textbook features of drift and shift one is tempted to enlarge on the orthomyxoviruses and their notorious promiscuity. Unabashed personal bias, however, leads us to focus on the Nidovirales, with the corona- and toroviruses as their most plausible representatives for illustrating evolution. With genome sizes between 26 and 32 kb, these are the largest RNA molecules with biological function and the concomitant highest probability for mutational changes (Horzinek, 1999). Although generally associated with acute, self-limiting enteric and respiratory infections, coronaviruses also establish persistent infections, as has been shown e.g. in mouse hepatitis virus models. Only a few studies have addressed the role of viral persistence during natural coronavirus infection and evolution, however, and the most impressive example would be the deadly Feline Infectious Peritonitis (FIP). Normally, feline coronaviruses cause mild enteric infections, with a high prevalence in almost any cat society worldwide. The low-virulence 'enteric' strains and the fatal FIP-strains are genetically very closely related, the latter being virulence variants of the former, which have arisen from the quasispecies swarm in individual, persistently infected animals (Vennema et al., 1998). This means that no two cases of FIP are caused by the same virus, and that horizontal transmission, i.e. natural cat-to-cat transfer is a rare exception rather than the rule. In other words: FIP is a non-contagious infectious disease, and viral evolution operates at the individual organismic level, without consequences for the population; we never see FIP epidemics. – Two feline coronavirus serotypes are recognized, both of which can cause FIP; they represent viral evolution by yet another mechanism, having arisen through RNA recombination between canine and feline coronaviruses. Various recombinant strains show cross-over points at different locations, showing independent mutational events, a consequence of biotope sharing between both domestic carnivores and repeated hijackings of heterologous hosts (Herrewegh et al., 1998). - SARS has made coronavirology presentable, but the vast amount of data from murine, porcine and feline infections still need to be perceived as important...

The genome organization, replication strategy (including the production of a 3'-coterminal nested set of subgenomic mRNAs), and key domains in the arteri-, corona- and torovirus replicases have been taken as evidence for their evolutionary relatedness, of a common ancestry. By serendipitous coincidence, viruses from these three groups were studied in the same laboratory in the early nineties (den Boon et al., 1991), and the second order in animal virus taxonomy was presented to the International Committee on Taxonomy of Viruses (ICTV) and eventually established. Taxonomy has been mocked as the bureaucracy of virology, and is explicitly not intended to "reflect any hierarchical or phylogenetic classification, but only a convenient order of presentation" (quoted from http://www.virustaxonomyonline.com). We politely disagree. Although a taxonomic structure above the level of family or genus has not been developed systematically (with the exception of the orders *Caudovirales*, *Mononegavirales* and *Nidovirales* – but note our heretic remarks about the Herpes- and Picornaviruses) the disposition of taxa is far from arbitrary and – in most virologists' minds – does reflect evolutionary dependencies. If you believe it, the world is ready for the 7th Kingdom of Viruses.

ACKNOWLEDGEMENTS

The work in Madrid quoted in this article has been supported by grant BFU 2005-00863, and by Fundación R. Areces.

REFERENCES

Cicin-Sain, L., Podlech, J., Messerle, M., Reddehase, M.J., and Koszinowski, U.H. (2005). Frequent coinfection of cells explains functional in vivo complementation between cytomegalovirus variants in the multiply infected host. J. Virol. *79*, 9492-9502.

de la Torre, J.C., and Holland, J.J. (1990). RNA virus quasispecies populations can suppress vastly superior mutant progeny. J. Virol. *64*, 6278-6281.

den Boon J.A., Snijder, E.J., Chirnside, E.D., de Vries, A.A., Horzinek, M.C., Spaan W.J. (1991). Equine arteritis virus is not a togavirus but belongs to the coronaviruslike superfamily. J. Virol. *65*, 2910-2920.

Domingo, E., ed. (2006). Quasispecies: Concepts and Implications for Virology. Curr. Top. Microbiol. Immunol., Vol. 299 (Berlin Heidelberg: Springer Verlag).

Domingo, E. 2007. Virus evolution. In Fields Virology 5th edition, D.M. Knipe, P.M. Howley et al., eds. (Philadelphia: Lippincott Williams and Wilkins), pp.389-421.

Grande-Pérez, A., Lazaro, E., Lowenstein, P., Domingo, E., and Manrubia, S.C. (2005). Suppression of viral infectivity through lethal defection. Proc. Natl. Acad. Sci. USA *102*, 4448-52.

Herrewegh A.A., Smeenk, I., Horzinek, M.C., Rottier, P.J., and de Groot, R.J. (1998). Feline coronavirus type II strains 79-1683 and 79-1146 originate from a double recombination between feline coronavirus type I and canine coronavirus. J. Virol. *72*, 4508-14.

Horzinek, M.C. (1999). Molecular evolution of corona- and toroviruses. Adv. Exp. Med. Biol. *473*, 61-72.

Iyer, L.M., Aravind, L., and Koonin, E.V. (2001). Common origin of four diverse families of large eukaryotic DNA viruses. J. Virol. *75*, 11720-11734.

Kahn, L.H. (2006). Confronting Zoonoses, Linking Human and Veterinary Medicine. Emerg. Infect. Dis. *12*, 556-561

Perales, C., Mateo, R., Mateu, M.G. and Domingo, E. (2007). Insights into RNA virus mutant spectrum and lethal mutagenesis events: replicative interference and complementation by multiple point mutants. J. Mol. Biol. *369*, 985-1000.

Pfeiffer, J.K., and Kirkegaard, K. (2005). Increased fidelity reduces poliovirus fitness under selective pressure in mice. PLoS Pathogens *1*, 102-110.

Vennema, H., Poland, A., Foley, J., Pedersen, N.C. 1998. Feline infectious peritonitis viruses arise by mutation from endemic feline enteric coronaviruses. Virology *243*, 150-157.

Vignuzzi, M., Stone, J.K., Arnold, J.J., Cameron, C.E., and Andino, R. (2006). Quasispecies diversity determines pathogenesis through cooperative interactions in a viral population. Nature *439*, 344-348.

Zinsstag, J., Schelling, E., Wyss, K., Mahamat, M.H. (2005). Potential of cooperation between human and animal health to strengthen health systems. Lancet *366*, 2142–2145

Other Books of Interest

www.ingramcontent.com/pod-product-compliance
Lightning Source LLC
LaVergne TN
LVHW011857200926
848484LV00002B/468
9781904455226